Global Economic History Series

VOLUME 21

The Quantitative Economic History of China

VOLUME 8/1

The titles published in this series are listed at *brill.com/qehc*

Chinese Household Registration Data of Wudu County (Gansu Province), 1947

Volume 1

Chinese Household Registration Data of Wudu County (Gansu Province), 1947

Volume 1
Text and Data

By

Minghui Li
Meimei Wang
Bas van Leeuwen
Jieli Li

BRILL

LEIDEN | BOSTON

The research leading to this book received funding from the European Research Council under the European Union's Horizon 2020 Programme / ERC-StG 637695 – HinDI, as part of the project "The Historical Dynamics of Industrialization in Northwestern Europe and China ca. 1800–2010: A Regional Interpretation."

Cover illustration: Part of the eighteenth-century remake of the twelfth-century original painting *Along the River During the Qingming Festival* by Zhang Zeduan. The remake is created by five Qing dynasty court painters (Chen Mu, Sun Hu, Jin Kun, Dai Hong and Cheng Zhidao) and is currently in the National Palace Museum in Taipei. The painting captures the daily life of people and the landscape of the capital, Bianjing (today's Kaifeng) from the Northern Song period. Licensed under public domain, via Wikimedia Commons.

The Library of Congress Cataloging-in-Publication Data is available online at https://catalog.loc.gov

Typeface for the Latin, Greek, and Cyrillic scripts: "Brill". See and download: brill.com/brill-typeface.

ISSN 2405-870X
ISBN 978-90-04-72737-3 (set)
ISBN 978-90-04-71569-1 (hardback, vol. 1)
ISBN 978-90-04-71570-7 (e-book, vol. 1)
DOI 10.1163/9789004715707
ISBN 978-90-04-72732-8 (hardback, vol. 2)
ISBN 978-90-04-72733-5 (e-book, vol. 2)
DOI 10.1163/9789004727335
ISBN 978-90-04-72734-2 (hardback, vol. 3)
ISBN 978-90-04-72735-9 (e-book, vol 3)
DOI 10.1163/9789004727359

This book is printed on acid-free paper and produced in a sustainable manner.

Contents

Volume 2

Volume 3

Figures and Tables

Figures

Tables

Key Periods in Chinese History

Qin state	(770 BCE–221 BCE)
Spring and Autumn Period	(770 BCE–476 BCE)
Warring States Period	(475 BCE–221 BCE)
Qin dynasty	(221 BCE–207 BCE)
Han dynasty	(202 BC–220 CE)
Three kingdoms	(220–280)
Western Jin dynasty	(266–317)
Eastern Jin dynasty	(317–420)
Sixteen kingdoms	(304–439)
Southern and Northern dynasties	(386–589)
Sui dynasty	(581–618)
Tang dynasty	(618–907)
Five dynasties and Ten kingdoms	(907–979)
North Song dynasty	(960–1127)
South Song dynasty	(1127–1279)
Yuan dynasty	(1271–1368)
Ming dynasty	(1368–1644)
Qing dynasty	(1644–1911)
Republic of China	(1912–1949)
Beiyang government period	(1912–1928)
Nationalist government period	(1928–1949)

Reigns of Emperors in the Late Qing Dynasty

Daoguang	1821–1850
Xianfeng	1851–1861
Tongzhi	1862–1874
Guangxu	1875–1908
Xuantong	1909–1911

Chinese Terms

Baojia system (保甲制度), a community self-defence system and an important type of county self-government organization, with the responsibility of building local security groups, levying taxes and collecting population statistics.

Buzheng Shisi (布政使司), common abbreviated reference to Chengxuan Buzheng Shisi (承宣布政使司), the provincial administration office during the Ming and Qing dynasties. Buzheng Shisi was informally called a province (行省) in Qing and was replaced by province in the Republican period.

Central Political Meeting (中央政治会议), the legislative authority during the Nationalist government period, responsible for making decisions and issuing ordinances concerning important subjects such as legislation, domestic affairs, economy, education, land use and transportation.

Constitution of the Self-governance for Cities, Towns and Villages (《城镇乡地方自治章程》), proposed by the Ministry of Civil Affairs and issued by the Qing government in 1908, endowing cities, towns and villages with the right to self-governance. The Constitution defines the spatial range and agencies of self-governance. It also elucidates the qualifications and duties of candidates from whom a local committee would be elected to make decisions about local public issues.

County (县), known as 'Xian' in Pinyin Romanization, was the basic formal unit in the hierarchy of territorial administration in China. In the Republican period, a county was the administrative unit under the supervision of a province while above a village or town.

Department (州), known as 'Zhou' in Pinyin Romanization, an intermediate agency of administrative supervision normally between a prefecture and its subordinate counties.

Di (氐), an ethnic group seen during ancient times in the north-western part of China. The first record of Di can be found on carapace-bone-scripts dated back to around 3,000 years ago. From the Qin dynasty onwards, the Di were mainly found in today's Sichuan, Gansu and Shaanxi provinces. It started to be integrated to the Han ethnic groups in the seventh century.

Ding (丁), officially referring to adult men aged between sixteen and sixty, was usually counted for the purpose of taxation and corvee labour. But during the Ming and Qing dynasties 'Ding' gradually failed to reflect the change in population. From the eighteenth century 'Ding' was no longer connected to real men, nor did it represent the number of adult men who paid tax. It became an abstract unit for taxation.

Household register (户籍登记), the register of households and inhabitants originating from the draft of the 'Law of Household Register' in 1911. Its contents had been constantly evolving and in the late 1940s it mainly contained three parts: the register for the origin of the household, the register for identity and the register for migration.

Law of Household Register (户籍法), a law to regulate the register of households and inhabitants, as well as the issuing of identity cards for civilians nationwide. It was first issued in 1931 but was not put into practice. In 1943, the Nationalist government announced the reimplementation of the law, and as a result the household register was regularly conducted to various degrees in different regions of China in the following years.

Ministry of Interior (内政部), founded in 1928 by the Nationalist government based on the former Ministry of Civil Affairs. It was responsible for managing issues such as civil affairs, land use, police, population statistics, customs, rituals, sanitation and healthcare.

Prefecture (府), known as 'Fu' in Pinyin Romanization, a unit of territorial administration coordinating counties and subordinated to province.

Qiang (羌), an ethnic group seen in ancient times in the western part of China. The first record of Qiang can be found on carapace-bone-scripts dated back to around 3,000 years ago. Historically, there were multiple factions of Qiang and most of them were distributed in today's Qinghai, Gansu and Sichuan provinces. Today's Tibetan ethnic group originated from one faction of Qiang.

Qu-cun-lü-lin system (区村闾邻制), administrative organization below the county level implemented by the Nationalist government for a short period after 1928. Compared with the Baojia system, its main purpose was self-governance. Yet it was replaced by the Baojia system before long as the Nationalists considered it urgent to implement an administrative organization that would be effective in self-defence.

Silk Road (丝绸之路), the historical network of routes linking China with Middle Asia, Western Asia and Europe in ancient times, particularly between the Han and Tang dynasties. The concept of a 'Silk Road' was first put forward by the German geographer Ferdinand von Richthofen (1833–1905) during the 1870s.

Sishu (私塾), traditional private primary school.

Sub-prefecture (厅), known as 'Ting' in Pinyin Romanization, a unit of territorial administration intermediating between a prefecture and its subordinate counties, or being directly responsible to provincial authorities.

Wang Anshi Reform (王安石变法) refers to the series of reforms proposed by Wang Anshi (1021–1086) and implemented by the central government during the Northern Song dynasty. The Reform triggered transformations in grain taxation, land taxation, agriculture and irrigation law, corvee labour, community administration (Baojia organization) and military training. However, as it infringed on the landlord class's interests, it was resisted intensively and eventually failed.

Xuantong census（宣统人口普查）, a nationwide census conducted by the Qing government during Emperor Xuantong's reign. It was initiated in order to determine the size of the population in different provinces, so as to calculate the quota for electing congressmen in different regions and to prepare the country for constitutional reform. It was terminated when the 1911 revolution to overthrow the Qing court took place.

Zhili Department (直隶州), known as 'Zhili Zhou' in Pinyin Romanization, an intermediate agency of administrative supervision normally between a province and its subordinate counties. In the Qing dynasty it was similar to a prefecture in the territorial administration hierarchy and was directly responsible to provincial authorities.

CHAPTER 1

Introduction to Wudu County

1.1 Introduction

Wudu, the county dealt with in this study, is located in northwestern China and has long been an important part of today's Gansu province. From the Han dynasty (202 BCE–220 CE) onwards, Gansu formed an integral part of the eastern section of the 'Silk Road',[1] which was the historical network of routes linking East Asia with Europe. Even though China's political and economic centres were mainly in the east and south of the country during the nineteenth and twentieth centuries, Gansu retained its significance as it was a crucial base for the central government to manage the vast northwestern frontiers. When the Second Sino-Japanese War (1937–1945) broke out, the central government moved the capital from the east to the west of China, and Gansu became a key area supporting and sustaining the country in various ways during this turbulent period.[2]

The size and composition of Gansu's population varied widely over different periods historically due to natural increases in population, changes of administrative boundaries, wars, migration and the ever-changing mix of various ethnicities. It is therefore challenging to clarify historical patterns of the region's demographic development. Wudu county, today being a district of Longnan city in Gansu province, also witnessed numerous changes in its population throughout history. In fact, Wudu's demographic development was closely connected to the general socio-economic transformations of Gansu and, at the same time, was linked with its neighbouring provinces.

Despite these challenges, scholars have noted the value of using sources concerning the Baojia (a system of community self-defence responsible for local police, militia work and collecting population statistics) to map the historical demographic development in China, particularly in the late imperial period. There have been a fair number of studies on the Baojia system in relation to Gansu during the late Qing (1644–1911) and Republican (1912–1949) eras. Cao Shuji's research on the history of China's population during the Qing dynasty, by collecting and analysing original demographic data from Gazetteers and a variety of other sources, estimates the populations of different provinces

1 Lu (2011).
2 Lu (2008), 321.

 | DOI:10.1163/9789004715707_002

and regions in selected periods.[3] In his study on the population of Republican China, Hou Yangfang estimated the national population using demographic data from different administrative units, which also include data for Gansu.[4] Both Cao's and Hou's studies provide a historical background of the administrative framework and policies regarding population (such as the Baojia system) in China, yet they have not given a more in-depth analysis on different provinces. Ran Jinhui and Li Huiyu's study on the Baojia system not only gives an overview of population administration policies throughout different Chinese dynasties, but also outlines the region-specific implementations of the Baojia system in various provinces during the Republican era. Among their region-specific discussions, they elaborate on the Baojia system in Gansu from the 1930s to the 1940s.[5] Fang Rong and Zhang Ruilan's study demonstrates the population administration policies and the corresponding demographic statistics of Gansu from the Qin dynasty to Republican China.[6] Due to the relatively abundant sources regarding the Republican period, their book details the evolution of Gansu's population and explains explicitly the quality and reliability of Gansu's demographic data for different years in the first half of the twentieth century. Although the authors extracted diverse and valuable demographic information such as sex ratio, marital status, education, and occupation from the sources, they have not further analyzed how the demographic characteristics were linked to Gansu's socio-economic developments.

In the past fifteen years, there have been important publications about the historical demography of Gansu using more analytical approaches. For example, Lu Weidong's study on the population of today's Gansu and Shaanxi provinces during Qing dynasty elaborates on the population institution, inter-province migration and the state of the Muslim (Hui) ethnicity group in this region.[7] In another book, Lu zooms in on the evolution of human settlements, urbanization and the pattern of migration caused by wars in Gansu during the late Qing period by using a set of village-level data derived from the Xuantong census. By applying GIS (Geographical Information System), he pictured the changing distributions of Gansu's inhabitants at the village level, to which previous studies pay insufficient attention.[8] Unfortunately, the Xuantong census of Wudu are lost and Lu's analysis thus excludes data on

3 Cao (2001).
4 Hou (2001).
5 Ran and Li (2005).
6 Fang and Zhang (2007).
7 Lu (2011).
8 Lu (2017).

Wudu county. In addition, Liu Dejun assesses the merits and problems of the Baojia data for Republican Gansu in his research. In particular, Liu examines the demographic data based on the provincial Baojia survey of 1935, and analyses the distribution of the population, the sex ratio, the area of different counties and regions, and the population density. Meanwhile, he demonstrates a household register form of towns and villages in Tai'an county that he retrieved from the Gansu Provincial Archives (甘肃省档案馆). However, he presents no household forms from other counties of Gansu.[9]

As Lu Weidong stated in his research, due to inadequate attention to certain types of sources and the lack of new approaches to analyse the massive, and sometimes messy, historical demographic data on China, many studies in the past are confined to estimating the demographic developments at the county, provincial and national levels. Consequently, the individuals who constituted the actual private and public spheres of the society are often left out.[10] Actually, there are a large number of forms, tables and reports encompassing demographic information about households and individuals from villages and towns in Republican China, which are surely helpful for understanding common people's life from a micro perspective. However, since such data are often problematic in quality and incomplete in quantity, they should be examined and analysed with caution.

Apart from focusing on population policy and long-term demographic changes, scholars have developed a keen interest in mapping out structural changes of the diverse aspects of Chinese society through socio-economic information obtained from population data. Over the past forty years, the well-known Lee-Campbell research group has built remarkable datasets about Chinese family, kinship and demographic behaviour that contain nearly three million records regarding around 490,000 persons from the mid-eighteenth to the early twentieth century. These records led to the production of a wealthy scholarship of fertility, mortality, marriage, social mobility, and inequality in Chinese history, though the data are limited to the Qing Imperial lineage and communities in Liaoning and Heilongjiang.[11] Since their studies are mostly about the Northeastern regions of China, and are focussed on the Imperial Lineage, some scholars have, at times, questioned the representativeness of these data.

---

9 Liu (2017).

10 Lu (2017), 3.

11 For a detailed discussion on the Lee-Campbell Research Group, see Campbell and Lee (2020).

The occupational structure from Qing dynasty to present day is another field that historical population data have contributed to. Guo and colleagues, for instance, used Baojia census, together with a wide range of sources such as gazetteers and surveys hosted by governments, to create a national occupational structure in 1640–1952. Although their work is innovative and insightful, the volume of the sample is relatively small and scattered.[12] Cheng Yang looks into the historical occupational structure by adopting an extensive collection of a different type of source, the trial papers for homicide prosecutions in China from 1734 to 1898. The occupational information from the trial papers, as Yang introduced, is overwhelmingly male-related. From his sample, he observed a stable male occupational structure in late Imperial China, which suggests a long-lasting stasis of the national economy.[13] Recent efforts by Ying Dai make use of the valuable lineage genealogies (家谱) to quantify the occupational structure in the Yangtze Valley in the twentieth century. Dai specifically compared the female labour force participation rate (FLFPR) among five localities of Shanghai and Chongqing by using household register data produced in 1948, finding a higher FLFPR in the rural areas. Particularly, the data she used for the FLFPR evaluation are the same type of data as our sample, attained from the household register enforced nationwide in the late 1940s.[14] Likewise, van Leeuwen and Li used the similar household statistics from Yugan, Jiangxi, to examine female agency within traditional Chinese households.[15] The latest studies indicate that a thorough look into local household register data in Republican China can reveal regional and gendered variations that past scholarship has paid insufficient attention to, and can add new knowledge to the broader occupational structure study.

Furthermore, historical population data are promising alternatives for quantitative research on education in the past. Wang and colleagues quantified such educational indicators as the enrolment rate, the ratio of female students, and students' family background in different periods throughout 1840s–2010s in China, using massive population surveys, census and other sources concerning a diversity of administrative regions.[16] However, they hardly touched upon the popularization of education below the county level due to the lack of sources. Nevertheless, their research is inspiring for this book in terms of interpreting education through demographic statistics.

12 Guo et al (2019).
13 Yang (2019); Yang (2022).
14 Ying Dai (2022).
15 Van Leeuwen and Li (2022).
16 Wang et al (2021).

The 1947 household register data of Wudu county that this book presents offers a good opportunity for a micro analysis, as these data are obtained from the original household register forms where the information of inhabitants was noted on the individual level, rather than from summarized reports. Although these data are problematic in certain ways, the experience of local Baojia organization during the 1930s–1940s enabled Gansu province to collect statistics more efficiently in the late 1940s. Also, the data uncover socio-economic and demographic characteristics of the common people in a couple of villages in the northwestern part of China right before the Communist Party established its centralized power. The data and the preliminary analyses of occupation, education, marriage, and the distribution of wealth in chapters 3 and 4 are worthy of further investigation for a macro history of China.

1.2 Historical Administrative Divisions of Wudu County

According to the Gazetteer, 'Wudu' as a place name originated from the Qin state during the Warring State period (475 BCE–221 BCE). The name and administration of this area changed through time. During the Ming dynasty (1368–1644), this region was called either 'Jiezhou department' (階州) or 'Jiexian county' (階县) in different emperors' reigns, depending on contemporary administrative policies. Despite the changes of name, it was constantly part of Shaanxi Buzheng Shisi (陕西布政使司). In the early years of the Qing dynasty, the western part of Shaanxi Buzheng Shisi was set apart as a single administrative unit, and was named Gansu Buzheng Shisi (甘肃布政使司), to which Jiezhou department belonged. In 1729, a larger administrative unit of Jiezhou zhili department (階州直隶州) was formed, consisting of the original Jiezhou department as well as the adjacent Wenxian county (文县) and Chengxian county (成县). In the meantime, the nearby Xigu thousand-household fortress (西固千户所), where military families were accommodated, was also merged into Jiezhou zhili department. In the wake of the rearrangement, Jiezhou zhili department became equal to a prefecture and was directly managed by the provincial government. From 1736 to 1913, the administrative area of Jiezhou zhili department remained stable.[17]

In 1913, the newly established Republican government required that all previous administrative units below province level, such as Prefecture (府), Zhili department (直隶州), Department (州) and Sub-prefecture (厅), be changed

17 Editorial Committee for Wudu County Gazetteer (1998), 87–91.

to county (县). In this context, Jiezhou was renamed Wudu county (武都县). At the same time, the area where Xigu households used to live was separated from Wudu to form Xigu county (today's Zhouqu county). In February 1929, the eastern part of Wudu county was split off to form Yongkang county, which was renamed Kangxian county in April 1929 (see the locations of Wudu, Wenxian, Chengxian, Kangxian and Xigu in 1939 in Figure 1.1). However, the administration of some areas of Wudu remained disputed. In 1941, one such area at the border with Xigu, including 294 households and 1,285 inhabitants, was formally merged into Xigu county. In 1942, the provincial government made further adjustments of disputed lands, resulting in the assignment of certain areas in the west of Wudu back to Xigu. There was a number of other administrative adjustments in the region afterwards, yet the boundary of Wudu county maintained relatively fixed until the establishment of the People's Republic of China in 1949.[18] In the past seven decades, there have been numerous changes in the territorial administration of Wudu. In 2004, the administrative unit of 'county' was relinquished and Wudu was set as a single district within the municipality of Longnan, Gansu province. This administrative system has retained its shape to the present day.

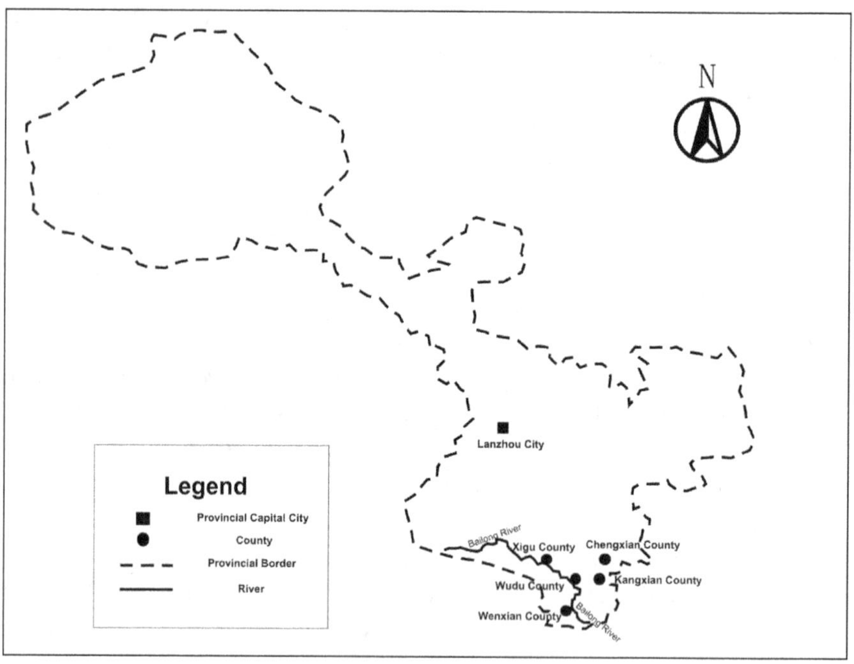

FIGURE 1.1 Sketch map of Gansu, 1939
SOURCE: OUYANG YING (1939), 25. GRAPHIC SCALE: 1:7,200,000

18 Ibid., 92–93.

1.3 The Geographical and Socio-economic Environment of Wudu County

Wudu county is located in the southeastern part of Gansu province and is surrounded by the Qinling mountains (秦岭) and the Min mountains (岷山). The terrain of Wudu is higher on the northwestern side and lower on the southeastern side, with tree-covered mountains encircling the county and the Bailong River (白龙江) crossing it from the higher to the lower areas. The highest point in Wudu county is 3,600 metres above sea level, and the height differences between mountains and valleys vary from 1,000 to 2,500 metres.[19]

According to the Gazetteer published during the 1990s, the height and the rather complicated terrains endow the county with sufficient rainfall, which enables the growth of a variety of crops and medicinal plants. Crops such as wheat, corn, potatoes, rapeseed and beets are grown widely. In the higher parts of the mountainous areas, the standard crops are soy beans, broad beans, green beans, barley and oats. In the semi-arid mountainous areas, the common crops include sorghum, grain (millet and grain), flax, and sesame. In the Bailong River valley, peanuts, cotton, rice, and sweet potatoes thrive.[20] Wudu county is also famous for its rich production of diverse types of precious medicinal drugs, including Hedysari Radix, Coptis Chinensis, Chinese Angelica, and Root of Pilose Asiabell.[21] It may be assumed that these natural conditions were seen in the 1940s as well.

Circa 2,000 years ago, Wudu was mainly settled by the Di (氐) and Qiang (羌) people; the former lived on both agriculture and husbandry while the latter lived essentially on husbandry. In fact, the Di's and Qiang's husbandry works had a great impact on local production, and husbandry is still an important part of the local economy today.[22] Nevertheless, the Di and Qiang groups lost their dominance in this region over the long run. By the Republican period, the Han Chinese group was already in the majority in this region, while the main minor ethnic groups were Muslim Hui and Tibetan.[23] It is noteworthy that the Northwest War during the Tongzhi emperor's reign (1862–1874), caused by conflicts between Han and Muslim Hui groups, resulted in tremendous population loss and a vast migration of the Hui group from Gansu to other areas.

19 Ibid., 143–144.
20 Ibid., 185.
21 Ibid., 263.
22 Ibid., 240.
23 For details see Table 2.4.

Yet research shows that the region of Jiezhou (the name of Wudu's region in Tongzhi's period) did not suffer much from the Northwest War.[24]

The long history of Di, Qiang, Hui, and Tibetan's co-residence in this region strongly affected local social traditions. Gazetteers composed during the Qing dynasty noted that the social customs in Wudu were very close to that of the ethnic minority groups. As described, local people were mostly brave, down to earth, polite and righteous, and usually lived a pure and non-luxurious life.[25]

---

24 Lu (2017), 257, 295.

25 Ye (1886), volume 8, Custom One.

CHAPTER 2

The Baojia System and the Household Register in Wudu County during the Republican Period

2.1 The Baojia (保甲) System and the Household Register (户籍登记)

The Baojia system was a grassroots administrative structure that the government applied to organize and unite households in specific periods of Chinese history. The organization of Baojia was founded on a decimal mechanism, in which normally ten households (户) were united as one Jia (甲), ten Jias formed one Bao (保), and ten Baos constituted one upper administrative unit such as a village, town, or district. It is widely thought that the Baojia system was not created until the 'Wang Anshi Reform' (王安石变法) in the eleventh century, and that the system was initially meant for military purposes. Specifically, it was devised for, first, investigating the population of every household to protect communities from suspicious people, and second, recruiting male residents from the communities to build a local defence corps.[1] In fact, similar administrative organizations based on decimalism had already taken shape as early as the Qin dynasty and transformed into various modes afterwards.[2] Such decimal systems were practised not only for surveying the size of the population, but more importantly, they were aimed at levying taxes, organizing corvee labour, and maintaining the societal order.[3]

In the early years of the Qing dynasty, the court inherited the policy legacy of the Ming dynasty, building its population register and taxation institution on a decimal administration called 'Lijia' (里甲). In practice, the major duty of Lijia was to assure the assignment of corvee and the levying of taxes that were linked to the number of Ding.[4] Ironically, however, officials never succeeded in counting a reliable number of 'Ding' or all of the inhabitants under Lijia. It became common that the wealthy escaped from reporting the increase of their household members to evade the corvee and tax, the burdens of which

1 Liu (1942), 153.

2 Ye (1936), 2–22; Ran and Li (2005), 17.

3 Ran and Li (2005), 3.

4 Officially 'Ding' (丁) referred to adult men aged between sixteen and sixty, and was counted for the purpose of taxation and corvee labour. But in the Qing dynasty 'Ding' was no longer connected to real men but became a unit for taxation. See Ho (1959).

 | DOI:10.1163/9789004715707_003

thus fell onto poor households. As a result, a large number of poor people ran away from their homes to avoid these burdens, thereafter making the population register exist in name only. To solve these problems, the Qing government issued an ordinance in 1712 that, from then on, the amount of tax would not be connected to the number of household inhabitants, meaning that the tax would not increase in accordance with the population growth. During the Yongzheng emperor's reign (1722–1735), taxes connected to 'Ding' were gradually relinquished in some provinces, and the size of lands became the major gauge for taxation. The changes in taxation to some extent eased people's fear of the population register. In the mid-eighteenth century, the Qing court started to enforce the Baojia system, which was much stricter than the Lijia, so as to curb conflicts associated with land taxation and to maintain the societal order.[5]

Entering the twentieth century, the empire faced various challenges both at home and from abroad. The Qing court decided to reform the country following the patterns of Western countries, and these changes included replacing the Baojia system with a self-governance system. In 1908, the government issued the *Constitution of the Self-governance for Cities, Towns and Villages* (《城镇乡地方自治章程》). According to the Constitution, administrative units within counties were categorized as cities, towns, and villages, depending on the size of the population. Members of the gentry in cities, towns, and villages would be elected to form local committees, which were supposed to supervise and make decisions about local public issues regarding education, public infrastructure, agriculture, industry, commerce and social relief.[6] The Constitution was not fully carried out as the Qing dynasty was soon overthrown in 1911, yet the three-layer (city-town-village) administration carried over to the early years of the Republic of China.

During the Beiyang government era (1912–1928), China was dominated by warlord conflicts between different provinces and regions, which obstructed the administrative organization below the county level from reaching a unified format. When the warlord conflicts were largely ended by the Nationalists led by Chiang Kai-shek in 1928, the city-town-village system was replaced by a four-layer system, namely Qu-cun-lü-lin (区村闾邻). In such a system, every five households formed one Lin (邻, literally meaning neighbour), every five Lins formed one Lü (闾, an administrative unit below the village), every four Lüs formed one Cun (村, village) and every twenty Cuns formed one Qu

5 Lu (2011), 20–21; Ran and Li (2005), 51–52.
6 Editorial Institute of the General Political Department (1911), 1–3.

(区, district).[7] The new four-layer administration was structured in a similar manner to the Baojia system, but it embraced the idea of self-governance that local administrations should govern their communities by themselves. This reflected what Sun Yat-sen (1866–1925), the founding father of the Republic of China, had advocated, namely that people were entitled to self-governance.[8] At the same time, the Communist Party was developing and was trying to expand its influence in rural areas of the provinces of Jiangxi, Hubei, Anhui and Henan in the central part of China. In 1931, the Communist Party established its reign in Jiangxi. Faced with the expansion of the Communists, the Nationalist government pondered over the necessity of implementing the Baojia system in the above-mentioned areas. As far as the Nationalists were concerned, the Baojia system would be efficient in investigating the inhabitants of every household, recruiting men for local military groups and enhancing social order and cohesion against the Communists. In addition, the households in such a system were required to oversee each other, making it easier to detect a suspicious person, especially a Communist member.[9]

In 1931, the Nationalist government initiated a pilot Baojia scheme in forty-three counties in Jiangxi province.[10] Before long, the pilot project was introduced to Hubei, Anhui and Henan provinces. The effectiveness of the Baojia system in population management and self-defence in the four above-mentioned provinces gave the Nationalists confidence that Baojia was more suitable for the whole country at that particular time. In fact, provincial governments of Jiangsu, Zhejiang, and Fujian began to consider the feasibility of practising Baojia in their territories shortly after the changes were made in the above-mentioned provinces. On the Central Political Meeting (中央政治会议) in November 1934, Baojia's efficiency in self-defence and its potential for being the foundation of self-governance was credited. Eventually, delegates reached the agreement that the Baojia system should be enforced nationwide.[11] Against this background, Baojia began to be organized across the country.

According to the Baojia regulation, the administrative arrangement proceeded in the following order. First, there should be surveys of the number of households and inhabitants (户口调查). The household was the most basic unit in this system, and a head should be set in every household. Each household was given a doorplate with a house number, and the doorplate was required

7 'The Law of the Organization of County' (1928).
8 Xin (1991), 187–188.
9 Ran and Li (2005), 59–64.
10 Editorial Committee for the Ministry of Interior Yearbook (1936), 373.
11 Ibid., 361.

to be hung at a visible site at the entrance of each residence. Ten households formed one Jia, and one head of the Jia was elected by the heads of households. Ten Jias formed one Bao, and a head of the Bao was elected by the heads of the Jia. A certain number of Baos formed a district, while a certain number of districts constituted a county. This structure was, however, not strictly followed in all provinces. In Gansu, for instance, ten Baos formed one village (乡) or town (镇) instead of a district.[12] Second, the heads of households, Jias and Baos ought to sign official documents to declare their duties for local security. The heads of households needed to report the birth, death and mobility of their household members, and were required to oversee other households within their own Jia. This was aimed at detecting people who might be associated with the Communists. The heads of Jias and Baos were responsible for local self-defence issues, such as reporting any changes in the local population, supervising the construction of defence infrastructure, and assisting the police in arresting suspected criminals.[13] Third, the heads of Baos and Jias should together work out sketch maps of the areas they took charge of, on which they should signify the names of their districts/towns/villages and the numbers of households and inhabitants. The maps should be first sent to the district magistrate, and later to the county magistrate.[14] Clearly, the self-defence function of Baojia was much appreciated in the 1930s, and was regarded as the primary reason for implementing Baojia countrywide.

In 1936, the Nationalist government announced that it was time to prepare for establishing a constitutional government via an election, prior to which they needed reliable statistics concerning the population of the country so that they could develop an idea of how to allocate the quota of election candidates to each province. In January, the Ministry of Interior (内政部) commanded that provinces and special cities submit reports of their Baojia households and inhabitants. Though far from perfect, the 1936 population statistics derived from the Baojia survey were of great importance in the demographic history of Republican China.[15]

While the Nationalist government emphasized the self-defence function of local administrative organization, the function of self-governance, much appreciated by Sun Yat-sen, followed as a close second. As the Second Sino-Japanese War broke out, a lot of territory in the coastal and central parts of China was occupied by Japan, and numerous Nationalist initiatives subsequently ground

12 Ibid., 395–394; Editorial Committee for Wudu County Gazetteer (1998), 91.

13 Editorial Committee for the Ministry of Interior Yearbook (1936), 361–363.

14 Ibid., 361–364.

15 Hou (2001), 73.

to a halt, severely eroding the role of Baojia. In October 1938, Japanese military invasions eased, and the Nationalist government thought it was high time to switch its focus from 'military fighting' to 'political construction of the country'. Under these new conditions, the Nationalist government began to implement the New County System (新县制) to consolidate its rule in the unoccupied, especially the rural, areas.[16] The focal point of the New County System was to enhance the self-governance function in local administration. Particularly, it highlighted a variety of self-governing duties, such as surveying households and inhabitants, setting the price of land, reclaiming wastelands, improving local health system and social relief, and abolishing backward traditions and customs. The government undoubtedly hoped that such works would strengthen people's faith in the Nationalist Party, and would mobilize people to actively participate in agricultural and industrial production that could sustain the country at war.[17]

There were adjustments to the administrative organization under the New County System, with Bao and Jia being the most fundamental building blocks. As before, both Baos and Jias were formed based on decimalism, yet now the practical organization seemingly became more flexible. To be specific, an upper administrative unit was allowed to encompass as many as fifteen lower units in densely populated areas, and as few as six units in sparsely populated areas. Baos and Jias constituted the villages or towns, both of which thereafter constituted a county.[18] The key roles of Bao and Jia in the New County System were characterized as surveying households and inhabitants, engaging in education and military training, regulating and reclaiming lands to enhance economic productivity, rendering medical and sanitary services, and offering military protection.[19] However, it is noteworthy that the processes of reorganizing Baojia varied by county: while some counties took it seriously, others did not.

The practice of the household register during Republican China was often mingled with the Baojia survey of households and inhabitants and (especially before the 1940s) was disorderly. In 1931 the first Law of Household Register (户籍法) was released, requiring information about the composition of households and the identification of individuals. Specifically, the required items on the register included the number of household members, sex, age, occupation,

16 Xin (1991), 184–185.
17 Ibid., 186–187.
18 Ibid., 189.
19 Ibid., 193.

date of birth, child adoption, marital status, custody, death and inheritance.[20] Nonetheless, a nationwide household register in accordance with the law was not possible at the time for various reasons. First, some provinces were suffering from floods or droughts that distracted them from allocating money or personnel to the register work. Second, other provinces had already arranged the survey of households and inhabitants when organizing their Baojia administration earlier, so they thought it unnecessary to compile a similar demographic register again. Additionally, the contents of the household register overlapped with that of the Baojia survey, which led people to doubt the justification for a household register. Contemporary experts perceived that the Baojia survey differed from the household register, in that the former aimed at organizing inhabitants in self-defence units, and the latter at collecting demographic information and assuring the identification of inhabitants. Nevertheless, it was confusing for many to separate the Baojia survey from the household register, and thus a nationwide register was suspended.[21]

As mentioned above, the Nationalist government started to conduct the Baojia survey in unoccupied areas a few years after the war with Japan began. However, most of the survey reports were unreliable for various reasons, such as the increasing mobility of inhabitants during the war and the underreporting of male adults who attempted to escape military service. In order to strengthen the quality of population statistics and to realize the wartime goal of 'political construction', the Nationalist government announced that it was to undertake the household register across the country again in 1943. It was stipulated that the register be divided into two parts, and be accomplished step by step within two years after the war ended. For the areas formerly occupied by the Japanese, the first step was to rearrange the administrative units of Bao and Jia on the basis of surveying local households and inhabitants. Specifically, the Baojia survey covered the inhabitant's sex, status of domicile, age, education, occupation and marital status. The population statistics and the surveys were later submitted to the county or municipal government.[22] The second step was to create household register books. In this phase, a formal form of each household would be filled in based on the Baojia survey, and these forms would be bound together to produce a register book. Aside from the register book, analytical reports based on the demographic statistics were supposed to be made regularly. For the areas not occupied by Japan, the local

20 'The Law of Household Register Issued in 1931' (1996), 29–53.

21 Hou (2001), 67–68.

22 Jiangxi Provincial Government (1943): 41–43; 'The Procedures of Implementing the Household and Identity Register based on Baojia Survey in Counties and Cities and Matters the Investigators Need Pay Attention to' (1944), 31.

government could conduct a household register directly without a Baojia survey. The registered information regarding counties and cities was meant to be submitted to provincial governments in the form of a report, which would eventually be sent to the Ministry of Interior.[23] By the end of 1946, nearly half of China's counties and cities had already reported their household register results.[24]

In order to achieve better-quality demographic statistics after the war, the government issued an amendment of the Law of Household Register in January 1946. According to the amendment, an arrangement of the Baojia administration and a survey of households and inhabitants had to be accomplished prior to the household register procedures, which now included three types. The first was the register for the origin of the household, which indicated whether the household originated from the place of registration. The second was the register for identity, which indicated demographic information such as birth, death, marital status and adoption. The third was the register for migration, where the state of immigration or emigration had to be noted. In finishing the household register, the inhabitants could then be issued with their identity card.[25] It was required that all counties and cities, whether they had already done the register or not, should register their inhabitants in accordance with the amendment and submit the results to the Ministry of Interior as soon as possible. From 1947 to early 1949, the household register was conducted countrywide, and updated reports were sent to the Ministry of Interior regularly.[26]

2.2 The Baojia Demographic Data of Gansu Province and Wudu County

The demographic data of Gansu province and Wudu county produced in the Republican period are massive in volume yet patchy in quality. During the Beiyang government period, Gansu province was controlled by different military factions with the administrative organization varying from place to place. Though local governors did carry out relatively strict household registers to enforce taxation and corvee assignment, relevant historical sources have not been preserved and thus detailed information about the household register in this period is scarce.[27] In 1934, the Nationalist government started to implement the Baojia system throughout the country, and Gansu was an active

23 Jiangxi Provincial Government (1943): 41–43.
24 Hou (2001), 78.
25 The News Bureau of the Executive Yuan (1947), 19.
26 Hou (2001), 80–89.
27 Fang and Zhang (2007), 419.

participant. Actually, the demographic statistics of Gansu from 1934 to 1949 were basically obtained through the Baojia system.[28]

In 1934, the Gansu government required that a complete Baojia administration should be arranged within one year across the province, following the Baojia ordinance issued by the central government. However, local officials anticipated that this work would not be easy. First, the territory of Gansu province was very large while a good proportion of villages and towns were embedded in the mountains; therefore, it was unlikely that surveyors would be able to travel from one place to another fast. Second, there were no specialized institutes to supervise the Baojia work within the whole province at the same time, so errors might not be discovered or corrected in a timely fashion if the work was to be conducted in different counties concurrently.[29] Considering these problems, and to work efficiently, the Gansu government divided all its counties into four groups, and commanded that the survey of households and inhabitants, as well as the arrangement of Bao and Jia administration, be carried out one after another among the four groups. Meanwhile, they expected all the work to be accomplished by the end of June 1935. Wudu county was assigned to the third group, whose Baojia work was done from January to May of 1935. The surveyed statistics of households and inhabitants were submitted to the Ministry of Interior at a later date, resulting in a publication in April 1936. According to the 1936 publication, there were 36,490 households and 173,938 inhabitants (91,229 males and 82,709 females) in Wudu county, and 1,060,387 households and 6,422,808 inhabitants (3,510,913 males and 2,911,895 females) in Gansu province as a whole.[30] According to Fang and Zhang, the quality of these statistics is high compared to the demographic data in the following war years, because they were obtained under strict governmental instructions and great political pressure to mobilize people against the Communists. Nonetheless, they are by no means precise. For example, the variable definitions of the types of households and the dissonant ways

28 Ibid., 455.

29 Lu (2008), 323.

30 The number of households and inhabitants of Gansu listed here were calculated by Fang and Zhang on the basis of the original publication by the Gansu government. In the original publication, statistics for Xiahe, Xihe and Liangdang counties are absent, because Xiahe had not finished the Baojia survey and the documents from Xihe and Liangdang did not arrive at the provincial government in time for publication. Fang and Zhang supplemented this gap with data published later. In 1935 the territory of Gansu province included Guyuan, Longde, Haiyuan and Huaping counties, which were incorporated into Ningxia Hui Autonomous Region in 1958. The provincial statistics of 1935 here exclude these four counties. See Fang and Zhang (2007), 465. 'The Overview Table of Households and Inhabitants Based on the Survey and Arrangement of Baojia in Counties of Gansu Province' (1936), 82–87.

of the counting in different counties blurred the quality.[31] In addition, the sex ratios in both Wudu and Gansu were unreasonably high (110 males versus 100 females for Wudu, and 121 males versus 100 females for Gansu), which means that a large number of women were unreported.

From 1937 to 1943, the Gansu government required a provincial Baojia survey of households and inhabitants roughly every year, but these demands were hardly met for two reasons. First, the outbreak of the Second Sino-Japanese War in 1937 brought about a massive influx of migrants from the eastern occupied provinces to the western unoccupied provinces including Gansu. This resulted in the increased mobility of inhabitants, which made the counting of residents more difficult. Second, the primary purpose of the Baojia survey during these years was to conscript soldiers for the army, recruit corvee labourers for building new airports and roads, and levy a grain tax for sustaining the war. Not only did ordinary people resent these duties, but officials at the county, Bao and Jia levels also resisted them because they were well aware that the local inhabitants would not cooperate in a population survey for such purposes. In fact, some inhabitants, especially young men, ran away from their home to escape the survey, while officials in some counties postponed or suspended the survey work, giving various excuses. For these reasons, obtaining reliable Baojia survey results was almost impossible. From the late 1938, the Gansu government tried to dispatch specialized survey supervisors to monitor the work in counties, and the supervisors did help detect numerous unreported cases. But this method did not essentially change the low quality of the Baojia data from 1937 to 1943.[32] Despite this, research shows that the Baojia survey in this period did play a positive role in identifying suitable young men and conscripting soldiers, which substantially supplied the Nationalist army with soldiers for the war.[33]

In 1946, the Nationalist government announced the implementation of the amended 'Law of Household Register', and the Gansu government immediately started to organize an overall Baojia survey and household register across the province accordingly. Regarding the Baojia arrangement, the amended law allowed a more flexible organization. Baos and Jias used to consist of ten smaller units. However, in practice, the number of the basic cells varied from six to fifteen depending on the population density. Now, according to the new law, Baos and Jias of cities could be combined in a larger structure consisting of ten to thirty smaller units. From the sample household register form attached to this amended law, it is known that the following information regarding

31 Fang and Zhang (2007), 461–462.
32 Ibid., 467–481.
33 Lu (2008), 325–326.

household members was required: name, sex, birthdate, status of domicile, date of first residence,[34] education, occupation, marital status, status of physical disability and identity card number.[35] Moreover, it was stated that new items could be added where necessary.[36] In Gansu, religion was supplemented in the register form. Although most counties had submitted their Baojia survey results by the end of 1946, only thirty-eight of the seventy-three counties managed to compile the household register books based on their Baojia survey forms.[37]

Since the survey and arrangement of Baojia was the central government's main focus in 1947, the Gansu government continued to undertake the provincial household register in this year. The household register work in different counties of Gansu was scheduled in three phases from April to October 1947, and that of Wudu county was completed in the third phase from June to August. By the end of October, the Baojia survey, the household register, and the issue of identity cards were basically completed in all counties, except for a few places where the Baojia system had not been practised for long (particularly in regions where Mongolian and Tibetan inhabitants dominated) and where societal order was not yet restored. The results were published as a report by the Gansu government in 1948.[38] The provincial officials regarded the quality of the 1947 survey and household register data as satisfactory, due to Gansu's relatively rich experience in organizing the Baojia system during the previous decade, and to the keen cooperation and participation of officials and gentry to different extent for the purpose of post-war construction.[39] The results show that there were 38,548 households and 223,942 inhabitants (114,783 males and 109,159 females) in Wudu county, and 1,222,170 households and 6,739,977 inhabitants (3,514,039 males and 3,225,938 females) in Gansu province as a whole in 1947.[40] Nevertheless, Fang and Zhang warned that the 1947 data are still biased, as the problem of underreporting was not entirely solved because of the political pressure regarding soldier recruitment, the labourer corvee and the grain tax. In particular, the female population across China was often underreported in this period.[41]

34 The starting date of residence in this county by the household.

35 Executive Yuan (1996), 87.

36 Ibid., 83.

37 Fang and Zhang (2007), 496.

38 Gansu Provincial Government (2015), 267–304.

39 Ibid., 303.

40 The original statistical report includes the demographic data of several counties that are not in the territory of Gansu today. The results from Gansu exclude the data for these counties. See Gansu Provincial Government (2015) and Fang and Zhang (2007), 502–509.

41 Fang and Zhang (2007), 503.

The Baojia survey of households and inhabitants was not held in Gansu during 1948–1949, because such work had already finished by the end of 1947. But the reporting of births, deaths and migration was getting increasingly institutionalized and relevant formats and statistical reports were created on a regular basis. The demographic statistics of Gansu in these two years are mainly built on such sources.

Nevertheless, Fang and Zhang point out that the quality of Gansu's demographic data from 1948 to 1949 is far from satisfying, despite the institutionalization of the household register. They compare the population data made by the Nationalist government in the mid-1949 with that of the Communist police department at the end of 1949, and find a great discrepancy. In terms of the Communist data, there were 1,808,505 households and 9,205,114 inhabitants in Gansu at the end of 1949, almost 30% more than the Nationalist data of the same year (see Table 2.1). They argue that the Communist government of Gansu eliminated the policies of conscripting soldiers, assigning the corvee and levying the grain tax in August 1949, which eased people's fear of the survey of households and inhabitants tremendously. Consequently, those who were previously invisible in the household register were now added to the register book, making the population results closer to reality.[42]

In general, the development of the population of Wudu and Gansu from the late Qing period to 1949 can be seen in Table 2.1.

2.3 The 1947 Household Register of Wudu County

From the forms where we obtain Wudu's demographic data, as presented at the end of this book, it can be inferred that these forms are part of the county's household register books, and are the result of the 1947 household register. Figures 2.1 and 2.2 show that our sources well match the format devised by the 1946 Law of Household Register. According to the regulation, for every 100 household forms, one register book would be bound. Meanwhile, books of domicile households were separated from those of lodging ones. All these books were produced as two copies, with one copy preserved by the village or town government and the other one sent to the county or municipal government.[43] Since the sources we have are books of Baos of several villages instead of one single village, we infer that these sources were copies gathered and preserved by the government of Wudu county. It is also noteworthy,

42 Ibid., 518–520.

43 'The Procedures of Implementing the Household and Identity Register Based on Baojia Survey in Counties and Cities and Matters the Investigators Need Pay Attention to' (1944).

TABLE 2.1 Population of Wudu and Gansu from late Qing to 1949

	Wudu county (Jiezhou)		Gansu province		China
	Number of households	Population	Number of households	Population	Population (million)
1851	–	–	–	16,573,000	436.1
1880	–	–	–	4,521,000	364.4
1907	18,433 (Jiezhou)	106,625	780,869	4,083,693	410.0*
1928	29,057 (Wudu)	137,811	948,133	5,052,467	–
1935	36,490 (Wudu)	173,938	1,060,387	6,422,808	530.0**
1946 (December)	38,157 (Wudu)	203,362	1,159,043	6,536,284	520.0
1947 (October)	38,548 (Wudu)	223,942	1,222,170	6,739,977	–
1949 (June)	38,596 (Wudu)	226,286	1,286,133	7,080,187	–
1949 (December)	–	–	1,808,505	9,205,114	541.7

SOURCES: DATA FOR WUDU (JIEZHOU) AND GANSU FROM 1907 TO 1949 ARE FROM FANG AND ZHANG (2007). DATA FOR GANSU AND CHINA FOR 1851 AND 1880 ARE FROM CAO (2001). DATA FOR CHINA AFTER 1911 ARE FROM HOU (2001)

Notes:

1. Although Fang and Zhang (2007) provide estimates of Gansu's population for the 1860s and 1890s, we tend to use statistics estimated by Cao Shuji when considering the population of both Gansu and China in the late Qing era. Cao specializes in the population history of the Qing dynasty and has conducted in-depth research on the essence of 'Ding', and therefore we prefer to rely on his data.
2. From 1862 to the mid-1870s a large-scale ethnic and religious war took place in the north-western provinces of the Qing empire. It was mainly related to the Muslim Hui groups in this region, and it resulted in a huge death toll and emigration. This explains why Gansu's population in 1880 was almost a quarter of that in 1851.
3. The numbers of households and inhabitants for 1935 in Wudu and Gansu are much larger than in 1928. In addition to natural population growth and immigration, this might be attributed to the relatively high quality of the 1935 Baojia survey, which was aimed at self-defence against the Communist force.
4. The data for Gansu cover the territory of today's Gansu province, and the data for Wudu cover the county's territory at the end of the 1940s. The data for Jiezhou exclude the populations of Xigu, Wenxian and Chengxian.

* 1911

** 1936

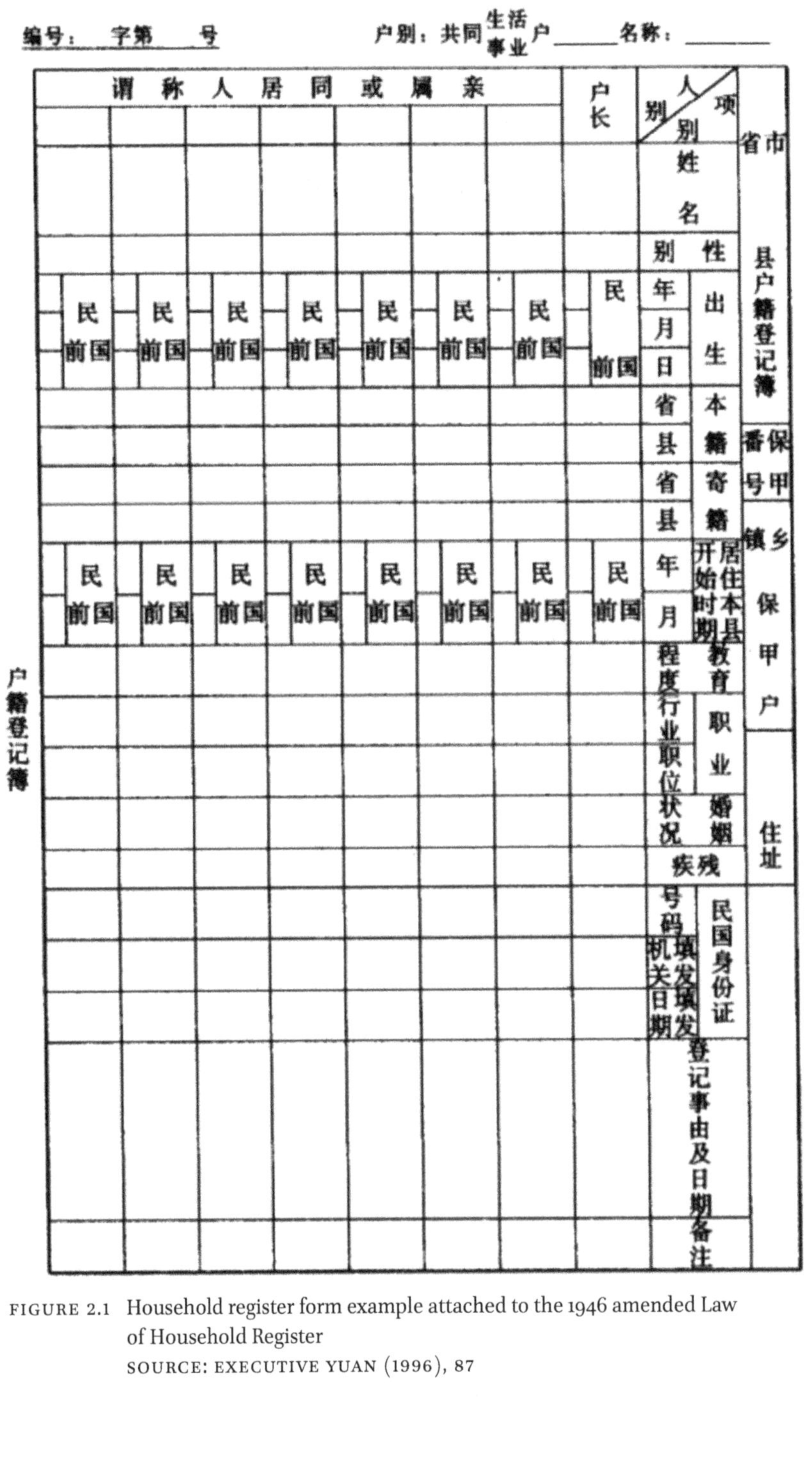

编号：字第　号　户别：共同生活/事业户____名称：____

户籍登记簿

省市县户籍登记簿

项别/人别	户长	亲属或同居人称谓						
姓名								
性别								
出生 年月日	民国前	民国前	民国前	民国前	民国前	民国前	民国前	民国前
本籍 省								
本籍 县								
寄籍 省								
寄籍 县								
居住本县开始时期 年月	民国前	民国前	民国前	民国前	民国前	民国前	民国前	民国前
教育程度								
职业 行业								
职业 职位								
婚姻状况								
残疾								
国民身份证 号码								
国民身份证 填发日期								
国民身份证 填发机关								
登记事由及日期								
备注								

保甲番号

住址：乡镇　保　甲　户

FIGURE 2.1 Household register form example attached to the 1946 amended Law of Household Register
SOURCE: EXECUTIVE YUAN (1996), 87

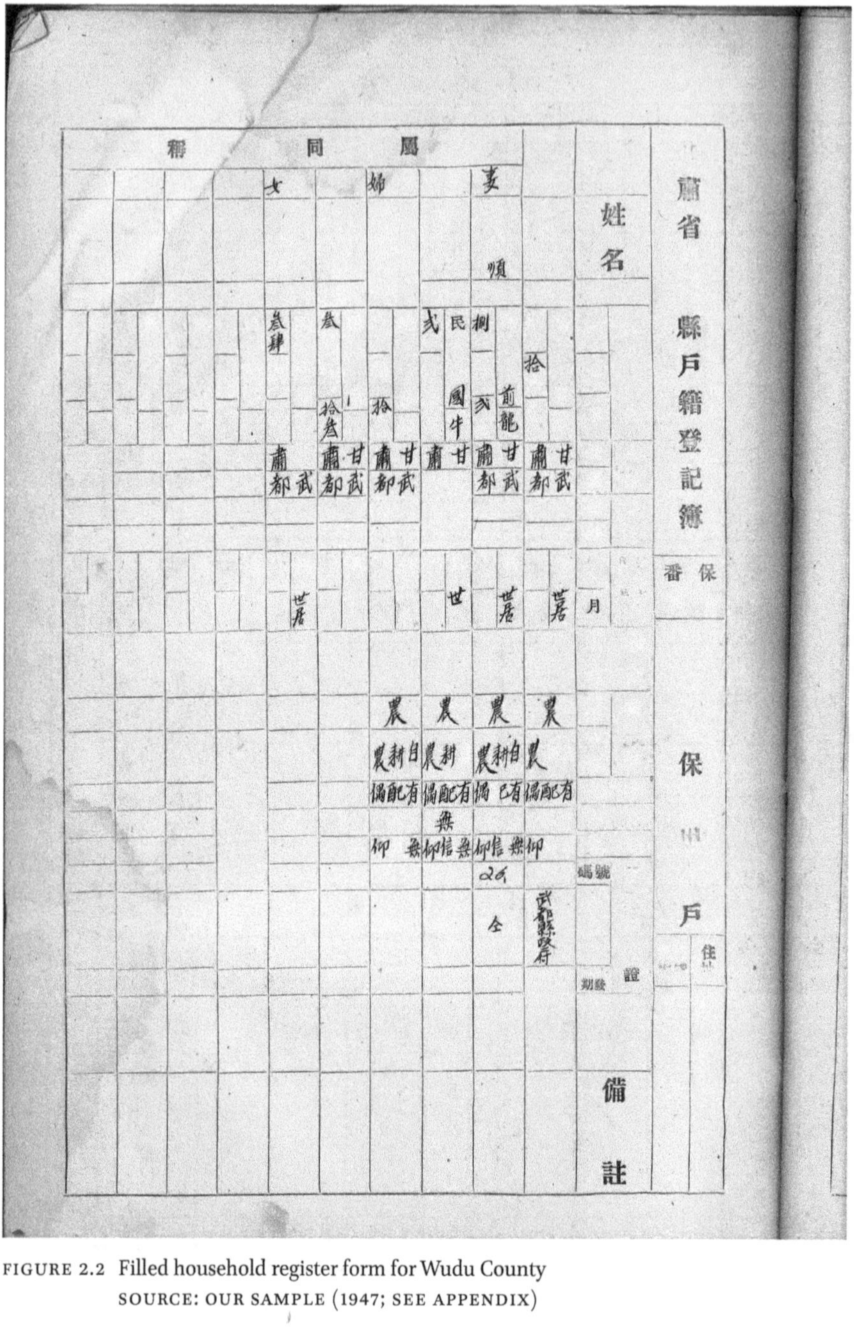

FIGURE 2.2 Filled household register form for Wudu County
SOURCE: OUR SAMPLE (1947; SEE APPENDIX)

however, that in a few cases new information after 1947 was added to the register books. For instance, in the sources one can find remarks that someone was born or died during 1948–1949. Yet it is not clear why and to what extent local officials added new information to the 1947 register books after 1947.

The data we present in the appendix of the book are part of the 1947 household register data of Wudu county. From Table 2.2 we know that our sample covers around two-fifths of all villages/towns of Wudu, yet the number of Baos, Jias, households and inhabitants cover around 7% of the total. The villages (Xiang in Pinyin, 乡) included in our sample are Muma Village (木马乡), Fujin Village (福津乡), Ganquan Village (甘泉乡), Pipa Village (琵琶乡), Anhua Village (安化乡), Housong Village (後宋乡), Liulin Village (柳林乡), Jiucheng Village (旧城乡) and Kongti Village (孔提乡). Apparently, some Baojia data regarding these nine villages are still missing. Moreover, the household register forms of one particular Bao do not have any indications about to which village this Bao belonged. In the appendix, we refer to this Bao as "village unknown 1st Bao". The locations of the nine known villages can be seen in Figure 2.3.

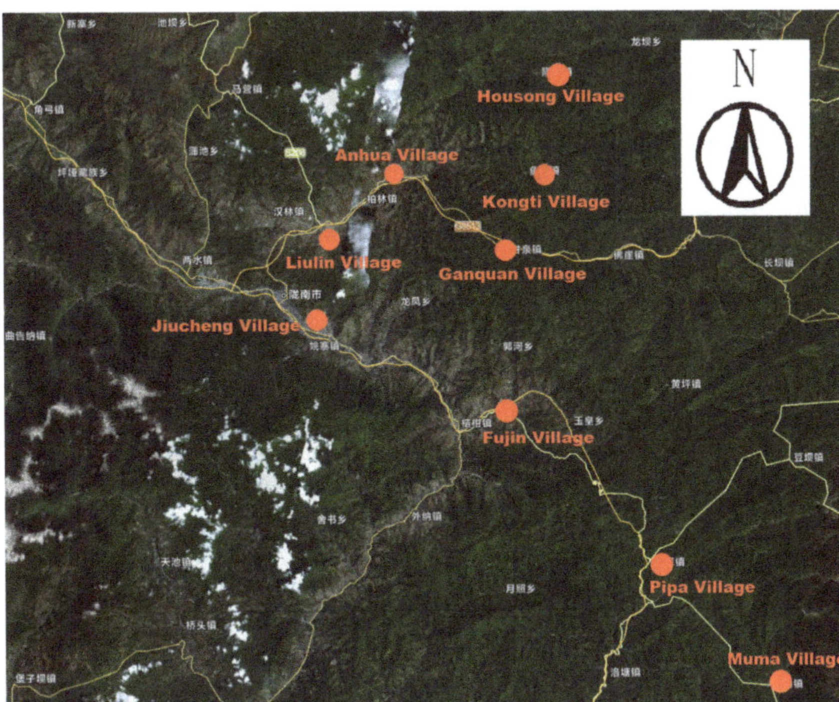

FIGURE 2.3 The locations of the nine villages of Wudu County in our sample
SOURCE: NATIONAL PLATFORM FOR COMMON GEOSPATIAL INFORMATION SERVICES (MAP WORLD). RED SPOTS ADDED BY THE AUTHORS

TABLE 2.2 The number of villages/towns, Baos, Jias, households and inhabitants in our sample and in Wudu county as a whole, 1947

	Number of villages	Number of Baos	Number of Jias	Number of households	Number of inhabitants
Our sample	9	14	185	2,605	15,913
Wudu county	22	195	2,655	38,548	223,942
The share of our sample (%)	40.9	7.2	7.0	6.8	7.1

In spite of the relatively small sample and the unavoidable problem of underreporting, the available data provide many new insights. On the one hand, they show the original format of the 1947 household register, helping us enhance our understanding of the institution of population management and its local practice in the late-1940s China. On the other hand, it renders richer records than the summarized statistical reports that researchers usually use, providing the opportunity for in-depth analysis that can add new knowledge to the demographic history of Wudu and Gansu in the mid-twentieth century. Moreover, the sample delivers abundant socio-economic information in relation to family structure, sex, marriage, education, occupation, and wealth distribution in an inland, underdeveloped, and less populated area of China. Hence, it is a precious source for scholars interested in this field to conduct comparative research between regions.

CHAPTER 3

The Socio-economic Position of Wudu County

3.1 Comparing Our 1947 Household Register of Wudu County to Alternative Sources

As mentioned in the previous chapters, our 1947 sample of Wudu county is part of a wider survey, the abbreviated results of which were published by the Nationalist government. From these results we know total population, education and occupation.[1] Yet many reference types of data are unavailable when relying solely on these official publications. This includes age-specific data, marital status, religion, relation to the head of household and various combinations of the previously mentioned indicators. Hence, even though the official publications are very informative, they have a limited connection with various branches of the socio-economic literature such as that on well-being.

3.2 Occupational Structure

In 1945 and 1947, the working age population (i.e. the population over twelve years old) in Wudu was about 70% of the total population.[2] Combined with an unemployment rate of 7.27%, as listed in Table 3.1, the total working population in Wudu county made up 64.9% of the total population.

Within this working population, it was the agricultural employment rate that was, at 93%, higher than the average level of the whole of Gansu (83%). However, almost all other economic sectors in Wudu lagged in occupational shares to Gansu. Indeed, due to the lack of promising job opportunities in other sectors, a large proportion of the population in Wudu had no other option but to stay in rural areas and engage in agricultural production for survival. These people, no matter how much/little time they spent on agricultural work each day, were categorized as agricultural labourers, which resulted in a higher overall employment rate in agriculture. In addition, in 1947, because of disruptions caused by the civil war, there was an economic depression in Gansu. This led to the shrinking of various economic sectors which thus accommodated fewer employed labourers than in 1945 and thus led to a

1 Cao (2013).
2 Fang and Zhang (2007), 632, 636.

 | DOI:10.1163/9789004715707_004

TABLE 3.1 Occupation in Wudu county and Gansu province in 1945 and 1947 (%)

	Unemployed population	Employed population	Employed population									
				Agriculture	Mining	Industry	Business	Transport	Civil services	Freelance	Personal services	Others
1945												
Gansu	15.91	84.09	100	80.73	0.24	6.16	4.35	0.37	2.51	1.05	2.15	2.43
Wudu	7.30	92.70	100	81.01	0.05	11.32	1.74	0.06	0.69	0.36	1.16	3.61
1947												
Gansu	18.17	81.23	100	82.72	0.10	8.51	3.95	0.26	1.68	0.52	0.88	1.37
Wudu	7.27	92.73	100	93.36	No data	0.71	1.17	0.01	0.36	0.17	0.92	3.29
Our sample (1947)	2.1	97.9	100	98.13		0.8	0.08	No data	0.17	0.05	0.06	0.71

SOURCE: FANG AND ZHANG (2007), 632–637. OUR SAMPLE (1947)

Note: The definitions of certain occupations and the unemployed/employed differed in different counties, and there were errors in reporting and copying relevant information at the county level.

TABLE 3.2 Employment by occupation and sex in Wudu county in 1947

Industry	Occupation	Men	Women
Agriculture	Landholding peasant	3,835	3,483
	Tenant farmer	1,029	928
	Hired labour	5	0
Industry	Stonemason/carpenter	4	0
	Weaver	3	58
Services	Civil government	5	0
	Businessperson	3	1
	Teacher	8	0
	Other	12	6

SOURCE: OUR SAMPLE (1947; SEE APPENDIX)

decreasing rate of employment. Yet this was counterbalanced by some of the unemployed population moving back to the agricultural sector, thus leading to higher agricultural employment numbers. Consequently, by and large overall employment remained constant.

As regards the (lack of) growth of industrial employment, Fang and Zhang argued that it does not mean the development of industry stagnated. They found that it was largely the result of more complete statistics on women engaging in traditional rural family handicraft manufacturing, which was largely included in agriculture.[3] The same economic decline occurred in Wudu, because of the notable expansion of the scale of agricultural labourers from around 80% in 1945 to 90% in 1947.

Looking at our sample, 69.12% consisted of employed working age people (10,999 out of 15,913). The ratio was very close to the statistics from the whole of Wudu mentioned above (64.9%). However, the proportion of the population engaged in agricultural work and the employment rate were relatively higher than the overall calculation made by Fang and Zhang. We have the occupational information of 10,244 people, and approximately 2% (215) of the total population above twelve years were unemployed. Among the entire employed population, 98.13% of people were in the agricultural sector. Only 0.8% were employed in industry. In addition to several carpenters, masons and blacksmiths, more than 85% of industrial labourers were composed of spinners and weavers (almost all female) (see Table 3.2). We should notice, as stated in

3 Ibid., 633.

Section 2.3, that our entire sample comes from rural areas. Hence the extremely high agricultural employment levels and the dominance of the textile industry in the industrial sector are understandable. Services appear with even less frequency, with most persons active as teachers, businesspeople or government employees.

3.3 Education

Besides occupation, a second variable strongly related to well-being is education. As shown in Table 3.3, the proportion of those who left education after spending some time in basic education (i.e. Sishu, primary school and junior high school in Gansu province during 1945–1947) declined. Conversely, the share of illiterate people and those pursuing higher education increased. The school age population also grew from 83.3% to 84.5%. In addition to natural demographic age structure fluctuation, Fang and Zhang argued that the migrants from other regions were the main driving force of the increasing demographic growth in Gansu after 1945. The polarization of the migrant population's educational participation played an important role in the changing demographic educational structure. In other words, the majority of those who migrated to Gansu were adults who were either illiterate or well educated.[4] In Wudu the illiteracy rate declined slightly during these years, but it was only the result of a decrease in the population below the age as most children went to school (i.e. below six years old).

Compared with Fang and Zhang's investigation, a slightly larger proportion of the population reached school age in our sample. Yet nearly all persons in our sample were illiterate as they were mostly agricultural workers. The lack of urban samples pulled up the average illiteracy rate. There were 13,317 people reaching school age out of our sample of 15,913 persons. Excluding 871 people who did not offer any information about their educational participation, we still have a valid sample of 12,445 persons. Among these 12,445 persons, 11,629 people were illiterate. The illiteracy rate was higher than the overall data from Wudu. Our data have a smaller share of persons obtaining primary, secondary and tertiary education. The only exception occurred in senior high school education. Our sample contains a higher percentage of the population who attended senior secondary education. We analysed their occupations and found that most of them either served in the services sector

4 Ibid., 641.

TABLE 3.3 Education in Wudu county and Gansu province in 1945 and 1947 (%)

	Population reaching school age		Population reaching school age													
				Illiterate		Sishu		Primary school		Junior high school		Senior high school		University	
	Wudu	Gansu	100	Wudu	Gansu	Wudu	Gansu	Wudu	Gansu	Wudu	Gansu	Wudu	Gansu	Wudu	Gansu
1945	83.45	83.33	100	91.70	82.60	3.10	5.35	4.70	10.56	0.40	1.09	0.08	0.29	0.04	0.10
1947	81.73	84.45	100	92.40	87.28	2.46	4.14	4.45	7.02	0.51	1.05	0.15	0.34	0.05	0.14
Our sample (1947)	83.68	–	100	93.44	–	2.11	–	4.14	–	0.44	–	0.19	–	0	–

SOURCE: FANG AND ZHANG (2007), 638–645. OUR SAMPLE (1947; SEE APPENDIX)

as headmasters or civil servants in townships and villages, or they were still enrolled as students. Therefore, we assume that rural areas were not economically attractive, so people with a better educational background tended to go to cities to work. Consequently, from an overall perspective of Wudu, the proportion of the population who obtained tertiary education was relatively higher, whereas in rural areas the gap caused by the lower percentage of people in tertiary education was filled by the higher proportion of those in senior secondary education.

The sex ratio among the large group of illiterate people in our sample was 5,750 males to 5,875 females (the other four people, not included in these numbers, did not offer information on their gender; see Table 3.4). Among those who once went to school, there were only fifteen females (twelve females attended primary school and three females studied in Sishu). Given the relatively large scale of modern primary education compared to the traditional Sishu, here we must acknowledge that, even in rural areas, modern education developed significantly. This was consistent with the general progress of modern education throughout the country as both central and local governments had been promoting the new type of education in the first half of the twentieth century.

TABLE 3.4 Educational enrolment rate by sex in Wudu county, 1860–1930 (%)

	Male					Female				
	Illiterate	**Sishu**	**Primary**	**Secondary**	**Higher**	**Illiterate**	**Sishu**	**Primary**	**Secondary**	**Higher**
1860s	100.0	0.0	0.0	0.0	0.0	100.0	0.0	0.0	0.0	0.0
1870s	96.9	3.1	0.0	0.0	0.0	100.0	0.0	0.0	0.0	0.0
1880s	94.8	5.0	0.2	0.0	0.0	99.8	0.2	0.0	0.0	0.0
1890s	91.6	8.3	0.1	0.1	0.0	99.7	0.2	0.2	0.2	0.0
1900s	93.1	6.7	0.2	0.1	0.0	100.0	0.0	0.0	0.0	0.0
1910s	92.3	5.6	2.1	1.1	0.0	100.0	0.0	0.0	0.0	0.0
1920s	90.3	2.2	7.5	1.7	0.0	99.6	0.1	0.3	0.0	0.0

SOURCE: OUR SAMPLE (1947; SEE APPENDIX)

3.4 Social Factors: Sex Ratio, Marriage Rates, Divorce and Life Expectancy

Social development is complicated. We focus on four aspects: sex ratio, marriage rate, divorce and life expectancy. First, we should notice from Tables 3.7 and 3.8 that the number of unmarried men was bigger than that of women at all ages but specifically below the age of forty. For example, for the age group 20–24, 2.3% of women were unmarried vs 26.9% of men. This must have largely been caused by a skewed sex ratio: only from age seventy did the total number of women exceeded that of men. Table 3.5, cited from Fang and Zhang's research, shows that the male to female sex ratio in Wudu was lower than that in Gansu province. The statistics in our sample shows that the sex ratio in 1947 was 105.8 (8,177 males and 7,729 females). This ratio matches Fang and Zhang's calculation.

If we investigate different age groups, as shown by Table 3.6, it's clear that the proportion of females decreased by age. This reflects the shorter average lifespan of females, arguably caused by their lower family status and worse living conditions.[5] There is a rich literature claiming that a higher female mortality rate compared to that of men could only be caused by either a statistical mistake or gender inequality.[6] According to our sample, there were more females than males only in the highest age groups, due to the natural difference in potential life expectancy.

TABLE 3.5 Sex ratio ((male/female)*100) in Wudu county and Gansu province in different years during the Republican period

	1928	1935	1946	1947	1949
Our sample	–	–	–	105.8	–
Wudu county	97.01	110.26	103.06	105.15	104.98
Gansu province	118.61	120.57	107.82	108.93	109.94

SOURCE: FANG AND ZHANG (2007), 526–528; OUR SAMPLE (1947; SEE APPENDIX)

5 Plus, it was very common in rural areas that women died from childbirth or gynecological diseases during their reproductive age.

6 Hao (1995), 2–11.

TABLE 3.6 Sex ratio ((male/female)*100) in different age groups in Wudu county and Gansu province, 1947

	< 18	18–44	45–55	> 55	**Overall sex ratio**
Our sample	103.44	107.1	121	98.06	105.8
Wudu county	104.17	105.44	116.3	98.95	105.15
Gansu province	110.13	100.11	133.27	114.48	108.93

SOURCE: OUR SAMPLE (1947; SEE APPENDIX). CALCULATED ACCORDING TO THE DATA FROM FANG AND ZHANG (2007), 536, 590–591

Second, marriage rates were very high, especially among women. From Table 3.7 we can see that, in the age class 15–19, 69% were or had been married. In the age group 20–24 this increased to 98%, moving towards 100% over time. For men (Table 3.8), the percentage of the corresponding age class was 42% and 71% respectively. Whereas women were basically universally married by the age of 30–40, for men this was true at ages 40–49.

The third category is divorce. This occurred, for women, mostly between the age of thirty and sixty (see Table 3.7). One could hypothesize that this was the busiest time for couples in work and with growing children, even though one could equally argue that couples reproducing and rearing children together can also strengthen family ties. Another reason for the increased number of divorces might be the changing marriage laws in the 1920s and 1930s, which offered women more freedom to divorce. This might be another reason as to why women aged 30–60 (i.e. married between ca. 1910 and 1940) witnessed the highest divorce share. Men (Table 3.8), however, divorced younger and more frequently. However, even though divorces occurred, it is noteworthy that the total number of divorces, either for men or women, was very small in all age classes. Divorce was rare before the twentieth century and was still not considered an option for most families in the early twentieth century. For women, especially rural and illiterate women, divorce meant the loss of economic support and was often linked with shame. This may explain the relatively small number of divorces in our sample.

Our fourth characteristic is life expectancy at age five. This cannot be directly estimated since we do not have age-specific mortality rates. Therefore, we calculated proxy life expectancy figures using marital status (see Tables 3.7 and 3.8). For each age we calculates the share of widowers/widows. This gives the group-specific mortality (of married persons) without having to calculate

TABLE 3.7 Marital status of women by age class in 1947

Age class	Widow	Married	Unmarried	Divorced	Total
90–99	0	0	0	0	0
80–89	17	2	0	0	19
70–79	84	37	0	2	123
60–69	204	218	0	5	427
50–59	145	434	3	4	586
40–49	84	752	1	10	847
35–39	22	342	2	5	371
30–34	22	540	7	1	570
25–29	14	506	11	1	532
20–24	16	501	12	3	532
15–19	5	310	142	1	458

SOURCE: OUR SAMPLE (1947; SEE APPENDIX)

TABLE 3.8 Marital status of men by age class in 1947

Age class	Widower	Married	Unmarried	Divorced	Total
90–99	1	0	0	0	1
80–89	15	3	0	0	18
70–79	50	49	4	1	104
60–69	134	301	0	7	442
50–59	132	524	5	4	665
40–49	87	807	17	14	925
35–39	22	378	15	10	425
30–34	29	539	50	11	629
25–29	18	454	87	8	567
20–24	12	385	162	6	565
15–19	5	175	255	2	437

SOURCE: OUR SAMPLE (1947; SEE APPENDIX)

the different number of persons by age group over time. Since almost 100% were married, this covers a large share of the population. Obviously, such an approach has several biases. For example, since in the oldest age group recent widows/widowers are included as well as those (in that age group) that were already widows/widowers, we must subtract the old mortality rate to get the age-specific mortality. This still leaves the problem that the survivors in the old age group might pass away at any age. So, we assume that their mortality rates are equal to the one we do observe. It is also possible that both partners pass away at the same time. Even though this increases mortality, we ignore it as it only has a minor effect.

Having age-specific mortality, we can calculate life expectancy from age fifteen (when significant shares of marriages occurred). We assume the result applies to those over the age of five by assuming no mortality between five and fifteen. This assumption applies because (1) mortality in that age group is limited anyway, and (2) because mortality as we calculated after age fifteen must be overestimated. After all, we use age groups where mean mortality might be higher than our estimate. Hence, we compensate for underestimating mortality at age 5–15 with overestimating it at age 15–90.

This results in estimates for male and female life expectancy. Please be aware that our female data report the life expectancy of men (women are, after all, often married to men so surviving women must have deceased husbands). Likewise, male data result in estimates of female life expectancy. Our life expectancy estimates for men and women at age five in Wudu in 1947 are thus 44.1 and 49.3 respectively. This compares to 'age at death' of ca. 58.9 in 1930 vs a life expectancy in Taiwan in 1930 of 52.[7] Yet in 1960 life expectancy in China at birth was 43.7 at birth (i.e. due to child mortality lower than the previous mentioned data) (World Bank 2022).

7 Van Zanden, Van Leeuwen and Xu (2021), 395.

CHAPTER 4

The Socio-economic Ranking: An Analysis of Surnames

4.1 Surnames and Clans in China

The patriarchal clan system is considered a crucial part of the foundation connecting villages and the political structure above village level. It played an important role in China's rural political system and rural social network.[1] In agricultural society, the relationship between the patriarchal clan system and the official Baojia system was very ambiguous. Clans had their own charters, organizations and various collective events, which helped to constrain and cultivate clan members to bring their behaviour in line with the government's moral imperatives. For the sake of the maintenance of basic moral order and effective grassroots governance, the government thus aimed to support and make use of the clan system.

However, the government never forgot to limit the overexpansion of the power of influential clans in order to prevent clans intervening too much in local administrative affairs.[2] To some extent, in a period during which there was strong state power, the Baojia system controlled by the government was intended to be disincentive to the power of local clans. But in fact, frequently the majority of households in a given, small area belonged to one or several clans. Therefore, in the Baojia system, the administrative area of a Bao or Li was usually divided based on existing linage organizations. When state power became weak, the Baojia system, which was established by the government for the convenience of its control over grassroots society, was gradually controlled by large local clans. Inevitably, cooperation and competition happened among different clans if they coexisted in the same village.[3] Having their power authorized by the government of the day, the heads of these clans exercised firmer control over their villages. The combination of the patriarchal clan system and the Baojia system thus became an important part of the whole political power network of the state.[4]

1 Duara (2003), 82.
2 Wang Liu Hui-chen (1975), 28–29.
3 Ahern (1973), 260–263.
4 Duara (2003), 102.

 | DOI:10.1163/9789004715707_005

There have been numerous studies about the significance of clans in local societies and economies. It was found that in a village with a strong clan force, informal relational contracts and reputational constraint, as existing within a clan, improved the trust among members and reduced coordination costs.[5] Supported by the clan system, formal institutions keeping a village under more effective governance were more easily implemented and maintained.[6] In addition, relatively abundant investment in public goods was much easier due to lower transaction costs during the process of raising funds.[7] But the theory of clan polarization claims that a lower diversity of clans in an area implied an increasing possibility of conflicts among clans. High coordination costs usually resulted in low investment in public goods.[8] Powerful clans preferred to form alliances with other clans of equivalent status to consolidate or even expand their economic networks and social influence.[9] Families without the background of a powerful clan usually tried various methods such as marriage to seek extensive cooperation and support from other clans which they could rely on. Even if they did not get a proper opportunity to establish close ties with powerful clans, they might have shifted their attention to uniting other small, less powerful clans in order to survive together. The network thus formed both within and between clans, and was an important factor affecting social mobility.

However, without clear records such as family trees or household division documents, it is not easy to identify each clan based on surnames alone, because people with the same surname did not necessarily belong to the same clan and members of a clan might have been scattered over different regions. Therefore, due to the lack of data, existing research about historical social mobility mainly focused on elite families.[10] To study social mobility in the more general sense of the whole society, Clark,[11] for example, tried the surname method. He measured relative representation among elites by comparing the proportion of a surname in the whole elite group with the proportion of a surname in the whole population. He assumed that if social mobility worked well, there should not be too much difference between these two proportions. Yu Hao used the surname method to test the social mobility from Qing China

5 Pan, Ning and Dai (2020).
6 Peng (2004).
7 Xu and Yao (2015); Olson (1965); Ostrom (1990).
8 Horowitz (1985); Montalvo and Reynal-Querol (2005).
9 Watson (1981).
10 For example, Pan and Fei (1947) and Ho (1962).
11 Clark (2015).

to the 2010s, and found different changing rates of elites' social status across time.[12] Nowadays, surname methods are widely accepted and frequently used to roughly identify elite groups and examine the change of their social status over time.

In the subsequent sections, we describe the socio-economic ranking among different surnames through the observation on the intermarriage and the wealth distribution among different surnames in Wudu county. Our sample is from fourteen Baos. These Baos were scattered across different villages/towns in Wudu. Despite the fact that members of the same clan might have been distributed among different villages, it was usual for them to be concentrated in one place. Before we discuss intermarriage (Section 4.3) and the wealth distribution (Section 4.4) among different surnames, we must take the degree of surname concentration into account (Section 4.2).

4.2 Surname Concentration

Due to geographic distance, the dominant surnames and surname concentration differed in each of the Baos in our sample. In a strict sense, surname concentration should be calculated based on the surnames of every person in the sample. But what we discuss here is the power of each clan and the intermarriages among them. To do so, we roughly calculate the concentration of clans by the surnames of those male residents who left information about the clans of their wives.

According to the statistics below, there were around 200–450 male residents with 10–30 different surnames in each Bao of our sample. The surname concentration differed. For example, Fujin Xiang 1st Bao and Kongti Xiang 1st Bao had both the largest samples and the biggest number of surnames. Similar to these two Baos, Fujin Xiang 5th Bao and Village Unknown 1st Bao had more than 400 male residents respectively, but they belonged to only approximately half the number of surnames. In the Anhua Xiang 14th Bao, the number of surnames was the same of that of Village Unknown 1st Bao, but the sample size was only 60% of the latter.

12 Hao (2021).

TABLE 4.1 Number of male residents and surnames in each Bao

	Number of male residents	Number of surnames		Number of male residents	Number of surnames
Muma Xiang 1st Bao	283	11	Pipa Xiang 8th Bao	349	19
Fujin Xiang 1st Bao	444	28	Anhua Xiang 14th Bao	261	18
Ganquan Xiang 7th Bao	359	26	Housong Xiang 3rd Bao	330	13
Liulin Xiang 2nd Bao	198	14	Liulin Xiang 1st Bao	317	15
Fujin Xiang 5th Bao	439	14	Housong Xiang 6th Bao	317	19
Fujin Xiang 8th Bao	253	11	Jiucheng Xiang 1st Bao	267	11
Village Unknown 1st Bao	423	18	Kongti Xiang 1st Bao	449	25

SOURCE: OUR SAMPLE (1947; SEE APPENDIX)

In order to show a complete description about surname concentration, we list the proportion of surnames that accounted for more than 5% in each Bao in Table 4.2.

Taking both the number of surnames and the top surnames accounting for more than 5% in each Bao into consideration makes the surname concentration clear. In Liulin Xiang 1st Bao, there were a total of fifteen surnames, but around 80% of the sample came from the same surname: Li. The second most dominant surname only accounted for 5%. The other 15% were scattered across thirteen clans. The Fujin Xiang 5th Bao, Housong Xiang 3rd Bao and the Anhua Xiang 14th Bao also had a high surname concentration. In the Fujin Xiang 5th Bao, 64% of male residents had the same family name – Shi – and three of fourteen surnames covered nearly 90% of all male residents. In the Housong Xiang 3rd Bao, 86% of the male residents was concentrated in three out of thirteen surnames relatively evenly and 94% of the population belonged to the top-four surnames. In the Anhua Xiang 14th Bao, there were

TABLE 4.2 Top surnames accounting for more than 5% in each Bao (%)

Muma Xiang 1st Bao		**Fujin Xiang 1st Bao**		**Ganquan Xiang 7th Bao**		**Liulin Xiang 2nd Bao**		**Fujin Xiang 5th Bao**	
Wang 王	35	Chen 陈	25	Wei 魏	17	Xi 郄	32	Shi 石	64
Bai 白	20	Zhao 赵	18	Dong 董	17	Hou 侯	18	Zhao 赵	12
He 何	18	Ma 马	10	Wang 王	9	Li 李	14	Li 李	11
Chen 陈	7	Dong 董	9	Hou 侯	8	Yang 杨	11	Huang 黄	6
Luo 罗	5	Han 韩	8	Yin 尹	8	Guo 郭	8		
Yang 杨	5	Li 李	7	Cao 曹	6	Wang 王	6		
				Yang 杨	6	Ran	5		

Fujin Xiang 8th Bao		**Village Unknown 1st Bao**		**Pipa Xiang 8th Bao**		**Anhua Xiang 14th Bao**		**Housong Xiang 3rd Bao**	
Yang 杨	28	Zhao 赵	20	Sun 孙	25	Zhang 张	45	Zhang 张	38
Zhang 张	20	Pan 潘	12	Dong 董	18	Jiao 焦	11	Fu 符	28
Zhao 赵	15	Lei 雷	10	Ma 马	9	Gao 高	9	Wang 王	20
Du 杜	13	Yang 杨	9	Shua 誜	7	Chen 陈	7	Li 李	8
Luo 罗	10	Wang 王	9	Gao 高	6	Wang 王	5		
Ding 丁	6	Li 李	8	Li 李	6				
Li 李	5	Jia 贾	6	Gou 垢	5				
		Deng 邓	6	Zhang 张	5				
		Liu 刘	5						

Liulin Xiang 1st Bao		**Housong Xiang 6th Bao**		**Jiucheng Xiang 1st Bao**		**Kongti Xiang 1st Bao**	
Li 李	80	Ma 马	20	Jia 贾	24	Yan 闫	14
You 尤	5	He 何	19	Yang 杨	18	Ma 马	13
		Yang 杨	12	Pang 庞	14	Yin 尹	10
		Wang 王	11	Zong 宗	13	Zhang 张	9
		Niu 牛	10	Zhang 张	10	Wang 王	8
		Sun 孙	9	Liu 刘	9	Yang 杨	7
		Zhang 张	9	Zhao 赵	9	Bao 包	6
						Xi 郄	6

SOURCE: OUR SAMPLE (1947; SEE APPENDIX)

eighteen surnames, but Zhang was the most popular, covering 45% of the male residents. However, in some Baos, such as the Kongti Xiang 1st Bao, Village Unknown 1st Bao, Jiucheng Xiang 1st Bao, Pipa Xiang 8th Bao and Ganquan Xiang 7th Bao, there was a much more balanced scale among different surnames. For instance, in the Pipa Xiang 8th Bao, the most popular surname only accounted for 14% and the top three surnames covered only around 70% of the male residents.

4.3 Intermarriage among Different Surnames

We list all the surnames of those male residents who left the information about the clan of their wives, the surnames of their wives and the number of people with each surname in each Bao in Table 4.3.

TABLE 4.3 Surnames of married male residents in Housong Xiang 3rd Bao and the major surnames of their wives

Surnames of husbands		Major surnames of wives and the number of people with each surname							
Zhang 张	124	Fu 符	23	Wang 王	19	Zhang 张	12	He 何	10
		Sun 孙	7	Li 李	6	Ma 马	6	Feng 冯	5
		Yang 杨	5	Qin 秦	4	Yin 尹	4	Liu 刘	3
Fu 符	92	Zhang 张	30	Wang 王	13	Sun 孙	12	He 何	7
		Yang 杨	6	Qin 秦	6	Li 李	3	Ma 马	3
Wang 王	65	Zhang 张	16	Fu 符	14	Wang 王	9	Qin 秦	5
		Sun 孙	5	Chen 陈	3	Yang 杨	3	He 何	2
Li 李	28	Zhang 张	13	Qin 秦	4	Wang 王	3	Fu 符	2
		He 何	2	Sun 孙	2				
Sun 孙	4	Zhang 张	3	He 何	1				
Jiao 焦	3	He 何	1	Fu 符	1	Du 杜	1		
Tan 谈	3	Zhang 张	2	Qiu 秋	1				
Xi 郊	3	Zhang 张	1	Wang 王	1	Yang 杨	1		
He 贺	2	Fu 符	1	Ma 马	1				
Qin 秦	2	Wang 王	1	Yang 杨	1				
Jin 靳	2	Zhang 张	1	Fu 符	1				
He 何	1	Zhang 张	1						
Ma 马	1	Yuan 袁	1						

SOURCE: OUR SAMPLE (1947; SEE APPENDIX)

We found that almost all the major surnames of wives in each Bao can also be found on the list of surnames of husbands in the same Bao. In other words, we can assume that there is a strong possibility that intermarriage mainly occurred within each Bao. Due to different surname structures, despite popular surnames appearing in many different Baos, the intermarriage of the same surname differed in different Baos. In Table 4.4 we take the surname Zhang as an example.

TABLE 4.4 The major surnames of wives in the Zhang family from different Baos

Number of husbands with the surname 'Zhang' in each Bao		**Major surnames of wives and the number of people with each surname**							
Housong Xiang 3rd Bao	124	Fu 符	23	Wang 王	19	Zhang 张	12	He 何	10
		Sun 孙	7	Li 李	6	Ma 马	6	Feng 冯	5
		Yang 杨	5	Qin 秦	4	Yin 尹	4	Liu 刘	3
Fujin Xiang 1st Bao	3	Li 李	1	He 贺	1	Shi 石	1		
Liulin Xiang 2nd Bao	1	Xi 郄	1						
Ganquan Xiang 7th Bao	2	Cui 崔	1	Zhao 赵	1				
Fujin Xiang 5th Bao	14	Shi 石	4	Zhang 张	3	Zhao 赵	2	Tan 谈	2
Fujin Xiang 8th Bao	50	Zhao 赵	10	Luo 罗	6	Zhang 张	6	Li 李	5
		Du 杜	3	Huang 黄	3	Ma 马	3		
Village Unknown 1st Bao	6	Deng 邓	2						
Pipa Xiang 8th Bao	16	Dong 董	5	Sun 孙	3	Li 李	2	Zhang 张	2
Anhua Xiang 14th Bao	118	Zhang 张	28	Wang 王	17	Li 李	9	Zeng 曾	6
		Gao 高	6	Jiao 焦	5	Pu 蒲	5	Tian 田	4
		Bao 包	3	Fan 樊	3	Jia 贾	3	Liu 刘	3
		Song 宋	3	Zhu 朱	3				
Liulin Xiang 1st Bao	4	Zhao 赵	2	Li 李	1	Feng 冯	1		

TABLE 4.4 The major surnames of wives (*cont.*)

Number of husbands with the surname 'Zhang' in each Bao		Major surnames of wives and the number of people with each surname							
Housong Xiang 6th Bao	29	Wang 王	12	Qin 秦	4	He 何	3	Ma 马	3
		Yuan 袁	3	Zhang 张	2				
Jiucheng Xiang 1st Bao	26	Zhang 张	5	Yang 杨	4	Zhu 朱	3	Jia 贾	2
		Li 黎	2	Liu 刘	2	Zhao 赵	2		
Kongti Xiang 1st Bao	40	Wang 王	12	Zhang 张	5	Yin 尹	4	Yang 杨	3
		Bao 包	2	Xu 许	2	Yan 闫	2		

SOURCE: OUR SAMPLE (1947; SEE APPENDIX)

In those Baos with a high surname concentration, the intermarriage of popular surnames happened more often within the group of major surnames. As shown in Table 4.3, Zhang, Fu and Wang were the top-three surnames in the Housong Xiang 3rd Bao; 19%, 15% and 10% of Zhang married Fu, Wang and Zhang respectively. These three surnames were those most frequently seen among the wives of Zhang families. In Fu families, Zhang (33%) and Wang (14%) were the major surnames of their wives, and 60% of the Wang married Fu, Wang and Zhang. In the Liulin Xiang 1st Bao, where there was extreme surname concentration, Li was the dominant surname. From Table 4.5, it is clear that the daughters from Li families accounted for the largest proportion among all surnames of wives in half of the total clans. Besides Li, Wang was another major surname of wives as it had the third most common female surname in the population.

If we observe the most common surnames, the intermarriage between Li and Wang in the Liulin Xiang 1st Bao is quite impressive. However, the overwhelming dominance of the top surname Li creates a problem, as there were many Li around, and obviously many other surnames must marry a Li while many Li simply cannot marry another surname because there were few of those around. For example, 12% of Li married a Wang wife, but 64% of Wang married a Li wife. This can be corrected using the Herfindahl – Hirschman Index (HHI).

TABLE 4.5 Surnames of married male residents in the Liulin Xiang 1st Bao and the major surnames of their wives

Surnames of husbands		Major surnames of wives and the number of people with each surname							
Li 李	253	Li 李	37	Wang 王	31	Zhang 张	20	Liu 刘	19
		Du 杜	15	Yang 杨	14	Tian 田	14	Zhao 赵	11
		Guo 郭	9	Fu 符	8	Yuan 袁	6	Jiao 焦	6
		Yin 尹	5	Pu 蒲	5	Lu 路	5	Fan 樊	4
		Xi 郄	4	Dang 党	3				
You 尤	16	Li 李	3	Wang 王	3	Liu 刘	2	Sun 孙	2
Wang 王	11	Li 李	7						
Huai 怀	8	Xi 郄	3						
Liu 刘	8	Li 李	3						
Yang 杨	4	Li 李	1	Wang 王	1	Yuan 袁	1	Zhao 赵	1
Zhang 张	4	Zhao 赵	2	Li 李	1	Feng 冯	1		
Guo 郭	4	Wang 王	2	Li 李	1	Zhang 张	1		
Fan 樊	2	Fan 樊	1	Jiao 焦	1				
Du 杜	2	Li 李	1	Wang 王	1				
Jiao 焦	1	Tian 田	1						
Dang 党	1	Zhang 张	1						
Pan 潘	1	Li 李	1						
Tian 田	1	Wang 王	1						
Yin 尹	1	Li 李	1						

SOURCE: OUR SAMPLE (1947; SEE APPENDIX)

From our calculation, the top-five surnames in the Liulin Xiang 1st Bao – Li, You, Wang, Huai and Liu – accounted for 80%, 5%, 3.5%, 2.5% and 2.5% of total married males respectively. The other, less frequent, surnames accounted for 6.5% in total. Using these numbers, the $\text{HHI} = 80\%^2 + 5\%^2 + 3.5\%^2 + 2.5\%^2 + 2.5\%^2 = 0.645$. This is high because of the dominance of Li. If marriages were completely random, the chance of being intermarried in the Liulin Xiang 1st Bao should be $1 - \text{HHI} = 35.5\%$. Simply said, the chance of intermarrying a given surname in this particular Bao was 35.5%. If the actual intermarriage rate between two surnames is 35.5% or higher, both surnames intermarry more than normal. If it is lower than 35.5%, the marriage rate is lower, and the marriage 'potential' is thus unfulfilled. In fact, only 12% of Li married Wang, meaning we get an intermarriage potential of 33.8% (12%/35.5%). This means that the Wang marry the Li much less frequently than might be expected, that is, they are far away from their full marriage potential. By doing the same calculation on other surnames, we can conclude that in Li families, the intermarriage potential of any surname is no more than 40% or, in other words, the intermarriage between males from the Li family and females from any surname was much less than normal.

On the contrary, if we look at those females with the surname Li who married Wang, we get an intermarriage potential of around 180% (64%/35.5%). This tells us that a daughter in a Li household was more likely to marry Wang family sons than those of other surnames.

As regards those male residents with infrequently seen surnames, most of them married popular surnames. For instance, in the Housong Xiang 3rd Bao, Sun, Tan, Xi, He and so forth were rare surnames but their wives mainly came from the most popular surnames: Fu, Wang and Zhang. In the Liulin Xiang 1st Bao, wives with the top surnames Li and Wang were also the mainstream in clans with the smallest population.

In those Baos with lower surname concentration, the intermarriage of both popular surnames and infrequently seen surnames had a much more random and discrete distribution. We take the Jiucheng Xiang 1st Bao as an example. For example, the sixty-four males with the surname Jia married about evenly with Pang, Sang, Yang, Zhao, etc. (see Table 4.6).

TABLE 4.6 Surnames of married male residents in the Jiucheng Xiang 1st Bao and the major surnames of their wives

Surnames of husbands		**Major surnames of wives and the number of people with each surname**							
Jia 贾	64	Jia 贾	8	Pang 庞	6	Sang 桑	4	Yang 杨	4
		Zhao 赵	4	Zhu 朱	4	Li 黎	3	Pan 潘	3
		Zong 宗	3	Shang 尚	3				
Yang 杨	49	Yang 杨	9	He 何	6	Li 李	4	Wang 王	4
		Chen 陈	3	Jia 贾	3	Zhao 赵	3	Zong 宗	3
		Wu 吴	2	Zhang 张	2				
Pang 庞	38	Li 李	7	Wang 王	4	Zhang 张	4	Zong 宗	3
		Yang 杨	3	Jia 贾	2	Pang 庞	2	Zhao 赵	2

TABLE 4.6 Surnames of married male residents (*cont.*)

Surnames of husbands		Major surnames of wives and the number of people with each surname							
Zong 宗	36	Pan 潘	5	Yang 杨	5	Zhu 朱	4	Jia 贾	3
		Li 李	2	Li 黎	2	Sang 桑	2	Zhang 张	2
		Zhao 赵	2						
Zhang 张	26	Zhang 张	5	Yang 杨	4	Zhu 朱	3	Jia 贾	2
		Li 黎	2	Liu 刘	2	Zhao 赵	2		
Liu 刘	25	Jia 贾	6	Zhao 赵	4	Zhang 张	3	Wang 王	3
		Li 黎	2	Zhu 朱	2				
Zhao 赵	23	Zong 宗	4	Jia 贾	3	Pang 庞	3	Zhao 赵	3
		Zhu 朱	2	Liu 刘	2				
Cao 曹	2	Jia 贾	2	Li 黎	1				
Wang 王	2	Zhao 赵	1	Zong 宗	1				
Gong 巩	1	Jia 贾	1						
Wei 魏	1	Pu 蒲	1						

SOURCE: OUR SAMPLE (1947; SEE APPENDIX)

To get a clear comparison among different types of Bao, we list the HHI and highest intermarriage potential in Table 4.7. For each type of Bao, we took two examples. We chose the Housong Xiang 3rd Bao and the Liulin Xiang 1st Bao as the representatives of Baos with a high surname concentration. The Jiucheng Xiang 1st Bao and the Ganquan Xiang 7th Bao are studied as the examples of Baos with a low surname concentration. The 1st and 5th Baos of Fujin Xiang are also included in Table 4.7 because they were the most common in the population covered by our sample. Moreover, we only calculate the highest intermarriage potential within those surnames that include more than five male residents. As a very simple example, in the Liulin Xiang 1st Bao, there was only one male resident with the surname Jiao and his wife was Tian, the pretty high Tian – Jiao intermarriage potential does not make much sense for our analysis.

TABLE 4.7 Comparison of Herfindahl Hirschman Index (HHI) and the highest intermarriage potential among different Baos

Characteristic place	Bao	HHI	1 – HHI (chance of random intermarriage)	Highest three intermarriage potentials (%)		
High surname concentration	Housong Xiang 3rd Bao	0.27	0.73	63.6 Zhang–Li	44.7 Zhang–Fu	33.7 Zhang–Wang
High surname concentration	Liulin Xiang 1st Bao	0.645	0.355	180 Li–Wang	106 Li–Liu	106 Xi–Huai
Low surname concentration	Jiucheng Xiang 1st Bao	0.156	0.844	28.4 Jia–Liu	23 Zhang–Zhang	21.8 Li–Pang
Low surname concentration	Ganquan Xiang 7th Bao	0.092	0.908	41 Liang–Jia	30 Zhang–Feng	25 Zhang–Cao
Largest population	Fujin Xiang 1st Bao	0.135	0.865	49.5 Li–Su	33 Yang–Duan	31.5 Feng–Liu
Largest population	Fujin Xiang 5th Bao	0.443	0.557	90 Shi–Huang	86 Shi–Li	63.2 Shi–Zhao

SOURCE: OUR SAMPLE (1947; SEE APPENDIX)

The Housong Xiang 3rd Bao and the Liulin Xiang 1st Bao had a much higher HHI and, consequently, the chance of random intermarriage (1 – HHI) was the lowest among all Baos. In other words, the higher the surname concentration, the higher the intermarriage rate among several popular surnames. Shifting our attention to the surnames with the highest intermarriage potential in different Baos, we find that in the Housong Xiang 3rd Bao and the Liulin Xiang 1st Bao, with a high surname concentration and among all wives, the most popular surnames were Zhang and Li respectively, and these two surnames were the most dominant in the population in the associated Bao. On the other hand, in the Jiucheng Xiang 1st Bao and the Ganquan Xiang 7th Bao, with a low surname concentration, there was no particularly dominant surname in the marriage market and even those who had a relatively high intermarriage potential were not from the families with the largest population share. This means that in those areas with a high surname concentration, dominant surnames were also popular in the marriage market, whereas in a region with a low surname concentration the intermarriages were random. Between the two largest Baos – the 1st and 5th Baos of Fujin Xiang – the former had a lower surname concentration. In the Fujin Xiang 1st Bao, there were 444 samples of male residents. The top-five surnames were Chen (109), Zhao (81), Ma (44), Dong (40) and Han (35). The top one and top five accounted for around 25% and 70% of all the male residents respectively. The Fujin Xiang 5th Bao had 439 male residents, among which 281 came from the Shi family. The top surname covered 64% of the male residents, while 93% of the male residents were from the top-four surnames: Shi, Zhao, Li and Huang. It's clear that our conclusion based on the comparison between the first two types of Baos is also applicable here. The Fujin Xiang 5th Bao, with a higher surname concentration, had a higher intermarriage potential and the largest clan Shi had an overwhelming privilege in the marriage market.

4.4 Wealth Distribution among Different Surnames

Farmland was a major part of household wealth in rural areas. Therefore, the status of land possession could be considered as a proxy to evaluate wealth distribution to a large extent. From our household registration form with the information about the occupation of each person, we can get an understanding about which surname was the dominant landowner and of course whether landholders had a higher marriage potential.

In the Muma Xiang 1st Bao, Fujin Xiang 1st Bao, Fujin Xiang 5th Bao and Fujin Xiang 8th Bao, all of the male residents were landholding peasants. In the Ganquan Xiang 7th Bao, Liulin Xiang 2nd Bao, Village Unknown 1st Bao, Pipa Xiang 8th Bao, Anhua Xiang 14th Bao, Housong Xiang 3rd Bao, Liulin Xiang 1st Bao, Housong Xiang 6th Bao, Jiucheng Xiang 1st Bao and Kongti Xiang 1st Bao, there were records of both landholding peasants and tenant farmers. Among these ten Baos, the Village Unknown 1st Bao, Pipa Xiang 8th Bao, Kongti Xiang 1st Bao and Ganquan Xiang 7th Bao only had a few tenant farmers registered and the group was too small to analyse. Therefore, we focus on the remaining six Baos.

As in Section 4.3, we calculate the top-five surnames with the largest number of landholding households in each Bao and compare them with the surname concentration.

From the comparison in Table 4.8, we found that the proportion of the most dominant surnames in the total population was very close to the proportion of the most dominant surnames of farmland owners in all landholding households. Moreover, this dominance in the population scale did not necessary lead to equivalent economic power. In Liulin Xiang 1st Bao, the surname Li covered 85% of the population; however, only around 80% of farmland-holding

TABLE 4.8 The top-five surnames with the largest population and the proportion of land-holding households in different Baos

Liulin Xiang 2nd Bao	The concentration of landownership	Li 36%	Xi 21%	Hou 14%	Yang 14%	Wang 7%
	Surname concentration	Xi 32%	Hou 18%	Li 14%	Yang 11%	Guo 8%
Anhua Xiang 14th Bao	The concentration of landownership	Zhang 42%	Jiao 14%	Gao 11%	Chen 8%	Ren 6%
	Surname concentration	Zhang 45%	Jiao 11%	Gao 9%	Chen 7%	Wang 5%
Housong Xiang 3rd Bao	The concentration of landownership	Zhang 35%	Fu 29%	Wang 19%	Li 12%	Sun 2%
	Surname concentration	Zhang 38%	Fu 28%	Wang 20%	Li 8%	Sun 1.2%
Liulin Xiang 1st Bao	The concentration of landownership	Li 85%	Wang 3.4%	You 2.3%	Huai 2.3%	Guo 2.3%
	Surname concentration	Li 80%	You 5%	Wang 3.5%	Huai 2.5%	Liu 2.5%
Housong Xiang 6th Bao	The concentration of landownership	He 28%	Ma 19.3%	Yang 17.5%	Zhang 12.3%	Wang 8.8%
	Surname concentration	Ma 20%	He 19%	Yang 12%	Wang 11%	Niu 10%
Kongti Xiang 1st Bao	The concentration of landownership	Ma 16%	Yan 12.2%	Yin 8.8%	Yang 8.8%	Xi 8.1%
	Surname concentration	Yan 14%	Ma 13%	Yin 10%	Zhang 9%	Wang 8%

SOURCE: OUR SAMPLE (1947; SEE APPENDIX)

households carried the family name Li. From the perspective of other surnames, their weight in the group of landowners exceeded their population scale. In Liulin Xiang 2nd Bao, Housong Xiang 6th Bao and Kongti Xiang 1st Bao, it was the surname ranking second in the population that had the majority of farmland-holding households. Even in Anhua Xiang 14th Bao and Housong Xiang 3rd Bao, where the top surname had the largest number of farmland owners, its proportion in farmland-holding households was only a little higher than its population scale. Therefore, we could conclude that the top surnames did not get much more farmland due to their dominance in the population scale.

Of course, one can argue that because of the convenience of the annexation of land within the same clan through internal trade and so forth, maybe the top surname still had the largest amount of farmland despite the fact that they did not account for a high-enough proportion in farmland-holding households. But if the most developed households with the top surnames really had economic power that was strong enough to concentrate massive amounts of farmland, it is not easy to explain the similarity of the proportion of top surnames in both the whole population and the group of farmland-holding households. Therefore, it is reasonable for us to guess that those households with the same surname in our male resident data were most likely to belong to many different clans.

To study whether landholders had a higher marriage potential, we split each top surname in landowning husband and non-landowning husband, and then calculated the highest intermarriage potential respectively. We chose the top-three surnames in Housong Xiang 3rd Bao, Anhua Xiang 14th Bao and Housong Xiang 6th Bao as examples.

From Table 4.9, in each Bao the top-three highest intermarriage potentials likely occurred for the same two surnames, namely Zhang and Fu in Housong Xiang 3rd Bao, Zhang and Chen in Anhua Xiang 14th Bao, and Wang and Ma in Housong Xiang 6th Bao. For these most popular surnames, their actual intermarriage rates with a landowning husband was higher than with tenant peasants. It shows that farmland owners had a higher marriage potential.

TABLE 4.9 The comparison of Herfindahl Hirschman Index (HHI) and the highest intermarriage potential among different Baos

	HHI	1 – HHI	Highest intermarriage potential (%)		
Housong Xiang 3rd Bao	0.27	0.73	76.1 Fu–Wang*	54.8 Zhang–Fu	44.2 Zhang–Fu*
Anhua Xiang 14th Bao	0.236	0.764	65 Zhang–Chen	43.2 Wang–Jiao	43.2 Zhang–Chen*
Housong Xiang 6th Bao	0.087	0.913	49.8 Wang–Ma	42.6 Wang–Ma*	36.5 Zhang–He*

SOURCE: OUR SAMPLE (1947; SEE APPENDIX)

Note: Surnames with * means non-landowning husband, surnames without * means landowning husband.

Appendix: Overview of Microdata for Wudu County

	The no. of household	Jia	Household	Relationship with the household head	Surname	First name	Sex	The Chinese zodiac	Birthdate	
									Year	Month
Muma Xiang 1st Bao	1	1	1	The household head	白	义理	Male	Horse	1905	8
Muma Xiang 1st Bao				The household head's wife	白	杜郭儿子	Female	Rabbit	1915	12
Muma Xiang 1st Bao				Father	白	发能	Male	Rooster	1896	6
Muma Xiang 1st Bao				Daughter	白	刘妹子	Female	Rooster	1933	6
Muma Xiang 1st Bao				The first son	白	马代子	Male	Horse	1943	1
Muma Xiang 1st Bao				The second son	白	国泰儿	Male	Rooster	1946	1
Muma Xiang 1st Bao				Uncle	白	发怀	Male	Ox	1913	8
Muma Xiang 1st Bao				Uncle's wife	白	陈孝莲子	Female	Ox	1913	6
Muma Xiang 1st Bao				Younger brother	白	代银儿	Male	Rat	1936	4
Muma Xiang 1st Bao	2	1	2	The household head	白	金富	Male	Rooster	1908	2
Muma Xiang 1st Bao				The household head's wife	白	杨润娃子	Female	Rooster	1908	8
Muma Xiang 1st Bao				Son	白	发德	Male	Rat	1936	4
Muma Xiang 1st Bao				Daughter	白	谢扬娃	Female	Dog	1946	11
Muma Xiang 1st Bao	3	1	3	The household head	白	发林	Male	Dog	1909	6
Muma Xiang 1st Bao				The household head's wife	白	陈石秀子	Female	Dog	1909	11
Muma Xiang 1st Bao				Daughter	白	同儿子	Female	Goat	1931	8

 | DOI:10.1163/9789004715707_006

Date	The beginning time of inhabitation in this county	Educational level	Profession sector	Occupation	Marital status	Physical disability	Religion
29	Inhabited for generations	Illiteracy	Agriculture	Land-holding peasant	Married	No	No religion
26	Inhabited for generations	Illiteracy	Agriculture	Land-holding peasant	Married	No	No religion
1	Inhabited for generations	Illiteracy	Agriculture	Land-holding peasant	Widowed	No	No religion
7	Inhabited for generations	Illiteracy	Agriculture	Land-holding peasant	Unmarried	Bald	No religion
18	Inhabited for generations					Dumb	
16	Inhabited for generations						
15	Inhabited for generations	Illiteracy	Agriculture	Land-holding peasant	Married	No	No religion
24	Inhabited for generations	Illiteracy	Agriculture	Land-holding peasant	Married	No	No religion
2	Inhabited for generations	Illiteracy					
21	Inhabited for generations	Illiteracy	Agriculture	Land-holding peasant	Married	No	No religion
5	Inhabited for generations	Illiteracy	Agriculture	Land-holding peasant	Married	No	No religion
11	Inhabited for generations	Primary school 1 year					
27	Inhabited for generations						
21	Inhabited for generations	Illiteracy	Agriculture	Land-holding peasant	Married	No	No religion
10	Inhabited for generations	Illiteracy	Agriculture	Land-holding peasant	Married	No	No religion
23	Inhabited for generations	Illiteracy	Agriculture	Land-holding peasant	Unmarried	Bald	No religion

Appendix: Overview of Microdata for Wudu County (*cont.*)

	The no. of household	Jia	Household	Relationship with the household head	Surname	First name	Sex	The Chinese zodiac	Birthdate	
									Year	Month
Muma Xiang 1st Bao				The first son	白	义海	Male	Ox	1938	1
Muma Xiang 1st Bao				The second son	白	三儿子	Male	Snake	1941	5
Muma Xiang 1st Bao				The third son	白	一娃子	Male	Monkey	1944	5
Muma Xiang 1st Bao	4	1	4	The household head	白	扬粹翠子	Female	Horse	1892	1
Muma Xiang 1st Bao				Daughter	白	四连子	Female	Tiger	1926	11
Muma Xiang 1st Bao	5	1	5	The household head	白	义合	Male	Dog	1909	5
Muma Xiang 1st Bao				The household head's wife	白	何么连子	Female	Tiger	1914	3
Muma Xiang 1st Bao				Daughter	白	董娃子	Female	Dragon	1940	9
Muma Xiang 1st Bao				Son	白	玉林子	Male	Dog	1946	4
Muma Xiang 1st Bao	6	1	6	The household head	白	义怀	Male	Horse	1904	1
Muma Xiang 1st Bao				The household head's wife	白	苟董银儿	Female	Snake	1904	10
Muma Xiang 1st Bao				Father	白	发钦	Male	Monkey	1882	1
Muma Xiang 1st Bao				Daughter	白	儿娃子	Female	Rooster	1945	10
Muma Xiang 1st Bao				Daughter	白	药儿娃	Female	Dragon	1940	9
Muma Xiang 1st Bao				Older brother's wife	白	杨董连子	Female	Rooster	1926	10
Muma Xiang 1st Bao	7	1	7	The household head	白	义德	Male	Dragon	1916	8
Muma Xiang 1st Bao				The household head's wife	白	陈粹儿子	Female	Dragon	1916	12
Muma Xiang 1st Bao				Mother	白	董张香子	Female	Goat	1894	3

Date	The beginning time of inhabitation in this county	Educational level	Profession sector	Occupation	Marital status	Physical disability	Religion
25	Inhabited for generations	Primary school 1 year					
6	Inhabited for generations					Bald	
3	Inhabited for generations						
16	Inhabited for generations	Illiteracy	Agriculture	Land-holding peasant	Widowed	No	No religion
28	Inhabited for generations	Illiteracy	Agriculture	Land-holding peasant	Divorced	No	No religion
19	Inhabited for generations	Illiteracy	Agriculture	Land-holding peasant	Married	No	No religion
7	Inhabited for generations	Illiteracy	Agriculture	Land-holding peasant	Married	No	No religion
12	Inhabited for generations	Illiteracy				Bald	
15	Inhabited for generations						
15	Inhabited for generations	Sishu 2 years	Agriculture	Land-holding peasant	Married	No	Buddhism
7	Inhabited for generations	Illiteracy	Agriculture	Land-holding peasant	Married	No	
23	Inhabited for generations	Illiteracy	Agriculture	Land-holding peasant	Widowed	No	
25	Inhabited for generations						
4	Inhabited for generations	Illiteracy					
9	Inhabited for generations	Illiteracy	Agriculture	Land-holding peasant	Widowed	Tuberculosis	
18	Inhabited for generations	Illiteracy	Agriculture	Land-holding peasant	Married	No	No religion
30	Inhabited for generations	Illiteracy	Agriculture	Land-holding peasant	Married	No	No religion
22	Inhabited for generations	Illiteracy	Agriculture	Land-holding peasant	Widowed	No	No religion

Appendix: Overview of Microdata for Wudu County (*cont.*)

	The no. of household	Jia	Household	Relationship with the household head	Surname	First name	Sex	The Chinese zodiac	Birthdate	
									Year	Month
Muma Xiang 1st Bao				Son	白	陈代儿	Male	Dragon	1940	5
Muma Xiang 1st Bao				Older brother's wife	白	陈真连子	Female	Horse	1918	3
Muma Xiang 1st Bao				Cohabitant	老	杨	Male	Dog	1897	8
Muma Xiang 1st Bao	8	1	8	The household head	白	义太	Male	Dog	1922	2
Muma Xiang 1st Bao				The household head's wife	白	谈董代子	Female	Monkey	1932	8
Muma Xiang 1st Bao				Father	白	发通	Male	Snake	1892	5
Muma Xiang 1st Bao				Mother	白	陈陈秀子	Female	Dragon	1903	8
Muma Xiang 1st Bao				Younger sister	白	黑女子	Female	Goat	1943	8
Muma Xiang 1st Bao				Younger brother	白	崴儿子	Male	Rooster	1945	8
Muma Xiang 1st Bao	9	1	9	The household head	白	发洪	Male	Pig	1923	10
Muma Xiang 1st Bao				The household head's wife	白	王良狗娃	Female	Dog	1922	5
Muma Xiang 1st Bao				Mother	白	何路儿子	Female	Pig	1886	6
Muma Xiang 1st Bao				Daughter	白	柏女子	Female	Dog	1946	12
Muma Xiang 1st Bao	10	1	10	The household head	白	发積	Male	Tiger	1914	6
Muma Xiang 1st Bao				The household head's wife	白	王翠连子	Female	Goat	1916	6
Muma Xiang 1st Bao				Father	白	金德	Male	Pig	1898	10
Muma Xiang 1st Bao				The first son	白	义双	Male	Tiger	1926	4

Date	The beginning time of inhabitation in this county	Educational level	Profession sector	Occupation	Marital status	Physical disability	Religion
7	Inhabited for generations	Primary school 1 year					
3	Inhabited for generations	Illiteracy	Agriculture	Land-holding peasant	Widowed	No	No religion
31	Inhabited for generations	Illiteracy	Agriculture	Hired labourer		No	No religion
28	Inhabited for generations	Illiteracy	Agriculture	Land-holding peasant	Married	No	No religion
15	Inhabited for generations	Illiteracy	Agriculture	Land-holding peasant	Married	No	No religion
27	Inhabited for generations	Illiteracy	Agriculture	Land-holding peasant	Married	No	No religion
25	Inhabited for generations	Illiteracy	Agriculture	Land-holding peasant	Married	No	No religion
15	Inhabited for generations						
22	Inhabited for generations						
3	Inhabited for generations	Illiteracy	Agriculture	Land-holding peasant	Married	No	No religion
11	Inhabited for generations	Illiteracy	Agriculture	Land-holding peasant	Married	No	No religion
12	Inhabited for generations	Illiteracy	Agriculture	Land-holding peasant	Widowed	No	No religion
24	Inhabited for generations						
6	Inhabited for generations	Illiteracy	Agriculture	Land-holding peasant	Married	No	No religion
23	Inhabited for generations	Illiteracy	Agriculture	Land-holding peasant	Married	No	No religion
1	Inhabited for generations	Illiteracy	Agriculture	Land-holding peasant		Mental disease	
26	Inhabited for generations	Illiteracy			Unmarried	No	No religion

Appendix: Overview of Microdata for Wudu County (*cont.*)

	The no. of household	Jia	Household	Relationship with the household head	Surname	First name	Sex	The Chinese zodiac	Birthdate	
									Year	Month
Muma Xiang 1st Bao				The second son	白	义仲	Male	Dog	1935	1
Muma Xiang 1st Bao				The third son	白	郭代子	Male	Tiger	1938	4
Muma Xiang 1st Bao				The fourth son	白	明安子	Male	Snake	1941	3
Muma Xiang 1st Bao				Grandson	白	兴世	Male	Tiger	1938	3
Muma Xiang 1st Bao	11	1	11	The household head	白	义清	Male	Rooster	1896	12
Muma Xiang 1st Bao				Son	白	兴德	Male	Pig	1910	11
Muma Xiang 1st Bao				Son's wife	白	王岩儿子	Female	Snake	1917	11
Muma Xiang 1st Bao				Granddaughter	白	董龄子	Female	Rat	1936	8
Muma Xiang 1st Bao				Grandson	白	无龄子	Male	Dog	1946	10
Muma Xiang 1st Bao	12	1	12	The household head	白	义明	Male	Ox	1900	8
Muma Xiang 1st Bao				The household head's wife	白	王代连子	Female	Dragon	1916	8
Muma Xiang 1st Bao				Daughter	白	素儿子	Female	Snake	1941	9
Muma Xiang 1st Bao				Older brother's wife	白	王腊娃子	Female	Dragon	1891	9
Muma Xiang 1st Bao	13	1	13	The household head	白	陈真儿子	Female	Tiger	1889	2
Muma Xiang 1st Bao				Son	白	兴章	Male	Rat	1924	6
Muma Xiang 1st Bao				Son's wife	白	何张连子	Female	Ox	1925	9

Date	The beginning time of inhabitation in this county	Educational level	Profession sector	Occupation	Marital status	Physical disability	Religion
26	Inhabited for generations	Primary school 3 years					
13	Inhabited for generations	Illiteracy					
19	Inhabited for generations	Illiteracy					
19	Inhabited for generations	Primary school 1 year				Bald	
27	Inhabited for generations	Illiteracy	Agriculture	Land-holding peasant	Widowed	No	No religion
17	Inhabited for generations	Illiteracy	Agriculture	Land-holding peasant	Married	No	No religion
11	Inhabited for generations	Illiteracy	Agriculture	Land-holding peasant	Married	No	No religion
31	Inhabited for generations	Illiteracy					
23	Inhabited for generations						
28	Inhabited for generations	Illiteracy	Agriculture	Land-holding peasant	Married	No	No religion
27	Inhabited for generations	Illiteracy	Agriculture	Land-holding peasant	Married	No	No religion
8	Inhabited for generations	Illiteracy					
6	Inhabited for generations	Illiteracy	Agriculture	Land-holding peasant	Widowed	No	No religion
1	Inhabited for generations	Illiteracy	Agriculture	Land-holding peasant	Widowed	No	No religion
21	Inhabited for generations	Illiteracy	Agriculture	Land-holding peasant	Married	No	No religion
3	Inhabited for generations	Illiteracy	Agriculture	Land-holding peasant	Married	No	No religion

Appendix: Overview of Microdata for Wudu County (*cont.*)

	The no. of household	Jia	Household	Relationship with the household head	Surname	First name	Sex	The Chinese zodiac	Birthdate	
									Year	Month
Muma Xiang 1st Bao				Daughter	白	肖儿子	Female	Rooster	1933	9
Muma Xiang 1st Bao	14	1	14	The household head	白	金先	Male	Monkey	1895	2
Muma Xiang 1st Bao				The household head's wife	白	玉连女子	Female	Dragon	1891	6
Muma Xiang 1st Bao				The first son	白	发鳌	Male	Snake	1917	8
Muma Xiang 1st Bao				The first son's wife	白	陈约儿子	Female	Tiger	1914	9
Muma Xiang 1st Bao				The second son	白	发乾	Male	Pig	1923	6
Muma Xiang 1st Bao				The second son's wife	白	王一儿娃	Female	Dragon	1928	8
Muma Xiang 1st Bao				Grandson	白	韩银儿	Male	Dog	1946	2
Muma Xiang 1st Bao	15	1	15	The household head	白	金顺	Male	Rooster	1908	8
Muma Xiang 1st Bao				The household head's wife	白	罗三儿子	Female	Goat	1906	8
Muma Xiang 1st Bao				Son	白	发升	Male	Tiger	1926	5
Muma Xiang 1st Bao				Son's wife	白	楮创连子	Female	Goat	1931	11
Muma Xiang 1st Bao	16	2	1	The household head	白	户谢娃子	Female	Dragon	1903	8
Muma Xiang 1st Bao				Grandson	白	五林子	Male	Dragon	1928	8
Muma Xiang 1st Bao	17	2	2	The household head	白	义虎	Male	Ox	1888	12
Muma Xiang 1st Bao				Older sister	白	马刘香子	Female	Rooster	1884	8
Muma Xiang 1st Bao	18	2	3	The household head	白	发清	Male	Horse	1881	7
Muma Xiang 1st Bao				Son	白	关成子	Male	Snake	1904	8

Date	The beginning time of inhabitation in this county	Educational level	Profession sector	Occupation	Marital status	Physical disability	Religion
4	Inhabited for generations	Illiteracy	Agriculture	Land-holding peasant	Unmarried	No	No religion
26	Inhabited for generations	Illiteracy	Agriculture	Land-holding peasant	Married	No	No religion
27	Inhabited for generations	Illiteracy	Agriculture	Land-holding peasant	Married	No	No religion
12	Inhabited for generations	Illiteracy	Agriculture	Land-holding peasant	Married	No	No religion
4	Inhabited for generations	Illiteracy	Agriculture	Land-holding peasant	Married	No	No religion
17	Inhabited for generations	Illiteracy	Agriculture	Land-holding peasant	Married	No	No religion
29	Inhabited for generations	Illiteracy	Agriculture	Land-holding peasant	Married	No	No religion
20	Inhabited for generations						
28	Inhabited for generations	Illiteracy	Agriculture	Land-holding peasant	Married	No	No religion
23	Inhabited for generations	Illiteracy	Agriculture	Land-holding peasant	Married	No	No religion
25	Inhabited for generations	Illiteracy	Agriculture	Land-holding peasant	Married	No	No religion
24	Inhabited for generations	Illiteracy	Agriculture	Land-holding peasant	Married	No	No religion
21	Inhabited for generations	Illiteracy	Agriculture	Land-holding peasant	Divorced	No	No religion
28	Inhabited for generations	Illiteracy	Agriculture	Land-holding peasant	Unmarried	No	No religion
14	Inhabited for generations	Sishu 2 years	Agriculture	Land-holding peasant	Widowed	No	No religion
24	Inhabited for generations	Illiteracy	Agriculture	Land-holding peasant	Widowed	No	No religion
16	Inhabited for generations	Illiteracy	Agriculture	Land-holding peasant	Widowed	No	No religion
15	Inhabited for generations	Illiteracy	Agriculture	Land-holding peasant	Married	No	No religion

Appendix: Overview of Microdata for Wudu County (*cont.*)

	The no. of household	Jia	Household	Relationship with the household head	Surname	First name	Sex	The Chinese zodiac	Birthdate	
									Year	Month
Muma Xiang 1st Bao				Son's wife	白	赵关秀子	Female	Pig	1898	9
Muma Xiang 1st Bao	19	2	4	The household head	白	义春	Male	Pig	1881	12
Muma Xiang 1st Bao				The household head's wife	白	孙刘儿子	Female	Dragon	1916	2
Muma Xiang 1st Bao				Son	白	一秀儿	Male	Goat	1943	12
Muma Xiang 1st Bao				Daughter	白	五秀儿	Female	Rooster	1945	8
Muma Xiang 1st Bao				Older sister	白	赵连子	Female	Monkey	1895	3
Muma Xiang 1st Bao	20	2	5	The household head	尹	怀玉	Male	Horse	1905	10
Muma Xiang 1st Bao				The household head's wife	尹	白秋儿子	Female	Dragon	1903	10
Muma Xiang 1st Bao				Father	尹	成通	Male	Rooster	1872	10
Muma Xiang 1st Bao				Mother	尹	王氏	Female	Pig	1875	2
Muma Xiang 1st Bao				Son	尹	天福	Male	Rat	1936	8
Muma Xiang 1st Bao	21	2	6	The household head	白	兴高	Male	Dragon	1903	6
Muma Xiang 1st Bao				The household head's wife	白	王郭儿子	Female	Snake	1904	12
Muma Xiang 1st Bao				Mother	白	王卢娃子	Female	Dog	1885	7
Muma Xiang 1st Bao				Son	白	六代子	Male	Monkey	1944	8
Muma Xiang 1st Bao				Younger brother	白	关银子	Male	Tiger	1926	11
Muma Xiang 1st Bao	22	2	7	The household head	白	金发	Male	Goat	1872	10

Date	The beginning time of inhabitation in this county	Educational level	Profession sector	Occupation	Marital status	Physical disability	Religion
18	Inhabited for generations	Illiteracy	Agriculture	Land-holding peasant	Married	No	No religion
16	Inhabited for generations	Illiteracy	Agriculture	Land-holding peasant	Married	No	No religion
17	Inhabited for generations	Illiteracy	Agriculture	Land-holding peasant	Married	No	No religion
9	Inhabited for generations						
22	Inhabited for generations						
25	Inhabited for generations					Blind	
16	Inhabited for generations	Illiteracy	Agriculture	Land-holding peasant	Married	No	No religion
21	Inhabited for generations	Illiteracy	Agriculture	Land-holding peasant	Married	No	No religion
2	Inhabited for generations				Married		
6	Inhabited for generations				Married		
31	Inhabited for generations	Primary school 1 year					
18	Inhabited for generations	Illiteracy	Agriculture	Land-holding peasant	Married	No	No religion
24	Inhabited for generations	Illiteracy	Agriculture	Land-holding peasant	Married	No	No religion
23	Inhabited for generations	Illiteracy	Agriculture	Land-holding peasant	Widowed	No	No religion
28	Inhabited for generations						
13	Inhabited for generations					Blind	
19	Inhabited for generations				Married		

Appendix: Overview of Microdata for Wudu County (*cont.*)

	The no. of household	Jia	Household	Relationship with the household head	Surname	First name	Sex	The Chinese zodiac	Birthdate Year	Month
Muma Xiang 1st Bao				The household head's wife	白	赵在在子	Female	Goat	1882	6
Muma Xiang 1st Bao				Son	白	发现	Male	Goat	1919	11
Muma Xiang 1st Bao				Granddaughter	白	催娃	Female	Dog	1946	8
Muma Xiang 1st Bao				Granddaughter	白	新娃子	Female	Pig	1947	3
Muma Xiang 1st Bao	23	2	8	The household head	白	金龙	Male	Rooster	1883	1
Muma Xiang 1st Bao				Son	白	发连	Male	Pig	1910	12
Muma Xiang 1st Bao				Son's wife	白	李卢娃子	Female	Rat	1912	12
Muma Xiang 1st Bao				The first grandson	白	义德	Male	Ox	1937	11
Muma Xiang 1st Bao				The second grandson	白	孟娃子	Male	Dragon	1940	2
Muma Xiang 1st Bao				Granddaughter	白	见娃子	Female	Dog	1946	8
Muma Xiang 1st Bao	24	2	9	The household head	白	发高	Male	Dog	1909	11
Muma Xiang 1st Bao				The household head's wife	白	杨银子娃	Female	Rat	1912	5
Muma Xiang 1st Bao				Mother	白	杨金连子	Female	Ox	1900	8
Muma Xiang 1st Bao				Son	白	桂娃子	Male	Rooster	1945	1
Muma Xiang 1st Bao				Older brother	白	发太	Male	Dragon	1903	10
Muma Xiang 1st Bao				Older brother's wife	白	林中秀子	Female	Horse	1905	9
Muma Xiang 1st Bao	25	2	10	The household head	陈	义福	Male	Rabbit	1902	6

Date	The beginning time of inhabitation in this county	Educational level	Profession sector	Occupation	Marital status	Physical disability	Religion
3	Inhabited for generations	Illiteracy	Agriculture	Land-holding peasant	Married	No	No religion
30	Inhabited for generations	Sishu 1 year	Agriculture	Land-holding peasant	Divorced	No	No religion
11	Inhabited for generations						
5	Inhabited for generations						
30	Inhabited for generations	Illiteracy	Agriculture	Land-holding peasant	Widowed	No	No religion
5	Inhabited for generations	Illiteracy	Agriculture	Land-holding peasant	Married	No	No religion
5	Inhabited for generations	Illiteracy	Agriculture	Land-holding peasant	Married	No	No religion
5	Inhabited for generations	Primary school half a year					
18	Inhabited for generations	Illiteracy					
21	Inhabited for generations						
16	Inhabited for generations	Illiteracy	Agriculture	Land-holding peasant	Married	No	No religion
14	Inhabited for generations	Illiteracy	Agriculture	Land-holding peasant	Married	No	No religion
28	Inhabited for generations	Illiteracy	Agriculture	Land-holding peasant	Widowed	No	No religion
27	Inhabited for generations						
9	Inhabited for generations	Illiteracy	Agriculture	Land-holding peasant	Married	No	No religion
3	Inhabited for generations	Illiteracy	Agriculture	Land-holding peasant	Married	No	No religion
7	Inhabited for generations	Illiteracy	Agriculture	Land-holding peasant	Married	No	No religion

Appendix: Overview of Microdata for Wudu County (*cont.*)

	The no. of household	Jia	Household	Relationship with the household head	Surname	First name	Sex	The Chinese zodiac	Birthdate Year	Month
Muma Xiang 1st Bao				The household head's wife	陈	白连儿子	Female	Ox	1900	11
Muma Xiang 1st Bao				Son	陈	兴隆	Male	Pig	1923	10
Muma Xiang 1st Bao				Son's wife	陈	谈中银儿	Female	Pig	1923	8
Muma Xiang 1st Bao				Daughter	陈	小娃子	Female	Rabbit	1939	7
Muma Xiang 1st Bao				Uncle	陈	顺学	Male	Ox	1870	5
Muma Xiang 1st Bao	26	2	11	The household head	陈	顺富	Male	Horse	1918	10
Muma Xiang 1st Bao				The household head's wife	陈	白赵儿子	Female	Tiger	1914	5
Muma Xiang 1st Bao				Daughter	陈	当银子	Female	Goat	1943	3
Muma Xiang 1st Bao				Son	陈	创代子	Male	Dog	1947	1
Muma Xiang 1st Bao	27	2	12	The household head	陈	义会	Male	Goat	1920	1
Muma Xiang 1st Bao				The household head's wife	陈	白二连子	Female	Goat	1919	3
Muma Xiang 1st Bao				Mother	陈	孙双连子	Female	Rooster	1896	3
Muma Xiang 1st Bao				Daughter	陈	安娃子	Female	Rooster	1945	4
Muma Xiang 1st Bao				Younger brother	陈	润儿子	Male	Ox	1925	4
Muma Xiang 1st Bao				Younger brother's wife	陈	白银翠儿	Female	Dragon	1928	6
Muma Xiang 1st Bao	28	2	13	The household head	陈	怀應	Male	Monkey	1920	12
Muma Xiang 1st Bao				The household head's wife	陈	白九银子	Female	Rooster	1921	10

Date	The beginning time of inhabitation in this county	Educational level	Profession sector	Occupation	Marital status	Physical disability	Religion
11	Inhabited for generations	Illiteracy	Agriculture	Land-holding peasant	Married	No	No religion
30	Inhabited for generations	Illiteracy	Agriculture	Land-holding peasant	Married	No	No religion
26	Inhabited for generations	Illiteracy	Agriculture	Land-holding peasant	Married	No	No religion
20	Inhabited for generations	Illiteracy				Bald	
3	Inhabited for generations				Widowed		
6	Inhabited for generations	Illiteracy	Agriculture	Land-holding peasant	Married	No	No religion
6	Inhabited for generations	Illiteracy	Agriculture	Land-holding peasant	Married	No	No religion
17	Inhabited for generations						
14	Inhabited for generations						
22	Inhabited for generations	Illiteracy	Agriculture	Land-holding peasant	Married	No	No religion
2	Inhabited for generations	Illiteracy	Agriculture	Land-holding peasant	Married	No	No religion
7	Inhabited for generations	Illiteracy	Agriculture	Land-holding peasant	Widowed	No	No religion
4	Inhabited for generations						
26	Inhabited for generations	Illiteracy	Agriculture	Land-holding peasant	Married	No	No religion
20	Inhabited for generations	Illiteracy	Agriculture	Land-holding peasant	Married	No	No religion
24	Inhabited for generations	Illiteracy	Agriculture	Land-holding peasant	Married	No	No religion
12	Inhabited for generations	Illiteracy	Agriculture	Land-holding peasant	Married	No	No religion

Appendix: Overview of Microdata for Wudu County (*cont.*)

	The no. of household	Jia	Household	Relationship with the household head	Surname	First name	Sex	The Chinese zodiac	Birthdate	
									Year	Month
Muma Xiang 1st Bao				Mother	陈	白连儿子	Female	Rat	1899	6
Muma Xiang 1st Bao				Son	陈	安德	Male	Dog	1934	12
Muma Xiang 1st Bao				Daughter	陈	金香子	Female	Dog	1946	5
Muma Xiang 1st Bao				Uncle's wife	陈	何长秀娃	Female	Monkey	1895	6
Muma Xiang 1st Bao	29	2	14	The household head	白	义现	Male	Horse	1905	5
Muma Xiang 1st Bao				Daughter	白	见娃子	Female	Monkey	1944	9
Muma Xiang 1st Bao				Younger brother	白	义先	Male	Monkey	1907	1
Muma Xiang 1st Bao				Younger brother's wife	白	马旦女子	Female	Rooster	1908	8
Muma Xiang 1st Bao	30	2	15	The household head	白	发勤	Male	Tiger	1901	10
Muma Xiang 1st Bao				The household head's wife	白	陈翠儿子	Female	Tiger	1901	11
Muma Xiang 1st Bao				Son	白	张娃子	Male	Rat	1936	9
Muma Xiang 1st Bao				Older brother's wife	白	陈连娃子	Female	Ox	1900	8
Muma Xiang 1st Bao	31	3	1	The household head	何	正银	Male	Tiger	1914	10
Muma Xiang 1st Bao				The household head's wife	何	陈腊儿子	Female	Rabbit	1916	1
Muma Xiang 1st Bao				Mother	何	白三娃子	Female	Horse	1893	8
Muma Xiang 1st Bao	32	3	2	The household head	何	正安	Male	Rat	1887	11
Muma Xiang 1st Bao				Son	何	孟子	Male	Monkey	1907	11
Muma Xiang 1st Bao				Son's wife	何	王润连子	Female	Rooster	1907	1

Date	The beginning time of inhabitation in this county	Educational level	Profession sector	Occupation	Marital status	Physical disability	Religion
14	Inhabited for generations	Illiteracy	Agriculture	Land-holding peasant	Widowed	No	No religion
21	Inhabited for generations	Primary school 2 years			Unmarried		
17	Inhabited for generations						
22	Inhabited for generations	Illiteracy	Agriculture	Land-holding peasant	Widowed	No	No religion
11	Inhabited for generations	Illiteracy	Agriculture	Land-holding peasant	Widowed	No	No religion
2	Inhabited for generations						
17	Inhabited for generations	Illiteracy	Agriculture	Land-holding peasant	Married	No	No religion
28	Inhabited for generations	Illiteracy	Agriculture	Land-holding peasant	Married	No	No religion
2	Inhabited for generations	Illiteracy	Agriculture	Land-holding peasant	Married	No	No religion
24	Inhabited for generations	Illiteracy	Agriculture	Land-holding peasant	Married	No	No religion
24	Inhabited for generations	Illiteracy					
28	Inhabited for generations	Illiteracy	Agriculture	Land-holding peasant	Widowed	No	No religion
28	Inhabited for generations	Sishu 1 year	Agriculture	Land-holding peasant	Married	No	No religion
8	Inhabited for generations	Illiteracy	Agriculture	Land-holding peasant	Married	No	No religion
15	Inhabited for generations	Illiteracy	Agriculture	Land-holding peasant	Widowed	No	No religion
4	Inhabited for generations	Illiteracy	Agriculture	Land-holding peasant	Widowed	No	No religion
2	Inhabited for generations	Illiteracy	Agriculture	Land-holding peasant	Married	No	No religion
6	Inhabited for generations	Illiteracy	Agriculture	Land-holding peasant	Married	No	No religion

Appendix: Overview of Microdata for Wudu County (*cont.*)

	The no. of household	Jia	Household	Relationship with the household head	Surname	First name	Sex	The Chinese zodiac	Birthdate Year	Month
Muma Xiang 1st Bao				Grandson	何	万章	Male	Tiger	1938	3
Muma Xiang 1st Bao	33	3	3	The household head	何	正太	Male	Ox	1913	3
Muma Xiang 1st Bao				The household head's wife	何	白蛾儿子	Female	Monkey	1907	9
Muma Xiang 1st Bao				Son	何	丙娃子	Male	Goat	1931	12
Muma Xiang 1st Bao				Daughter	何	卢娃子	Female	Dragon	1928	6
Muma Xiang 1st Bao	34	3	4	The household head	何	正宝	Male	Rat	1898	1
Muma Xiang 1st Bao				The household head's wife	何	白桃香子	Female	Ox	1900	6
Muma Xiang 1st Bao				Son	何	五代子	Male	Dog	1934	6
Muma Xiang 1st Bao				Son's wife	何	陈九翠子	Female	Dragon	1928	10
Muma Xiang 1st Bao	35	3	5	The household head	何	廷顺	Male	Dragon	1890	1
Muma Xiang 1st Bao				The household head's wife	何	王连翠子	Female	Ox	1900	10
Muma Xiang 1st Bao				Son	何	万银	Male	Rabbit	1939	3
Muma Xiang 1st Bao				Daughter	何	九旦娃	Female	Dog	1922	11
Muma Xiang 1st Bao				Daughter	何	崴娃子	Female	Goat	1943	8
Muma Xiang 1st Bao	36	3	6	The household head	谈	学清	Male	Goat	1894	9
Muma Xiang 1st Bao				The household head's wife	谈	陈赵儿子	Female	Ox	1900	4
Muma Xiang 1st Bao				Son	谈	朝银	Male	Dog	1922	6

Date	The beginning time of inhabitation in this county	Educational level	Profession sector	Occupation	Marital status	Physical disability	Religion
15	Inhabited for generations	Primary school 1 year					
6	Inhabited for generations	Illiteracy	Agriculture	Land-holding peasant	Married	No	No religion
5	Inhabited for generations	Illiteracy	Agriculture	Land-holding peasant	Married	No	No religion
15	Inhabited for generations	Illiteracy	Agriculture	Land-holding peasant	Unmarried	No	No religion
2	Inhabited for generations	Illiteracy	Agriculture	Land-holding peasant	Divorced	Right eye blind	No religion
5	Inhabited for generations	Illiteracy	Agriculture	Land-holding peasant	Married	No	No religion
20	Inhabited for generations	Illiteracy	Agriculture	Land-holding peasant	Married	No	No religion
19	Inhabited for generations	Illiteracy	Agriculture	Land-holding peasant	Married	No	No religion
26	Inhabited for generations	Illiteracy	Agriculture	Land-holding peasant	Married	No	No religion
11	Inhabited for generations	Illiteracy	Agriculture	Land-holding peasant	Married	No	No religion
20	Inhabited for generations	Illiteracy	Agriculture	Land-holding peasant	Married	No	No religion
22	Inhabited for generations	Primary school 1 year					
7	Inhabited for generations	Illiteracy	Agriculture	Land-holding peasant	Unmarried	Lame	No religion
25	Inhabited for generations						
3	Inhabited for generations	Illiteracy	Agriculture	Land-holding peasant	Married	No	No religion
24	Inhabited for generations	Illiteracy	Agriculture	Land-holding peasant	Married	No	No religion
4	Inhabited for generations	Illiteracy	Agriculture	Land-holding peasant	Married	No	No religion

Appendix: Overview of Microdata for Wudu County (*cont.*)

	The no. of house-hold	Jia	House-hold	Relation-ship with the house-hold head	Surname	First name	Sex	The Chinese zodiac	Birthdate	
									Year	Month
Muma Xiang 1st Bao				Son's wife	谈	白代翠子	Female	Rat	1924	12
Muma Xiang 1st Bao	37	3	7	The house-hold head	陈	何九娃子	Female	Horse	1881	9
Muma Xiang 1st Bao				Daughter	陈	苏儿子	Female	Rat	1912	6
Muma Xiang 1st Bao				Grandson	陈	代管子	Male	Goat	1943	11
Muma Xiang 1st Bao	38	3	8	The house-hold head	陈	方董儿子	Female	Tiger	1900	1
Muma Xiang 1st Bao				Son	陈	进隆	Male	Ox	1926	1
Muma Xiang 1st Bao				Son's wife	陈	白营连子	Female	Rooster	1921	6
Muma Xiang 1st Bao	39	3	9	The house-hold head	陈	尚德	Male	Rooster	1908	11
Muma Xiang 1st Bao				The house-hold head's wife	陈	王创连子	Female	Rabbit	1915	6
Muma Xiang 1st Bao				Son	陈	进金	Male	Horse	1930	9
Muma Xiang 1st Bao				Son's wife	陈	白四季子	Female	Snake	1929	7
Muma Xiang 1st Bao	40	3	10	The house-hold head	陈	进宽	Male	Rooster	1908	11
Muma Xiang 1st Bao				The house-hold head's wife	陈	谈五儿子	Female	Dog	1909	6
Muma Xiang 1st Bao				Daughter	陈	创娃子	Female	Dog	1947	1
Muma Xiang 1st Bao	41	3	11	The house-hold head	陈	尚虎	Male	Snake	1892	11
Muma Xiang 1st Bao				The house-hold head's wife	陈	刘星连子	Female	Ox	1888	2
Muma Xiang 1st Bao				Son	陈	进发	Male	Tiger	1915	1
Muma Xiang 1st Bao				Son's wife	陈	何罗代子	Female	Ox	1913	12

Date	The beginning time of inhabitation in this county	Educational level	Profession sector	Occupation	Marital status	Physical disability	Religion
3	Inhabited for generations	Illiteracy	Agriculture	Land-holding peasant	Married	No	No religion
14	Inhabited for generations	Illiteracy	Agriculture	Land-holding peasant	Widowed	No	No religion
24	Inhabited for generations	Illiteracy	Agriculture	Land-holding peasant	Divorced	No	No religion
8	Inhabited for generations						
9	Inhabited for generations	Illiteracy	Agriculture	Land-holding peasant	Widowed	No	No religion
26	Inhabited for generations	Illiteracy	Agriculture	Land-holding peasant	Married	No	No religion
10	Inhabited for generations	Illiteracy	Agriculture	Land-holding peasant	Married	No	No religion
4	Inhabited for generations	Sishu 2 years	Agriculture	Land-holding peasant	Married	No	No religion
17	Inhabited for generations	Illiteracy	Agriculture	Land-holding peasant	Married	No	No religion
7	Inhabited for generations	Illiteracy	Agriculture	Land-holding peasant	Married	No	No religion
16	Inhabited for generations	Illiteracy	Agriculture	Land-holding peasant	Married	No	No religion
5	Inhabited for generations	Illiteracy	Agriculture	Land-holding peasant	Married	No	No religion
11	Inhabited for generations	Illiteracy	Agriculture	Land-holding peasant	Married	No	No religion
9	Inhabited for generations						
21	Inhabited for generations	Illiteracy	Agriculture	Land-holding peasant	Married	No	No religion
12	Inhabited for generations	Illiteracy	Agriculture	Land-holding peasant	Married	No	No religion
25	Inhabited for generations	Illiteracy	Agriculture	Land-holding peasant	Married	No	No religion
8	Inhabited for generations	Illiteracy	Agriculture	Land-holding peasant	Married	No	No religion

Appendix: Overview of Microdata for Wudu County (*cont.*)

	The no. of household	Jia	Household	Relationship with the household head	Surname	First name	Sex	The Chinese zodiac	Birthdate Year	Month
Muma Xiang 1st Bao				Cohabitant	何	白关花子	Female	Rat	1887	6
Muma Xiang 1st Bao	42	3	12	The household head	陈	尚富	Male	Tiger	1901	3
Muma Xiang 1st Bao				The household head's wife	陈	何连儿子	Female	Horse	1905	3
Muma Xiang 1st Bao				Son	陈	润娃子	Male	Dog	1946	5
Muma Xiang 1st Bao				Daughter	陈	孝娃子	Female	Rabbit	1940	1
Muma Xiang 1st Bao	43	3	13	The household head	陈	奉登	Male	Rat	1875	1
Muma Xiang 1st Bao				Son's wife	陈	白连女子	Female	Pig	1886	4
Muma Xiang 1st Bao				Son's wife	陈	白四儿娃	Female	Dog	1909	5
Muma Xiang 1st Bao				Grandson	陈	进满	Male	Goat	1906	10
Muma Xiang 1st Bao				Grandson's wife	陈	白金翠子	Female	Dragon	1928	6
Muma Xiang 1st Bao				Grand daughter	陈	当翠儿	Female	Monkey	1944	4
Muma Xiang 1st Bao				Cohabitant	白	白连女子	Female	Rat	1887	4
Muma Xiang 1st Bao	44	3	14	The household head	谈	朝奉	Male	Dragon	1903	3
Muma Xiang 1st Bao				The household head's wife	谈	王翠女子	Female	Monkey	1907	7
Muma Xiang 1st Bao				Father	谈	学宝	Male	Dog	1878	2
Muma Xiang 1st Bao				Son	谈	廷槐	Male	Rat	1936	6
Muma Xiang 1st Bao				Daughter	谈	女子	Female	Pig	1947	6
Muma Xiang 1st Bao				Older brother's wife	谈	王保香子	Female	Rat	1899	8

Date	The beginning time of inhabitation in this county	Educational level	Profession sector	Occupation	Marital status	Physical disability	Religion
14	Inhabited for generations	Illiteracy	Agriculture	Land-holding peasant	Widowed	No	No religion
23	Inhabited for generations	Sishu 1 year	Agriculture	Land-holding peasant	Married	No	No religion
28	Inhabited for generations	Illiteracy	Agriculture	Land-holding peasant	Married	No	No religion
17	Inhabited for generations						
29	Inhabited for generations	Illiteracy					
15	Inhabited for generations						
27	Inhabited for generations	Illiteracy	Agriculture	Land-holding peasant	Widowed	No	No religion
23	Inhabited for generations	Illiteracy	Agriculture	Land-holding peasant	Divorced	No	No religion
27	Inhabited for generations	Illiteracy	Agriculture	Land-holding peasant	Married	Blind	No religion
14	Inhabited for generations	Illiteracy	Agriculture	Land-holding peasant	Married	No	No religion
26	Inhabited for generations						
12	Inhabited for generations	Illiteracy	Agriculture	Land-holding peasant	Widowed	No	No religion
30	Inhabited for generations	Illiteracy	Agriculture	Land-holding peasant	Married	No	No religion
18	Inhabited for generations	Illiteracy	Agriculture	Land-holding peasant	Married	No	No religion
24	Inhabited for generations				Widowed		
4	Inhabited for generations	Primary school 2 years					
19	Inhabited for generations						
9	Inhabited for generations	Illiteracy	Agriculture	Land-holding peasant	Widowed	No	No religion

Appendix: Overview of Microdata for Wudu County (*cont.*)

	The no. of household	Jia	Household	Relationship with the household head	Surname	First name	Sex	The Chinese zodiac	Birthdate Year	Month
Muma Xiang 1st Bao	45	4	1	The household head	陈	义进	Male	Ox	1925	10
Muma Xiang 1st Bao				The household head's wife	陈	董冯儿子	Female	Rat	1924	8
Muma Xiang 1st Bao				Son	陈	创娃子	Male	Rooster	1945	12
Muma Xiang 1st Bao	46	4	2	The household head	白	兴玉	Male	Rat	1899	9
Muma Xiang 1st Bao				The household head's wife	白	谈关女子	Female	Tiger	1901	8
Muma Xiang 1st Bao				Father	白	义才	Male	Tiger	1877	11
Muma Xiang 1st Bao				Son	白	斗全	Male	Dog	1934	8
Muma Xiang 1st Bao	47	4	3	The household head	白	兴太	Male	Rabbit	1902	4
Muma Xiang 1st Bao				The household head's wife	白	谈见娃子	Female	Horse	1905	1
Muma Xiang 1st Bao				Father	白	义思	Male	Dragon	1879	12
Muma Xiang 1st Bao				Mother	白	李四娃子	Female	Rabbit	1878	9
Muma Xiang 1st Bao				Son	白	赵娃子	Male	Ox	1925	10
Muma Xiang 1st Bao				Son's wife	白	孙燕女子	Female	Pig	1923	9
Muma Xiang 1st Bao				The second son	白	斗明	Male	Rooster	1933	12
Muma Xiang 1st Bao	48	4	4	The household head	白	金升	Male	Tiger	1900	1
Muma Xiang 1st Bao				The household head's wife	白	李桃女子	Female	Tiger	1901	8
Muma Xiang 1st Bao				Grandson	白	义银	Male	Rat	1924	5

Date	The beginning time of inhabitation in this county	Educational level	Profession sector	Occupation	Marital status	Physical disability	Religion
2	Inhabited for generations	Illiteracy	Agriculture	Land-holding peasant	Married		No religion
15	Inhabited for generations	Illiteracy	Agriculture	Land-holding peasant	Married	No	No religion
21	Inhabited for generations						
6	Inhabited for generations	Illiteracy	Agriculture	Land-holding peasant	Married	No	No religion
8	Inhabited for generations	Illiteracy	Agriculture	Land-holding peasant	Married	No	No religion
2	Inhabited for generations						
24	Inhabited for generations	Primary school 2 years				Bald	
12	Inhabited for generations	Illiteracy	Agriculture	Land-holding peasant	Married	No	No religion
26	Inhabited for generations	Illiteracy	Agriculture	Land-holding peasant	Married	No	No religion
22	Inhabited for generations	Illiteracy	Agriculture	Land-holding peasant	Married	No	No religion
9	Inhabited for generations	Illiteracy	Agriculture	Land-holding peasant	Married	No	No religion
15	Inhabited for generations	Illiteracy	Agriculture	Land-holding peasant	Married	No	No religion
29	Inhabited for generations	Illiteracy	Agriculture	Land-holding peasant	Married	No	No religion
18	Inhabited for generations	Primary school 2 years					
19	Inhabited for generations	Illiteracy	Agriculture	Land-holding peasant	Married	No	No religion
18	Inhabited for generations	Illiteracy	Agriculture	Land-holding peasant	Married	No	No religion
28	Inhabited for generations	Illiteracy	Agriculture	Land-holding peasant	Married	No	No religion

Appendix: Overview of Microdata for Wudu County (*cont.*)

	The no. of household	Jia	Household	Relationship with the household head	Surname	First name	Sex	The Chinese zodiac	Birthdate Year	Month
Muma Xiang 1st Bao				Grandson's wife	白	谈张二娃	Female	Rabbit	1927	9
Muma Xiang 1st Bao	49	4	5	The household head	白	义清	Male	Tiger	1901	2
Muma Xiang 1st Bao				Son	白	兴位	Male	Ox	1925	10
Muma Xiang 1st Bao				Son's wife	白	刘苏儿子	Female	Rat	1924	7
Muma Xiang 1st Bao				Grand daughter	白	男娃子	Female	Dog	1946	2
Muma Xiang 1st Bao	50	4	6	The household head	白	兴银	Male	Dog	1909	10
Muma Xiang 1st Bao				The household head's wife	白	杨真儿子	Female	Goat	1906	3
Muma Xiang 1st Bao				Son	白	斗金	Male	Pig	1935	7
Muma Xiang 1st Bao				Uncle	白	义钦	Male	Monkey	1907	7
Muma Xiang 1st Bao	51	4	7	The household head	白	义忠	Male	Ox	1888	9
Muma Xiang 1st Bao				The household head's wife	白	李楼娃子	Female	Snake	1892	6
Muma Xiang 1st Bao				Daughter	白	连儿子	Female	Rabbit	1915	9
Muma Xiang 1st Bao				Son	白	兴林	Male	Monkey	1902	12
Muma Xiang 1st Bao				Son's wife	白	何石娃子	Female	Rooster	1921	6
Muma Xiang 1st Bao				The first grandson	白	刘儿娃	Male	Horse	1943	1
Muma Xiang 1st Bao				The second son's wife	白	高儿娃	Male	Rooster	1945	6
Muma Xiang 1st Bao	52	4	8	The household head	白	义钧	Male	Horse	1918	7
Muma Xiang 1st Bao				The household head's wife	白	陈三儿子	Female	Dragon	1916	6

Date	The beginning time of inhabitation in this county	Educational level	Profession sector	Occupation	Marital status	Physical disability	Religion
9	Inhabited for generations	Illiteracy	Agriculture	Land-holding peasant	Married	No	No religion
14	Inhabited for generations	Sishu 1 year	Agriculture	Land-holding peasant	Widowed	No	Buddhism
3	Inhabited for generations	Illiteracy	Agriculture	Land-holding peasant	Married	No	
15	Inhabited for generations	Illiteracy	Agriculture	Land-holding peasant	Married	No	
13	Inhabited for generations						
25	Inhabited for generations	Illiteracy	Agriculture	Land-holding peasant	Married	No	No religion
31	Inhabited for generations	Illiteracy	Agriculture	Land-holding peasant	Married	No	No religion
23	Inhabited for generations	Primary school 1 year					
1	Inhabited for generations	Illiteracy	Agriculture	Land-holding peasant	Widowed	No	No religion
11	Inhabited for generations	Illiteracy	Agriculture	Land-holding peasant	Married	No	No religion
30	Inhabited for generations	Illiteracy	Agriculture	Land-holding peasant	Married	No	No religion
23	Inhabited for generations	Illiteracy	Agriculture	Land-holding peasant	Married	No	No religion
22	Inhabited for generations	Illiteracy	Agriculture	Land-holding peasant	Married	No	No religion
8	Inhabited for generations	Illiteracy	Agriculture	Land-holding peasant	Married	No	No religion
12	Inhabited for generations						
21	Inhabited for generations						
15	Inhabited for generations	Illiteracy	Agriculture	Land-holding peasant	Married	Bald	No religion
15	Inhabited for generations	Illiteracy	Agriculture	Land-holding peasant	Married	No	No religion

Appendix: Overview of Microdata for Wudu County (*cont.*)

	The no. of household	Jia	Household	Relationship with the household head	Surname	First name	Sex	The Chinese zodiac	Birthdate Year	Month
Muma Xiang 1st Bao				Daughter	白	四儿娃	Female	Rooster	1945	11
Muma Xiang 1st Bao	53	4	9	The household head	白	兴通	Male	Goat	1882	2
Muma Xiang 1st Bao				The household head's wife	白	李兴连娃	Female	Monkey	1883	12
Muma Xiang 1st Bao				Son	白	斗富	Male	Ox	1913	11
Muma Xiang 1st Bao				Son's wife	白	王蛇儿娃	Female	Snake	1917	8
Muma Xiang 1st Bao				Grand daughter	白	东翠娃	Female	Rooster	1945	12
Muma Xiang 1st Bao	54	4	10	The household head	白	兴隆	Male	Tiger	1901	3
Muma Xiang 1st Bao				The household head's wife	白	谈桃秀子	Female	Rabbit	1901	1
Muma Xiang 1st Bao				Father	白	义彩	Male	Goat	1882	4
Muma Xiang 1st Bao				Son	白	斗星	Male	Pig	1935	3
Muma Xiang 1st Bao				Daughter	白	罗代子	Female	Rooster	1933	9
Muma Xiang 1st Bao	55	4	11	The household head	白	兴虎	Male	Snake	1917	1
Muma Xiang 1st Bao				The household head's wife	白	王崴儿娃	Female	Snake	1918	2
Muma Xiang 1st Bao				Son	白	斗魁	Male	Rabbit	1939	9
Muma Xiang 1st Bao				Daughter	银	翠儿	Female	Horse	1942	3
Muma Xiang 1st Bao	56	4	12	The household head	王	顺江	Male	Pig	1909	2
Muma Xiang 1st Bao				The household head's wife	王	李见女子	Female	Rooster	1908	3

Date	The beginning time of inhabitation in this county	Educational level	Profession sector	Occupation	Marital status	Physical disability	Religion
19	Inhabited for generations					Bald	
12	Inhabited for generations	Illiteracy	Agriculture	Land-holding peasant	Married	No	No religion
16	Inhabited for generations	Illiteracy	Agriculture	Land-holding peasant	Married	No	No religion
9	Inhabited for generations	Illiteracy	Agriculture	Land-holding peasant	Married	No	No religion
22	Inhabited for generations	Illiteracy	Agriculture	Land-holding peasant	Married	No	No religion
18	Inhabited for generations						
25	Inhabited for generations	Sishu 1 year	Agriculture	Land-holding peasant	Married	No	Buddhism
7	Inhabited for generations	Illiteracy	Agriculture	Land-holding peasant	Married	No	
4	Inhabited for generations	Illiteracy	Agriculture	Land-holding peasant	Widowed	No	
13	Inhabited for generations	Primary school 3 years			Unmarried	No	
4	Inhabited for generations	Illiteracy	Agriculture	Land-holding peasant	Unmarried	No	
23	Inhabited for generations	Illiteracy	Agriculture	Land-holding peasant	Married	No	No religion
6	Inhabited for generations	Illiteracy	Agriculture	Land-holding peasant	Married	No	No religion
23	Inhabited for generations	Primary school 1 year					
31	Inhabited for generations						
11	Inhabited for generations	Illiteracy	Agriculture	Land-holding peasant	Married	No	No religion
6	Inhabited for generations	Illiteracy	Agriculture	Land-holding peasant	Married	No	No religion

Appendix: Overview of Microdata for Wudu County (*cont.*)

	The no. of household	Jia	Household	Relationship with the household head	Surname	First name	Sex	The Chinese zodiac	Birthdate Year	Month
Muma Xiang 1st Bao				Mother	白	褚关秀子	Female	Ox	1888	8
Muma Xiang 1st Bao				Son	王	白斗强	Male	Ox	1937	2
Muma Xiang 1st Bao				Daughter	王	当翠儿	Female	Rabbit	1939	11
Muma Xiang 1st Bao				Cohabitant	白	兴元	Male	Horse	1905	5
Muma Xiang 1st Bao					白	王南儿子	Female	Rooster	1908	8
Muma Xiang 1st Bao	57	4	13	The household head	白	发成	Male	Pig	1898	11
Muma Xiang 1st Bao				The household head's wife	白	陈称儿子	Female	Ox	1900	9
Muma Xiang 1st Bao				Son	白	胖儿娃	Male	Horse	1942	7
Muma Xiang 1st Bao	58	5	1	The household head	王	顺富	Male	Rabbit	1902	10
Muma Xiang 1st Bao				The household head's wife	王	韩买连子	Female	Horse	1905	11
Muma Xiang 1st Bao				Mother	王	谈董秀子	Female	Dog	1885	9
Muma Xiang 1st Bao				Son	王	安太	Male	Ox	1937	5
Muma Xiang 1st Bao				The first daughter	王	永林子	Female	Dragon	1940	10
Muma Xiang 1st Bao				The second daughter	王	徐娃子	Female	Horse	1942	5
Muma Xiang 1st Bao				The third daughter	王	蔡娃子	Female	Goat	1943	11
Muma Xiang 1st Bao	59	5	2	The household head	杨	才兴	Male	Rooster	1908	11
Muma Xiang 1st Bao				The household head's wife	杨	陈胡儿子	Female	Dog	1909	8
Muma Xiang 1st Bao				Father	杨	世通	Male	Dragon	1890	1

Date	The beginning time of inhabitation in this county	Educational level	Profession sector	Occupation	Marital status	Physical disability	Religion
11	Inhabited for generations	Illiteracy	Agriculture	Land-holding peasant	Widowed	No	No religion
17	Inhabited for generations	Primary school 2 years					
22	Inhabited for generations	Illiteracy					
5	Inhabited for generations	Illiteracy	Agriculture	Land-holding peasant	Married	No	No religion
28	Inhabited for generations	Illiteracy	Agriculture	Land-holding peasant	Married	No	No religion
29	Inhabited for generations	Illiteracy	Agriculture	Land-holding peasant	Married	No	No religion
29	Inhabited for generations	Illiteracy	Agriculture	Land-holding peasant	Married	No	No religion
1	Inhabited for generations						
2	Inhabited for generations	Sishu 1 year	Agriculture	Land-holding peasant	Married	No	Buddhism
30	Inhabited for generations	Illiteracy	Agriculture	Land-holding peasant	Married	No	
9	Inhabited for generations	Illiteracy	Agriculture	Land-holding peasant	Widowed	No	
5	Inhabited for generations	Primary school 1 year					
7	Inhabited for generations	Illiteracy					
23	Inhabited for generations						
30	Inhabited for generations						
26	Inhabited for generations	Illiteracy	Agriculture	Land-holding peasant	Married	No	No religion
19	Inhabited for generations	Illiteracy	Agriculture	Land-holding peasant	Married	No	No religion
27	Inhabited for generations	Illiteracy	Agriculture	Land-holding peasant	Widowed	No	No religion

Appendix: Overview of Microdata for Wudu County (*cont.*)

	The no. of household	Jia	Household	Relationship with the household head	Surname	First name	Sex	The Chinese zodiac	Birthdate Year	Month
Muma Xiang 1st Bao				Grandmother	贾	氏	Female	Goat	1870	2
Muma Xiang 1st Bao				Son	杨	银成娃	Male	Rooster	1933	9
Muma Xiang 1st Bao	60	5	3	The household head	杨	才钦	Male	Tiger	1901	10
Muma Xiang 1st Bao				The household head's wife	杨	白刘儿子	Female	Snake	1904	6
Muma Xiang 1st Bao				Mother	杨	贾翠子	Female	Rooster	1884	2
Muma Xiang 1st Bao				Grandmother	何	氏	Female	Pig	1862	1
Muma Xiang 1st Bao	61	5	4	The household head	王	應会	Male	Rat	1887	10
Muma Xiang 1st Bao				The household head's wife	王	何虎银儿	Female	Tiger	1889	12
Muma Xiang 1st Bao				Son	王	何娃子	Male	Rat	1912	8
Muma Xiang 1st Bao				Son's wife	王	何刘儿子	Female	Tiger	1914	6
Muma Xiang 1st Bao				The second son	王	顺福	Male	Monkey	1932	3
Muma Xiang 1st Bao				Grandson	王	海娃子	Male	Snake	1941	7
Muma Xiang 1st Bao	62	5	5	The household head	王	應连	Male	Dog	1909	3
Muma Xiang 1st Bao				The household head's wife	王	杨冯儿子	Female	Monkey	1907	4
Muma Xiang 1st Bao				Father	王	碧福	Male	Rat	1887	10
Muma Xiang 1st Bao				Son	王	顺林	Male	Rabbit	1939	3
Muma Xiang 1st Bao	63	5	6	The household head	王	應智	Male	Goat	1920	1

Date	The beginning time of inhabitation in this county	Educational level	Profession sector	Occupation	Marital status	Physical disability	Religion
18	Inhabited for generations						
4	Inhabited for generations	Primary school 1 year				Bald	
11	Inhabited for generations	Illiteracy	Agriculture	Land-holding peasant	Married	No	No religion
8	Inhabited for generations	Illiteracy	Agriculture	Land-holding peasant	Married	No	No religion
22	Inhabited for generations	Illiteracy	Agriculture	Land-holding peasant	Widowed	No	No religion
2	Inhabited for generations						
19	Inhabited for generations	Illiteracy	Agriculture	Land-holding peasant	Married	No	No religion
8	Inhabited for generations	Illiteracy	Agriculture	Land-holding peasant	Married	No	No religion
27	Inhabited for generations	Illiteracy	Agriculture	Land-holding peasant	Married	No	No religion
3	Inhabited for generations	Illiteracy	Agriculture	Land-holding peasant	Married	No	No religion
20	Inhabited for generations	Primary school 3 years			Unmarried		
6	Inhabited for generations	Illiteracy					
6	Inhabited for generations	Sishu 2 years	Agriculture	Land-holding peasant	Married	No	No religion
13	Inhabited for generations	Illiteracy	Agriculture	Land-holding peasant	Married	No	No religion
19	Inhabited for generations	Illiteracy	Agriculture	Land-holding peasant	Married	No	No religion
14	Inhabited for generations	Primary school 1 year					
2	Inhabited for generations	Illiteracy	Agriculture	Land-holding peasant	Married	No	No religion

Appendix: Overview of Microdata for Wudu County (*cont.*)

									Birthdate	
	The no. of household	**Jia**	**Household**	**Relationship with the household head**	**Surname**	**First name**	**Sex**	**The Chinese zodiac**	**Year**	**Month**
Muma Xiang 1st Bao				The household head's wife	王	白冯翠子	Female	Monkey	1920	8
Muma Xiang 1st Bao				Son	王	代林子	Male	Dragon	1940	12
Muma Xiang 1st Bao				Daughter	王	秦当儿	Female	Monkey	1944	6
Muma Xiang 1st Bao	64	5	7	The household head	王	顺连	Male	Rooster	1908	10
Muma Xiang 1st Bao				The household head's wife	王	何卢银子	Female	Rat	1912	4
Muma Xiang 1st Bao				Father	王	應德	Male	Dog	1885	8
Muma Xiang 1st Bao				The first daughter	王	贤儿娃	Female	Rooster	1933	1
Muma Xiang 1st Bao				The second daughter	王	六娃子	Female	Dragon	1940	7
Muma Xiang 1st Bao	65	5	8	The household head	王	顺正	Male	Ox	1900	7
Muma Xiang 1st Bao				The household head's wife	王	白蔡儿子	Female	Dragon	1903	11
Muma Xiang 1st Bao				Daughter	王	代线子	Female	Rabbit	1939	5
Muma Xiang 1st Bao				The first son	王	张代儿	Male	Goat	1931	4
Muma Xiang 1st Bao				The second son	王	银代儿	Male	Horse	1942	3
Muma Xiang 1st Bao	66	5	9	The household head	王	顺弟	Male	Rabbit	1915	7
Muma Xiang 1st Bao				The household head's wife	王	褚么女子	Female	Ox	1913	7
Muma Xiang 1st Bao				Father	王	應中	Male	Rat	1887	9
Muma Xiang 1st Bao				Daughter	王	胖女子	Female	Rabbit	1939	6

Date	The beginning time of inhabitation in this county	Educational level	Profession sector	Occupation	Marital status	Physical disability	Religion
28	Inhabited for generations	Illiteracy	Agriculture	Land-holding peasant	Married	No	No religion
9	Inhabited for generations	Illiteracy					
4	Inhabited for generations						
14	Inhabited for generations	Illiteracy	Agriculture	Land-holding peasant	Married	No	No religion
4	Inhabited for generations	Illiteracy	Agriculture	Land-holding peasant	Married	No	No religion
14	Inhabited for generations	Illiteracy	Agriculture	Land-holding peasant	Married	No	No religion
14	Inhabited for generations	Illiteracy	Agriculture	Land-holding peasant	Unmarried	No	No religion
16	Inhabited for generations	Illiteracy					
30	Inhabited for generations	Illiteracy	Agriculture	Land-holding peasant	Married	No	No religion
24	Inhabited for generations	Illiteracy	Agriculture	Land-holding peasant	Married	No	No religion
7	Inhabited for generations	Illiteracy		Land-holding peasant			
7	Inhabited for generations	Illiteracy	Agriculture	Land-holding peasant	Unmarried	No	No religion
3	Inhabited for generations	Illiteracy					
25	Inhabited for generations	Illiteracy	Agriculture	Land-holding peasant	Married	No	No religion
2	Inhabited for generations	Illiteracy	Agriculture	Land-holding peasant	Married	No	No religion
2	Inhabited for generations	Illiteracy	Agriculture	Land-holding peasant	Widowed	No	No religion
23	Inhabited for generations	Illiteracy					

Appendix: Overview of Microdata for Wudu County (*cont.*)

	The no. of household	Jia	Household	Relationship with the household head	Surname	First name	Sex	The Chinese zodiac	Birthdate Year	Month
Muma Xiang 1st Bao	67	5	10	The household head	王	顺义	Male	Horse	1905	2
Muma Xiang 1st Bao				Daughter	王	赵儿娃	Female	Monkey	1932	5
Muma Xiang 1st Bao	68	5	11	The household head	王	顺魁	Male	Dog	1908	1
Muma Xiang 1st Bao				The household head's wife	王	马左儿子	Female	Pig	1910	10
Muma Xiang 1st Bao				Mother	王	高腊银子	Female	Ox	1876	12
Muma Xiang 1st Bao				Daughter	王	路娃子	Female	Pig	1935	9
Muma Xiang 1st Bao	69	5	12	The household head	王	顺玉	Male	Rat	1899	8
Muma Xiang 1st Bao				The household head's wife	王	白董儿子	Female	Ox	1900	9
Muma Xiang 1st Bao				Son	王	安才	Male	Dog	1934	8
Muma Xiang 1st Bao				The first daughter	王	二连子	Female	Dragon	1940	10
Muma Xiang 1st Bao				The second son	王	没儿子	Female	Goat	1943	6
Muma Xiang 1st Bao	70	5	13	The household head	王	顺贵	Male	Rabbit	1915	1
Muma Xiang 1st Bao				The household head's wife	王	杨赵儿子	Female	Ox	1914	1
Muma Xiang 1st Bao				Mother	王	高星楼儿	Female	Goat	1870	7
Muma Xiang 1st Bao				Son	王	安成	Male	Dragon	1940	2
Muma Xiang 1st Bao	71	5	14	The household head	王	何冯女子	Female	Monkey	1895	2

Date	The beginning time of inhabitation in this county	Educational level	Profession sector	Occupation	Marital status	Physical disability	Religion
5	Inhabited for generations	Illiteracy	Agriculture	Land-holding peasant	Widowed	No	No religion
1	Inhabited for generations	Illiteracy	Agriculture	Land-holding peasant	Unmarried	No	No religion
1	Inhabited for generations	Illiteracy	Agriculture	Land-holding peasant	Married	Right eye blind	No religion
25	Inhabited for generations	Illiteracy	Agriculture	Land-holding peasant	Married	No	No religion
15	Inhabited for generations						
23	Inhabited for generations	Illiteracy	Agriculture	Land-holding peasant	Unmarried	No	No religion
6	Inhabited for generations	Illiteracy	Agriculture	Land-holding peasant	Married	No	No religion
21	Inhabited for generations	Illiteracy	Agriculture	Land-holding peasant	Married	No	No religion
12	Inhabited for generations	Primary school 2 years					
2	Inhabited for generations	Illiteracy					
7	Inhabited for generations						
16	Inhabited for generations	Illiteracy	Agriculture	Land-holding peasant	Married	No	No religion
5	Inhabited for generations	Illiteracy	Agriculture	Land-holding peasant	Married	No	No religion
21	Inhabited for generations						
9	Inhabited for generations	Primary school 1 year					
22	Inhabited for generations	Illiteracy	Agriculture	Land-holding peasant	Widowed	No	No religion

Appendix: Overview of Microdata for Wudu County (*cont.*)

	The no. of household	Jia	Household	Relationship with the household head	Surname	First name	Sex	The Chinese zodiac	Birthdate Year	Month
Muma Xiang 1st Bao				Son	王	顺代	Male	Goat	1931	7
Muma Xiang 1st Bao				Son's wife	王	何筛娃子	Female	Goat	1931	8
Muma Xiang 1st Bao	72	6	1	The household head	何	正福	Male	Rabbit	1902	7
Muma Xiang 1st Bao				The household head's wife	何	王毛旦	Female	Rabbit	1902	6
Muma Xiang 1st Bao	73	6	2	The household head	何	万英	Male	Ox	1925	9
Muma Xiang 1st Bao				The household head's wife	何	刘桃连子	Female	Dragon	1928	8
Muma Xiang 1st Bao				Mother	何	王么女子	Female	Rabbit	1890	5
Muma Xiang 1st Bao				Younger brother	何	润林娃	Male	Snake	1941	10
Muma Xiang 1st Bao	74	6	3	The household head	何	廷武	Male	Rooster	1896	10
Muma Xiang 1st Bao				The household head's wife	何	白称儿子	Female	Monkey	1895	5
Muma Xiang 1st Bao				Father	何	正秀	Male	Ox	1876	10
Muma Xiang 1st Bao				Son	何	万福	Male	Pig	1910	4
Muma Xiang 1st Bao				Son's wife	何	王翠儿子	Female	Monkey	1920	8
Muma Xiang 1st Bao				Grandson	何	玉林娃	Male	Snake	1941	1
Muma Xiang 1st Bao				Grand daughter	何	玉秀儿	Female	Goat	1943	6
Muma Xiang 1st Bao				Grand daughter	何	玉英儿	Female	Dog	1946	3
Muma Xiang 1st Bao				Uncle	何	正應	Male	Rabbit	1890	7

Date	The beginning time of inhabitation in this county	Educational level	Profession sector	Occupation	Marital status	Physical disability	Religion
30	Inhabited for generations	Primary school 2 years			Married	Bald	
28	Inhabited for generations	Illiteracy	Agriculture	Land-holding peasant	Married	No	No religion
15	Inhabited for generations	Illiteracy	Agriculture	Land-holding peasant	Married	No	No religion
8	Inhabited for generations	Illiteracy	Agriculture	Land-holding peasant	Married	No	No religion
14	Inhabited for generations	Illiteracy		Land-holding peasant	Married	Tuberculosis	No religion
29	Inhabited for generations	Illiteracy		Land-holding peasant	Married	No	No religion
4	Inhabited for generations	Illiteracy		Land-holding peasant	Widowed	Nose	No religion
1	Inhabited for generations	Illiteracy					
15	Inhabited for generations	Sishu 2 years	Agriculture	Land-holding peasant	Married	No	No religion
15	Inhabited for generations	Illiteracy	Agriculture	Land-holding peasant	Married	No	No religion
16	Inhabited for generations	Illiteracy			Widowed		
1	Inhabited for generations	Illiteracy	Agriculture	Land-holding peasant	Married	No	No religion
25	Inhabited for generations	Illiteracy		Land-holding peasant	Married	No	No religion
24	Inhabited for generations	Illiteracy					
7	Inhabited for generations						
3	Inhabited for generations						
15	Inhabited for generations	Illiteracy	Agriculture	Land-holding peasant	Widowed	Lame	No religion

Appendix: Overview of Microdata for Wudu County (*cont.*)

									Birthdate	
	The no. of household	Jia	Household	Relationship with the household head	Surname	First name	Sex	The Chinese zodiac	Year	Month
Muma Xiang 1st Bao	75	6	4	The household head	何	廷虎	Male	Pig	1909	12
Muma Xiang 1st Bao				The household head's wife	何	李当翠儿	Female	Goat	1906	8
Muma Xiang 1st Bao				Mother	何	苟买秀子	Female	Dragon	1890	1
Muma Xiang 1st Bao				Son	何	王定娃	Male	Dragon	1940	10
Muma Xiang 1st Bao	76	6	5	The household head	何	正发	Male	Rooster	1895	1
Muma Xiang 1st Bao				The household head's wife	何	陈有香子	Female	Tiger	1889	8
Muma Xiang 1st Bao				Father	何	文祥	Male	Dragon	1867	2
Muma Xiang 1st Bao				Son	何	廷福	Male	Rooster	1933	4
Muma Xiang 1st Bao				Son's wife	何	白马儿娃	Female	Horse	1930	9
Muma Xiang 1st Bao	77	6	6	The household head	何	正宽	Male	Pig	1910	5
Muma Xiang 1st Bao				Mother	何	白么女子	Female	Goat	1882	12
Muma Xiang 1st Bao				Son	何	长儿娃	Male	Dragon	1940	8
Muma Xiang 1st Bao	78	6	7	The household head	何	正礼	Male	Pig	1910	8
Muma Xiang 1st Bao				The household head's wife	何	刘银翠儿	Female	Horse	1905	4
Muma Xiang 1st Bao				Son	何	双代儿	Male	Goat	1931	10
Muma Xiang 1st Bao				Daughter	何	毛儿娃	Female	Dragon	1940	11
Muma Xiang 1st Bao	79	6	8	The household head	何	廷泰	Male	Horse	1919	1

Date	The beginning time of inhabitation in this county	Educational level	Profession sector	Occupation	Marital status	Physical disability	Religion
12	Inhabited for generations	Illiteracy	Agriculture	Land-holding peasant	Married	No	No religion
27	Inhabited for generations	Illiteracy	Agriculture	Land-holding peasant	Married	No	No religion
17	Inhabited for generations	Illiteracy	Agriculture	Land-holding peasant	Married	No	No religion
29	Inhabited for generations	Illiteracy					
8	Inhabited for generations	Illiteracy	Agriculture	Land-holding peasant	Married	No	No religion
29	Inhabited for generations	Illiteracy	Agriculture	Land-holding peasant	Married	No	No religion
7	Inhabited for generations	Illiteracy			Widowed		
10	Inhabited for generations	Illiteracy	Agriculture	Land-holding peasant	Married	No	No religion
7	Inhabited for generations	Illiteracy	Agriculture	Land-holding peasant	Married	No	No religion
17	Inhabited for generations	Illiteracy	Agriculture	Land-holding peasant	Widowed	No	No religion
1	Inhabited for generations	Illiteracy	Agriculture	Land-holding peasant	Widowed	No	No religion
18	Inhabited for generations	Illiteracy					
27	Inhabited for generations	Illiteracy	Agriculture	Land-holding peasant	Married	No	No religion
6	Inhabited for generations	Illiteracy	Agriculture	Land-holding peasant	Married	No	No religion
28	Inhabited for generations	Illiteracy	Agriculture	Land-holding peasant	Married	No	No religion
9	Inhabited for generations	Illiteracy					
25	Inhabited for generations	Sishu 1 year	Agriculture	Land-holding peasant	Married	No	No religion

Appendix: Overview of Microdata for Wudu County (*cont.*)

	The no. of house-hold	Jia	House-hold	Relation-ship with the house-hold head	Surname	First name	Sex	The Chinese zodiac	Birthdate	
									Year	Month
Muma Xiang 1st Bao				The house-hold head's wife	何	杨么连子	Female	Horse	1918	7
Muma Xiang 1st Bao				Grand-mother	何	陈龙秀女	Female	Goat	1858	6
Muma Xiang 1st Bao				Father	何	正钦	Male	Pig	1886	9
Muma Xiang 1st Bao				Daughter	何	良翠儿	Female	Pig	1947	3
Muma Xiang 1st Bao	80	6	9	The house-hold head	何	正海	Male	Snake	1895	5
Muma Xiang 1st Bao				The house-hold head's wife	何	刘桥连子	Female	Tiger	1901	12
Muma Xiang 1st Bao				Son	何	廷裕	Male	Ox	1937	2
Muma Xiang 1st Bao				Grand daughter	何	赵女子	Female	Snake	1941	2
Muma Xiang 1st Bao				Older brother's wife	何	王买娃子	Female	Tiger	1877	9
Muma Xiang 1st Bao	81	6	10	The house-hold head	何	文有	Male	Snake	1892	12
Muma Xiang 1st Bao				Son	何	正堂	Male	Rooster	1908	11
Muma Xiang 1st Bao				Grand daughter	何	义儿娃	Female	Rabbit	1939	9
Muma Xiang 1st Bao				Grandson	何	富林娃	Male	Snake	1941	6
Muma Xiang 1st Bao				Nephew	何	正举	Male	Snake	1917	2
Muma Xiang 1st Bao				Cohabitant	刘	娃子	Male	Snake	1929	8
Muma Xiang 1st Bao	82	6	11	The house-hold head	何	廷先	Male	Dragon	1903	5
Muma Xiang 1st Bao				The house-hold head's wife	何	郝蔡秀子	Female	Tiger	1901	11

Date	The beginning time of inhabitation in this county	Educational level	Profession sector	Occupation	Marital status	Physical disability	Religion
9	Inhabited for generations	Illiteracy	Agriculture	Land-holding peasant	Married	No	No religion
14	Inhabited for generations				Widowed		
10	Inhabited for generations	Illiteracy	Agriculture	Land-holding peasant	Widowed	No	No religion
28	Inhabited for generations						
22	Inhabited for generations	Illiteracy	Agriculture	Land-holding peasant	Married	No	No religion
15	Inhabited for generations	Illiteracy	Agriculture	Land-holding peasant	Married	No	No religion
11	Inhabited for generations	Primary school 1 year					
28	Inhabited for generations	Illiteracy					
15	Inhabited for generations				Widowed	Blind	
28	Inhabited for generations	Illiteracy	Agriculture	Land-holding peasant	Married	No	No religion
5	Inhabited for generations	Illiteracy	Agriculture	Land-holding peasant	Married	No	No religion
13	Inhabited for generations	Illiteracy					
27	Inhabited for generations	Illiteracy					
23	Inhabited for generations	Illiteracy	Agriculture	Land-holding peasant	Unmarried	No	No religion
19	Inhabited for generations	Illiteracy	Agriculture		Unmarried	Bald	No religion
26	Inhabited for generations	Illiteracy	Agriculture	Land-holding peasant	Married	No	No religion
16	Inhabited for generations	Illiteracy	Agriculture	Land-holding peasant	Married	No	No religion

Appendix: Overview of Microdata for Wudu County (*cont.*)

	The no. of house-hold	Jia	House-hold	Relation-ship with the house-hold head	Surname	First name	Sex	The Chinese zodiac	Birthdate	
									Year	Month
Muma Xiang 1st Bao				Mother	何	郝冬秀子	Female	Horse	1893	12
Muma Xiang 1st Bao				Son	何	万贵	Male	Pig	1935	8
Muma Xiang 1st Bao				Daughter	何	节翠子	Female	Horse	1942	3
Muma Xiang 1st Bao	83	6	12	The household head	何	正选	Male	Goat	1893	2
Muma Xiang 1st Bao				The household head's wife	何	贾五秀子	Female	Rabbit	1890	12
Muma Xiang 1st Bao				Father	何	文通	Male	Dragon	1867	7
Muma Xiang 1st Bao				The first son	何	廷成	Male	Pig	1923	7
Muma Xiang 1st Bao				The first son's wife	何	董连儿子	Female	Horse	1918	6
Muma Xiang 1st Bao				The second son	何	廷明	Male	Goat	1931	8
Muma Xiang 1st Bao				The second son's wife	何	王全儿娃	Female	Horse	1930	9
Muma Xiang 1st Bao				Younger brother	何	正强	Male	Tiger	1901	7
Muma Xiang 1st Bao	84	6	13	The household head	何	廷官	Male	Tiger	1901	3
Muma Xiang 1st Bao				The household head's wife	何	王天夫子	Female	Pig	1898	2
Muma Xiang 1st Bao				Father	何	正德	Male	Rabbit	1877	6
Muma Xiang 1st Bao				Mother	何	白石秀子	Female	Rabbit	1881	10
Muma Xiang 1st Bao				Son	何	万禄	Male	Snake	1929	8
Muma Xiang 1st Bao				Son's wife	何	李长儿娃	Female	Tiger	1926	6

Date	The beginning time of inhabitation in this county	Educational level	Profession sector	Occupation	Marital status	Physical disability	Religion
9	Inhabited for generations	Illiteracy	Agriculture	Land-holding peasant	Widowed	No	No religion
13	Inhabited for generations	Primary school 1 year			Unmarried		
2	Inhabited for generations	Illiteracy					
3	Inhabited for generations	Sishu 2 years	Agriculture	Land-holding peasant	Married	No	No religion
20	Inhabited for generations	Illiteracy	Agriculture	Land-holding peasant	Married	No	No religion
22	Inhabited for generations				Widowed	No	No religion
2	Inhabited for generations	Illiteracy	Agriculture	Land-holding peasant	Married	No	No religion
4	Inhabited for generations	Illiteracy	Agriculture	Land-holding peasant	Married	No	No religion
20	Inhabited for generations	Primary school graduated			Married	No	
7	Inhabited for generations	Illiteracy	Agriculture	Land-holding peasant	Married	No	
28	Inhabited for generations	Illiteracy	Agriculture	Land-holding peasant	Widowed	No	
23	Inhabited for generations	Illiteracy	Agriculture	Land-holding peasant	Married	No	No religion
14	Inhabited for generations	Illiteracy	Agriculture	Land-holding peasant	Married	No	No religion
24	Inhabited for generations	Illiteracy	Agriculture	Land-holding peasant	Married	No	No religion
20	Inhabited for generations				Married		
19	Inhabited for generations	Illiteracy	Agriculture	Land-holding peasant	Married	No	No religion
26	Inhabited for generations	Illiteracy	Agriculture	Land-holding peasant	Married	No	No religion

Appendix: Overview of Microdata for Wudu County (*cont.*)

	The no. of household	Jia	Household	Relationship with the household head	Surname	First name	Sex	The Chinese zodiac	Birthdate	
									Year	Month
Muma Xiang 1st Bao				Grandson	何	根代儿	Male	Dog	1946	10
Muma Xiang 1st Bao	85	6	14	The household head	何	正圳	Male	Rat	1913	1
Muma Xiang 1st Bao				Father	何	文元	Male	Snake	1880	10
Muma Xiang 1st Bao				Mother	何	刘银香子	Female	Rat	1887	5
Muma Xiang 1st Bao				Nephew	何	廷富	Male	Pig	1910	10
Muma Xiang 1st Bao				Son's wife	何	刘同女子	Female	Horse	1918	7
Muma Xiang 1st Bao				Grandson	何	张林娃	Male	Dragon	1940	4
Muma Xiang 1st Bao	86	6	15	The household head	何	廷宝	Male	Goat	1906	10
Muma Xiang 1st Bao				The household head's wife	何	杨关儿子	Female	Goat	1906	11
Muma Xiang 1st Bao				Mother	何	李郭香儿	Female	Pig	1882	1
Muma Xiang 1st Bao				Son	何	万升	Male	Horse	1930	10
Muma Xiang 1st Bao				Son's wife	何	刘代儿娃	Female	Snake	1929	8
Muma Xiang 1st Bao				Daughter	何	路儿娃	Female	Dragon	1940	10
Muma Xiang 1st Bao				Older brother's wife	何	董刘儿子	Female	Rabbit	1902	6
Muma Xiang 1st Bao	87	7	1	The household head	何	廷荣	Male	Monkey	1907	10
Muma Xiang 1st Bao				The household head's wife	何	杨见儿子	Female	Dog	1909	8
Muma Xiang 1st Bao				Father	何	正利	Male	Rabbit	1866	1

Date	The beginning time of inhabitation in this county	Educational level	Profession sector	Occupation	Marital status	Physical disability	Religion
3	Inhabited for generations						
14	Inhabited for generations	Illiteracy	Agriculture	Land-holding peasant	Divorced	No	No religion
29	Inhabited for generations	Illiteracy	Agriculture	Land-holding peasant	Married	No	No religion
23	Inhabited for generations	Illiteracy	Agriculture	Land-holding peasant	Married	No	No religion
6	Inhabited for generations	Illiteracy	Agriculture	Land-holding peasant	Married	No	No religion
15	Inhabited for generations	Illiteracy	Agriculture	Land-holding peasant	Married	No	No religion
2	Inhabited for generations	Illiteracy					
20	Inhabited for generations	Illiteracy	Agriculture	Land-holding peasant	Married	No	No religion
25	Inhabited for generations	Illiteracy	Agriculture	Land-holding peasant	Married	No	No religion
24	Inhabited for generations	Illiteracy	Agriculture	Land-holding peasant	Widowed	No	No religion
13	Inhabited for generations	Illiteracy	Agriculture	Land-holding peasant	Married	No	No religion
18	Inhabited for generations	Illiteracy	Agriculture	Land-holding peasant	Married	No	No religion
18	Inhabited for generations	Illiteracy		Land-holding peasant			
5	Inhabited for generations	Illiteracy	Agriculture	Land-holding peasant	Widowed	No	No religion
20	Inhabited for generations	Illiteracy	Agriculture	Land-holding peasant	Married	No	No religion
19	Inhabited for generations	Illiteracy	Agriculture	Land-holding peasant	Married	No	No religion
23	Inhabited for generations				Widowed		

Appendix: Overview of Microdata for Wudu County (*cont.*)

	The no. of household	Jia	Household	Relationship with the household head	Surname	First name	Sex	The Chinese zodiac	Birthdate Year	Month
Muma Xiang 1st Bao				Daughter	何	银秀儿	Female	Dog	1947	1
Muma Xiang 1st Bao	88	7	2	The household head	王	顺宽	Male	Ox	1914	1
Muma Xiang 1st Bao				The household head's wife	王	李贤儿子	Female	Goat	1919	8
Muma Xiang 1st Bao				Mother	王	褚金香子	Female	Pig	1886	11
Muma Xiang 1st Bao				Daughter	王	四儿娃	Female	Horse	1942	5
Muma Xiang 1st Bao	89	7	3	The household head	何	贾胡儿子	Female	Tiger	1901	3
Muma Xiang 1st Bao				Mother	何	王秋娃子	Female	Rat	1875	5
Muma Xiang 1st Bao				Daughter	何	约儿娃	Female	Rooster	1933	9
Muma Xiang 1st Bao	90	6	4	The household head	何	正彦	Male	Monkey	1907	12
Muma Xiang 1st Bao				The household head's wife	何	杨秋儿子	Female	Ox	1925	11
Muma Xiang 1st Bao				Father	何	文玉	Male	Snake	1892	3
Muma Xiang 1st Bao				Mother	何	王翠儿子	Female	Monkey	1895	6
Muma Xiang 1st Bao				Daughter	何	玉娃子	Female	Rooster	1945	5
Muma Xiang 1st Bao	91	7	5	The household head	何	廷宰	Male	Dog	1909	3
Muma Xiang 1st Bao				The household head's wife	何	王三儿子	Female	Ox	1913	8
Muma Xiang 1st Bao				Younger brother	何	廷相	Male	Rat	1912	5
Muma Xiang 1st Bao				Younger brother's wife	何	白买儿子	Female	Ox	1913	2

Date	The beginning time of inhabitation in this county	Educational level	Profession sector	Occupation	Marital status	Physical disability	Religion
12	Inhabited for generations						
17	Inhabited for generations	Illiteracy	Agriculture	Land-holding peasant	Married	No	No religion
3	Inhabited for generations	Illiteracy	Agriculture	Land-holding peasant	Married	No	No religion
29	Inhabited for generations	Illiteracy	Agriculture	Land-holding peasant	Widowed	No	No religion
18	Inhabited for generations	Illiteracy					
16	Inhabited for generations	Illiteracy	Agriculture	Land-holding peasant	Divorced	No	No religion
2	Inhabited for generations						
28	Inhabited for generations	Illiteracy	Agriculture	Land-holding peasant	Unmarried	No	No religion
13	Inhabited for generations	Illiteracy	Agriculture	Land-holding peasant	Married	No	No religion
20	Inhabited for generations	Illiteracy	Agriculture	Land-holding peasant	Married	No	No religion
19	Inhabited for generations	Illiteracy	Agriculture	Land-holding peasant	Married	No	No religion
7	Inhabited for generations	Illiteracy	Agriculture	Land-holding peasant	Married	No	No religion
26	Inhabited for generations						
22	Inhabited for generations	Illiteracy	Agriculture	Land-holding peasant	Married	No	No religion
15	Inhabited for generations	Illiteracy	Agriculture	Land-holding peasant	Married	No	No religion
15	Inhabited for generations	Illiteracy	Agriculture	Land-holding peasant	Married	No	No religion
10	Inhabited for generations	Illiteracy	Agriculture	Land-holding peasant	Married	No	No religion

Appendix: Overview of Microdata for Wudu County (*cont.*)

	The no. of household	Jia	Household	Relationship with the household head	Surname	First name	Sex	The Chinese zodiac	Birthdate Year	Month
Muma Xiang 1st Bao				Nephew	何	二娃子	Male	Dragon	1940	3
Muma Xiang 1st Bao	92	7	6	The household head	何	廷春	Male	Dog	1897	2
Muma Xiang 1st Bao				The household head's wife	何	王王儿子	Female	Dog	1897	8
Muma Xiang 1st Bao				Son	何	万清	Male	Rabbit	1915	8
Muma Xiang 1st Bao				Son's wife	何	王双代儿	Female	Dog	1922	1
Muma Xiang 1st Bao	93	7	7	The household head	何	正隆	Male	Dragon	1903	4
Muma Xiang 1st Bao				The household head's wife	何	王何儿子	Female	Snake	1904	6
Muma Xiang 1st Bao				Son	何	廷银	Male	Tiger	1938	12
Muma Xiang 1st Bao				Daughter	何	代银儿	Female	Dragon	1940	5
Muma Xiang 1st Bao	94	7	8	The household head	王	安银	Male	Tiger	1915	1
Muma Xiang 1st Bao				The household head's wife	王	白罗家娃	Female	Snake	1929	4
Muma Xiang 1st Bao				Mother	马	中儿子	Female	Ox	1900	4
Muma Xiang 1st Bao	95	7	9	The household head	何	正礼	Male	Ox	1900	7
Muma Xiang 1st Bao				The household head's wife	何	李见儿子	Female	Dog	1909	3
Muma Xiang 1st Bao				Father	何	文永	Male	Rooster	1884	3
Muma Xiang 1st Bao				The first son	何	丙丁子	Male	Dragon	1928	8

Date	The beginning time of inhabitation in this county	Educational level	Profession sector	Occupation	Marital status	Physical disability	Religion
30	Inhabited for generations	Illiteracy					
6	Inhabited for generations	Illiteracy	Agriculture	Land-holding peasant	Married	No	No religion
2	Inhabited for generations	Illiteracy	Agriculture	Land-holding peasant	Married	No	No religion
27	Inhabited for generations	Illiteracy	Agriculture	Land-holding peasant	Married	No	No religion
28	Inhabited for generations	Illiteracy	Agriculture	Land-holding peasant	Married	No	No religion
18	Inhabited for generations	Illiteracy	Agriculture	Land-holding peasant	Married	No	No religion
17	Inhabited for generations	Illiteracy	Agriculture	Land-holding peasant	Married	No	No religion
6	Inhabited for generations	Primary school 1 year					
21	Inhabited for generations	Illiteracy					
24	Inhabited for generations	Illiteracy	Agriculture	Land-holding peasant	Married	No	No religion
19	Inhabited for generations	Illiteracy	Agriculture	Land-holding peasant	Married	No	No religion
28	Inhabited for generations	Illiteracy	Agriculture	Land-holding peasant	Widowed	No	No religion
14	Inhabited for generations	Illiteracy	Agriculture	Land-holding peasant	Married	No	No religion
28	Inhabited for generations	Illiteracy	Agriculture	Land-holding peasant	Married	No	No religion
24	Inhabited for generations	Illiteracy	Agriculture	Land-holding peasant	Widowed	Looking ill	No religion
29	Inhabited for generations	Illiteracy	Agriculture	Land-holding peasant	Unmarried	No	No religion

Appendix: Overview of Microdata for Wudu County (*cont.*)

	The no. of household	Jia	Household	Relationship with the household head	Surname	First name	Sex	The Chinese zodiac	Birthdate Year	Month
Muma Xiang 1st Bao				The second son	何	廷朝	Male	Rooster	1933	2
Muma Xiang 1st Bao				The first daughter	何	夭女子	Female	Monkey	1945	1
Muma Xiang 1st Bao				The second daughter	何	桂女子	Female	Pig	1941	6
Muma Xiang 1st Bao				Uncle's wife	何	王桥连子	Female	Pig	1898	8
Muma Xiang 1st Bao				Grandson	何	万连	Male	Rabbit	1927	9
Muma Xiang 1st Bao				Grandson's wife	何	刘蔡连子	Female	Pig	1923	8
Muma Xiang 1st Bao	96	7	10	The household head	何	廷福	Male	Dog	1909	10
Muma Xiang 1st Bao				Mother	何	王谢娃子	Female	Rooster	1884	12
Muma Xiang 1st Bao				Son	何	代银子	Male	Rabbit	1940	1
Muma Xiang 1st Bao				Younger sister	何	郝娃子	Female	Rabbit	1915	8
Muma Xiang 1st Bao	97	7	11	The household head	何	正玉	Male	Pig	1886	10
Muma Xiang 1st Bao				Younger brother's wife	何	王冯儿子	Female	Monkey	1895	4
Muma Xiang 1st Bao				Nephew	何	秦秦娃	Female	Pig	1923	8
Muma Xiang 1st Bao				Grandson	何	万李	Male	Ox	1913	10
Muma Xiang 1st Bao				Grandson's wife	何	李刘连子	Female	Dragon	1928	8
Muma Xiang 1st Bao	98	7	12	The household head	杨	世德	Male	Rooster	1908	10
Muma Xiang 1st Bao				The household head's wife	杨	何蜡儿子	Female	Pig	1909	2

Date	The beginning time of inhabitation in this county	Educational level	Profession sector	Occupation	Marital status	Physical disability	Religion
24	Inhabited for generations	Primary school 2 years					
11	Inhabited for generations						
20	Inhabited for generations						
20	Inhabited for generations	Illiteracy	Agricul-ture	Land-holding peasant	Widowed	No	No religion
10	Inhabited for generations	Illiteracy	Agricul-ture	Land-holding peasant	Married	No	No religion
26	Inhabited for generations	Illiteracy	Agricul-ture	Land-holding peasant	Married	No	No religion
20	Inhabited for generations	Illiteracy	Agricul-ture	Land-holding peasant	Widowed	Left eye blind	No religion
25	Inhabited for generations	Illiteracy	Agricul-ture	Land-holding peasant	Widowed	No	No religion
26	Inhabited for generations	Illiteracy					
25	Inhabited for generations	Illiteracy	Agricul-ture	Land-holding peasant	Unmarried	No	No religion
25	Inhabited for generations	Illiteracy	Agricul-ture	Land-holding peasant	Widowed	No	No religion
16	Inhabited for generations	Illiteracy	Agricul-ture	Land-holding peasant	Widowed	No	No religion
22	Inhabited for generations	Illiteracy	Agricul-ture	Land-holding peasant	Widowed	No	No religion
24	Inhabited for generations	Illiteracy	Agricul-ture	Land-holding peasant	Married	No	No religion
19	Inhabited for generations	Illiteracy	Agricul-ture	Land-holding peasant	Married	No	No religion
18	Inhabited for generations	Illiteracy	Agricul-ture	Land-holding peasant	Married	No	No religion
11	Inhabited for generations	Illiteracy	Agricul-ture	Land-holding peasant	Married	No	No religion

Appendix: Overview of Microdata for Wudu County (*cont.*)

									Birthdate	
	The no. of household	Jia	Household	Relationship with the household head	Surname	First name	Sex	The Chinese zodiac	Year	Month
Muma Xiang 1st Bao				Mother	杨	何金连子	Female	Dog	1885	2
Muma Xiang 1st Bao				Younger brother	杨	世凤	Male	Goat	1919	7
Muma Xiang 1st Bao				Younger brother's wife	杨	何连儿子	Female	Ox	1913	8
Muma Xiang 1st Bao	99	7	13	The household head	杨	福贵	Male	Snake	1918	1
Muma Xiang 1st Bao				The household head's wife	杨	李胖女子	Female	Goat	1919	8
Muma Xiang 1st Bao				Father	杨	金才	Male	Goat	1903	8
Muma Xiang 1st Bao				Mother	杨	王冯娃子	Female	Horse	1881	3
Muma Xiang 1st Bao				Grandfather	杨	世清	Male	Dragon	1885	2
Muma Xiang 1st Bao				Uncle's wife	杨	何万儿子	Female	Rat	1912	8
Muma Xiang 1st Bao	100	7	14	The household head	杨	福寿	Male	Goat	1906	12
Muma Xiang 1st Bao				The household head's wife	杨	何何翠娃	Female	Goat	1919	3
Muma Xiang 1st Bao				Mother	杨	王楼翠子	Female	Rooster	1938	3
Muma Xiang 1st Bao				Son	杨	双义	Male	Dog	1934	6
Muma Xiang 1st Bao				Daughter	杨	叩娃子	Female	Dragon	1940	10
Muma Xiang 1st Bao	101	7	15	The household head	杨	廷申	Male	Horse	1876	8
Muma Xiang 1st Bao				The first son	杨	世林	Male	Goat	1906	5
Muma Xiang 1st Bao				The first son's wife	杨	何冬冬娃	Female	Dragon	1903	8

Date	The beginning time of inhabitation in this county	Educational level	Profession sector	Occupation	Marital status	Physical disability	Religion
11	Inhabited for generations	Illiteracy	Agriculture	Land-holding peasant	Widowed	No	No religion
21	Inhabited for generations	Illiteracy	Agriculture	Land-holding peasant	Married	No	No religion
25	Inhabited for generations	Illiteracy	Agriculture	Land-holding peasant	Married	No	No religion
22	Inhabited for generations	Illiteracy	Agriculture	Land-holding peasant	Married	No	No religion
10	Inhabited for generations	Illiteracy	Agriculture	Land-holding peasant	Married	No	No religion
10	Inhabited for generations	Illiteracy					
5	Inhabited for generations	Illiteracy		Land-holding peasant			
6	Inhabited for generations	Illiteracy	Agriculture	Land-holding peasant	Widowed	No	No religion
27	Inhabited for generations	Illiteracy	Agriculture	Land-holding peasant	Divorced	No	No religion
19	Inhabited for generations	Illiteracy	Agriculture	Land-holding peasant	Married	No	No religion
28	Inhabited for generations	Illiteracy	Agriculture	Land-holding peasant	Married	No	No religion
26	Inhabited for generations	Illiteracy	Agriculture	Land-holding peasant	Widowed	No	No religion
16	Inhabited for generations	Primary school 2 years					
15	Inhabited for generations	Illiteracy					
12	Inhabited for generations	Illiteracy	Agriculture	Land-holding peasant	Widowed	No	No religion
7	Inhabited for generations	Illiteracy	Agriculture	Land-holding peasant	Married	No	No religion
25	Inhabited for generations	Illiteracy	Agriculture	Land-holding peasant	Married	No	No religion

Appendix: Overview of Microdata for Wudu County (*cont.*)

	The no. of household	Jia	Household	Relationship with the household head	Surname	First name	Sex	The Chinese zodiac	Birthdate	
									Year	Month
Muma Xiang 1st Bao				The second son	杨	世庆	Male	Rat	1912	7
Muma Xiang 1st Bao				The second son's wife	杨	何蔡儿子	Female	Dog	1913	8
Muma Xiang 1st Bao				Grand daughter	杨	创银子	Female	Snake	1941	10
Muma Xiang 1st Bao				Grand daughter	杨	真娃子	Female	Horse	1942	2
Muma Xiang 1st Bao	102	8	1	The household head	王	碧凤	Male	Horse	1881	4
Muma Xiang 1st Bao				The household head's wife	王	孟润叩儿	Female	Pig	1886	9
Muma Xiang 1st Bao				Son	王	應洪	Male	Rooster	1908	6
Muma Xiang 1st Bao				Son's wife	王	陈四娃子	Female	Dog	1909	3
Muma Xiang 1st Bao	103	8	2	The household head	白	义福	Male	Rabbit	1915	3
Muma Xiang 1st Bao				The household head's wife	白	王南儿娃	Female	Horse	1918	11
Muma Xiang 1st Bao				Mother	白	王郭女子	Female	Tiger	1888	1
Muma Xiang 1st Bao				Son	白	童儿娃	Male	Snake	1941	3
Muma Xiang 1st Bao				Daughter	白	五儿	Female	Dog	1946	6
Muma Xiang 1st Bao	104	8	3	The household head	高	明德	Male	Horse	1905	4
Muma Xiang 1st Bao				The household head's wife	高	张创采儿	Female	Rat	1912	8
Muma Xiang 1st Bao				Father	高	发王	Male	Dragon	1879	12
Muma Xiang 1st Bao				Mother	高	王三代儿	Female	Rooster	1883	1

Date	The beginning time of inhabitation in this county	Educational level	Profession sector	Occupation	Marital status	Physical disability	Religion
18	Inhabited for generations	Illiteracy	Agriculture	Land-holding peasant	Married	No	No religion
29	Inhabited for generations	Illiteracy	Agriculture	Land-holding peasant	Married	No	No religion
21	Inhabited for generations	Illiteracy					
19	Inhabited for generations						
27	Inhabited for generations	Illiteracy	Agriculture	Land-holding peasant	Married	No	No religion
5	Inhabited for generations	Illiteracy	Agriculture	Land-holding peasant	Married	No	No religion
22	Inhabited for generations	Illiteracy	Agriculture	Land-holding peasant	Married	No	No religion
12	Inhabited for generations	Illiteracy	Agriculture	Land-holding peasant	Married	No	No religion
17	Inhabited for generations	Illiteracy	Agriculture	Land-holding peasant	Married	No	No religion
18	Inhabited for generations	Illiteracy	Agriculture	Land-holding peasant	Married	No	No religion
30	Inhabited for generations	Illiteracy	Agriculture	Land-holding peasant	Widowed	No	No religion
23	Inhabited for generations	Illiteracy					
24	Inhabited for generations	Illiteracy					
6	Inhabited for generations	Illiteracy	Agriculture	Land-holding peasant	Married	No	No religion
27	Inhabited for generations	Illiteracy	Agriculture	Land-holding peasant	Married	No	No religion
15	Inhabited for generations	Illiteracy	Agriculture	Land-holding peasant	Married	No	No religion
28	Inhabited for generations	Illiteracy	Agriculture	Land-holding peasant	Married	No	No religion

Appendix: Overview of Microdata for Wudu County (*cont.*)

	The no. of household	Jia	Household	Relationship with the household head	Surname	First name	Sex	The Chinese zodiac	Birthdate	
									Year	Month
Muma Xiang 1st Bao	105	8	4	The household head	高	明宝	Male	Dragon	1891	12
Muma Xiang 1st Bao				The first son	高	朝海	Male	Horse	1918	2
Muma Xiang 1st Bao				The household head's wife	高	王五连子	Female	Goat	1891	4
Muma Xiang 1st Bao				The second son	高	朝银	Male	Goat	1931	4
Muma Xiang 1st Bao				The second son's wife	高	李王目娃	Female	Dragon	1928	10
Muma Xiang 1st Bao	106	8	5	The household head	高	明学	Male	Horse	1905	1
Muma Xiang 1st Bao				The household head's wife	高	石爱娃	Female	Goat	1906	6
Muma Xiang 1st Bao				Daughter	创	银子	Female	Rat	1936	8
Muma Xiang 1st Bao	107	8	6	The household head	王	應珍	Male	Dragon	1916	7
Muma Xiang 1st Bao				The household head's wife	王	赵路儿子	Female	Ox	1913	3
Muma Xiang 1st Bao				Son	王	兴娃	Male	Horse	1942	12
Muma Xiang 1st Bao	108	8	7	The household head	白	金揔	Male	Dragon	1880	2
Muma Xiang 1st Bao				The first son	白	发义	Male	Dog	1909	10
Muma Xiang 1st Bao				The household head's wife	白	王石儿娃	Female	Dragon	1916	6
Muma Xiang 1st Bao				The second son	白	发兴	Male	Horse	1930	9
Muma Xiang 1st Bao				The second son's wife	白	王创娃子	Female	Dragon	1928	10
Muma Xiang 1st Bao	109	8	8	The household head	高	明发	Male	Pig	1898	11

Date	The beginning time of inhabitation in this county	Educational level	Profession sector	Occupation	Marital status	Physical disability	Religion
15	Inhabited for generations	Illiteracy	Agriculture	Land-holding peasant	Widowed	No	No religion
15	Inhabited for generations	Illiteracy	Agriculture	Land-holding peasant	Married	No	No religion
1	Inhabited for generations	Illiteracy			Married	Hands lost	No religion
4	Inhabited for generations	Illiteracy	Agriculture	Land-holding peasant	Married	No	No religion
25	Inhabited for generations	Illiteracy	Agriculture	Land-holding peasant	Married	No	No religion
30	Inhabited for generations	Illiteracy	Agriculture	Land-holding peasant	Married	No	No religion
6	Inhabited for generations	Illiteracy	Agriculture	Land-holding peasant	Married	No	No religion
30	Inhabited for generations	Illiteracy					
8	Inhabited for generations	Illiteracy	Agriculture	Land-holding peasant	Married	No	No religion
9	Inhabited for generations	Illiteracy	Agriculture	Land-holding peasant	Married	No	No religion
23	Inhabited for generations						
12	Inhabited for generations	Illiteracy	Agriculture	Land-holding peasant	Widowed	No	No religion
19	Inhabited for generations	Illiteracy	Agriculture	Land-holding peasant	Married	No	No religion
5	Inhabited for generations	Illiteracy	Agriculture	Land-holding peasant	Married	No	No religion
4	Inhabited for generations	Illiteracy	Agriculture	Land-holding peasant	Married	No	No religion
22	Inhabited for generations	Illiteracy	Agriculture	Land-holding peasant	Married	No	No religion
2	Inhabited for generations	Illiteracy	Agriculture	Land-holding peasant	Widowed	No	No religion

Appendix: Overview of Microdata for Wudu County (*cont.*)

	The no. of household	Jia	Household	Relationship with the household head	Surname	First name	Sex	The Chinese zodiac	Birthdate	
									Year	Month
Muma Xiang 1st Bao				The first son	高	五娃子	Male	Rooster	1933	6
Muma Xiang 1st Bao				The second son	高	夭来娃	Male	Rooster	1946	1
Muma Xiang 1st Bao				The first daughter	高	欠娃子	Female	Pig	1935	11
Muma Xiang 1st Bao				The second daughter	高	贤娃子	Female	Goat	1943	10
Muma Xiang 1st Bao	110	8	9	The household head	高	明福	Male	Rat	1899	5
Muma Xiang 1st Bao				The household head's wife	高	王桥儿子	Female	Goat	1906	9
Muma Xiang 1st Bao				Father	高	廷路	Male	Ox	1886	3
Muma Xiang 1st Bao				Mother	高	白成成子	Female	Ox	1876	2
Muma Xiang 1st Bao				Son	高	六来儿	Male	Tiger	1926	8
Muma Xiang 1st Bao				Son's wife	高	王夫女子	Female	Snake	1929	8
Muma Xiang 1st Bao	111	8	10	The household head	白	义清	Male	Rabbit	1902	2
Muma Xiang 1st Bao				The household head's wife	白	王玉儿子	Female	Rat	1912	12
Muma Xiang 1st Bao				Mother	白	谈来娃子	Female	Horse	1881	7
Muma Xiang 1st Bao				Daughter	白	秦根儿	Female	Ox	1937	7
Muma Xiang 1st Bao	112	8	11	The household head	高	朝宽	Male	Snake	1929	6
Muma Xiang 1st Bao				The household head's wife	高	赵贤儿娃	Female	Snake	1929	8
Muma Xiang 1st Bao				Son	高	才儿娃	Male	Horse	1942	3

Date	The beginning time of inhabitation in this county	Educational level	Profession sector	Occupation	Marital status	Physical disability	Religion
17	Inhabited for generations	Illiteracy	Agriculture	Land-holding peasant	Unmarried	No	No religion
8	Inhabited for generations						
3	Inhabited for generations	Illiteracy					
15	Inhabited for generations						
22	Inhabited for generations	Illiteracy	Agriculture	Land-holding peasant	Married	No	No religion
8	Inhabited for generations	Illiteracy	Agriculture	Land-holding peasant	Married	No	No religion
14	Inhabited for generations	Illiteracy	Agriculture	Land-holding peasant	Married	No	No religion
11	Inhabited for generations	Illiteracy			Married		No religion
5	Inhabited for generations	Illiteracy	Agriculture	Land-holding peasant	Married	No	No religion
19	Inhabited for generations	Illiteracy	Agriculture	Land-holding peasant	Married	No	No religion
4	Inhabited for generations	Illiteracy	Agriculture	Land-holding peasant	Married	No	No religion
11	Inhabited for generations	Illiteracy	Agriculture	Land-holding peasant	Married	No	No religion
15	Inhabited for generations	Illiteracy	Agriculture	Land-holding peasant	Widowed	No	No religion
29	Inhabited for generations	Illiteracy					
3	Inhabited for generations	Illiteracy	Agriculture	Land-holding peasant	Married	No	No religion
29	Inhabited for generations	Illiteracy	Agriculture	Land-holding peasant	Married	No	No religion
18	Inhabited for generations						

Appendix: Overview of Microdata for Wudu County (*cont.*)

	The no. of household	Jia	Household	Relationship with the household head	Surname	First name	Sex	The Chinese zodiac	Birthdate Year	Month
Muma Xiang 1st Bao				Daughter	高	才英儿	Female	Rooster	1945	6
Muma Xiang 1st Bao	113	8	12	The household head	马	华南	Male	Horse	1905	3
Muma Xiang 1st Bao				The household head's wife	马	谈跳银子	Female	Dragon	1903	8
Muma Xiang 1st Bao				Father	马	荣学	Male	Horse	1881	5
Muma Xiang 1st Bao				Mother	马	谈冬冬子	Female	Horse	1881	11
Muma Xiang 1st Bao				Son	马	桂生儿	Male	Snake	1928	7
Muma Xiang 1st Bao				Son's wife	马	谈毛娃	Female	Rabbit	1927	2
Muma Xiang 1st Bao				The second son	马	福龙	Male	Dog	1934	9
Muma Xiang 1st Bao	114	8	13	The household head	谈	学武	Male	Horse	1893	10
Muma Xiang 1st Bao				Younger brother	谈	学文	Male	Dog	1897	2
Muma Xiang 1st Bao				Younger brother's wife	谈	李何儿子	Female	Rooster	1896	8
Muma Xiang 1st Bao				Nephew	谈	正正娃	Male	Rooster	1921	6
Muma Xiang 1st Bao				Son's wife	谈	赵青果儿	Female	Goat	1919	10
Muma Xiang 1st Bao	115	8	14	The household head	马	华廷	Male	Rabbit	1902	6
Muma Xiang 1st Bao				The household head's wife	马	杨路儿子	Female	Rat	1912	4
Muma Xiang 1st Bao				Father	马	荣李	Male	Rabbit	1878	7

Date	The beginning time of inhabitation in this county	Educational level	Profession sector	Occupation	Marital status	Physical disability	Religion
14	Inhabited for generations						
27	Inhabited for generations	Illiteracy	Agriculture	Land-holding peasant	Married	No	No religion
25	Inhabited for generations	Illiteracy	Agriculture	Land-holding peasant	Married	No	No religion
5	Inhabited for generations	Illiteracy	Agriculture	Land-holding peasant	Married	No	No religion
23	Inhabited for generations	Illiteracy	Agriculture	Land-holding peasant	Married	No	No religion
12	Inhabited for generations	Illiteracy			Married	Hands broken	
2	Inhabited for generations	Illiteracy	Agriculture	Land-holding peasant	Married	No	No religion
1	Inhabited for generations	Primary school 2 years					
7	Inhabited for generations	Illiteracy	Agriculture	Land-holding peasant	Widowed	No	No religion
22	Inhabited for generations	Illiteracy	Agriculture	Land-holding peasant	Married	No	No religion
12	Inhabited for generations	Illiteracy	Agriculture	Land-holding peasant	Married	No	No religion
20	Inhabited for generations	Illiteracy	Agriculture	Land-holding peasant	Married	No	No religion
9	Inhabited for generations	Illiteracy	Agriculture	Land-holding peasant	Married	No	No religion
14		Illiteracy	Agriculture	Land-holding peasant	Married	No	No religion
19		Illiteracy	Agriculture	Land-holding peasant	Married	No	No religion
15		Illiteracy	Agriculture	Land-holding peasant	Married	No	No religion

Appendix: Overview of Microdata for Wudu County (*cont.*)

	The no. of household	Jia	Household	Relationship with the household head	Surname	First name	Sex	The Chinese zodiac	Birthdate	
									Year	Month
Muma Xiang 1st Bao				Mother	马	李罗香子	Female	Pig	1883	10
Muma Xiang 1st Bao				The first son	马	娃子	Male	Rabbit	1939	7
Muma Xiang 1st Bao				The second son	马	福清	Male	Snake	1941	4
Muma Xiang 1st Bao				The first daughter	马	女子	Female	Ox	1937	8
Muma Xiang 1st Bao				The second daughter	马	崴女子	Female	Dog	1946	2
Muma Xiang 1st Bao				Uncle	马	荣禄	Male	Horse	1893	8
Muma Xiang 1st Bao	116	8	15	The household head	马	华兴	Male	Horse	1904	1
Muma Xiang 1st Bao				The household head's wife	马	谈创银儿	Female	Goat	1916	5
Muma Xiang 1st Bao				The first daughter	马	石娃	Female	Rat	1936	11
Muma Xiang 1st Bao				The second daughter	马	小女儿	Female	Dog	1946	10
Muma Xiang 1st Bao	117	9	1	The household head	马	华高	Male	Horse	1905	11
Muma Xiang 1st Bao				The household head's wife	马	高唐银子	Female	Snake	1904	3
Muma Xiang 1st Bao				The first daughter	马	麦花儿	Female	Goat	1943	12
Muma Xiang 1st Bao				The second daughter	马	明当儿	Female	Dog	1946	9
Muma Xiang 1st Bao				Daughter	马	麦儿娃	Female	Dragon	1940	8
Muma Xiang 1st Bao	118	9	2	The household head	王	應德	Male	Tiger	1914	3

Date	The beginning time of inhabitation in this county	Educational level	Profession sector	Occupation	Marital status	Physical disability	Religion
3		Illiteracy	Agriculture	Land-holding peasant	Married	No	No religion
22	Inhabited for generations	Illiteracy	Agriculture				
6	Inhabited for generations	Primary school 1 year					
10	Inhabited for generations	Illiteracy					
16	Inhabited for generations					Left eye blind	
15	Inhabited for generations	Illiteracy	Agriculture	Land-holding peasant	Widowed	No	No religion
25	Inhabited for generations	Illiteracy	Agriculture	Land-holding peasant	Married	No	No religion
16	Inhabited for generations	Illiteracy	Agriculture	Land-holding peasant	Married	No	No religion
18	Inhabited for generations	Illiteracy					
19	Inhabited for generations						
22	Inhabited for generations	Illiteracy	Agriculture	Land-holding peasant	Married	No	No religion
21	Inhabited for generations	Illiteracy	Agriculture	Land-holding peasant	Married	No	No religion
28	Inhabited for generations						
17	Inhabited for generations						
19	Inhabited for generations	Illiteracy					
4	Inhabited for generations	Illiteracy	Agriculture	Land-holding peasant	Married	No	No religion

Appendix: Overview of Microdata for Wudu County (*cont.*)

	The no. of household	Jia	Household	Relationship with the household head	Surname	First name	Sex	The Chinese zodiac	Birthdate Year	Month
Muma Xiang 1st Bao				The household head's wife	王	褚连女子	Female	Dragon	1916	3
Muma Xiang 1st Bao				Father	王	碧现	Male	Monkey	1883	2
Muma Xiang 1st Bao				Mother	王	李崴连儿	Female	Ox	1887	1
Muma Xiang 1st Bao	119	9	3	The household head	王	應宽	Male	Horse	1904	8
Muma Xiang 1st Bao				The household head's wife	王	白王儿子	Female	Goat	1919	11
Muma Xiang 1st Bao				Mother	王	赵连娃子	Female	Rooster	1885	3
Muma Xiang 1st Bao				Daughter	王	称儿子	Female	Rooster	1945	3
Muma Xiang 1st Bao	120	9	4	The household head	王	應见	Male	Snake	1904	7
Muma Xiang 1st Bao				The household head's wife	王	谈苟儿子	Female	Horse	1905	4
Muma Xiang 1st Bao				Son	王	顺要	Male	Rabbit	1939	12
Muma Xiang 1st Bao				Daughter	王	代女子	Female	Dog	1946	12
Muma Xiang 1st Bao	121	9	5	The household head	王	豆林儿子	Male	Ox	1913	8
Muma Xiang 1st Bao				Son	王	顺海	Male	Rooster	1933	6
Muma Xiang 1st Bao	122	9	6	The household head	王	碧德	Male	Goat	1928	4
Muma Xiang 1st Bao				The household head's wife	王	何见儿子	Female	Ox	1913	8
Muma Xiang 1st Bao				Son	王	石儿娃	Male	Tiger	1926	11

Date	The beginning time of inhabitation in this county	Educational level	Profession sector	Occupation	Marital status	Physical disability	Religion
22	Inhabited for generations	Illiteracy	Agriculture	Land-holding peasant	Married	No	No religion
8	Inhabited for generations	Illiteracy	Agriculture	Land-holding peasant	Married	No	No religion
24	Inhabited for generations	Illiteracy	Agriculture	Land-holding peasant	Married	No	No religion
11	Inhabited for generations	Illiteracy	Agriculture	Land-holding peasant	Married	No	No religion
1	Inhabited for generations	Illiteracy	Agriculture	Land-holding peasant	Married	Ulcer on the face	No religion
3	Inhabited for generations	Illiteracy	Agriculture	Land-holding peasant	Widowed	No	No religion
5	Inhabited for generations						
10	Inhabited for generations	Illiteracy	Agriculture	Land-holding peasant	Married	No	No religion
27	Inhabited for generations	Illiteracy	Agriculture	Land-holding peasant	Married	No	No religion
18	Inhabited for generations						
9	Inhabited for generations	Primary school 1 year					
15	Inhabited for generations	Illiteracy	Agriculture	Land-holding peasant	Divorced	No	No religion
7	Inhabited for generations	Primary school 3 years					
13	Inhabited for generations	Illiteracy	Agriculture	Land-holding peasant	Married	No	No religion
15	Inhabited for generations	Illiteracy	Agriculture	Land-holding peasant	Married	No	No religion
24	Inhabited for generations	Illiteracy	Agriculture	Land-holding peasant	Unmarried	No	No religion

Appendix: Overview of Microdata for Wudu County (*cont.*)

	The no. of household	Jia	Household	Relationship with the household head	Surname	First name	Sex	The Chinese zodiac	Birthdate	
									Year	Month
Muma Xiang 1st Bao	123	9	7	The household head	王	碧富	Male	Dog	1897	11
Muma Xiang 1st Bao				The household head's wife	王	杨翠翠子	Female	Rooster	1896	8
Muma Xiang 1st Bao				Daughter	王	四娃子	Female	Dragon	1940	5
Muma Xiang 1st Bao				Older brother	王	碧應	Male	Goat	1894	6
Muma Xiang 1st Bao	124	9	8	The household head	王	碧双	Male	Rooster	1908	2
Muma Xiang 1st Bao				The household head's wife	王	刘翠银儿	Female	Monkey	1907	10
Muma Xiang 1st Bao				The first daughter	王	蔡儿娃	Female	Snake	1941	5
Muma Xiang 1st Bao				The second daughter	王	三儿娃	Female	Rooster	1945	4
Muma Xiang 1st Bao	125	9	9	The household head	王	應中	Male	Rooster	1933	9
Muma Xiang 1st Bao				Mother	王	何九连子	Female	Rooster	1881	10
Muma Xiang 1st Bao				Grandmother	王	何春香子	Female	Monkey	1882	2
Muma Xiang 1st Bao	126	9	10	The household head	王	應德	Male	Ox	1900	9
Muma Xiang 1st Bao				The household head's wife	王	藤官女子	Female	Rooster	1908	8
Muma Xiang 1st Bao				Mother	王	尹真娃子	Female	Goat	1870	1
Muma Xiang 1st Bao				The first son	王	孝崇儿	Male	Rooster	1921	4
Muma Xiang 1st Bao				The first son's wife	王	罗约儿子	Female	Horse	1930	2
Muma Xiang 1st Bao				The second son	王	双成儿	Male	Dog	1946	5

Date	The beginning time of inhabitation in this county	Educational level	Profession sector	Occupation	Marital status	Physical disability	Religion
24	Inhabited for generations	Illiteracy	Agriculture	Land-holding peasant	Married	No	No religion
12	Inhabited for generations	Illiteracy	Agriculture	Land-holding peasant	Married	No	No religion
20	Inhabited for generations	Illiteracy					
7	Inhabited for generations	Illiteracy	Agriculture	Land-holding peasant	Widowed	Larynx disease	No religion
27	Inhabited for generations	Illiteracy	Agriculture	Land-holding peasant	Married	No	No religion
25	Inhabited for generations	Illiteracy	Agriculture	Land-holding peasant	Married	No	No religion
15	Inhabited for generations	Illiteracy	Agriculture	Land-holding peasant			
17	Inhabited for generations						
5	Inhabited for generations	Primary school 2 years					
1	Inhabited for generations	Illiteracy	Agriculture	Land-holding peasant	Divorced	No	No religion
2	Inhabited for generations	Illiteracy	Agriculture	Land-holding peasant	Widowed	No	No religion
27	Inhabited for generations	Illiteracy	Agriculture	Land-holding peasant	Married	Ulcer on the face	No religion
15	Inhabited for generations	Illiteracy	Agriculture	Land-holding peasant	Married	No	No religion
25	Inhabited for generations				Widowed		
22	Inhabited for generations	Illiteracy	Agriculture	Land-holding peasant	Married	No	No religion
10	Inhabited for generations	Illiteracy	Agriculture	Land-holding peasant	Married	No	No religion
25	Inhabited for generations						

Appendix: Overview of Microdata for Wudu County (*cont.*)

	The no. of household	Jia	Household	Relationship with the household head	Surname	First name	Sex	The Chinese zodiac	Birthdate Year	Month
Muma Xiang 1st Bao	127	9	11	The household head	王	碧银	Male	Goat	1882	4
Muma Xiang 1st Bao				The household head's wife	王	李代银子	Female	Monkey	1895	5
Muma Xiang 1st Bao	128	9	12	The household head	王	白夭香子	Female	Monkey	1883	3
Muma Xiang 1st Bao				Son's wife	王	郭桂连子	Female	Pig	1910	4
Muma Xiang 1st Bao				Grandson	王	應通	Male	Ox	1937	11
Muma Xiang 1st Bao	129	9	13	The household head	王	顺怀	Male	Goat	1919	11
Muma Xiang 1st Bao				The household head's wife	王	何桃连子	Female	Rat	1898	8
Muma Xiang 1st Bao				Father	王	應顺	Male	Dragon	1891	6
Muma Xiang 1st Bao				Mother	王	郭真儿子	Female	Rat	1899	2
Muma Xiang 1st Bao				Younger sister	王	楼儿娃	Female	Rat	1936	6
Muma Xiang 1st Bao	130	9	14	The household head	王	應斗	Male	Dog	1909	5
Muma Xiang 1st Bao				The household head's wife	王	白夭连子	Female	Snake	1917	10
Muma Xiang 1st Bao				The first son	王	马儿娃	Male	Horse	1942	9
Muma Xiang 1st Bao				The second son	王	保代儿	Male	Pig	1941	2
Muma Xiang 1st Bao	131	9	15	The household head	王	應珍	Male	Pig	1910	9
Muma Xiang 1st Bao				The household head's wife	王	谈冬儿子	Female	Monkey	1907	12

Date	The beginning time of inhabitation in this county	Educational level	Profession sector	Occupation	Marital status	Physical disability	Religion
15	Inhabited for generations	Illiteracy	Agriculture	Land-holding peasant	Married	No	No religion
17	Inhabited for generations	Illiteracy	Agriculture	Land-holding peasant	Married	No	No religion
18	Inhabited for generations	Illiteracy	Agriculture	Land-holding peasant	Widowed	No	No religion
19	Inhabited for generations	Illiteracy	Agriculture	Land-holding peasant	Widowed	No	No religion
15	Inhabited for generations	Primary school 1 year					
3	Inhabited for generations	Illiteracy	Agriculture	Land-holding peasant	Married	Left hand lost	No religion
15	Inhabited for generations	Illiteracy	Agriculture	Land-holding peasant	Married	No	No religion
9	Inhabited for generations	Illiteracy	Agriculture	Land-holding peasant	Married	No	No religion
7	Inhabited for generations	Illiteracy	Agriculture	Land-holding peasant	Married	No	No religion
4	Inhabited for generations	Illiteracy					
27	Inhabited for generations	Illiteracy	Agriculture	Land-holding peasant	Married	No	No religion
3	Inhabited for generations	Illiteracy	Agriculture	Land-holding peasant	Married	No	No religion
7	Inhabited for generations						
25	Inhabited for generations						
21	Inhabited for generations	Sishu 1 year	Agriculture	Land-holding peasant	Married	No	No religion
13	Inhabited for generations	Illiteracy	Agriculture	Land-holding peasant	Married	No	No religion

Appendix: Overview of Microdata for Wudu County (*cont.*)

	The no. of household	Jia	Household	Relationship with the household head	Surname	First name	Sex	The Chinese zodiac	Birthdate	
									Year	Month
Muma Xiang 1st Bao				Mother	王	何官女子	Female	Monkey	1883	9
Muma Xiang 1st Bao				Son	王	顺明	Male	Dog	1934	8
Muma Xiang 1st Bao				Uncle	王	碧会	Male	Ox	1888	11
Muma Xiang 1st Bao				Uncle's wife	王	罗三香子	Female	Horse	1893	3
Muma Xiang 1st Bao				Younger sister	王	孝银儿	Female	Rat	1936	8
Muma Xiang 1st Bao	132	10	1	The household head	王	應官	Male	Monkey	1907	12
Muma Xiang 1st Bao				The household head's wife	王	白四儿娃	Female	Rooster	1921	6
Muma Xiang 1st Bao				Daughter	王	赵儿子	Female	Dog	1946	7
Muma Xiang 1st Bao				Younger brother	王	五娃子	Male	Ox	1925	9
Muma Xiang 1st Bao	133	10	2	The household head	王	大情	Male	Dog	1882	2
Muma Xiang 1st Bao				The household head's wife	王	刘刘秀子	Female	Dog	1885	9
Muma Xiang 1st Bao				Son	王	廷兴	Male	Rooster	1896	4
Muma Xiang 1st Bao				Son's wife	王	贾玉连儿	Female	Dog	1909	8
Muma Xiang 1st Bao	134	10	3	The household head	王	廷先	Male	Dragon	1916	8
Muma Xiang 1st Bao				The household head's wife	王	何孝连儿	Female	Tiger	1914	12
Muma Xiang 1st Bao				Older brother	王	树保子	Male	Rabbit	1902	2
Muma Xiang 1st Bao	135	10	4	The household head	王	廷武	Male	Ox	1899	1

Date	The beginning time of inhabitation in this county	Educational level	Profession sector	Occupation	Marital status	Physical disability	Religion
28	Inhabited for generations	Illiteracy	Agriculture	Land-holding peasant	Widowed	No	No religion
24	Inhabited for generations	Primary school 2 years					
12	Inhabited for generations	Illiteracy	Agriculture	Land-holding peasant	Married	No	No religion
27	Inhabited for generations	Illiteracy	Agriculture	Land-holding peasant	Married	No	No religion
31	Inhabited for generations	Illiteracy					
28	Inhabited for generations	Illiteracy	Agriculture	Land-holding peasant	Married	No	No religion
4	Inhabited for generations	Illiteracy	Agriculture	Land-holding peasant	Married	No	No religion
7	Inhabited for generations						
13	Inhabited for generations	Illiteracy	Agriculture	Land-holding peasant	Unmarried	Dumb	No religion
6	Inhabited for generations	Illiteracy	Agriculture	Land-holding peasant	Married	No	No religion
8	Inhabited for generations	Illiteracy	Agriculture	Land-holding peasant	Married	No	No religion
4	Inhabited						
19	Inhabited for generations	Illiteracy	Agriculture	Land-holding peasant	Married	No	No religion
29	Inhabited for generations	Illiteracy	Agriculture	Land-holding peasant	Married	No	No religion
12	Inhabited for generations	Illiteracy	Agriculture	Land-holding peasant	Married	No	No religion
9	Inhabited for generations	Illiteracy	Agriculture	Land-holding peasant	Widowed	Dumb	No religion
4	Inhabited for generations	Illiteracy	Agriculture	Land-holding peasant	Married	No	No religion

Appendix: Overview of Microdata for Wudu County (*cont.*)

	The no. of household	Jia	Household	Relationship with the household head	Surname	First name	Sex	The Chinese zodiac	Birthdate Year	Month
Muma Xiang 1st Bao				The household head's wife	王	贾锅官儿	Female	Ox	1900	8
Muma Xiang 1st Bao				Son	王	碧未	Male	Snake	1917	7
Muma Xiang 1st Bao				Son's wife	王	何一儿子	Female	Rabbit	1915	8
Muma Xiang 1st Bao	136	10	5	The household head	王	廷元	Male	Dog	1909	3
Muma Xiang 1st Bao				The household head's wife	王	何桥连子	Female	Rat	1912	8
Muma Xiang 1st Bao				Son	王	碧喜	Male	Dragon	1940	6
Muma Xiang 1st Bao	137	10	6	The household head	王	碧乾	Male	Rat	1912	4
Muma Xiang 1st Bao				The household head's wife	王	何四儿子	Female	Rabbit	1915	5
Muma Xiang 1st Bao				Son	王	應海	Male	Ox	1937	10
Muma Xiang 1st Bao	138	10	7	The household head	王	杨马儿子	Male	Horse	1893	10
Muma Xiang 1st Bao				Son	撞	成儿	Male	Goat	1919	6
Muma Xiang 1st Bao				Son's wife	王	罗楼银儿	Female	Dragon	1916	8
Muma Xiang 1st Bao				Grandson	王	孝代子	Male	Goat	1943	10
Muma Xiang 1st Bao				Granddaughter	王	孝女子	Female	Pig	1947	3
Muma Xiang 1st Bao	139	10	8	The household head	王	官成	Male	Rabbit	1915	8
Muma Xiang 1st Bao	140	10	9	The household head	王	廷来	Male	Monkey	1883	12

Date	The beginning time of inhabitation in this county	Educational level	Profession sector	Occupation	Marital status	Physical disability	Religion
28	Inhabited for generations	Illiteracy	Agriculture	Land-holding peasant	Married	No	No religion
3	Inhabited for generations	Illiteracy	Agriculture	Land-holding peasant	Married	No	No religion
25	Inhabited for generations	Illiteracy	Agriculture	Land-holding peasant	Married	No	No religion
25	Inhabited for generations	Illiteracy	Agriculture	Land-holding peasant	Married	No	No religion
27	Inhabited for generations	Illiteracy	Agriculture	Land-holding peasant	Married	No	No religion
20	Inhabited for generations	Primary school 1 year					
1	Inhabited for generations	Illiteracy	Agriculture	Land-holding peasant	Married	No	No religion
21	Inhabited for generations	Illiteracy	Agriculture	Land-holding peasant	Married	No	No religion
8	Inhabited for generations	Primary school 1 year					
9	Inhabited for generations	Illiteracy	Agriculture	Land-holding peasant	Widowed	No	No religion
18	Inhabited for generations	Illiteracy	Agriculture	Land-holding peasant	Married	No	No religion
29	Inhabited for generations	Illiteracy	Agriculture	Land-holding peasant	Married	No	No religion
27	Inhabited for generations						
1	Inhabited for generations						
25	Inhabited for generations	Illiteracy	Agriculture	Land-holding peasant	Unmarried	No	No religion
19	Inhabited for generations	Illiteracy	Agriculture	Land-holding peasant	Married	No	No religion

Appendix: Overview of Microdata for Wudu County (*cont.*)

	The no. of household	Jia	Household	Relationship with the household head	Surname	First name	Sex	The Chinese zodiac	Birthdate	
									Year	Month
Muma Xiang 1st Bao				The household head's wife	王	李附银子	Female	Ox	1888	4
Muma Xiang 1st Bao	141	10	10	The household head	王	碧连	Male	Ox	1913	7
Muma Xiang 1st Bao				The household head's wife	王	刘创银子	Female	Ox	1913	4
Muma Xiang 1st Bao				Son	王	應国	Male	Ox	1937	11
Muma Xiang 1st Bao	142	10	11	The household head	王	廷合	Male	Rabbit	1878	4
Muma Xiang 1st Bao				The household head's wife	王	王金秀子	Female	Goat	1892	11
Muma Xiang 1st Bao				The first son	王	碧德	Male	Snake	1904	11
Muma Xiang 1st Bao				Son's wife	王	白桥连子	Female	Rat	1913	5
Muma Xiang 1st Bao				The second son	王	碧堂	Male	Rabbit	1902	2
Muma Xiang 1st Bao				The household head's wife	王	王创儿子	Female	Snake	1917	2
Muma Xiang 1st Bao				Granddaughter	王	当女子	Female	Monkey	1944	4
Muma Xiang 1st Bao	143	10	12	The household head	王	應文	Male	Horse	1919	1
Muma Xiang 1st Bao				The household head's wife	王	刘银连子	Female	Monkey	1920	8
Muma Xiang 1st Bao				Mother	王	陈苏秀子	Female	Dog	1885	3
Muma Xiang 1st Bao				Daughter	王	桂儿子	Female	Goat	1943	10
Muma Xiang 1st Bao	144	10	13	The household head	罗	世堂	Male	Rabbit	1878	11

Date	The beginning time of inhabitation in this county	Educational level	Profession sector	Occupation	Marital status	Physical disability	Religion
26	Inhabited for generations	Illiteracy	Agriculture	Land-holding peasant	Married	No	No religion
27	Inhabited for generations	Illiteracy	Agriculture	Land-holding peasant	Married	No	No religion
12	Inhabited for generations	Illiteracy	Agriculture	Land-holding peasant	Married	No	No religion
17	Inhabited for generations	Primary school 1 year					
24	Inhabited for generations	Illiteracy	Agriculture	Land-holding peasant	Married	No	No religion
27	Inhabited for generations	Illiteracy	Agriculture	Land-holding peasant	Married	No	No religion
28	Inhabited for generations	Illiteracy	Agriculture	Land-holding peasant	Married	Dwarf	No religion
1	Inhabited for generations	Illiteracy	Agriculture	Land-holding peasant	Married	No	No religion
6	Inhabited for generations	Illiteracy	Agriculture	Land-holding peasant	Married	No	No religion
10	Inhabited for generations	Illiteracy	Agriculture	Land-holding peasant	Married	No	No religion
18	Inhabited for generations						
5	Inhabited for generations	Illiteracy	Agriculture	Land-holding peasant	Married	No	No religion
20	Inhabited for generations	Illiteracy	Agriculture	Land-holding peasant	Married	No	No religion
3	Inhabited for generations	Illiteracy	Agriculture	Land-holding peasant	Widowed	No	No religion
13	Inhabited for generations					Left eye blind	
7	Inhabited for generations				Widowed		

Appendix: Overview of Microdata for Wudu County (*cont.*)

	The no. of household	Jia	Household	Relationship with the household head	Surname	First name	Sex	The Chinese zodiac	Birthdate	
									Year	Month
Muma Xiang 1st Bao				Son's wife	罗	王四娃子	Female	Rooster	1884	9
Muma Xiang 1st Bao				Grandson	罗	兴贵	Male	Dragon	1916	7
Muma Xiang 1st Bao				Grandson's wife	罗	王根女子	Female	Dog	1909	8
Muma Xiang 1st Bao				Great grandson	罗	发富	Male	Rabbit	1927	6
Muma Xiang 1st Bao				Great grandson	罗	麦连子	Female	Rat	1936	6
Muma Xiang 1st Bao	145	10	14	The household head	罗	世珍	Male	Rat	1912	6
Muma Xiang 1st Bao				The household head's wife	罗	白百儿子	Female	Dragon	1916	8
Muma Xiang 1st Bao				Father	罗	义进	Male	Goat	1893	1
Muma Xiang 1st Bao				Mother	罗	花儿子	Female	Rabbit	1890	4
Muma Xiang 1st Bao				Son	罗	仲一	Male	Tiger	1938	7
Muma Xiang 1st Bao				Daughter	罗	小女子	Female	Dog	1946	6
Muma Xiang 1st Bao				Older brother	罗	世乾	Male	Pig	1910	2
Muma Xiang 1st Bao				Older brother's wife	罗	白金连子	Female	Monkey	1913	4
Muma Xiang 1st Bao				Niece	罗	一代儿	Female	Snake	1942	1
Muma Xiang 1st Bao				Nephew	罗	同代儿	Male	Goat	1943	6
Muma Xiang 1st Bao				Younger brother	罗	世洪	Male	Snake	1929	11
Muma Xiang 1st Bao	146	10	15	The household head	罗	世才	Male	Dog	1909	12

Date	The beginning time of inhabitation in this county	Educational level	Profession sector	Occupation	Marital status	Physical disability	Religion
23	Inhabited for generations	Illiteracy	Agriculture	Land-holding peasant	Widowed	No	No religion
20	Inhabited for generations	Illiteracy	Agriculture	Land-holding peasant	Married	No	No religion
11	Inhabited for generations	Illiteracy	Agriculture	Land-holding peasant	Married	No	No religion
4	Inhabited for generations	Illiteracy	Agriculture	Land-holding peasant	Unmarried	No	No religion
27	Inhabited for generations	Illiteracy	Agriculture	Land-holding peasant		No	No religion
25	Inhabited for generations	Illiteracy	Agriculture	Land-holding peasant	Married	Bald	No religion
13	Inhabited for generations	Illiteracy	Agriculture	Land-holding peasant	Married	No	No religion
22	Inhabited for generations	Illiteracy	Agriculture	Land-holding peasant	Married	No	No religion
3	Inhabited for generations	Illiteracy	Agriculture	Land-holding peasant	Married	No	No religion
24	Inhabited for generations	Primary school 1 year					
14	Inhabited for generations						
12	Inhabited for generations	Illiteracy	Agriculture	Land-holding peasant	Married	No	No religion
3	Inhabited for generations	Illiteracy	Agriculture	Land-holding peasant	Married	No	No religion
1	Inhabited for generations						
7	Inhabited for generations						
15	Inhabited for generations	Illiteracy	Agriculture	Land-holding peasant	Unmarried	No	No religion
16	Inhabited for generations	Illiteracy	Agriculture	Land-holding peasant	Married	No	No religion

Appendix: Overview of Microdata for Wudu County (*cont.*)

	The no. of household	Jia	Household	Relationship with the household head	Surname	First name	Sex	The Chinese zodiac	Birthdate Year	Month
Muma Xiang 1st Bao				The household head's wife	罗	何桂银儿	Female	Rabbit	1902	4
Muma Xiang 1st Bao				Father	罗	义虎	Male	Horse	1900	1
Muma Xiang 1st Bao				Mother	罗	董润连儿	Female	Snake	1880	1
Muma Xiang 1st Bao				Son	罗	兴儿	Male	Goat	1943	7
Muma Xiang 1st Bao				Daughter	罗	创银儿	Female	Goat	1931	4
Muma Xiang 1st Bao	147	11	1	The household head	罗	义富	Male	Horse	1893	8
Muma Xiang 1st Bao				Son	罗	世兴	Male	Rabbit	1927	11
Muma Xiang 1st Bao				Son's wife	罗	王扣娃	Female	Goat	1931	8
Muma Xiang 1st Bao	148	11	2	The household head	罗	世银	Male	Rat	1912	3
Muma Xiang 1st Bao				The household head's wife	罗	白董儿子	Female	Horse	1919	1
Muma Xiang 1st Bao				Son	罗	九代儿	Male	Horse	1942	10
Muma Xiang 1st Bao				Daughter	罗	冯代儿	Female	Ox	1937	4
Muma Xiang 1st Bao	149	11	3	The household head	罗	世奎	Male	Pig	1910	4
Muma Xiang 1st Bao				The household head's wife	罗	王真连儿	Female	Snake	1917	10
Muma Xiang 1st Bao				Mother	罗	何花女子	Female	Pig	1874	10
Muma Xiang 1st Bao				The first son	罗	仲礼	Male	Dog	1934	6
Muma Xiang 1st Bao				The second son	银	代儿	Male	Snake	1941	5

Date	The beginning time of inhabitation in this county	Educational level	Profession sector	Occupation	Marital status	Physical disability	Religion
11	Inhabited for generations	Illiteracy	Agriculture	Land-holding peasant	Married	No	No religion
21	Inhabited for generations	Illiteracy	Agriculture	Land-holding peasant	Married	No	No religion
17	Inhabited for generations	Illiteracy	Agriculture	Land-holding peasant	Married	No	No religion
21	Inhabited for generations	Illiteracy					
20	Inhabited for generations	Illiteracy	Agriculture	Land-holding peasant	Unmarried	No	No religion
25	Inhabited for generations	Illiteracy	Agriculture	Land-holding peasant	Widowed	No	No religion
17	Inhabited for generations	Illiteracy	Agriculture	Land-holding peasant	Married	No	No religion
28	Inhabited for generations	Illiteracy	Agriculture	Land-holding peasant	Married	No	No religion
17	Inhabited for generations	Illiteracy	Agriculture	Land-holding peasant	Married	Tuberculosis	No religion
19	Inhabited for generations	Illiteracy	Agriculture	Land-holding peasant	Married	No	No religion
17	Inhabited for generations						
13	Inhabited for generations	Illiteracy					
1	Inhabited for generations	Illiteracy	Agriculture	Land-holding peasant	Married	No	No religion
20	Inhabited for generations	Illiteracy	Agriculture	Land-holding peasant	Married	No	No religion
2	Inhabited for generations				Widowed		
8	Inhabited for generations	Primary school 3 years					
31	Inhabited for generations	Illiteracy					

Appendix: Overview of Microdata for Wudu County (*cont.*)

	The no. of household	Jia	Household	Relationship with the household head	Surname	First name	Sex	The Chinese zodiac	Birthdate Year	Month
Muma Xiang 1st Bao	150	11	4	The household head	罗	世德	Male	Ox	1888	8
Muma Xiang 1st Bao				The household head's wife	罗	王翠儿子	Female	Rooster	1896	6
Muma Xiang 1st Bao				Father	罗	义太	Male	Dragon	1879	1
Muma Xiang 1st Bao				Son	罗	仲明	Male	Rat	1924	8
Muma Xiang 1st Bao				Son's wife	罗	贾同女子	Female	Ox	1925	12
Muma Xiang 1st Bao	151	11	5	The household head	罗	义清	Male	Snake	1880	12
Muma Xiang 1st Bao				Son	罗	世明	Male	Dragon	1903	8
Muma Xiang 1st Bao				Son's wife	罗	董董连子	Female	Rabbit	1902	12
Muma Xiang 1st Bao				Grandson	罗	仲贵	Male	Pig	1924	1
Muma Xiang 1st Bao				Grandson's wife	罗	王良女子	Female	Rooster	1921	11
Muma Xiang 1st Bao				Great grandson	银	生儿	Male	Monkey	1944	2
Muma Xiang 1st Bao	152	11	6	The household head	王	廷现	Male	Snake	1904	4
Muma Xiang 1st Bao				The household head's wife	王	罗金连子	Female	Tiger	1901	7
Muma Xiang 1st Bao				The first son	王	刘代娃	Male	Pig	1935	8
Muma Xiang 1st Bao				The second son	王	碧飞	Male	Tiger	1938	6
Muma Xiang 1st Bao				Daughter	王	真香娃	Female	Monkey	1944	5
Muma Xiang 1st Bao	153	11	7	The household head	王	廷仲	Male	Goat	1882	7

Date	The beginning time of inhabitation in this county	Educational level	Profession sector	Occupation	Marital status	Physical disability	Religion
11	Inhabited for generations	Illiteracy	Agriculture	Land-holding peasant	Married	No	No religion
30	Inhabited for generations	Illiteracy	Agriculture	Land-holding peasant	Married	No	No religion
28	Inhabited for generations	Illiteracy	Agriculture	Land-holding peasant	Widowed	No	No religion
29	Inhabited for generations	Illiteracy	Agriculture	Land-holding peasant	Married	No	No religion
1	Inhabited for generations	Illiteracy	Agriculture	Land-holding peasant	Married	No	No religion
15	Inhabited for generations	Illiteracy	Agriculture	Land-holding peasant	Widowed	No	No religion
25	Inhabited for generations	Illiteracy	Agriculture	Land-holding peasant	Married	No	No religion
7	Inhabited for generations	Illiteracy	Agriculture	Land-holding peasant	Married	No	No religion
19	Inhabited for generations	Illiteracy	Agriculture	Land-holding peasant	Married	No	No religion
10	Inhabited for generations	Illiteracy	Agriculture	Land-holding peasant	Married	No	No religion
14	Inhabited for generations						
19	Inhabited for generations	Illiteracy	Agriculture	Land-holding peasant	Married	No	No religion
1	Inhabited for generations	Illiteracy	Agriculture	Land-holding peasant	Married	No	No religion
15	Inhabited for generations	Illiteracy	Agriculture	Land-holding peasant	Unmarried	No	No religion
12	Inhabited for generations	Primary school 1 year					
21	Inhabited for generations						
15	Inhabited for generations	Illiteracy	Agriculture	Land-holding peasant	Widowed	No	No religion

Appendix: Overview of Microdata for Wudu County (*cont.*)

	The no. of household	Jia	Household	Relationship with the household head	Surname	First name	Sex	The Chinese zodiac	Birthdate Year	Month
Muma Xiang 1st Bao				Son's wife	王	谈官银子	Female	Dog	1909	9
Muma Xiang 1st Bao				Son	王	九旦娃	Male	Dragon	1928	11
Muma Xiang 1st Bao	154	11	8	The household head	王	應龙	Male	Tiger	1901	10
Muma Xiang 1st Bao				The household head's wife	王	高董连子	Female	Tiger	1900	1
Muma Xiang 1st Bao				Father	王	廷白	Male	Dog	1872	10
Muma Xiang 1st Bao				Mother	王	何玉娃子	Female	Rabbit	1878	10
Muma Xiang 1st Bao				Son	王	顺谈	Male	Horse	1918	3
Muma Xiang 1st Bao				Son's wife	王	谈刘娃	Female	Pig	1923	6
Muma Xiang 1st Bao				Grandson	王	长娃	Male	Snake	1941	7
Muma Xiang 1st Bao	155	11	9	The household head	王	應洪	Male	Pig	1898	2
Muma Xiang 1st Bao				The household head's wife	王	何石儿子	Female	Snake	1903	1
Muma Xiang 1st Bao				The first daughter	王	官翠娃	Female	Dragon	1940	8
Muma Xiang 1st Bao				The second daughter	王	九翠娃	Female	Dog	1946	10
Muma Xiang 1st Bao	156	11	10	The household head	王	谈肖儿子	Female	Horse	1905	12
Muma Xiang 1st Bao				Son	王	石生娃	Male	Rabbit	1939	9
Muma Xiang 1st Bao	157	11	11	The household head	王	谈小娃子	Female	Rat	1887	1
Muma Xiang 1st Bao				Son's wife	王	谈连娃子	Female	Monkey	1907	5
Muma Xiang 1st Bao				Granddaughter	王	四银子	Female	Goat	1946	8

Date	The beginning time of inhabitation in this county	Educational level	Profession sector	Occupation	Marital status	Physical disability	Religion
18	Inhabited for generations	Illiteracy	Agriculture	Land-holding peasant	Divorced	No	No religion
1	Inhabited for generations	Illiteracy	Agriculture	Land-holding peasant	Unmarried	No	No religion
2	Inhabited for generations	Illiteracy	Agriculture	Land-holding peasant	Married	No	No religion
5	Inhabited for generations	Illiteracy	Agriculture	Land-holding peasant	Married	No	No religion
16	Inhabited for generations						
6	Inhabited for generations						
28	Inhabited for generations	Illiteracy	Agriculture	Land-holding peasant	Married	Left vision poor	No religion
18	Inhabited for generations	Illiteracy	Agriculture	Land-holding peasant	Married	No	No religion
15	Inhabited for generations						
24	Inhabited for generations	Illiteracy	Agriculture	Land-holding peasant	Married	No	No religion
1	Inhabited for generations	Illiteracy	Agriculture	Land-holding peasant	Married	No	No religion
11	Inhabited for generations	Illiteracy					
13	Inhabited for generations						
27	Inhabited for generations	Illiteracy	Agriculture	Land-holding peasant	Divorced	Deaf	No religion
27	Inhabited for generations	Illiteracy				Bald	
26	Inhabited for generations	Illiteracy	Agriculture	Land-holding peasant	Divorced	No	No religion
3	Inhabited for generations	Illiteracy	Agriculture	Land-holding peasant	Divorced	No	No religion
5	Inhabited for generations					Bald	

Appendix: Overview of Microdata for Wudu County (*cont.*)

	The no. of household	Jia	Household	Relationship with the household head	Surname	First name	Sex	The Chinese zodiac	Birthdate Year	Month
Muma Xiang 1st Bao	158	11	12	The household head	王	白桃连子	Female	Ox	1913	12
Muma Xiang 1st Bao				Son	王	双林子	Male	Dragon	1928	8
Muma Xiang 1st Bao				Daughter	王	蔡银子	Female	Goat	1931	4
Muma Xiang 1st Bao	159	11	13	The household head	王	碧才	Male	Goat	1894	5
Muma Xiang 1st Bao				Son	王	應龙	Male	Rat	1924	8
Muma Xiang 1st Bao				Son's wife	王	白郭儿娃	Female	Rat	1924	12
Muma Xiang 1st Bao				Grandson	附	兴娃	Male	Rooster	1945	5
Muma Xiang 1st Bao				Uncle's wife	王	白润娃子	Female	Rat	1899	5
Muma Xiang 1st Bao	160	11	14	The household head	王	應勤	Male	Goat	1919	11
Muma Xiang 1st Bao				The household head's wife	王	董苏连子	Female	Rat	1912	4
Muma Xiang 1st Bao				Mother	王	贾楼儿子	Female	Pig	1898	7
Muma Xiang 1st Bao				Son	王	顺强	Male	Pig	1935	5
Muma Xiang 1st Bao	161	11	15	The household head	王	碧福	Male	Pig	1898	10
Muma Xiang 1st Bao				The household head's wife	王	白石秀子	Female	Pig	1898	11
Muma Xiang 1st Bao				Son	王	應海	Male	Goat	1919	2
Muma Xiang 1st Bao				Son's wife	王	何冯翠子	Female	Pig	1923	8
Muma Xiang 1st Bao	162	12	1	The household head	王	碧益	Male	Horse	1893	2

Date	The beginning time of inhabitation in this county	Educational level	Profession sector	Occupation	Marital status	Physical disability	Religion
2	Inhabited for generations	Illiteracy	Agriculture	Land-holding peasant	Widowed	No	No religion
29	Inhabited for generations	Illiteracy	Agriculture	Land-holding peasant	Unmarried	No	No religion
20	Inhabited for generations	Illiteracy	Agriculture	Land-holding peasant	Unmarried	No	No religion
2	Inhabited for generations	Sishu 1 year	Agriculture	Land-holding peasant	Widowed	No	Buddhism
15	Inhabited for generations	Illiteracy	Agriculture	Land-holding peasant	Married	No	
23	Inhabited for generations	Illiteracy	Agriculture	Land-holding peasant	Married	No	
7	Inhabited for generations						
15	Inhabited for generations	Illiteracy	Agriculture	Land-holding peasant	Widowed	No	
6	Inhabited for generations	Illiteracy	Agriculture	Land-holding peasant	Married	No	No religion
21	Inhabited for generations	Illiteracy	Agriculture	Land-holding peasant	Married	No	No religion
16	Inhabited for generations	Illiteracy	Agriculture	Land-holding peasant	Widowed	No	No religion
6	Inhabited for generations	Primary school 1 year					
17	Inhabited for generations	Illiteracy	Agriculture	Land-holding peasant	Married	No	No religion
13	Inhabited for generations	Illiteracy	Agriculture	Land-holding peasant	Married	No	No religion
21	Inhabited for generations	Illiteracy	Agriculture	Land-holding peasant	Married	No	No religion
26	Inhabited for generations	Illiteracy	Agriculture	Land-holding peasant	Married	No	No religion
2	Inhabited for generations	Illiteracy	Agriculture	Land-holding peasant	Married	No	No religion

Appendix: Overview of Microdata for Wudu County (*cont.*)

	The no. of household	Jia	Household	Relationship with the household head	Surname	First name	Sex	The Chinese zodiac	Birthdate Year	Month
Muma Xiang 1st Bao				The household head's wife	王	何连女子	Female	Rooster	1900	2
Muma Xiang 1st Bao				Son	王	應伦	Male	Pig	1910	9
Muma Xiang 1st Bao				Son's wife	王	谈翠女子	Female	Tiger	1915	1
Muma Xiang 1st Bao				Grandson	王	顺兴	Male	Tiger	1938	10
Muma Xiang 1st Bao	163	12	2	The household head	王	碧杰	Male	Rat	1887	10
Muma Xiang 1st Bao				Son	王	應双	Male	Rat	1924	10
Muma Xiang 1st Bao				Son's wife	王	何蔡连子	Female	Tiger	1922	11
Muma Xiang 1st Bao	164	12	3	The household head	王	應成	Male	Rooster	1884	3
Muma Xiang 1st Bao				The first son	王	顺德	Male	Tiger	1901	4
Muma Xiang 1st Bao				The first son's wife	王	陈同儿子	Female	Tiger	1901	10
Muma Xiang 1st Bao				The second son	王	顺怀	Male	Rat	1924	12
Muma Xiang 1st Bao				The second son's wife	王	高同儿子	Female	Rooster	1921	12
Muma Xiang 1st Bao				Grandson	王	安兴	Male	Rabbit	1939	11
Muma Xiang 1st Bao	165	12	4	The household head	王	顺玉	Male	Rabbit	1902	6
Muma Xiang 1st Bao				The household head's wife	王	杨王代儿	Female	Tiger	1901	9
Muma Xiang 1st Bao				Mother	王	谈许女子	Female	Tiger	1889	2

Date	The beginning time of inhabitation in this county	Educational level	Profession sector	Occupation	Marital status	Physical disability	Religion
17	Inhabited for generations	Illiteracy	Agriculture	Land-holding peasant	Married	No	No religion
7	Inhabited for generations	Illiteracy	Agriculture	Land-holding peasant	Married	No	No religion
7	Inhabited for generations	Illiteracy	Agriculture	Land-holding peasant	Married	No	No religion
11	Inhabited for generations	Primary school 1 year					
24	Inhabited for generations	Illiteracy	Agriculture	Land-holding peasant	Widowed	No	No religion
27	Inhabited for generations	Illiteracy	Agriculture	Land-holding peasant	Married	No	No religion
25	Inhabited for generations	Illiteracy	Agriculture	Land-holding peasant	Widowed	No	No religion
26	Inhabited for generations	Illiteracy	Agriculture	Land-holding peasant	Married	No	No religion
19	Inhabited for generations	Illiteracy	Agriculture	Land-holding peasant	Married	No	No religion
9	Inhabited for generations	Illiteracy	Agriculture	Land-holding peasant	Married	No	No religion
29	Inhabited for generations	Illiteracy	Agriculture	Land-holding peasant	Married	No	No religion
30	Inhabited for generations	Illiteracy	Agriculture	Land-holding peasant	Married	No	No religion
12	Inhabited for generations	Primary school 1 year					
8	Inhabited for generations	Illiteracy	Agriculture	Land-holding peasant	Married	No	No religion
18	Inhabited for generations	Illiteracy	Agriculture	Land-holding peasant	Married	No	No religion
1	Inhabited for generations	Illiteracy	Agriculture	Land-holding peasant	Widowed	No	No religion

Appendix: Overview of Microdata for Wudu County (*cont.*)

	The no. of household	Jia	Household	Relationship with the household head	Surname	First name	Sex	The Chinese zodiac	Birthdate Year	Month
Muma Xiang 1st Bao				Daughter	王	有娃	Female	Snake	1941	10
Muma Xiang 1st Bao	166	12	5	The household head	王	顺高	Male	Rabbit	1902	6
Muma Xiang 1st Bao				The household head's wife	王	贾约家娃	Female	Dog	1909	10
Muma Xiang 1st Bao	167	12	6	The household head	王	碧星	Male	Pig	1910	12
Muma Xiang 1st Bao				Older brother's wife	王	陈银翠子	Female	Goat	1906	9
Muma Xiang 1st Bao				Niece	王	代女子	Female	Ox	1937	3
Muma Xiang 1st Bao	168	12	7	The household head	王	顺恩	Male	Goat	1919	10
Muma Xiang 1st Bao				Mother	王	白六女子	Female	Snake	1892	7
Muma Xiang 1st Bao	169	12	8	The household head	王	顺益	Male	Dragon	1917	1
Muma Xiang 1st Bao				The household head's wife	王	白见娃子	Female	Tiger	1914	9
Muma Xiang 1st Bao				Mother	王	白蔡香子	Female	Ox	1886	6
Muma Xiang 1st Bao				Daughter	王	双代娃	Female	Snake	1941	7
Muma Xiang 1st Bao	170	12	9	The household head	王	顺成	Male	Ox	1913	12
Muma Xiang 1st Bao				The household head's wife	王	来儿子	Female	Rabbit	1927	5
Muma Xiang 1st Bao				Father	王	應星	Male	Rooster	1884	2
Muma Xiang 1st Bao				Mother	王	谈李润儿	Female	Rooster	1884	6
Muma Xiang 1st Bao				Daughter	王	谢连子	Female	Rat	1936	9

Date	The beginning time of inhabitation in this county	Educational level	Profession sector	Occupation	Marital status	Physical disability	Religion
5	Inhabited for generations	Illiteracy					
5	Inhabited for generations	Illiteracy	Agriculture	Land-holding peasant	Married	No	No religion
15	Inhabited for generations	Illiteracy	Agriculture	Land-holding peasant	Married	No	No religion
3	Inhabited for generations	Illiteracy	Agriculture	Land-holding peasant	Divorced	No	No religion
12	Inhabited for generations	Illiteracy	Agriculture	Land-holding peasant	Divorced	No	No religion
26	Inhabited for generations	Illiteracy					
18	Inhabited for generations	Illiteracy	Agriculture	Land-holding peasant	Unmarried	Left eye blind	No religion
28	Inhabited for generations	Illiteracy	Agriculture	Land-holding peasant	Widowed	No	No religion
29	Inhabited for generations	Illiteracy	Agriculture	Land-holding peasant	Married	No	No religion
5	Inhabited for generations	Illiteracy	Agriculture	Land-holding peasant	Married	No	No religion
14	Inhabited for generations	Illiteracy	Agriculture	Land-holding peasant	Widowed	No	No religion
1	Inhabited for generations	Illiteracy					
10	Inhabited for generations	Illiteracy	Agriculture	Land-holding peasant	Married	No	No religion
11	Inhabited for generations	Illiteracy	Agriculture	Land-holding peasant	Married	No	No religion
15	Inhabited for generations	Illiteracy	Agriculture	Land-holding peasant	Married	No	No religion
25	Inhabited for generations	Illiteracy	Agriculture	Land-holding peasant	Married	No	No religion
18	Inhabited for generations	Illiteracy					

Appendix: Overview of Microdata for Wudu County (*cont.*)

	The no. of house-hold	Jia	House-hold	Relation-ship with the house-hold head	Surname	First name	Sex	The Chinese zodiac	Birthdate	
									Year	Month
Muma Xiang 1st Bao				Uncle's wife	王	孙崴连子	Female	Rooster	1884	8
Muma Xiang 1st Bao	171	12	10	The house-hold head	王	應富	Male	Tiger	1908	4
Muma Xiang 1st Bao				Son's wife	王	陈银连子	Female	Rat	1899	4
Muma Xiang 1st Bao				Grandson	王	安仁	Male	Rat	1936	10
Muma Xiang 1st Bao				Grand-daughter	王	桂儿娃	Female	Goat	1931	7
Muma Xiang 1st Bao	172	12	11	The house-hold head	王	碧见	Male	Monkey	1920	4
Muma Xiang 1st Bao				The house-hold head's wife	王	董毛儿娃	Female	Snake	1917	5
Muma Xiang 1st Bao				Mother	王	李代娃子	Female	Snake	1892	4
Muma Xiang 1st Bao				Son	王	刘成子	Male	Dragon	1928	8
Muma Xiang 1st Bao	173	12	12	The house-hold head	王	廷元	Male	Ox	1888	2
Muma Xiang 1st Bao				The house-hold head's wife	王	李创女子	Female	Pig	1898	7
Muma Xiang 1st Bao				Son	王	一生儿	Male	Rat	1936	3
Muma Xiang 1st Bao	174	12	13	The house-hold head	王	碧乾	Male	Tiger	1901	2
Muma Xiang 1st Bao				The house-hold head's wife	王	何官娃子	Female	Dog	1909	4
Muma Xiang 1st Bao				Father	王	廷魁	Male	Ox	1888	9
Muma Xiang 1st Bao				Daughter	王	创连子	Female	Rooster	1945	1
Muma Xiang 1st Bao	175	12	14	The house-hold head	王	廷富	Male	Goat	1905	1

Date	The beginning time of inhabitation in this county	Educational level	Profession sector	Occupation	Marital status	Physical disability	Religion
26	Inhabited for generations	Illiteracy	Agriculture	Land-holding peasant	Widowed	No	No religion
12	Inhabited for generations	Illiteracy	Agriculture	Land-holding peasant	Divorced	No	No religion
15	Inhabited for generations	Illiteracy	Agriculture	Land-holding peasant	Divorced	No	No religion
8	Inhabited for generations	Primary school 1 year					
9	Inhabited for generations	Illiteracy	Agriculture	Land-holding peasant	Unmarried	No	No religion
29	Inhabited for generations	Illiteracy	Agriculture	Land-holding peasant	Married	No	No religion
15	Inhabited for generations	Illiteracy	Agriculture	Land-holding peasant	Married	No	No religion
5	Inhabited for generations	Illiteracy	Agriculture	Land-holding peasant	Widowed	No	No religion
29	Inhabited for generations	Illiteracy	Agriculture	Land-holding peasant	Unmarried	No	No religion
19	Inhabited for generations	Illiteracy	Agriculture	Land-holding peasant	Married	No	No religion
21	Inhabited for generations	Illiteracy	Agriculture	Land-holding peasant	Married	No	No religion
11	Inhabited for generations	Illiteracy					
14	Inhabited for generations	Illiteracy	Agriculture	Land-holding peasant	Married	No	No religion
16	Inhabited for generations	Illiteracy	Agriculture	Land-holding peasant	Married	No	No religion
9	Inhabited for generations	Illiteracy	Agriculture	Land-holding peasant	Widowed	No	No religion
13	Inhabited for generations						
23	Inhabited for generations	Illiteracy	Agriculture	Land-holding peasant	Married	No	No religion

Appendix: Overview of Microdata for Wudu County (*cont.*)

	The no. of household	Jia	Household	Relationship with the household head	Surname	First name	Sex	The Chinese zodiac	Birthdate	
									Year	Month
Muma Xiang 1st Bao				The household head's wife	王	尹银女子	Female	Goat	1905	1
Muma Xiang 1st Bao				Daughter	王	买生儿	Female	Dog	1946	9
Muma Xiang 1st Bao	176	12	15	The household head	王	碧洪	Male	Rat	1899	4
Muma Xiang 1st Bao				The household head's wife	王	高秋儿子	Female	Dragon	1903	4
Muma Xiang 1st Bao				Son	王	九来儿	Male	Rabbit	1927	10
Muma Xiang 1st Bao				Daughter	王	良娃	Female	Monkey	1932	7
Muma Xiang 1st Bao	177	13	1	The household head	王	廷尉	Male	Dog	1897	8
Muma Xiang 1st Bao				The household head's wife	王	董跳银儿	Female	Rooster	1896	4
Muma Xiang 1st Bao	178	13	2	The household head	王	大均	Male	Ox	1888	2
Muma Xiang 1st Bao				Son	王	廷喜	Male	Monkey	1907	4
Muma Xiang 1st Bao				Son's wife	王	王兆翠儿	Female	Snake	1917	7
Muma Xiang 1st Bao				Grandson	王	约儿子	Female	Snake	1941	11
Muma Xiang 1st Bao	179	13	3	The household head	王	大情	Male	Snake	1880	11
Muma Xiang 1st Bao				The household head's wife	王	马冬女子	Female	Snake	1880	11
Muma Xiang 1st Bao				Son	王	廷禄	Male	Dragon	1916	8
Muma Xiang 1st Bao	180	13	4	The household head	王	碧虎	Male	Tiger	1914	3

Date	The beginning time of inhabitation in this county	Educational level	Profession sector	Occupation	Marital status	Physical disability	Religion
20	Inhabited for generations	Illiteracy	Agriculture	Land-holding peasant	Married	No	No religion
10	Inhabited for generations						
10	Inhabited for generations	Illiteracy	Agriculture	Land-holding peasant	Married	No	No religion
3	Inhabited for generations	Illiteracy	Agriculture	Land-holding peasant	Married	No	No religion
8	Inhabited for generations	Illiteracy	Agriculture	Land-holding peasant	Divorced	No	No religion
14	Inhabited for generations	Illiteracy	Agriculture	Land-holding peasant	Unmarried	No	No religion
12	Inhabited for generations	Sishu 2 years	Agriculture	Land-holding peasant	Married	No	Buddhism
4	Inhabited for generations	Illiteracy	Agriculture	Land-holding peasant	Married	No	
16	Inhabited for generations	Sishu 1 year	Agriculture	Land-holding peasant	Widowed	No	Buddhism
1	Inhabited for generations	Illiteracy	Agriculture	Land-holding peasant	Married	Injured	
22	Inhabited for generations	Illiteracy	Agriculture	Land-holding peasant	Married	No	
7	Inhabited for generations	Illiteracy					
8	Inhabited for generations	Illiteracy	Agriculture	Land-holding peasant	Married	No	No religion
8	Inhabited for generations	Illiteracy	Agriculture	Land-holding peasant	Married	No	No religion
23	Inhabited for generations	Illiteracy	Agriculture	Land-holding peasant	Unmarried	Bald	No religion
11	Inhabited for generations	Illiteracy	Agriculture	Land-holding peasant	Married	No	No religion

Appendix: Overview of Microdata for Wudu County (*cont.*)

	The no. of household	Jia	Household	Relationship with the household head	Surname	First name	Sex	The Chinese zodiac	Birthdate Year	Month
Muma Xiang 1st Bao				The household head's wife	王	罗三女子	Female	Tiger	1914	3
Muma Xiang 1st Bao				Son	王	應科	Male	Tiger	1938	9
Muma Xiang 1st Bao				Daughter	王	九贤子	Female	Pig	1935	10
Muma Xiang 1st Bao	181	13	5	The household head	王	碧明	Male	Rat	1924	11
Muma Xiang 1st Bao				The household head's wife	王	白当娃子	Female	Dog	1922	4
Muma Xiang 1st Bao				Son	王	杨兴娃	Male	Rooster	1945	11
Muma Xiang 1st Bao				Daughter	王	贤儿娃	Female	Horse	1942	9
Muma Xiang 1st Bao	182	13	6	The household head	王	碧怀	Male	Rat	1912	11
Muma Xiang 1st Bao				The household head's wife	王	白李代子	Female	Tiger	1914	7
Muma Xiang 1st Bao				Mother	王	王连女子	Female	Rooster	1883	1
Muma Xiang 1st Bao				Son	王	代官成	Male	Horse	1942	1
Fujin Xiang 1st Bao	183	1	1	The household head	陈	青山	Male	Rabbit	1902	3
Fujin Xiang 1st Bao				The household head's wife	李	蔡娃	Female	Rabbit	1902	4
Fujin Xiang 1st Bao				Mother	赵	张女	Female	Rabbit	1865	10
Fujin Xiang 1st Bao				Son	陈	国民	Male	Rat	1926	4
Fujin Xiang 1st Bao				The second son	陈	庚辰	Male	Goat	1943	2

Date	The beginning time of inhabitation in this county	Educational level	Profession sector	Occupation	Marital status	Physical disability	Religion
27	Inhabited for generations	Illiteracy	Agriculture	Land-holding peasant	Married	No	No religion
28	Inhabited for generations	Primary school 1 year					
20	Inhabited for generations	Illiteracy	Agriculture	Land-holding peasant	Unmarried	No	No religion
21	Inhabited for generations	Illiteracy	Agriculture	Land-holding peasant	Married	No	No religion
19	Inhabited for generations	Illiteracy	Agriculture	Land-holding peasant	Married	No	No religion
19	Inhabited for generations						
28	Inhabited for generations						
11	Inhabited for generations	Illiteracy	Agriculture	Land-holding peasant	Married	Blind	No religion
6	Inhabited for generations	Illiteracy	Agriculture	Land-holding peasant	Married	No	No religion
17	Inhabited for generations	Illiteracy	Agriculture	Land-holding peasant	Widowed	No	No religion
17	Inhabited for generations						
8	Inhabited for generations	Illiteracy	Agriculture	Land-holding peasant	Married	No	No religion
17	Inhabited for generations	Illiteracy	Agriculture	Land-holding peasant	Married	No	No religion
15	Inhabited for generations						
18	Inhabited for generations	Illiteracy	Agriculture	Land-holding peasant	Unmarried	No	No religion
28	Inhabited for generations						

Appendix: Overview of Microdata for Wudu County (*cont.*)

	The no. of household	Jia	Household	Relationship with the household head	Surname	First name	Sex	The Chinese zodiac	Birthdate	
									Year	Month
Fujin Xiang 1st Bao				Uncle	陈	啟琴	Male	Dragon	1879	2
Fujin Xiang 1st Bao	184	1	2	The household head	陈	义连	Male	Horse	1905	5
Fujin Xiang 1st Bao				The household head's wife	李	马秀	Female	Horse	1905	9
Fujin Xiang 1st Bao				Daughter	陈	桃香	Female	Horse	1930	5
Fujin Xiang 1st Bao				Son	陈	王三义	Male	Dog	1934	10
Fujin Xiang 1st Bao	185	1	3	The household head	赵	学謙	Male	Monkey	1895	4
Fujin Xiang 1st Bao				The household head's wife	赵	怀娃	Female	Rooster	1908	7
Fujin Xiang 1st Bao				Son	赵	磁儿	Male	Rooster	1945	2
Fujin Xiang 1st Bao	186	1	4	The household head	陈	国财	Male	Horse	1918	5
Fujin Xiang 1st Bao				The household head's wife	李	登香	Female	Monkey	1920	11
Fujin Xiang 1st Bao				Daughter	陈	女儿	Female	Monkey	1944	2
Fujin Xiang 1st Bao				Son	陈	代娃	Male	Rooster	1945	1
Fujin Xiang 1st Bao	187	1	5	The household head	陈	田生	Male	Monkey	1883	6
Fujin Xiang 1st Bao				The household head's wife	赵	成娃	Female	Horse	1893	1
Fujin Xiang 1st Bao				Son	陈	国喜	Male	Rat	1924	2
Fujin Xiang 1st Bao				Son's wife	张	怀娃	Female	Tiger	1926	4
Fujin Xiang 1st Bao				The second son	陈	虎儿	Male	Ox	1937	8

Date	The beginning time of inhabitation in this county	Educational level	Profession sector	Occupation	Marital status	Physical disability	Religion
28	Inhabited for generations						
3	Inhabited for generations	Illiteracy	Agriculture	Land-holding peasant	Married	No	No religion
27	Inhabited for generations	Illiteracy	Agriculture	Land-holding peasant	Married	No	No religion
4	Inhabited for generations	Illiteracy	Agriculture	Land-holding peasant	Married	No	No religion
17	Inhabited for generations	Bao school 4 years	Agriculture	Land-holding peasant	Married	No	No religion
7	Inhabited for generations	Illiteracy	Agriculture	Land-holding peasant	Married	No	No religion
1	Inhabited for generations	Illiteracy	Agriculture	Land-holding peasant	Married	No	No religion
22	Inhabited for generations						
27	Inhabited for generations	Illiteracy	Agriculture	Land-holding peasant	Married	No	No religion
5	Inhabited for generations	Illiteracy	Agriculture	Land-holding peasant	Married	No	No religion
8	Inhabited for generations	Illiteracy	Agriculture				
14	Inhabited for generations						
8	Inhabited for generations	Illiteracy	Agriculture	Land-holding peasant	Married	No	No religion
21	Inhabited for generations	Illiteracy	Agriculture	Land-holding peasant	Married	No	No religion
7	Inhabited for generations	Illiteracy	Agriculture	Land-holding peasant	Married	No	No religion
12	Inhabited for generations	Illiteracy	Agriculture	Land-holding peasant	Married	No	No religion
24	Inhabited for generations						

Appendix: Overview of Microdata for Wudu County (*cont.*)

	The no. of household	Jia	Household	Relationship with the household head	Surname	First name	Sex	The Chinese zodiac	Birthdate	
									Year	Month
Fujin Xiang 1st Bao				Granddaughter	陈	润女儿	Female	Dog	1946	10
Fujin Xiang 1st Bao	188	1	6	The household head	卢	凤英	Male	Ox	1913	1
Fujin Xiang 1st Bao				The household head's wife	岳	王家	Female	Rat	1912	11
Fujin Xiang 1st Bao				Father	卢	彦明	Male	Tiger	1889	6
Fujin Xiang 1st Bao				Mother	赵	冯女	Female	Goat	1894	12
Fujin Xiang 1st Bao				Son	卢	长生	Male	Pig	1947	2
Fujin Xiang 1st Bao				Daughter	卢	月香	Female	Rooster	1933	10
Fujin Xiang 1st Bao				The second daughter	卢	岁月	Female	Tiger	1938	4
Fujin Xiang 1st Bao				The third daughter	卢	菊香	Female	Snake	1941	9
Fujin Xiang 1st Bao				Younger brother	卢	凤财	Male	Rabbit	1927	9
Fujin Xiang 1st Bao				Younger sister	卢	银柱	Female	Pig	1935	12
Fujin Xiang 1st Bao				The second younger sister	卢	崴柱	Female	Rabbit	1939	7
Fujin Xiang 1st Bao	189	1	7	The household head	卢	崴平	Female	Rabbit	1915	2
Fujin Xiang 1st Bao				Son	王	才儿	Male	Rabbit	1939	3
Fujin Xiang 1st Bao				The household head's huband	陈	国隆	Male	Ox	1913	3
Fujin Xiang 1st Bao	190	1	8	The household head	陈	义忠	Male	Rat	1899	9
Fujin Xiang 1st Bao				The household head's wife	赵	苟女	Female	Monkey	1907	8

Date	The beginning time of inhabitation in this county	Educational level	Profession sector	Occupation	Marital status	Physical disability	Religion
26	Inhabited for generations						
23	Inhabited for generations	Illiteracy	Agriculture	Land-holding peasant	Married	No	No religion
1	Inhabited for generations	Illiteracy	Agriculture	Land-holding peasant	Married	No	No religion
27	Inhabited for generations	Illiteracy	Agriculture	Land-holding peasant	Married	No	No religion
4	Inhabited for generations	Illiteracy	Agriculture	Land-holding peasant	Married	No	No religion
14	Inhabited for generations						
9	Inhabited for generations	Illiteracy	Agriculture	Land-holding peasant	Unmarried	No	No religion
19	Inhabited for generations						
5	Inhabited for generations						
27	Inhabited for generations	Illiteracy	Agriculture	Land-holding peasant	Unmarried	No	No religion
13	Inhabited for generations						
17	Inhabited for generations						
27	Inhabited for generations	Illiteracy	Agriculture	Land-holding peasant	Unmarried	No	No religion
6	Inhabited for generations	Illiteracy					
8	Inhabited for generations	Illiteracy	Agriculture	Land-holding peasant	Married	No	No religion
2	Inhabited for generations	Illiteracy	Agriculture	Land-holding peasant	Married	No	No religion
3	Inhabited for generations	Illiteracy	Agriculture	Land-holding peasant	Married	No	No religion

Appendix: Overview of Microdata for Wudu County (*cont.*)

	The no. of house-hold	Jia	House-hold	Relation-ship with the house-hold head	Surname	First name	Sex	The Chinese zodiac	Birthdate Year	Month
Fujin Xiang 1st Bao				Father	陈	啟转	Male	Monkey	1871	11
Fujin Xiang 1st Bao				Mother	王	银翠娃	Female	Rat	1875	9
Fujin Xiang 1st Bao				Son	陈	田发	Male	Ox	1925	2
Fujin Xiang 1st Bao				The second son	陈	田丰	Male	Dog	1934	10
Fujin Xiang 1st Bao				The third son	陈	三成	Male	Horse	1942	3
Fujin Xiang 1st Bao				The fourth son	陈	保儿	Male	Rooster	1945	11
Fujin Xiang 1st Bao				Daughter	陈	女儿	Female	Pig	1935	9
Fujin Xiang 1st Bao	191	1	9	The house-hold head	陈	守林	Male	Rooster	1921	5
Fujin Xiang 1st Bao				The house-hold head's wife	张	让娃	Female	Tiger	1926	5
Fujin Xiang 1st Bao				Daughter	陈	伯英	Female	Goat	1943	11
Fujin Xiang 1st Bao				The second daughter	陈	东英	Female	Rooster	1945	12
Fujin Xiang 1st Bao				Father	陈	张送子	Male	Dog	1885	2
Fujin Xiang 1st Bao	192	1	10	The house-hold head	陈	田义	Male	Rabbit	1915	5
Fujin Xiang 1st Bao				The house-hold head's wife	李	柱娃	Female	Goat	1919	3
Fujin Xiang 1st Bao				Son	陈	菱儿	Male	Dog	1946	10
Fujin Xiang 1st Bao				Uncle's wife	韩	有连	Female	Monkey	1883	10
Fujin Xiang 1st Bao	193	1	11	The house-hold head	陈	义吉	Male	Rooster	1908	11

Date	The beginning time of inhabitation in this county	Educational level	Profession sector	Occupation	Marital status	Physical disability	Religion
12	Inhabited for generations						
8	Inhabited for generations						
8	Inhabited for generations	Illiteracy	Agriculture	Land-holding peasant	Unmarried	No	No religion
12	Inhabited for generations	Bao school 3 years					
24	Inhabited for generations						
18	Inhabited for generations						
14	Inhabited for generations	Illiteracy					
4	Inhabited for generations	Illiteracy	Agriculture	Land-holding peasant	Married	No	No religion
12	Inhabited for generations	Illiteracy	Agriculture	Land-holding peasant	Married	No	No religion
3	Inhabited for generations						
22	Inhabited for generations						
28	Inhabited for generations	Illiteracy	Agriculture	Land-holding peasant	Widowed	No	No religion
26	Inhabited for generations	Illiteracy	Agriculture	Land-holding peasant	Married	Hemorrhoids	No religion
5	Inhabited for generations	Illiteracy	Agriculture	Land-holding peasant	Married	No	No religion
1	Inhabited for generations						
1	Inhabited for generations	Illiteracy	Agriculture	Land-holding peasant	Widowed	No	No religion
9	Inhabited for generations	Illiteracy	Agriculture	Land-holding peasant	Married	No	No religion

Appendix: Overview of Microdata for Wudu County (*cont.*)

	The no. of household	Jia	Household	Relationship with the household head	Surname	First name	Sex	The Chinese zodiac	Birthdate	
									Year	Month
Fujin Xiang 1st Bao				The household head's wife	卢	银环	Female	Dragon	1916	11
Fujin Xiang 1st Bao				Mother	崔	富姐	Female	Dog	1874	9
Fujin Xiang 1st Bao				Son	陈	所娃	Male	Monkey	1944	2
Fujin Xiang 1st Bao				The second son	陈	七十四	Male	Pig	1947	8
Fujin Xiang 1st Bao	194	1	12	The household head	陈	田仓	Male	Monkey	1920	11
Fujin Xiang 1st Bao				The household head's wife	韩	进娃	Female	Horse	1918	4
Fujin Xiang 1st Bao				Daughter	陈	智娃	Female	Dog	1946	2
Fujin Xiang 1st Bao	195	1	13	The household head	陈	田喜	Male	Ox	1913	8
Fujin Xiang 1st Bao				The household head's wife	李	鸡娃	Female	Rooster	1908	9
Fujin Xiang 1st Bao				Mother	刘	马姐	Female	Dog	1885	4
Fujin Xiang 1st Bao				Younger brother	陈	才	Male	Tiger	1926	6
Fujin Xiang 1st Bao				Younger brother's wife	李	改娃	Female	Goat	1931	1
Fujin Xiang 1st Bao	196	1	14	The household head	陈	义明	Male	Pig	1898	11
Fujin Xiang 1st Bao				The household head's wife	岳	顺女	Female	Pig	1898	8
Fujin Xiang 1st Bao				Son	陈	国华	Male	Monkey	1932	8
Fujin Xiang 1st Bao				The second son	陈	谢生	Male	Ox	1937	5

Date	The beginning time of inhabitation in this county	Educational level	Profession sector	Occupation	Marital status	Physical disability	Religion
1	Inhabited for generations	Illiteracy	Agriculture	Land-holding peasant	Married	No	No religion
1	Inhabited for generations						
7	Inhabited for generations						
2	Inhabited for generations						
12	Inhabited for generations	Illiteracy	Agriculture	Land-holding peasant	Married	No	No religion
5	Inhabited for generations	Illiteracy	Agriculture	Land-holding peasant	Married	No	No religion
18	Inhabited for generations						
24	Inhabited for generations	Illiteracy	Agriculture	Land-holding peasant	Married	No	No religion
25	Inhabited for generations	Illiteracy	Agriculture	Land-holding peasant	Married	Eye disease	No religion
4	Inhabited for generations	Illiteracy	Agriculture	Land-holding peasant	Widowed	No	No religion
5	Inhabited for generations	Illiteracy	Agriculture	Land-holding peasant	Married	No	No religion
19	Inhabited for generations	Illiteracy	Agriculture	Land-holding peasant	Married	No	No religion
17	Inhabited for generations	Illiteracy	Agriculture	Land-holding peasant	Married	No	No religion
15	Inhabited for generations	Illiteracy	Agriculture	Land-holding peasant	Married	No	No religion
3	Inhabited for generations	Illiteracy	Agriculture	Land-holding peasant	Unmarried	No	No religion
14	Inhabited for generations	Illiteracy	Agriculture	Land-holding peasant	Unmarried	No	No religion

Appendix: Overview of Microdata for Wudu County (*cont.*)

									Birthdate	
	The no. of household	Jia	Household	Relationship with the household head	Surname	First name	Sex	The Chinese zodiac	Year	Month
Fujin Xiang 1st Bao				The third son	陈	具卯儿	Male	Horse	1942	1
Fujin Xiang 1st Bao				Daughter	陈	桂香	Female	Dog	1946	2
Fujin Xiang 1st Bao	197	1	15	The household head	卢	凤发	Male	Snake	1930	11
Fujin Xiang 1st Bao				The household head's wife	李	柱环	Female	Pig	1910	1
Fujin Xiang 1st Bao				Son	卢	正才儿	Male	Dog	1933	8
Fujin Xiang 1st Bao				The second son	卢	崴才儿	Male	Ox	1937	1
Fujin Xiang 1st Bao				The third son	卢	保娃	Male	Rooster	1945	3
Fujin Xiang 1st Bao				Daughter	卢	怀女儿	Female	Goat	1943	5
Fujin Xiang 1st Bao				Daughter	卢	存女儿	Female	Pig	1946	1
Fujin Xiang 1st Bao	198	2	1	The household head	卢	凤鸣	Male	Rabbit	1902	12
Fujin Xiang 1st Bao				The household head's wife	李	路儿	Female	Rat	1912	7
Fujin Xiang 1st Bao				Son	卢	崴俱	Male	Rat	1936	1
Fujin Xiang 1st Bao				Daughter	卢	具女儿	Female	Tiger	1936	6
Fujin Xiang 1st Bao				The second daughter	卢	崴女儿	Female	Pig	1947	9
Fujin Xiang 1st Bao	199	2	2	The household head	卢	彦章	Male	Horse	1881	2
Fujin Xiang 1st Bao				Son	徐	芦家	Male	Horse	1905	2
Fujin Xiang 1st Bao				Son's wife	三	女子	Female	Rooster	1908	7
Fujin Xiang 1st Bao				Grandson	卢	聚财	Male	Rat	1936	1

Date	The beginning time of inhabitation in this county	Educational level	Profession sector	Occupation	Marital status	Physical disability	Religion
26	Inhabited for generations						
16	Inhabited for generations						
23	Inhabited for generations	Illiteracy	Agriculture	Land-holding peasant	Married	No	No religion
3	Inhabited for generations	Illiteracy	Agriculture	Land-holding peasant	Married	No	No religion
13	Inhabited for generations	Illiteracy	Agriculture	Land-holding peasant	Unmarried		
1	Inhabited for generations	Illiteracy	Agriculture				
17	Inhabited for generations						
3	Inhabited for generations						
11	Inhabited for generations						
3	Inhabited for generations	Illiteracy	Agriculture	Land-holding peasant	Married	No	No religion
31	Inhabited for generations	Illiteracy	Others	Weaving	Married	No	No religion
1	Inhabited for generations	Bao school 1 year					
9	Inhabited for generations	Illiteracy					
3	Inhabited for generations						
5	Inhabited for generations	Illiteracy	Agriculture	Land-holding peasant	Widowed	No	No religion
7	Inhabited for generations	Illiteracy	Agriculture	Land-holding peasant	Married	No	No religion
31	Inhabited for generations	Illiteracy	Others	Weaving	Married	No	No religion
25	Inhabited for generations	Bao school 2 years					

Appendix: Overview of Microdata for Wudu County (*cont.*)

	The no. of household	Jia	Household	Relationship with the household head	Surname	First name	Sex	The Chinese zodiac	Birthdate Year	Month
Fujin Xiang 1st Bao				The second grandson	卢	六十四	Male	Rooster	1945	12
Fujin Xiang 1st Bao				Granddaughter	卢	桃小	Female	Goat	1931	7
Fujin Xiang 1st Bao	200	2	3	The household head	陈	田福	Male	Pig	1898	4
Fujin Xiang 1st Bao				The household head's wife	赵	月连	Female	Rat	1899	12
Fujin Xiang 1st Bao				Mother	徐	赛娃	Female	Ox	1873	6
Fujin Xiang 1st Bao				Son	陈	国林	Male	Horse	1930	4
Fujin Xiang 1st Bao				Daughter	陈	崴学女	Female	Snake	1941	6
Fujin Xiang 1st Bao	201	2	4	The household head	王	啟祥	Male	Dog	1897	7
Fujin Xiang 1st Bao				The household head's wife	陈	换换	Female	Dog	1897	2
Fujin Xiang 1st Bao				Son	王	同姓	Male	Horse	1930	10
Fujin Xiang 1st Bao				The second son	王	時娃	Male	Monkey	1944	12
Fujin Xiang 1st Bao	202	2	5	The household head	陈	学德	Male	Dog	1909	2
Fujin Xiang 1st Bao				The household head's wife	赵	香香	Female	Pig	1900	4
Fujin Xiang 1st Bao				Son	陈	全生	Male	Dragon	1940	7
Fujin Xiang 1st Bao				The second son	陈	射全	Male	Horse	1942	4
Fujin Xiang 1st Bao				The third son	陈	明全	Male	Pig	1947	3
Fujin Xiang 1st Bao				Daughter	陈	桂英	Female	Monkey	1932	8

Date	The beginning time of inhabitation in this county	Educational level	Profession sector	Occupation	Marital status	Physical disability	Religion
26	Inhabited for generations						
31	Inhabited for generations	Illiteracy	Agriculture	Land-holding peasant	Unmarried	No	No religion
2	Inhabited for generations	Illiteracy	Agriculture	Land-holding peasant	Married	No	No religion
6	Inhabited for generations	Illiteracy	Agriculture	Land-holding peasant	Married	No	No religion
14	Inhabited for generations						
5	Inhabited for generations	Illiteracy	Agriculture	Land-holding peasant	Unmarried	No	No religion
26	Inhabited for generations						
14	Inhabited for generations	Illiteracy	Agriculture	Land-holding peasant	Married	No	No religion
27	Inhabited for generations	Illiteracy	Others	Weaving	Married	No	No religion
9	Inhabited for generations	Illiteracy	Agriculture	Land-holding peasant	Unmarried	No	No religion
26	Inhabited for generations						
17	Inhabited for generations	Sishu 3 years	Agriculture	Land-holding peasant	Married	No	No religion
6	Inhabited for generations	Illiteracy	Agriculture	Land-holding peasant	Married	No	No religion
26	Inhabited for generations	Bao school 1 year					
2	Inhabited for generations						
9	Inhabited for generations						
24	Inhabited for generations	Illiteracy	Agriculture	Land-holding peasant	Unmarried	No	No religion

Appendix: Overview of Microdata for Wudu County (*cont.*)

	The no. of household	Jia	Household	Relationship with the household head	Surname	First name	Sex	The Chinese zodiac	Birthdate	
									Year	Month
Fujin Xiang 1st Bao	203	2	6	The household head	刘	占荣	Male	Rooster	1908	1
Fujin Xiang 1st Bao				The household head's wife	岳	柱娃	Female	Dog	1909	10
Fujin Xiang 1st Bao	204	2	7	The household head	陈	学明	Male	Horse	1893	7
Fujin Xiang 1st Bao				The household head's wife	成	钱旦	Female	Rabbit	1902	3
Fujin Xiang 1st Bao				Son	陈	赵射	Male	Dragon	1940	2
Fujin Xiang 1st Bao	205	2	8	The household head	陈	仲山	Male	Ox	1913	6
Fujin Xiang 1st Bao				The household head's wife	罗	李娃	Female	Dog	1913	11
Fujin Xiang 1st Bao				Son	陈	永洪	Male	Monkey	1932	10
Fujin Xiang 1st Bao				Daughter	陈	双连	Female	Pig	1935	10
Fujin Xiang 1st Bao				The second daughter	陈	双秀	Female	Rabbit	1939	4
Fujin Xiang 1st Bao				The third daughter	陈	双桂	Female	Dog	1936	6
Fujin Xiang 1st Bao	206	2	9	The household head	岳	登堂	Male	Goat	1882	9
Fujin Xiang 1st Bao				The household head's wife	贾	二女	Female	Goat	1894	9
Fujin Xiang 1st Bao				Son	岳	文忠	Male	Pig	1923	9
Fujin Xiang 1st Bao				Daughter	岳	柱女	Female	Snake	1941	7
Fujin Xiang 1st Bao	207	2	10	The household head	岳	文才	Male	Ox	1900	6

Date	The beginning time of inhabitation in this county	Educational level	Profession sector	Occupation	Marital status	Physical disability	Religion
17	Inhabited for generations	Illiteracy	Agriculture	Land-holding peasant	Married	No	No religion
17	Inhabited for generations	Illiteracy	Agriculture	Land-holding peasant	Married	No	No religion
4	Inhabited for generations	Illiteracy	Agriculture	Land-holding peasant	Married	No	No religion
13	Inhabited for generations	Illiteracy	Agriculture	Land-holding peasant	Married	No	No religion
12	Inhabited for generations	Illiteracy					
18	Inhabited for generations	Illiteracy	Agriculture	Land-holding peasant	Married	No	No religion
27	Inhabited for generations	Illiteracy	Agriculture	Land-holding peasant	Married	No	No religion
25	Inhabited for generations	Bao school 4 years	Agriculture	Land-holding peasant	Unmarried	No	No religion
7	Inhabited for generations						
27	Inhabited for generations						
3	Inhabited for generations						
23	Inhabited for generations	Illiteracy	Agriculture	Land-holding peasant	Married	Blind	No religion
12	Inhabited for generations	Illiteracy	Agriculture	Land-holding peasant	Married	No	No religion
11	Inhabited for generations	Illiteracy	Agriculture	Land-holding peasant	Widowed	No	No religion
20	Inhabited for generations						
3	Inhabited for generations	Illiteracy	Agriculture	Land-holding peasant	Married	No	No religion

Appendix: Overview of Microdata for Wudu County (*cont.*)

	The no. of household	Jia	Household	Relationship with the household head	Surname	First name	Sex	The Chinese zodiac	Birthdate	
									Year	Month
Fujin Xiang 1st Bao				The household head's wife	陈	扣娃	Female	Monkey	1907	3
Fujin Xiang 1st Bao				Son	岳	兴怀	Male	Monkey	1920	10
Fujin Xiang 1st Bao				Son's wife	雷	建香	Female	Monkey	1920	12
Fujin Xiang 1st Bao				Grandson	岳	全才	Male	Dog	1946	7
Fujin Xiang 1st Bao				Granddaughter	岳	平见儿	Female	Snake	1941	8
Fujin Xiang 1st Bao				The second granddaughter	岳	崴具娃	Female	Rooster	1945	3
Fujin Xiang 1st Bao	208	2	11	The household head	刘	青荣	Male	Dog	1922	5
Fujin Xiang 1st Bao				The household head's wife	贺	时明	Female	Ox	1925	12
Fujin Xiang 1st Bao				Daughter	刘	秋英	Female	Rooster	1945	10
Fujin Xiang 1st Bao				The second daughter	刘	具英	Female	Pig	1947	3
Fujin Xiang 1st Bao	209	2	12	The household head	岳	兴禄	Male	Rooster	1884	3
Fujin Xiang 1st Bao				Son	岳	昭娃	Male	Rabbit	1927	1
Fujin Xiang 1st Bao				Uncle	岳	文禄	Male	Rabbit	1878	1
Fujin Xiang 1st Bao	210	2	13	The household head	岳	兴才	Male	Monkey	1920	11
Fujin Xiang 1st Bao				The household head's wife	浩	黑女	Female	Dog	1923	4
Fujin Xiang 1st Bao				Mother	雷	富娃	Female	Tiger	1901	7
Fujin Xiang 1st Bao				Daughter	岳	凤英	Female	Pig	1947	2

Date	The beginning time of inhabitation in this county	Educational level	Profession sector	Occupation	Marital status	Physical disability	Religion
3	Inhabited for generations	Illiteracy	Agriculture	Land-holding peasant	Married	No	No religion
20	Inhabited for generations	Illiteracy	Agriculture	Land-holding peasant	Married	No	No religion
20	Inhabited for generations	Illiteracy	Agriculture	Land-holding peasant	Married	No	No religion
1	Inhabited for generations						
24	Inhabited for generations						
13	Inhabited for generations						
30	Inhabited for generations	Illiteracy	Agriculture	Land-holding peasant	Married	No	No religion
29	Inhabited for generations	Illiteracy	Agriculture	Land-holding peasant	Married	No	No religion
20	Inhabited for generations						
21	Inhabited for generations						
18	Inhabited for generations	Illiteracy	Agriculture	Land-holding peasant	Widowed	No	No religion
11	Inhabited for generations	Illiteracy	Agriculture	Land-holding peasant	Unmarried	No	No religion
10	Inhabited for generations						
18	Inhabited for generations	Illiteracy	Agriculture	Land-holding peasant	Married	No	No religion
20	Inhabited for generations	Illiteracy	Agriculture	Land-holding peasant	Married	No	No religion
16	Inhabited for generations	Illiteracy	Agriculture	Land-holding peasant	Widowed	No	No religion
5	Inhabited for generations						

Appendix: Overview of Microdata for Wudu County (*cont.*)

	The no. of household	Jia	Household	Relationship with the household head	Surname	First name	Sex	The Chinese zodiac	Birthdate	
									Year	Month
Fujin Xiang 1st Bao				Younger sister	岳	全桃儿	Female	Monkey	1932	4
Fujin Xiang 1st Bao	211	2	14	The household head	岳	文德	Male	Dragon	1891	4
Fujin Xiang 1st Bao				The household head's wife	富	儿	Female	Tiger	1889	8
Fujin Xiang 1st Bao				Son	岳	李生	Male	Rooster	1908	9
Fujin Xiang 1st Bao				Son's wife	赵	雪连	Female	Horse	1918	11
Fujin Xiang 1st Bao				Grandson	岳	早儿	Male	Snake	1941	1
Fujin Xiang 1st Bao				Granddaughter	岳	双哇	Female	Goat	1943	3
Fujin Xiang 1st Bao	212	2	15	The household head	岳	登才	Male	Monkey	1920	9
Fujin Xiang 1st Bao				The household head's wife	陈	须子	Female	Snake	1917	9
Fujin Xiang 1st Bao				Father	岳	万田	Male	Dragon	1891	4
Fujin Xiang 1st Bao				Mother	陈	跳连	Female	Rooster	1884	7
Fujin Xiang 1st Bao	213	3	1	The household head	岳	万林	Male	Snake	1892	8
Fujin Xiang 1st Bao				The household head's wife	李	青顺	Female	Monkey	1895	10
Fujin Xiang 1st Bao				Son's wife	陈	杜女	Female	Monkey	1920	9
Fujin Xiang 1st Bao				Granddaughter	岳	二女	Female	Pig	1947	7
Fujin Xiang 1st Bao				Daughter	岳	四全	Female	Monkey	1932	11
Fujin Xiang 1st Bao				The second daughter	岳	谢女	Female	Rabbit	1939	8

Date	The beginning time of inhabitation in this county	Educational level	Profession sector	Occupation	Marital status	Physical disability	Religion
1	Inhabited for generations	Illiteracy	Agriculture	Land-holding peasant	Unmarried	No	No religion
5	Inhabited for generations	Illiteracy	Agriculture	Land-holding peasant	Married	No	No religion
24	Inhabited for generations	Illiteracy	Agriculture	Land-holding peasant	Married	No	No religion
9	Inhabited for generations	Illiteracy	Agriculture	Land-holding peasant	Married	No	No religion
11	Inhabited for generations	Illiteracy	Agriculture	Land-holding peasant	Married	No	No religion
23	Inhabited for generations						
13	Inhabited for generations						
30	Inhabited for generations	Illiteracy	Agriculture	Land-holding peasant	Married	No	No religion
8	Inhabited for generations	Illiteracy	Others	Weaving	Married	No	No religion
1	Inhabited for generations	Illiteracy	Agriculture	Land-holding peasant	Married	No	No religion
31	Inhabited for generations	Illiteracy	Others	Weaving	Married	No	No religion
3	Inhabited for generations	Illiteracy	Agriculture	Land-holding peasant	Married	No	No religion
22	Inhabited for generations	Illiteracy	Others	Weaving	Married	No	No religion
26	Inhabited for generations	Illiteracy	Others	Weaving	Widowed	No	No religion
10	Inhabited for generations						
20	Inhabited for generations	Illiteracy	Agriculture	Land-holding peasant	Unmarried	No	No religion
9	Inhabited for generations						

Appendix: Overview of Microdata for Wudu County (*cont.*)

	The no. of household	Jia	Household	Relationship with the household head	Surname	First name	Sex	The Chinese zodiac	Birthdate	
									Year	Month
Fujin Xiang 1st Bao	214	3	2	The household head	岳	登寿	Male	Dog	1922	5
Fujin Xiang 1st Bao				The household head's wife	雷	才娃	Female	Goat	1919	3
Fujin Xiang 1st Bao				Son	岳	富儿	Male	Snake	1941	9
Fujin Xiang 1st Bao				The second son	刘	贵儿	Male	Dog	1946	9
Fujin Xiang 1st Bao				Grandmother	马	转姐	Female	Pig	1874	12
Fujin Xiang 1st Bao				Cohabitant	赵	做娃	Female	Dog	1909	4
Fujin Xiang 1st Bao				Cohabitant's son	转	玉儿	Male	Rat	1936	12
Fujin Xiang 1st Bao	215	3	3	The household head	岳	常寿	Female	Rat	1899	8
Fujin Xiang 1st Bao				Daughter	岳	英娃	Female	Horse	1918	3
Fujin Xiang 1st Bao				Grandson	岳	桂儿	Male	Rat	1936	12
Fujin Xiang 1st Bao				The second grandson	岳	林林	Male	Monkey	1944	11
Fujin Xiang 1st Bao				Granddaughter	岳	桂女	Female	Pig	1947	5
Fujin Xiang 1st Bao				The second daughter	岳	么女	Female	Rabbit	1939	10
Fujin Xiang 1st Bao				Family member	陈	黄拜	Male	Snake	1892	9
Fujin Xiang 1st Bao				Family member	成	禄儿	Male	Rabbit	1915	7
Fujin Xiang 1st Bao	216	3	4	The household head	岳	兴贤	Male	Ox	1900	5
Fujin Xiang 1st Bao				The household head's wife	成	柱女	Female	Horse	1918	3
Fujin Xiang 1st Bao				Mother	张	银娃	Female	Rooster	1871	12

Date	The beginning time of inhabitation in this county	Educational level	Profession sector	Occupation	Marital status	Physical disability	Religion
16	Inhabited for generations	Illiteracy	Agriculture	Land-holding peasant	Married	No	No religion
1	Inhabited for generations	Illiteracy	Agriculture	Land-holding peasant	Married	No	No religion
26	Inhabited for generations	Illiteracy					
12	Inhabited for generations						
22	Inhabited for generations						
19	Inhabited for generations	Illiteracy	Agriculture	Land-holding peasant	Widowed	No	No religion
24	Inhabited for generations						
30	Inhabited for generations	Illiteracy	Agriculture	Land-holding peasant	Married	No	No religion
28	Inhabited for generations	Illiteracy	Agriculture	Land-holding peasant	Married	No	No religion
25	Inhabited for generations	Bao school 2 years					
22	Inhabited for generations						
10	Inhabited for generations						
24	Inhabited for generations						
29	Inhabited for generations	Illiteracy	Agriculture	Land-holding peasant	Married	No	No religion
16	Inhabited for generations	Illiteracy	Agriculture	Land-holding peasant	Married	No	No religion
25	Inhabited for generations	Illiteracy	Agriculture	Land-holding peasant	Married	No	No religion
15	Inhabited for generations	Illiteracy	Agriculture	Land-holding peasant	Married	No	No religion
3	Inhabited for generations						

Appendix: Overview of Microdata for Wudu County (*cont.*)

									Birthdate	
	The no. of household	Jia	Household	Relationship with the household head	Surname	First name	Sex	The Chinese zodiac	Year	Month
Fujin Xiang 1st Bao				Son	岳	谢有	Male	Snake	1941	1
Fujin Xiang 1st Bao				The second son	岳	王贵	Male	Rooster	1945	11
Fujin Xiang 1st Bao				Younger brother	岳	兴虎	Male	Dragon	1916	10
Fujin Xiang 1st Bao				Younger brother's wife	李	怀儿	Female	Snake	1929	11
Fujin Xiang 1st Bao				Nephew	岳	得时	Male	Pig	1947	2
Fujin Xiang 1st Bao	217	3	5	The household head	陈	国益	Male	Dog	1909	4
Fujin Xiang 1st Bao				The household head's wife	赵	东香	Female	Ox	1913	6
Fujin Xiang 1st Bao				Son	陈	张娃	Male	Rat	1924	12
Fujin Xiang 1st Bao				Son's wife	何	举娃	Female	Rabbit	1927	2
Fujin Xiang 1st Bao				Grandson	陈	桃娃	Male	Monkey	1944	8
Fujin Xiang 1st Bao	218	3	6	The household head	陈	义生	Male	Dragon	1891	2
Fujin Xiang 1st Bao				The household head's wife	何	鸡娃	Female	Ox	1888	11
Fujin Xiang 1st Bao				Son	陈	平娃	Male	Tiger	1926	3
Fujin Xiang 1st Bao				Son's wife	何	二女	Female	Monkey	1920	12
Fujin Xiang 1st Bao				Daughter	陈	林娃	Female	Snake	1941	11
Fujin Xiang 1st Bao	219	3	7	The household head	陈	义富	Male	Tiger	1889	9
Fujin Xiang 1st Bao				The household head's wife	赵	良娃	Female	Monkey	1895	2

Date	The beginning time of inhabitation in this county	Educational level	Profession sector	Occupation	Marital status	Physical disability	Religion
3	Inhabited for generations						
3	Inhabited for generations						
29	Inhabited for generations	Illiteracy	Agriculture	Land-holding peasant	Married	No	No religion
5	Inhabited for generations	Illiteracy	Agriculture	Land-holding peasant	Married	No	No religion
1	Inhabited for generations						
1	Inhabited for generations	Illiteracy	Agriculture	Land-holding peasant	Married	No	No religion
1	Inhabited for generations	Illiteracy	Others	Weaving	Married	No	No religion
11	Inhabited for generations	Illiteracy	Agriculture	Land-holding peasant	Married	No	No religion
17	Inhabited for generations	Illiteracy	Others	Weaving	Married	No	No religion
24	Inhabited for generations						
1	Inhabited for generations	Illiteracy	Agriculture	Land-holding peasant	Married	No	No religion
6	Inhabited for generations	Illiteracy	Others	Weaving	Married	No	No religion
1	Inhabited for generations	Illiteracy	Agriculture	Land-holding peasant	Married	No	No religion
24	Inhabited for generations	Illiteracy	Others	Weaving	Married	No	No religion
16	Inhabited for generations						
30	Inhabited for generations	Illiteracy	Agriculture	Land-holding peasant	Married	No	No religion
15	Inhabited for generations	Illiteracy	Others	Weaving	Married	No	No religion

Appendix: Overview of Microdata for Wudu County (*cont.*)

	The no. of house-hold	Jia	House-hold	Relation-ship with the house-hold head	Surname	First name	Sex	The Chinese zodiac	Birthdate	
									Year	Month
Fujin Xiang 1st Bao				Son	陈	国仓	Male	Tiger	1914	8
Fujin Xiang 1st Bao				Son's wife	罗	鸡娃	Female	Tiger	1914	12
Fujin Xiang 1st Bao				Grandson	陈	根义	Male	Horse	1942	8
Fujin Xiang 1st Bao	220	3	8	The house-hold head	陈	义德	Male	Rooster	1908	6
Fujin Xiang 1st Bao				The house-hold head's wife	菊	娃	Female	Pig	1910	10
Fujin Xiang 1st Bao				Son	陈	国琴	Male	Tiger	1938	12
Fujin Xiang 1st Bao				The second son	陈	月杨	Male	Snake	1941	12
Fujin Xiang 1st Bao				Daughter	陈	琴娃	Female	Pig	1935	10
Fujin Xiang 1st Bao	221	3	9	The house-hold head	陈	起富	Male	Goat	1894	12
Fujin Xiang 1st Bao				The house-hold head's wife	董	玉儿	Female	Snake	1904	10
Fujin Xiang 1st Bao				Son	陈	义鸿	Male	Ox	1937	12
Fujin Xiang 1st Bao				Daughter	陈	核秀	Female	Monkey	1932	4
Fujin Xiang 1st Bao				The second daughter	陈	才秀	Female	Dragon	1940	12
Fujin Xiang 1st Bao	222	3	10	The house-hold head	陈	义高	Male	Horse	1880	4
Fujin Xiang 1st Bao				Son	陈	国全	Male	Snake	1917	5
Fujin Xiang 1st Bao				Son's wife	杨	跳娃	Female	Rabbit	1915	5
Fujin Xiang 1st Bao				Grandson	陈	仓娃	Male	Horse	1942	7
Fujin Xiang 1st Bao				Grand-daughter	陈	仓女	Female	Rooster	1945	2

Date	The beginning time of inhabitation in this county	Educational level	Profession sector	Occupation	Marital status	Physical disability	Religion
1	Inhabited for generations	Illiteracy	Agriculture	Land-holding peasant	Married	No	No religion
1	Inhabited for generations	Illiteracy	Others	Weaving	Married	No	No religion
13	Inhabited for generations						
11	Inhabited for generations	Illiteracy	Agriculture	Land-holding peasant	Married	No	No religion
30	Inhabited for generations	Illiteracy	Others	Weaving	Married	No	No religion
24	Inhabited for generations	Bao school 3 years					
24	Inhabited for generations						
7	Inhabited for generations	Illiteracy					
6	Inhabited for generations	Sishu 3 years	Agriculture	Land-holding peasant	Married	No	No religion
18	Inhabited for generations	Illiteracy	Others	Weaving	Married	No	No religion
2	Inhabited for generations	Bao school 3 years	Agriculture				
22	Inhabited for generations	Illiteracy					
3	Inhabited for generations						
21	Inhabited for generations						
2	Inhabited for generations	Illiteracy	Agriculture	Land-holding peasant	Married	No	No religion
30	Inhabited for generations	Illiteracy	Others	Weaving	Married	No	No religion
29	Inhabited for generations						
26	Inhabited for generations						

Appendix: Overview of Microdata for Wudu County (*cont.*)

	The no. of household	Jia	Household	Relationship with the household head	Surname	First name	Sex	The Chinese zodiac	Birthdate	
									Year	Month
Fujin Xiang 1st Bao	223	3	11	The household head	王	治世	Male	Horse	1905	2
Fujin Xiang 1st Bao				The household head's wife	董	秀花女	Female	Dog	1908	1
Fujin Xiang 1st Bao				Son	王	生	Male	Ox	1937	12
Fujin Xiang 1st Bao				The second son	王	四斤娃	Male	Monkey	1945	11
Fujin Xiang 1st Bao	224	3	12	The household head	赵	大婚娃	Male	Goat	1906	8
Fujin Xiang 1st Bao				The household head's wife	陈	斤娃	Female	Dog	1909	2
Fujin Xiang 1st Bao				Daughter	赵	香儿	Female	Dragon	1940	6
Fujin Xiang 1st Bao				The second daughter	赵	桃香儿	Female	Rooster	1945	7
Fujin Xiang 1st Bao	225	3	13	The household head	董	其才	Male	Ox	1925	2
Fujin Xiang 1st Bao				The household head's wife	岳	崴英	Female	Rat	1924	4
Fujin Xiang 1st Bao				Mother	赵	筱娃	Female	Dragon	1903	1
Fujin Xiang 1st Bao				Younger brother	董	二牛	Male	Tiger	1938	5
Fujin Xiang 1st Bao				Younger sister	董	荣娃	Female	Monkey	1932	9
Fujin Xiang 1st Bao				The second younger sister	董	贵女	Female	Dog	1946	4
Fujin Xiang 1st Bao				Daughter	董	秋女	Female	Rooster	1945	10
Fujin Xiang 1st Bao	226	3	14	The household head	董	大贵	Male	Horse	1918	5

Date	The beginning time of inhabitation in this county	Educational level	Profession sector	Occupation	Marital status	Physical disability	Religion
18	Inhabited for generations	Illiteracy	Agriculture	Land-holding peasant	Married	No	No religion
15	Inhabited for generations	Illiteracy	Others	Weaving	Married	No	No religion
30	Inhabited for generations	Illiteracy	Agriculture				
12	Inhabited for generations						
20	Inhabited for generations	Illiteracy	Agriculture	Land-holding peasant	Married	No	No religion
17	Inhabited for generations	Illiteracy	Others	Weaving	Married	No	No religion
8	Inhabited for generations						
3	Inhabited for generations						
2	Inhabited for generations	Illiteracy	Agriculture	Land-holding peasant	Married	No	No religion
27	Inhabited for generations	Illiteracy	Agriculture	Land-holding peasant	Married	No	No religion
7	Inhabited for generations	Illiteracy	Agriculture	Land-holding peasant	Widowed	No	No religion
28	Inhabited for generations	Bao school 2 years					
28	Inhabited for generations	Illiteracy	Agriculture		Unmarried		
20	Inhabited for generations						
10	Inhabited for generations						
23	Inhabited for generations	Illiteracy	Agriculture	Land-holding peasant	Married	No	No religion

Appendix: Overview of Microdata for Wudu County (*cont.*)

	The no. of household	Jia	Household	Relationship with the household head	Surname	First name	Sex	The Chinese zodiac	Birthdate Year	Month
Fujin Xiang 1st Bao				The household head's wife	陈	蒋秀	Female	Tiger	1914	11
Fujin Xiang 1st Bao				Son	董	新房	Male	Dragon	1940	5
Fujin Xiang 1st Bao				The second son	董	二房	Male	Horse	1942	11
Fujin Xiang 1st Bao				Daughter	陈	翠英	Female	Rooster	1945	9
Fujin Xiang 1st Bao	227	3	15	The household head	董	其德	Male	Tiger	1926	2
Fujin Xiang 1st Bao				The household head's wife	王	柱连	Female	Ox	1925	11
Fujin Xiang 1st Bao				Mother	赵	蜡女	Female	Dragon	1903	1
Fujin Xiang 1st Bao	228	4	1	The household head	董	吹元	Male	Rabbit	1878	4
Fujin Xiang 1st Bao				The household head's wife	李	银唤	Female	Rooster	1884	5
Fujin Xiang 1st Bao				Daughter	董	发亥	Female	Pig	1923	7
Fujin Xiang 1st Bao	229	4	2	The household head	张	谢元	Male	Tiger	1914	12
Fujin Xiang 1st Bao				The household head's wife	李	全秀	Female	Rabbit	1915	5
Fujin Xiang 1st Bao				Father	张	万有	Male	Horse	1880	9
Fujin Xiang 1st Bao				Son	张	顺緒	Male	Rooster	1945	10
Fujin Xiang 1st Bao				Uncle	张	万富	Male	Rabbit	1890	3
Fujin Xiang 1st Bao				Uncle's wife	贺	蔡女	Female	Rabbit	1890	11
Fujin Xiang 1st Bao	230	4	3	The household head	董	治兴	Male	Snake	1880	6

Date	The beginning time of inhabitation in this county	Educational level	Profession sector	Occupation	Marital status	Physical disability	Religion
27	Inhabited for generations	Illiteracy	Agriculture	Land-holding peasant	Married	No	No religion
31	Inhabited for generations	Bao school 1 year					
24	Inhabited for generations						
3	Inhabited for generations						
17	Inhabited for generations	Illiteracy	Agriculture	Land-holding peasant	Married	No	No religion
27	Inhabited for generations	Illiteracy	Others	Weaving	Married	No	No religion
10	Inhabited for generations	Illiteracy	Others	Weaving	Widowed	No	No religion
15	Inhabited for generations	Illiteracy	Agriculture	Land-holding peasant	Married	No	No religion
1	Inhabited for generations	Illiteracy	Others	Weaving	Married	No	No religion
28	Inhabited for generations	Illiteracy	Others	Weaving	Widowed	No	No religion
1	Inhabited for generations	Illiteracy	Agriculture	Land-holding peasant	Married	No	No religion
18	Inhabited for generations	Illiteracy	Agriculture	Land-holding peasant	Married	No	No religion
15	Inhabited for generations						
4	Inhabited for generations						
10	Inhabited for generations	Illiteracy	Agriculture	Land-holding peasant	Married	No	No religion
1	Inhabited for generations	Illiteracy	Agriculture	Land-holding peasant	Married	No	No religion
23	Inhabited for generations		Agriculture		Married		

Appendix: Overview of Microdata for Wudu County (*cont.*)

	The no. of household	Jia	Household	Relationship with the household head	Surname	First name	Sex	The Chinese zodiac	Birthdate Year	Month
Fujin Xiang 1st Bao				The household head's wife	王	苏家	Female	Rat	1887	2
Fujin Xiang 1st Bao				Daughter	董	崴菊	Female	Tiger	1914	12
Fujin Xiang 1st Bao				Grandson	董	朝朝	Male	Horse	1942	7
Fujin Xiang 1st Bao				Family member	马	润福	Male	Snake	1917	5
Fujin Xiang 1st Bao	231	4	4	The household head	马	福喜	Male	Tiger	1901	3
Fujin Xiang 1st Bao				The household head's wife	成	董娃	Female	Monkey	1907	2
Fujin Xiang 1st Bao				Daughter	马	转娃	Female	Tiger	1926	5
Fujin Xiang 1st Bao				The second daughter	马	香儿	Female	Dragon	1940	4
Fujin Xiang 1st Bao				Family member	贺	福娃	Male	Dog	1909	3
Fujin Xiang 1st Bao	232	4	5	The household head	马	济安	Male	Rabbit	1915	3
Fujin Xiang 1st Bao				The household head's wife	石	顺娃	Female	Snake	1917	12
Fujin Xiang 1st Bao				Mother	许	仓娃	Female	Ox	1888	2
Fujin Xiang 1st Bao				Son	马	福成	Male	Rooster	1945	10
Fujin Xiang 1st Bao				Daughter	马	福女	Female	Goat	1943	2
Fujin Xiang 1st Bao				Younger brother	马	娘招	Male	Rat	1924	8

Date	The beginning time of inhabitation in this county	Educational level	Profession sector	Occupation	Marital status	Physical disability	Religion	
27	Inhabited for generations		Agriculture		Married			
10	Inhabited for generations	Illiteracy	Agriculture	Land-holding peasant	Married	No	No religion	
18	Inhabited for generations	Bao school 1 year						
18	Inhabited for generations	Illiteracy	Agriculture	Land-holding peasant	Married	No	No religion	
19	Inhabited for generations	Illiteracy	Agriculture	Land-holding peasant	Married	No	No religion	
12	Inhabited for generations	Illiteracy	Agriculture	Land-holding peasant	Married	No	No religion	
17	Inhabited for generations	Illiteracy	Agriculture	Land-holding peasant	Married	No	No religion	
4	Inhabited for generations	Illiteracy						
20	Inhabited for generations	Illiteracy	Agriculture	Land-holding peasant	Married	No	No religion	Matrilocal husband of the household head's first daughter
1	Inhabited for generations	Illiteracy	Agriculture	Land-holding peasant	Married	No	No religion	
21	Inhabited for generations	Illiteracy	Agriculture	Land-holding peasant	Married	No	No religion	
20	Inhabited for generations	Illiteracy	Agriculture	Land-holding peasant	Widowed	No	No religion	
31	Inhabited for generations							
2	Inhabited for generations							
15	Inhabited for generations	Illiteracy	Agriculture	Land-holding peasant	Married	No	No religion	

Appendix: Overview of Microdata for Wudu County (*cont.*)

	The no. of household	Jia	Household	Relationship with the household head	Surname	First name	Sex	The Chinese zodiac	Birthdate Year	Month
Fujin Xiang 1st Bao				Younger brother's wife	谈	崴润	Female	Ox	1925	3
Fujin Xiang 1st Bao	233	4	6	The household head	董	治全	Male	Goat	1906	3
Fujin Xiang 1st Bao				The household head's wife	张	福连	Female	Monkey	1907	5
Fujin Xiang 1st Bao				Father	董	仲生	Male	Snake	1880	5
Fujin Xiang 1st Bao				Daughter	董	三娃	Female	Snake	1929	12
Fujin Xiang 1st Bao				The second daughter	董	桂英	Female	Dog	1934	1
Fujin Xiang 1st Bao				The third daughter	董	玉英	Female	Pig	1935	10
Fujin Xiang 1st Bao				The fourth daughter	董	四英	Female	Dragon	1940	2
Fujin Xiang 1st Bao				The fifth daughter	董	五英	Female	Goat	1943	2
Fujin Xiang 1st Bao				Family member	赵	双娃	Male	Rabbit	1927	5
Fujin Xiang 1st Bao	234	4	7	The household head	马	济富	Male	Rabbit	1927	3
Fujin Xiang 1st Bao				The household head's wife	赵	崴菊	Female	Dragon	1928	10
Fujin Xiang 1st Bao				Father	马	福荣	Male	Ox	1900	12
Fujin Xiang 1st Bao				Mother	崔	柱娃	Female	Goat	1906	3
Fujin Xiang 1st Bao				Younger brother	马	次富	Male	Horse	1945	1
Fujin Xiang 1st Bao				Younger sister	马	反女	Female	Rat	1936	8
Fujin Xiang 1st Bao				Nephew	马	世华	Male	Horse	1918	1

Date	The beginning time of inhabitation in this county	Educational level	Profession sector	Occupation	Marital status	Physical disability	Religion
30	Inhabited for generations	Illiteracy	Agriculture	Land-holding peasant	Married	No	No religion
28	Inhabited for generations	Illiteracy	Agriculture	Land-holding peasant	Married	No	No religion
10	Inhabited for generations	Illiteracy	Agriculture	Land-holding peasant	Married	No	No religion
11	Inhabited for generations						
13	Inhabited for generations	Illiteracy	Agriculture	Land-holding peasant	Married	No	No religion
10	Inhabited for generations	Illiteracy					
27	Inhabited for generations						
15	Inhabited for generations						
17	Inhabited for generations						
1	Inhabited for generations	Illiteracy	Agriculture	Land-holding peasant	Married	No	No religion
5	Inhabited for generations	Illiteracy	Agriculture	Land-holding peasant	Married	No	No religion
2	Inhabited for generations	Illiteracy	Agriculture	Land-holding peasant	Married	No	No religion
1	Inhabited for generations	Illiteracy	Agriculture	Land-holding peasant	Married	No	No religion
31	Inhabited for generations	Illiteracy	Agriculture	Land-holding peasant	Married	No	No religion
3	Inhabited for generations						
16	Inhabited for generations						
26	Inhabited for generations	Illiteracy	Agriculture	Land-holding peasant	Married	No	No religion

Appendix: Overview of Microdata for Wudu County (*cont.*)

									Birthdate	
	The no. of house-hold	Jia	House-hold	Relation-ship with the house-hold head	Surname	First name	Sex	The Chinese zodiac	Year	Month
Fujin Xiang 1st Bao				Nephew's wife	刘	赛女	Female	Dragon	1916	7
Fujin Xiang 1st Bao				Nephew's son	马	成富	Male	Pig	1947	5
Fujin Xiang 1st Bao	235	4	8	The house-hold head	马	福隆	Male	Tiger	1914	2
Fujin Xiang 1st Bao				The house-hold head's wife	陈	柱娃	Female	Dog	1913	9
Fujin Xiang 1st Bao				Mother	陈	亥儿	Female	Rooster	1884	4
Fujin Xiang 1st Bao				Son	马	顺緒	Male	Snake	1945	6
Fujin Xiang 1st Bao				Younger brother	马	福万	Male	Tiger	1926	1
Fujin Xiang 1st Bao				Younger brother's wife	赵	良娃	Female	Monkey	1920	5
Fujin Xiang 1st Bao				Cousin	马	福成	Male	Goat	1894	7
Fujin Xiang 1st Bao				Cousin's wife	谈	胖儿	Female	Dragon	1891	5
Fujin Xiang 1st Bao	236	4	9	The house-hold head	马	平娃	Male	Pig	1910	4
Fujin Xiang 1st Bao				The house-hold head's wife	徐	其娃	Female	Pig	1910	5
Fujin Xiang 1st Bao				Father	马	大清	Male	Dragon	1891	11
Fujin Xiang 1st Bao				Younger brother	马	治奎	Male	Monkey	1920	6
Fujin Xiang 1st Bao				Younger brother's wife	成	桃林	Female	Horse	1918	8
Fujin Xiang 1st Bao	237	4	10	The house-hold head	马	进财	Male	Ox	1937	6
Fujin Xiang 1st Bao				Mother	高	香娃	Female	Tiger	1914	5

Date	The beginning time of inhabitation in this county	Educational level	Profession sector	Occupation	Marital status	Physical disability	Religion
30	Inhabited for generations	Illiteracy	Agriculture	Land-holding peasant	Married	No	No religion
10	Inhabited for generations						
7	Inhabited for generations	Illiteracy	Agriculture	Land-holding peasant	Married	No	No religion
27	Inhabited for generations	Illiteracy	Agriculture	Land-holding peasant	Married	No	No religion
27	Inhabited for generations				Widowed		
11	Inhabited for generations	Illiteracy	Agriculture				
26	Inhabited for generations	Illiteracy	Agriculture	Land-holding peasant	Married	No	No religion
5	Inhabited for generations	Illiteracy	Agriculture	Land-holding peasant	Married	No	No religion
1	Inhabited for generations				Married		
1	Inhabited for generations				Married		
20	Inhabited for generations	Illiteracy	Agriculture	Land-holding peasant	Married	No	No religion
9	Inhabited for generations	Illiteracy	Others	Weaving	Married	No	No religion
9	Inhabited for generations	Illiteracy	Agriculture	Land-holding peasant	Widowed	No	No religion
1	Inhabited for generations	Illiteracy	Agriculture	Land-holding peasant	Married	No	No religion
1	Inhabited for generations	Illiteracy	Others	Weaving	Married	No	No religion
16	Inhabited for generations	Illiteracy	Agriculture	Land-holding peasant	Unmarried	No	No religion
21	Inhabited for generations	Illiteracy	Agriculture	Land-holding peasant	Widowed	No	No religion

Appendix: Overview of Microdata for Wudu County (*cont.*)

									Birthdate	
	The no. of household	Jia	Household	Relationship with the household head	Surname	First name	Sex	The Chinese zodiac	Year	Month
Fujin Xiang 1st Bao				Younger brother	马	朱凤	Male	Horse	1942	6
Fujin Xiang 1st Bao				Younger sister	马	进女	Female	Snake	1941	4
Fujin Xiang 1st Bao				The second younger sister	马	宗女	Female	Dog	1946	8
Fujin Xiang 1st Bao	238	4	11	The household head	马	济清	Male	Rooster	1908	10
Fujin Xiang 1st Bao				The household head's wife	赵	东连	Female	Rabbit	1902	11
Fujin Xiang 1st Bao				Father	马	福有	Male	Rabbit	1878	9
Fujin Xiang 1st Bao				Mother	赵	时	Female	Horse	1881	3
Fujin Xiang 1st Bao				Son	马	治田	Male	Monkey	1932	9
Fujin Xiang 1st Bao				Daughter	马	明香	Female	Pig	1935	4
Fujin Xiang 1st Bao				The second daughter	马	柱明	Female	Goat	1943	5
Fujin Xiang 1st Bao				The third daughter	马	进香	Female	Dog	1946	6
Fujin Xiang 1st Bao				Cohabitant	马	治全	Male	Pig	1913	6
Fujin Xiang 1st Bao				Cohabitant's wife	陈	苏娘	Female	Rat	1924	5
Fujin Xiang 1st Bao				Cohabitant	马	谢德	Male	Pig	1886	5
Fujin Xiang 1st Bao	239	4	12	The household head	段	相印	Male	Dragon	1891	2
Fujin Xiang 1st Bao				The household head's wife	杨	林女	Female	Snake	1904	4
Fujin Xiang 1st Bao				Son	陈	月园	Male	Dog	1922	2

Date	The beginning time of inhabitation in this county	Educational level	Profession sector	Occupation	Marital status	Physical disability	Religion
1	Inhabited for generations						
1	Inhabited for generations						
1	Inhabited for generations						
12	Inhabited for generations	Illiteracy	Agriculture	Land-holding peasant	Married	No	No religion
7	Inhabited for generations	Illiteracy	Agriculture	Land-holding peasant	Married	No	No religion
29	Inhabited for generations				Married		
3	Inhabited for generations				Married		
29	Inhabited for generations	Primary school 5 years			Unmarried		
3	Inhabited for generations	Illiteracy					
8	Inhabited for generations						
1	Inhabited for generations						
1	Inhabited for generations	Illiteracy	Agriculture	Land-holding peasant	Married	No	No religion
1	Inhabited for generations	Illiteracy	Agriculture	Land-holding peasant	Married	No	No religion
1	Inhabited for generations	Illiteracy	Agriculture	Land-holding peasant	Widowed	No	No religion
6	Inhabited for generations	Illiteracy	Agriculture	Land-holding peasant	Married	No	No religion
27	Inhabited for generations	Illiteracy	Agriculture	Weaving	Married	No	No religion
1	Inhabited for generations	Illiteracy	Agriculture	Land-holding peasant	Married	No	No religion

Appendix: Overview of Microdata for Wudu County (*cont.*)

	The no. of household	Jia	Household	Relationship with the household head	Surname	First name	Sex	The Chinese zodiac	Birthdate Year	Month
Fujin Xiang 1st Bao				Son's wife	陈	六女	Female	Horse	1930	3
Fujin Xiang 1st Bao	240	4	13	The household head	段	芝理	Male	Rooster	1908	6
Fujin Xiang 1st Bao				The household head's wife	王	金连	Female	Dragon	1903	4
Fujin Xiang 1st Bao				Son	段	明贵	Male	Rooster	1933	10
Fujin Xiang 1st Bao				Daughter	段	称女	Female	Dragon	1940	8
Fujin Xiang 1st Bao				Cohabitant	王	五儿子	Male	Dog	1925	7
Fujin Xiang 1st Bao	241	4	14	The household head	马	济奎	Male	Ox	1925	12
Fujin Xiang 1st Bao				Mother	张	嘉女	Female	Ox	1887	10
Fujin Xiang 1st Bao				Younger brother	马	四生	Male	Rat	1936	2
Fujin Xiang 1st Bao				Uncle	张	富清	Male	Horse	1893	6
Fujin Xiang 1st Bao				Cousin	张	全钰	Male	Horse	1918	7
Fujin Xiang 1st Bao				Cousin's wife	石	崴秀	Female	Dragon	1916	11
Fujin Xiang 1st Bao				Nephew	张	富保	Male	Horse	1942	5
Fujin Xiang 1st Bao				Daughter	马	香娃	Female	Rabbit	1939	10
Fujin Xiang 1st Bao	242	4	15	The household head	汪	培元	Male	Monkey	1920	11
Fujin Xiang 1st Bao				The household head's wife	杨	崴香	Female	Ox	1925	1
Fujin Xiang 1st Bao				Father	汪	金城	Male	Horse	1893	11

Date	The beginning time of inhabitation in this county	Educational level	Profession sector	Occupation	Marital status	Physical disability	Religion
13	Inhabited for generations	Illiteracy	Agriculture	Weaving	Married	No	No religion
1	Inhabited for generations	Illiteracy	Agriculture	Land-holding peasant	Married	No	No religion
4	Inhabited for generations	Illiteracy	Agriculture	Land-holding peasant	Married	No	No religion
15	Inhabited for generations	Illiteracy					
11	Inhabited for generations	Illiteracy					
1	Inhabited for generations	Illiteracy	Agriculture	Land-holding peasant	Unmarried	No	No religion
11	Inhabited for generations	Illiteracy	Agriculture	Land-holding peasant	Widowed	No	No religion
28	Inhabited for generations	Illiteracy	Others	Weaving	Widowed	No	No religion
8	Inhabited for generations	Illiteracy	Agriculture				
16	Inhabited for generations	Illiteracy	Agriculture	Land-holding peasant	Widowed	No	No religion
6	Inhabited for generations	Illiteracy	Agriculture	Land-holding peasant	Married	No	No religion
7	Inhabited for generations	Illiteracy	Others	Weaving	Married	No	No religion
5	Inhabited for generations						
11	Inhabited for generations						
29	Inhabited for generations	Lower primary school graduated	Agriculture	Land-holding peasant	Married	No	No religion
2	Inhabited for generations	Illiteracy	Others	Weaving	Married	No	No religion
8	Inhabited for generations	Sishu 5 years	Agriculture	Land-holding peasant	Married	No	No religion

Appendix: Overview of Microdata for Wudu County (*cont.*)

	The no. of household	Jia	Household	Relationship with the household head	Surname	First name	Sex	The Chinese zodiac	Birthdate	
									Year	Month
Fujin Xiang 1st Bao				Mother	马	喜娃	Female	Tiger	1901	4
Fujin Xiang 1st Bao	243	5	1	The household head	段	福田	Male	Goat	1882	10
Fujin Xiang 1st Bao				The household head's wife	赵	菜女	Female	Snake	1880	8
Fujin Xiang 1st Bao				Son	段	芝明	Male	Snake	1917	10
Fujin Xiang 1st Bao				Son's wife	换	娃	Female	Snake	1917	9
Fujin Xiang 1st Bao				Grandson	段	郭郭	Male	Tiger	1938	10
Fujin Xiang 1st Bao				Concubine	李	跳娃	Female	Dragon	1891	11
Fujin Xiang 1st Bao				Daughter	段	将娃	Female	Goat	1931	12
Fujin Xiang 1st Bao				Hired labourer	福	忠	Male	Horse	1893	10
Fujin Xiang 1st Bao	244	5	2	The household head	马	学德	Male	Tiger	1901	1
Fujin Xiang 1st Bao				The household head's wife	汪	秋娃	Female	Pig	1910	10
Fujin Xiang 1st Bao				Son	马	送成	Male	Dog	1946	12
Fujin Xiang 1st Bao				Daughter	马	玉香	Female	Goat	1934	2
Fujin Xiang 1st Bao	245	5	3	The household head	马	犁牛	Male	Horse	1893	8
Fujin Xiang 1st Bao				The household head's wife	马	董秀	Female	Horse	1893	10
Fujin Xiang 1st Bao				Son	马	具娃	Male	Rabbit	1927	4
Fujin Xiang 1st Bao				Son's wife	赵	注其娃	Female	Snake	1929	7

Date	The beginning time of inhabitation in this county	Educational level	Profession sector	Occupation	Marital status	Physical disability	Religion
15	Inhabited for generations	Illiteracy	Others	Weaving	Married	No	No religion
13	Inhabited for generations	Illiteracy	Agriculture	Land-holding peasant	Married	No	No religion
13	Inhabited for generations	Illiteracy	Others	Weaving	Married	No	No religion
7	Inhabited for generations	Illiteracy	Agriculture	Land-holding peasant	Married	No	No religion
1	Inhabited for generations	Illiteracy	Others	Weaving	Married	No	No religion
24	Inhabited for generations	Bao school 1 year					
1	Inhabited for generations	Illiteracy	Others	Weaving	Married	No	No religion
21	Inhabited for generations	Illiteracy	Others	Weaving	Unmarried	No	No religion
16	Inhabited for generations	Illiteracy	Agriculture	Tenant farmer	Widowed	No	No religion
11	Inhabited for generations	Illiteracy	Agriculture	Land-holding peasant	Married	Blind	No religion
1	Inhabited for generations	Illiteracy	Others	Weaving	Married	No	No religion
23	Inhabited for generations						
21	Inhabited for generations	Illiteracy			Unmarried		
30	Inhabited for generations	Illiteracy	Agriculture	Land-holding peasant	Married	No	No religion
18	Inhabited for generations	Illiteracy	Others	Weaving	Married	No	No religion
12	Inhabited for generations	Illiteracy	Agriculture	Land-holding peasant	Married	No	No religion
4	Inhabited for generations	Illiteracy	Others	Weaving	Married	No	No religion

Appendix: Overview of Microdata for Wudu County (*cont.*)

	The no. of household	Jia	Household	Relationship with the household head	Surname	First name	Sex	The Chinese zodiac	Birthdate Year	Birthdate Month
Fujin Xiang 1st Bao				The second son	马	具生	Male	Rooster	1933	9
Fujin Xiang 1st Bao	246	5	4	The household head	余	子清	Male	Rat	1887	11
Fujin Xiang 1st Bao				The household head's wife	李	连秀	Female	Rooster	1896	4
Fujin Xiang 1st Bao	247	5	5	The household head	马	玉明	Male	Monkey	1907	3
Fujin Xiang 1st Bao				The household head's wife	殷	翠环	Female	Rabbit	1902	9
Fujin Xiang 1st Bao				Son	马	济发	Male	Rat	1924	4
Fujin Xiang 1st Bao				Son's wife	张	润娃	Female	Rabbit	1927	3
Fujin Xiang 1st Bao				The second son	马	平平	Male	Horse	1942	1
Fujin Xiang 1st Bao				The third son	马	崴平	Male	Dog	1946	12
Fujin Xiang 1st Bao				Daughter	马	菊香	Female	Goat	1931	11
Fujin Xiang 1st Bao				Older brother's wife	董	双秀	Female	Monkey	1907	12
Fujin Xiang 1st Bao				Nephew	马	聚财	Male	Rooster	1933	10
Fujin Xiang 1st Bao				The second nephew	马	崴财	Male	Rooster	1945	5
Fujin Xiang 1st Bao	248	5	6	The household head	马	玉山	Male	Pig	1910	12
Fujin Xiang 1st Bao				The household head's wife	李	登林	Female	Ox	1913	1
Fujin Xiang 1st Bao				Son	马	四居	Male	Rat	1936	8

Date	The beginning time of inhabitation in this county	Educational level	Profession sector	Occupation	Marital status	Physical disability	Religion
9	Inhabited for generations	Bao school 3 years			Widowed	No	No religion
23	Inhabited for generations	Illiteracy	Agriculture	Land-holding peasant	Married	No	No religion
15	Inhabited for generations	Illiteracy	Others	Weaving	Married	No	No religion
16	Inhabited for generations	Illiteracy	Agriculture	Land-holding peasant	Married	No	No religion
4	Inhabited for generations	Illiteracy	Agriculture	Land-holding peasant	Married	No	No religion
25	Inhabited for generations	Lower primary school 3 years	Agriculture	Land-holding peasant	Married	No	No religion
1	Inhabited for generations	Illiteracy	Agriculture	Land-holding peasant	Married	No	No religion
1	Inhabited for generations						
18	Inhabited for generations						
16	Inhabited for generations	Illiteracy	Agriculture		Unmarried		
13	Inhabited for generations	Illiteracy	Agriculture	Land-holding peasant	Widowed	No	No religion
13	Inhabited for generations	Illiteracy	Agriculture	Land-holding peasant	Unmarried	No	No religion
23	Inhabited for generations						
31	Inhabited for generations	Illiteracy	Agriculture	Land-holding peasant	Married	No	No religion
7	Inhabited for generations	Illiteracy	Others	Weaving	Married	No	No religion
28	Inhabited for generations	Bao school 3 years					

Appendix: Overview of Microdata for Wudu County (*cont.*)

	The no. of household	Jia	Household	Relationship with the household head	Surname	First name	Sex	The Chinese zodiac	Birthdate Year	Month
Fujin Xiang 1st Bao	249	5	7	The household head	马	福生	Male	Dragon	1903	5
Fujin Xiang 1st Bao				The household head's wife	董	秀儿	Female	Pig	1898	11
Fujin Xiang 1st Bao				Son	马	三贵	Male	Pig	1923	9
Fujin Xiang 1st Bao				Son's wife	李	双玄	Female	Ox	1925	8
Fujin Xiang 1st Bao				Grandson	马	吴家	Male	Dog	1946	12
Fujin Xiang 1st Bao				Granddaughter	马	佐娃	Female	Monkey	1944	8
Fujin Xiang 1st Bao	250	5	8	The household head	马	福祥	Male	Horse	1905	12
Fujin Xiang 1st Bao				The household head's wife	赵	羊姐	Female	Goat	1906	9
Fujin Xiang 1st Bao				Son	马	曾娃	Male	Rat	1923	1
Fujin Xiang 1st Bao				Son's wife	赵	扣儿	Female	Horse	1930	10
Fujin Xiang 1st Bao				The second son	马	曾娃	Male	Snake	1941	9
Fujin Xiang 1st Bao				Daughter	马	润月	Female	Ox	1937	5
Fujin Xiang 1st Bao	251	5	9	The household head	孙	得福	Male	Dragon	1903	3
Fujin Xiang 1st Bao				The household head's wife	郭	菊香	Female	Tiger	1914	11
Fujin Xiang 1st Bao				Son	孙	桂林	Male	Tiger	1938	1
Fujin Xiang 1st Bao				Daughter	孙	顺益	Female	Horse	1942	8
Fujin Xiang 1st Bao	252	5	10	The household head	马	治义	Male	Tiger	1901	10

Date	The beginning time of inhabitation in this county	Educational level	Profession sector	Occupation	Marital status	Physical disability	Religion
10	Inhabited for generations	Illiteracy	Agriculture	Land-holding peasant	Married	No	No religion
1	Inhabited for generations	Illiteracy	Agriculture	Land-holding peasant	Married	No	No religion
17	Inhabited for generations	Illiteracy	Agriculture	Land-holding peasant	Married	No	No religion
1	Inhabited for generations	Illiteracy	Agriculture	Land-holding peasant	Married	No	No religion
8	Inhabited for generations						
25	Inhabited for generations						
12	Inhabited for generations	Illiteracy	Agriculture	Land-holding peasant	Married	No	No religion
29	Inhabited for generations	Illiteracy	Others	Weaving	Married	No	No religion
19	Inhabited for generations	Illiteracy	Agriculture	Land-holding peasant	Married	No	No religion
24	Inhabited for generations	Illiteracy	Others	Weaving	Married	No	No religion
6	Inhabited for generations						
25	Inhabited for generations						
28	Inhabited for generations	Illiteracy	Agriculture	Land-holding peasant	Married	No	No religion
19	Inhabited for generations	Illiteracy	Others	Weaving	Married	No	No religion
23	Inhabited for generations	Bao school 2 years					
3	Inhabited for generations						
26	Inhabited for generations	Illiteracy	Agriculture	Land-holding peasant	Widowed	No	No religion

Appendix: Overview of Microdata for Wudu County (*cont.*)

	The no. of household	Jia	Household	Relationship with the household head	Surname	First name	Sex	The Chinese zodiac	Birthdate	
									Year	Month
Fujin Xiang 1st Bao				Nephew	马	济川	Male	Rabbit	1902	7
Fujin Xiang 1st Bao				Nephew's wife	王	郭儿	Female	Tiger	1906	8
Fujin Xiang 1st Bao				Nephew's son	马	治明	Male	Rabbit	1927	11
Fujin Xiang 1st Bao				Grandson's wife	王	舍娃	Female	Rabbit	1927	9
Fujin Xiang 1st Bao				The second grandson	马	治国	Male	Rooster	1933	5
Fujin Xiang 1st Bao	253	5	11	The household head	马	福登	Male	Goat	1919	9
Fujin Xiang 1st Bao				The household head's wife	李	月儿	Female	Horse	1918	7
Fujin Xiang 1st Bao				Father	马	世德	Male	Dog	1885	5
Fujin Xiang 1st Bao				Mother	高	妇女	Female	Dragon	1891	10
Fujin Xiang 1st Bao				Son	董	李成	Male	Rooster	1945	1
Fujin Xiang 1st Bao				Older brother	马	师娃	Male	Rat	1912	3
Fujin Xiang 1st Bao				Older brother's wife	郭	射娃	Female	Rat	1912	3
Fujin Xiang 1st Bao				Nephew	马	刘柏	Male	Dog	1946	8
Fujin Xiang 1st Bao				Younger brother	马	四娃	Male	Rooster	1921	5
Fujin Xiang 1st Bao				The third younger brother	马	福科	Male	Snake	1929	6
Fujin Xiang 1st Bao				Younger brother's wife	崔	腰女	Female	Rooster	1917	6
Fujin Xiang 1st Bao				Niece	马	玉英	Female	Goat	1945	2

Date	The beginning time of inhabitation in this county	Educational level	Profession sector	Occupation	Marital status	Physical disability	Religion	
31	Inhabited for generations	Sishu 4 years	Agriculture	Land-holding peasant	Married	No	No religion	
19	Inhabited for generations	Illiteracy	Agriculture	Land-holding peasant	Married	No	No religion	
11	Inhabited for generations	Illiteracy	Agriculture	Land-holding peasant	Married	No	No religion	
1	Inhabited for generations	Illiteracy	Agriculture	Land-holding peasant	Married	No	No religion	
31	Inhabited for generations	Bao school 3 years	Agriculture	Land-holding peasant	Unmarried	No	No religion	
24	Inhabited for generations	Illiteracy	Agriculture	Land-holding peasant	Married	No	No religion	
18	Inhabited for generations	Illiteracy	Others	Weaving	Married	No	No religion	
15	Inhabited for generations	Illiteracy	Agriculture	Land-holding peasant	Married	No	No religion	
1	Inhabited for generations	Illiteracy	Others	Weaving	Married	No	No religion	
10	Inhabited for generations							
26	Inhabited for generations	Illiteracy	Agriculture	Land-holding peasant	Married	No	No religion	
2	Inhabited for generations	Illiteracy	Others	Weaving	Married	No	No religion	
29	Inhabited for generations							
18	Inhabited for generations					Blind		
1	Inhabited for generations	Illiteracy	Agriculture	Land-holding peasant	Unmarried	No	No religion	Joined the army
6	Inhabited for generations	Illiteracy	Agriculture	Land-holding peasant	Married	No	No religion	
8	Inhabited for generations							

Appendix: Overview of Microdata for Wudu County (*cont.*)

	The no. of household	Jia	Household	Relationship with the household head	Surname	First name	Sex	The Chinese zodiac	Birthdate Year	Month
Fujin Xiang 1st Bao	254	5	12	The household head	马	福明	Male	Ox	1913	4
Fujin Xiang 1st Bao				Younger sister	马	富儿	Female	Rooster	1933	7
Fujin Xiang 1st Bao	255	5	13	The household head	段	福起	Male	Horse	1875	11
Fujin Xiang 1st Bao				The household head's wife	张	红花	Female	Dragon	1879	11
Fujin Xiang 1st Bao				Son's wife	马	坡娃	Female	Snake	1917	5
Fujin Xiang 1st Bao				Grandson	段	西虎	Male	Tiger	1938	9
Fujin Xiang 1st Bao				Granddaughter	段	早女儿	Female	Dog	1934	2
Fujin Xiang 1st Bao				Nephew's son	段	寡儿	Male	Horse	1918	4
Fujin Xiang 1st Bao	256	5	14	The household head	段	福有	Male	Rat	1875	8
Fujin Xiang 1st Bao				The household head's wife	岳	成姐	Female	Goat	1882	4
Fujin Xiang 1st Bao				Son	谢	生娃	Male	Horse	1918	10
Fujin Xiang 1st Bao	257	5	15	The household head	马	福寿	Male	Rooster	1908	3
Fujin Xiang 1st Bao				The household head's wife	陈	招妇儿	Female	Pig	1912	6
Fujin Xiang 1st Bao				Son	马	齐桂	Male	Tiger	1938	12
Fujin Xiang 1st Bao				The second son	马	登桂	Male	Horse	1942	2
Fujin Xiang 1st Bao				Daughter	马	桂女	Female	Rooster	1945	2
Fujin Xiang 1st Bao	258	6	1	The household head	汪	金有	Male	Rabbit	1915	1

Date	The beginning time of inhabitation in this county	Educational level	Profession sector	Occupation	Marital status	Physical disability	Religion
6	Inhabited for generations	Illiteracy	Agriculture	Land-holding peasant	Widowed	No	No religion
2	Inhabited for generations	Illiteracy					
8	Inhabited for generations	Illiteracy	Agriculture	Land-holding peasant	Married	No	No religion
18	Inhabited for generations	Illiteracy	Others	Weaving	Married	No	No religion
22	Inhabited for generations	Illiteracy	Others	Weaving	Widowed	No	No religion
29	Inhabited for generations	Bao school 1 year					
13	Inhabited for generations	Illiteracy					
1	Inhabited for generations	Illiteracy	Agriculture	Land-holding peasant	Unmarried	No	No religion
15	Inhabited for generations	Illiteracy	Agriculture	Land-holding peasant	Married	Left eye disease	No religion
20	Inhabited for generations	Illiteracy	Agriculture	Land-holding peasant	Married	No	No religion
9	Inhabited for generations	Illiteracy	Agriculture	Land-holding peasant	Unmarried	No	No religion
3	Inhabited for generations	Illiteracy	Agriculture	Land-holding peasant	Married	Right eye injuried	No religion
21	Inhabited for generations	Illiteracy	Agriculture	Land-holding peasant	Married	No	No religion
23	Inhabited for generations	Illiteracy	Agriculture	Land-holding peasant	Unmarried	No	No religion
24	Inhabited for generations	Illiteracy	Agriculture	Land-holding peasant	Unmarried	No	No religion
25	Inhabited for generations						
10	Inhabited for generations	Illiteracy	Agriculture	Land-holding peasant	Married	No	No religion

Appendix: Overview of Microdata for Wudu County (*cont.*)

	The no. of household	Jia	Household	Relationship with the household head	Surname	First name	Sex	The Chinese zodiac	Birthdate	
									Year	Month
Fujin Xiang 1st Bao				The household head's wife	雷	益娃	Female	Ox	1913	5
Fujin Xiang 1st Bao				Father	汪	治叶	Male	Dragon	1891	3
Fujin Xiang 1st Bao				Mother	冯	儿子	Female	Goat	1894	4
Fujin Xiang 1st Bao				Son	汪	忠信	Male	Pig	1935	10
Fujin Xiang 1st Bao				The second son	汪	崴忠	Male	Tiger	1938	12
Fujin Xiang 1st Bao				Younger brother	汪	文科	Male	Horse	1918	8
Fujin Xiang 1st Bao				Younger brother's wife	张	三路	Female	Snake	1917	5
Fujin Xiang 1st Bao				Nephew	汪	富德	Male	Horse	1942	7
Fujin Xiang 1st Bao				Niece	汪	富香	Female	Dog	1946	11
Fujin Xiang 1st Bao				The second niece	汪	桂香	Female	Dog	1946	1
Fujin Xiang 1st Bao				Daughter	汪	成香	Female	Pig	1947	5
Fujin Xiang 1st Bao	259	6	2	The household head	段	福义	Male	Dragon	1891	12
Fujin Xiang 1st Bao				The household head's wife	杨	卑儿	Female	Rabbit	1902	8
Fujin Xiang 1st Bao				Daughter	段	四女子	Female	Rat	1924	5
Fujin Xiang 1st Bao	260	6	3	The household head	马	福贵	Male	Horse	1893	1
Fujin Xiang 1st Bao				Daughter	马	成明儿	Female	Rooster	1933	4
Fujin Xiang 1st Bao				Younger brother	马	福林	Male	Monkey	1907	6

Date	The beginning time of inhabitation in this county	Educational level	Profession sector	Occupation	Marital status	Physical disability	Religion
26	Inhabited for generations	Illiteracy	Agriculture	Land-holding peasant	Married	No	No religion
27	Inhabited for generations	Illiteracy	Agriculture	Land-holding peasant	Married	No	No religion
24	Inhabited for generations	Illiteracy	Agriculture	Land-holding peasant	Married	No	No religion
6	Inhabited for generations	Bao school 4 years				Right leg injured	
3	Inhabited for generations						
31	Inhabited for generations	Illiteracy	Agriculture	Land-holding peasant	Married	No	No religion
25	Inhabited for generations	Illiteracy	Agriculture	Land-holding peasant	Married	No	No religion
13	Inhabited for generations						
5	Inhabited for generations						
31	Inhabited for generations						
19	Inhabited for generations						
2	Inhabited for generations	Illiteracy	Agriculture	Land-holding peasant	Married	No	No religion
7	Inhabited for generations	Illiteracy	Agriculture	Land-holding peasant	Married	No	No religion
1	Inhabited for generations	Illiteracy	Agriculture	Land-holding peasant	Widowed	No	No religion
22	Inhabited for generations	Illiteracy	Agriculture	Land-holding peasant	Widowed	No	No religion
5	Inhabited for generations	Illiteracy	Others	Weaving	Unmarried	No	No religion
10	Inhabited for generations	Illiteracy	Agriculture	Land-holding peasant	Widowed	No	No religion

Appendix: Overview of Microdata for Wudu County (*cont.*)

	The no. of household	Jia	Household	Relationship with the household head	Surname	First name	Sex	The Chinese zodiac	Birthdate Year	Month
Fujin Xiang 1st Bao	261	6	4	The household head	贾	得全	Male	Horse	1918	1
Fujin Xiang 1st Bao				The household head's wife	李	谢女	Female	Rat	1924	4
Fujin Xiang 1st Bao				Son	贾	信娃	Male	Horse	1942	1
Fujin Xiang 1st Bao	262	6	5	The household head	雷	治中	Male	Ox	1913	5
Fujin Xiang 1st Bao				The household head's wife	金	娃	Female	Snake	1917	6
Fujin Xiang 1st Bao				Daughter	雷	王女	Female	Horse	1942	7
Fujin Xiang 1st Bao				Son	雷	四才	Male	Pig	1947	5
Fujin Xiang 1st Bao	263	6	6	The household head	董	治仓	Male	Goat	1894	10
Fujin Xiang 1st Bao				The household head's wife	杨	毕娃	Female	Dragon	1903	8
Fujin Xiang 1st Bao				Son	董	其斌	Male	Ox	1913	7
Fujin Xiang 1st Bao				Son's wife	赵	成娃	Female	Ox	1925	4
Fujin Xiang 1st Bao	264	6	7	The household head	马	进才	Male	Ox	1913	2
Fujin Xiang 1st Bao				The household head's wife	贾	跳娃	Female	Rabbit	1915	6
Fujin Xiang 1st Bao				Son	马	东成	Male	Rooster	1945	6
Fujin Xiang 1st Bao	265	6	8	The household head	董	稀器娃	Female	Rooster	1921	1
Fujin Xiang 1st Bao				Son	董	松柏	Male	Snake	1941	1
Fujin Xiang 1st Bao				The second son	董	长青	Male	Rooster	1945	1

Date	The beginning time of inhabitation in this county	Educational level	Profession sector	Occupation	Marital status	Physical disability	Religion
24	Inhabited for generations	Illiteracy	Agriculture	Land-holding peasant	Married	No	No religion
30	Inhabited for generations	Illiteracy	Others	Land-holding peasant	Married	No	No religion
17	Inhabited for generations	Illiteracy					
29	Inhabited for generations	Illiteracy	Agriculture	Land-holding peasant	Married	No	No religion
10	Inhabited for generations	Illiteracy	Agriculture	Land-holding peasant	Married	No	No religion
27	Inhabited for generations						
11	Inhabited for generations						
29	Inhabited for generations	Illiteracy	Agriculture	Land-holding peasant	Married	No	No religion
4	Inhabited for generations	Illiteracy	Others	Land-holding peasant	Married	No	No religion
29	Inhabited for generations	Illiteracy	Agriculture	Land-holding peasant	Married	No	No religion
29	Inhabited for generations	Illiteracy	Others	Land-holding peasant	Married	No	No religion
12	Inhabited for generations	Illiteracy	Agriculture	Land-holding peasant	Married	No	No religion
2	Inhabited for generations	Illiteracy	Agriculture	Land-holding peasant	Married	No	No religion
13	Inhabited for generations						
16	Inhabited for generations	Illiteracy	Agriculture	Land-holding peasant	Married	No	No religion
3	Inhabited for generations						
8	Inhabited for generations						

Appendix: Overview of Microdata for Wudu County (*cont.*)

									Birthdate	
	The no. of household	Jia	Household	Relationship with the household head	Surname	First name	Sex	The Chinese zodiac	Year	Month
Fujin Xiang 1st Bao				Family member	陈	万海	Unknown	Rat	1912	1
Fujin Xiang 1st Bao	266	6	9	The household head	董	其法	Male	Monkey	1895	3
Fujin Xiang 1st Bao				The household head's wife	郑	银女	Female	Tiger	1901	3
Fujin Xiang 1st Bao				Son	董	润德	Male	Horse	1930	2
Fujin Xiang 1st Bao				Son's wife	赵	元英	Female	Snake	1929	3
Fujin Xiang 1st Bao				Daughter	董	根贵	Female	Dog	1934	9
Fujin Xiang 1st Bao				The second daughter	董	崴贵	Female	Tiger	1938	9
Fujin Xiang 1st Bao				Granddaughter	董	吕香	Female	Dog	1946	1
Fujin Xiang 1st Bao				Cohabitant	郑	炳雲	Male	Ox	1888	3
Fujin Xiang 1st Bao	267	6	10	The household head	董	治忠	Male	Pig	1886	2
Fujin Xiang 1st Bao				The household head's wife	马	翠儿	Female	Rooster	1884	3
Fujin Xiang 1st Bao				Son	董	其玉	Male	Tiger	1914	4
Fujin Xiang 1st Bao				Son's wife	杨	悟儿	Female	Rabbit	1915	1
Fujin Xiang 1st Bao	268	6	11	The household head	董	其富	Male	Tiger	1926	2
Fujin Xiang 1st Bao				The household head's wife	赵	翠娃	Female	Tiger	1926	1
Fujin Xiang 1st Bao				Father	董	治锡	Male	Rabbit	1902	11

Date	The beginning time of inhabitation in this county	Educational level	Profession sector	Occupation	Marital status	Physical disability	Religion	
1	Inhabited for generations	Illiteracy	Agriculture	Land-holding peasant	Married	No	No religion	
14	Inhabited for generations	Illiteracy	Agriculture	Land-holding peasant	Married	No	No religion	
28	Inhabited for generations	Illiteracy	Others	Land-holding peasant	Married	No	No religion	
16	Inhabited for generations	Primary school 5 years			Married	No	No religion	
13	Inhabited for generations	Illiteracy	Others	Land-holding peasant	Married	No	No religion	
29	Inhabited for generations	Illiteracy	Agriculture					
23	Inhabited for generations	Illiteracy	Agriculture					
4	Inhabited for generations							
30	Inhabited for generations	Illiteracy	Agriculture	Land-holding peasant	Widowed	No	No religion	
9	Inhabited for generations	Illiteracy	Agriculture	Land-holding peasant	Married	No	No religion	
3	Inhabited for generations	Illiteracy	Agriculture	Land-holding peasant	Married	No	No religion	
1	Inhabited for generations	Illiteracy	Agriculture	Land-holding peasant	Married	No	No religion	
9	Inhabited for generations	Illiteracy	Agriculture	Land-holding peasant	Married	No	No religion	
25	Inhabited for generations	Illiteracy	Agriculture	Land-holding peasant	Married	No	No religion	
1	Inhabited for generations	Illiteracy	Agriculture	Land-holding peasant	Married	No	No religion	
1	Inhabited for generations	Illiteracy	Agriculture	Land-holding peasant	Married	No	No religion	Died in August 1947

Appendix: Overview of Microdata for Wudu County (*cont.*)

	The no. of household	Jia	Household	Relationship with the household head	Surname	First name	Sex	The Chinese zodiac	Birthdate Year	Month
Fujin Xiang 1st Bao				Mother	李	岁秀	Female	Rabbit	1902	11
Fujin Xiang 1st Bao				Son	董	吉庆	Male	Dog	1947	3
Fujin Xiang 1st Bao				Younger brother	董	其贵	Male	Rooster	1933	4
Fujin Xiang 1st Bao	269	6	12	The household head	董	治花	Male	Monkey	1907	2
Fujin Xiang 1st Bao				The household head's wife	赵	女儿	Female	Ox	1909	5
Fujin Xiang 1st Bao				Father	董	仲义	Male	Rat	1875	9
Fujin Xiang 1st Bao				Mother	赵	芦儿	Female	Ox	1876	9
Fujin Xiang 1st Bao				Cousin	董	治甲	Male	Horse	1893	12
Fujin Xiang 1st Bao				Cousin's wife	谈	茶娃	Female	Snake	1904	10
Fujin Xiang 1st Bao				Nephew	董	和贵儿	Male	Monkey	1944	12
Fujin Xiang 1st Bao				Niece	董	和儿	Female	Dog	1934	7
Fujin Xiang 1st Bao	270	6	13	The household head	谈	学琴	Male	Rooster	1884	9
Fujin Xiang 1st Bao				The household head's wife	杨	性娃	Female	Goat	1881	2
Fujin Xiang 1st Bao				Daughter	谈	玉贵秀	Female	Dragon	1916	7
Fujin Xiang 1st Bao				The second daughter	谈	玉娃子	Female	Goat	1919	8
Fujin Xiang 1st Bao				Family member	董	治叶	Male	Rooster	1908	2
Fujin Xiang 1st Bao				Family member	赵	佛保儿	Male	Rat	1923	7
Fujin Xiang 1st Bao	271	6	14	The household head	麻	玉贵	Male	Rooster	1921	9

Date	The beginning time of inhabitation in this county	Educational level	Profession sector	Occupation	Marital status	Physical disability	Religion
2	Inhabited for generations	Illiteracy	Agriculture	Land-holding peasant	Married	No	No religion
20	Inhabited for generations						
27	Inhabited for generations	Bao school 4 years			Unmarried		
15	Inhabited for generations	Illiteracy	Agriculture	Land-holding peasant	Married	No	No religion
15	Inhabited for generations	Illiteracy	Others	Weaving	Married	No	No religion
30	Inhabited for generations				Married		
5	Inhabited for generations				Married		
11	Inhabited for generations	Illiteracy	Agriculture	Land-holding peasant	Married	No	No religion
1	Inhabited for generations	Illiteracy	Others	Weaving	Married	No	No religion
26	Inhabited for generations						
17	Inhabited for generations	Illiteracy					
8	Inhabited for generations	Illiteracy	Agriculture	Land-holding peasant	Married	No	No religion
4	Inhabited for generations	Illiteracy	Others	Weaving	Married	No	No religion
9	Inhabited for generations	Illiteracy	Agriculture	Weaving	Married	No	No religion
4	Inhabited for generations	Illiteracy	Others	Weaving	Married	No	No religion
27	Inhabited for generations	Illiteracy	Agriculture	Land-holding peasant	Married	No	No religion
8	Inhabited for generations	Illiteracy	Agriculture	Land-holding peasant	Married	No	No religion
26	Inhabited for generations	Illiteracy	Agriculture	Land-holding peasant	Married	No	No religion

Appendix: Overview of Microdata for Wudu County (*cont.*)

									Birthdate	
	The no. of house-hold	Jia	House-hold	Relation-ship with the house-hold head	Surname	First name	Sex	The Chinese zodiac	Year	Month
Fujin Xiang 1st Bao				The house-hold head's wife	徐	巧全儿	Female	Pig	1923	8
Fujin Xiang 1st Bao				Mother	徐	房生	Female	Dragon	1891	5
Fujin Xiang 1st Bao				Daughter	麻	信女	Female	Rooster	1945	3
Fujin Xiang 1st Bao				Niece	桃	信女	Female	Tiger	1938	4
Fujin Xiang 1st Bao	272	6	15	The house-hold head	董	其福	Male	Dragon	1903	9
Fujin Xiang 1st Bao				The house-hold head's wife	马	禧桂	Female	Dragon	1903	11
Fujin Xiang 1st Bao				Mother	李	分金	Female	Goat	1881	3
Fujin Xiang 1st Bao				Son	董	润林	Male	Horse	1930	2
Fujin Xiang 1st Bao				Son's wife	马	崴明	Female	Dragon	1928	9
Fujin Xiang 1st Bao				The second son	马	长生	Male	Pig	1947	3
Fujin Xiang 1st Bao				Younger brother	董	其三	Male	Ox	1913	3
Fujin Xiang 1st Bao				Younger brother's wife	岳	论娃	Female	Dragon	1916	12
Fujin Xiang 1st Bao	273	7	1	The house-hold head	赵	培华	Male	Horse	1918	10
Fujin Xiang 1st Bao				The house-hold head's wife	苏	玉娃	Female	Rooster	1921	1
Fujin Xiang 1st Bao				Father	赵	永祥	Male	Ox	1888	6
Fujin Xiang 1st Bao				Daughter	赵	成香	Female	Rabbit	1939	1
Fujin Xiang 1st Bao				The second daughter	赵	二女	Female	Horse	1942	8

Date	The beginning time of inhabitation in this county	Educational level	Profession sector	Occupation	Marital status	Physical disability	Religion
26	Inhabited for generations	Illiteracy	Others	Land-holding peasant	Married	No	No religion
11	Inhabited for generations	Illiteracy	Others	Land-holding peasant	Married	No	No religion
23	Inhabited for generations						
3	Inhabited for generations	Illiteracy					
4	Inhabited for generations	Illiteracy	Agricul-ture	Land-holding peasant	Married	No	No religion
29	Inhabited for generations	Illiteracy	Agricul-ture	Land-holding peasant	Married	No	No religion
13	Inhabited for generations				Widowed		
23	Inhabited for generations	Bao school 3 years			Married	No	No religion
26	Inhabited for generations	Illiteracy	Others	Weaving	Married	No	No religion
9	Inhabited for generations						
13	Inhabited for generations	Illiteracy	Agricul-ture	Land-holding peasant	Married	No	No religion
7	Inhabited for generations	Illiteracy	Others	Weaving	Married	No	No religion
26	Inhabited for generations	Sishu 3 years	Agricul-ture	Land-holding peasant	Married	No	No religion
19	Inhabited for generations	Illiteracy	Others	Land-holding peasant	Married	No	No religion
3	Inhabited for generations	Illiteracy	Agricul-ture	Land-holding peasant	Widowed	No	No religion
25	Inhabited for generations						
23	Inhabited for generations						

Appendix: Overview of Microdata for Wudu County (*cont.*)

	The no. of household	Jia	Household	Relationship with the household head	Surname	First name	Sex	The Chinese zodiac	Birthdate	
									Year	Month
Fujin Xiang 1st Bao				The third daughter	赵	菜香	Female	Rooster	1945	6
Fujin Xiang 1st Bao				The fourth daughter	赵	进香	Female	Dog	1946	2
Fujin Xiang 1st Bao	274	7	2	The household head	董	治连	Male	Rooster	1908	1
Fujin Xiang 1st Bao				The household head's wife	赵	玉连	Female	Monkey	1907	12
Fujin Xiang 1st Bao				Mother	徐	周文	Female	Ox	1888	9
Fujin Xiang 1st Bao				Grandmother	赵	四女	Female	Ox	1864	2
Fujin Xiang 1st Bao				Son	董	佑禧	Male	Monkey	1932	8
Fujin Xiang 1st Bao				Daughter	董	贵女	Female	Snake	1929	9
Fujin Xiang 1st Bao				Family member	朱	青云	Male	Rat	1924	5
Fujin Xiang 1st Bao				Grandson	董	庆婚	Male	Dog	1946	8
Fujin Xiang 1st Bao	275	7	3	The household head	赵	永财	Male	Snake	1917	2
Fujin Xiang 1st Bao				The household head's wife	李	跳秀	Female	Pig	1910	12
Fujin Xiang 1st Bao				Mother	赵	岁住	Female	Tiger	1876	10
Fujin Xiang 1st Bao				Daughter	赵	东香	Female	Tiger	1926	12
Fujin Xiang 1st Bao				The second daughter	赵	江明	Female	Rat	1936	11
Fujin Xiang 1st Bao	276	7	4	The household head	董	其昌	Male	Monkey	1907	4
Fujin Xiang 1st Bao				The household head's wife	杜	成娃	Female	Goat	1906	8

Date	The beginning time of inhabitation in this county	Educational level	Profession sector	Occupation	Marital status	Physical disability	Religion	
5	Inhabited for generations							
26	Inhabited for generations							Died in June
15	Inhabited for generations	Sishu 3 years	Agriculture	Land-holding peasant	Married	No	No religion	
6	Inhabited for generations	Illiteracy	Others	Land-holding peasant	Married	No	No religion	
16	Inhabited for generations					Two eyes blind		
2	Inhabited for generations		Unemployed					
10	Inhabited for generations	Higher primary school 5 years						
24	Inhabited for generations	Illiteracy	Others	Land-holding peasant	Married	No	No religion	
1	Inhabited for generations	Illiteracy	Agriculture	Land-holding peasant	Married	No	No religion	
16	Inhabited for generations							
21	Inhabited for generations	Illiteracy	Agriculture	Land-holding peasant	Married	No	No religion	
31	Inhabited for generations	Illiteracy	Agriculture	Land-holding peasant	Married	No	No religion	
1	Inhabited for generations				Widowed			
6	Inhabited for generations	Illiteracy	Agriculture	Land-holding peasant	Widowed	No	No religion	
16	Inhabited for generations							
15	Inhabited for generations	Illiteracy	Agriculture	Land-holding peasant	Married	No	No religion	
19	Inhabited for generations	Illiteracy	Agriculture	Land-holding peasant	Married	No	No religion	

Appendix: Overview of Microdata for Wudu County (*cont.*)

	The no. of household	Jia	Household	Relationship with the household head	Surname	First name	Sex	The Chinese zodiac	Birthdate Year	Month
Fujin Xiang 1st Bao				Daughter	董	有女	Female	Rooster	1933	1
Fujin Xiang 1st Bao	277	7	5	The household head	董	仲常	Male	Pig	1886	2
Fujin Xiang 1st Bao				The household head's wife	常	早姐	Female	Rooster	1896	11
Fujin Xiang 1st Bao				Son	董	富娃	Male	Tiger	1914	4
Fujin Xiang 1st Bao	278	7	6	The household head	董	三娃	Female	Dragon	1916	11
Fujin Xiang 1st Bao				Father	董	仲和	Male	Monkey	1871	6
Fujin Xiang 1st Bao				Mother	陈	贵连	Female	Goat	1881	6
Fujin Xiang 1st Bao				Older sister	董	保娃	Female	Tiger	1914	8
Fujin Xiang 1st Bao				Nephew	董	爱娃	Male	Dog	1934	2
Fujin Xiang 1st Bao				Niece	董	玄娃	Female	Goat	1943	11
Fujin Xiang 1st Bao	279	7	7	The household head	董	治发	Male	Tiger	1901	4
Fujin Xiang 1st Bao				The household head's wife	李	爱娃	Female	Goat	1906	10
Fujin Xiang 1st Bao				Son	董	四斤半	Male	Dog	1934	6
Fujin Xiang 1st Bao				Daughter	董	送家女	Female	Goat	1943	6
Fujin Xiang 1st Bao				The second daughter	董	送银女	Female	Monkey	1944	8
Fujin Xiang 1st Bao	280	7	8	The household head	董	其送	Male	Goat	1919	5
Fujin Xiang 1st Bao				The household head's wife	李	润女	Female	Rooster	1919	8

Date	The beginning time of inhabitation in this county	Educational level	Profession sector	Occupation	Marital status	Physical disability	Religion
28	Inhabited for generations	Illiteracy	Agriculture	Land-holding peasant	Unmarried	No	No religion
15	Inhabited for generations	Illiteracy	Agriculture	Land-holding peasant	Married	No	No religion
1	Inhabited for generations	Illiteracy	Agriculture	Land-holding peasant	Married	No	No religion
13	Inhabited for generations	Illiteracy	Agriculture	Land-holding peasant	Widowed	Deaf	No religion
5	Inhabited for generations	Illiteracy	Agriculture	Land-holding peasant	Widowed	No	No religion
13	Inhabited for generations				Married	Leg broken	
13	Inhabited for generations				Married	No	
1	Inhabited for generations	Illiteracy	Agriculture	Land-holding peasant	Widowed	No	No religion
1	Inhabited for generations	Bao school 2 years			Unmarried		
8	Inhabited for generations						
25	Inhabited for generations	Illiteracy	Agriculture	Land-holding peasant	Married	No	No religion
31	Inhabited for generations	Illiteracy	Agriculture	Land-holding peasant	Married	No	No religion
4	Inhabited for generations	Illiteracy					
27	Inhabited for generations						
1	Inhabited for generations						
23	Inhabited for generations	Illiteracy	Agriculture	Land-holding peasant	Married	No	No religion
21	Inhabited for generations	Illiteracy	Others	Land-holding peasant	Married	No	No religion

Appendix: Overview of Microdata for Wudu County (*cont.*)

	The no. of household	Jia	Household	Relationship with the household head	Surname	First name	Sex	The Chinese zodiac	Birthdate	
									Year	Month
Fujin Xiang 1st Bao				Father	董	正邦	Male	Rabbit	1902	3
Fujin Xiang 1st Bao				Mother	三	斤女	Female	Goat	1882	3
Fujin Xiang 1st Bao	281	7	9	The household head	朱	海元	Male	Rabbit	1890	3
Fujin Xiang 1st Bao				The household head's wife	李	双喜	Female	Rabbit	1902	3
Fujin Xiang 1st Bao				Daughter	朱	生娃	Female	Rooster	1934	12
Fujin Xiang 1st Bao	282	7	10	The household head	董	治清	Male	Snake	1892	6
Fujin Xiang 1st Bao				The household head's wife	徐	叩娃	Female	Rabbit	1890	6
Fujin Xiang 1st Bao	283	7	11	The household head	董	治德	Male	Rat	1924	8
Fujin Xiang 1st Bao				The household head's wife	岳	菊娃	Female	Pig	1922	7
Fujin Xiang 1st Bao				Son	董	早儿	Male	Monkey	1944	3
Fujin Xiang 1st Bao				Daughter	董	早女	Female	Pig	1947	4
Fujin Xiang 1st Bao	284	7	12	The household head	董	其才	Male	Snake	1904	7
Fujin Xiang 1st Bao				The household head's wife	赵	学足	Female	Snake	1917	2
Fujin Xiang 1st Bao	285	7	13	The household head	董	崴杜	Male	Goat	1919	9
Fujin Xiang 1st Bao				The household head's wife	汪	玉明	Female	Rooster	1921	4
Fujin Xiang 1st Bao				Mother	陈	徐儿	Female	Snake	1902	6

Date	The beginning time of inhabitation in this county	Educational level	Profession sector	Occupation	Marital status	Physical disability	Religion
23	Inhabited for generations	Illiteracy	Agriculture	Land-holding peasant	Married	No	No religion
16	Inhabited for generations	Illiteracy	Others	Land-holding peasant	Married	No	No religion
12	Inhabited for generations	Illiteracy	Agriculture	Land-holding peasant	Married	No	No religion
5	Inhabited for generations	Illiteracy	Agriculture	Land-holding peasant	Married	No	No religion
3	Inhabited for generations	Illiteracy					
5	Inhabited for generations	Sishu 3 years	Agriculture	Land-holding peasant	Married	No	No religion
29	Inhabited for generations	Illiteracy	Others	Land-holding peasant	Married	Deaf	No religion
24	Inhabited for generations	Illiteracy	Agriculture	Land-holding peasant	Married	No	No religion
20	Inhabited for generations	Illiteracy	Others	Land-holding peasant	Married	No	No religion
25	Inhabited for generations						
28	Inhabited for generations						
8	Inhabited for generations	Illiteracy	Agriculture	Land-holding peasant	Married	No	No religion
23	Inhabited for generations	Illiteracy	Others	Land-holding peasant	Married	No	No religion
25	Inhabited for generations	Illiteracy	Agriculture	Land-holding peasant	Married	No	No religion
24	Inhabited for generations	Illiteracy	Others	Land-holding peasant	Married	No	No religion
3	Inhabited for generations	Illiteracy	Others	Land-holding peasant	Widowed	No	No religion

Appendix: Overview of Microdata for Wudu County (*cont.*)

	The no. of household	Jia	Household	Relationship with the household head	Surname	First name	Sex	The Chinese zodiac	Birthdate	
									Year	Month
Fujin Xiang 1st Bao				Son	董	县落	Male	Horse	1942	12
Fujin Xiang 1st Bao				Daughter	董	足香	Female	Rabbit	1939	9
Fujin Xiang 1st Bao				The second daughter	董	金香	Female	Rooster	1945	4
Fujin Xiang 1st Bao	286	7	14	The household head	董	仲庭	Male	Dog	1885	5
Fujin Xiang 1st Bao				The household head's wife	徐	耕儿	Female	Horse	1893	9
Fujin Xiang 1st Bao				Son	董	崴炳	Male	Rooster	1921	9
Fujin Xiang 1st Bao				Son's wife	李	春明	Female	Ox	1925	11
Fujin Xiang 1st Bao				Grandson	董	双喜	Male	Dog	1946	7
Fujin Xiang 1st Bao				The second son	董	韩才	Male	Rooster	1933	1
Fujin Xiang 1st Bao	287	7	15	The household head	董	仲裕	Male	Horse	1880	12
Fujin Xiang 1st Bao				Son	董	治国	Male	Dragon	1928	4
Fujin Xiang 1st Bao				Son's wife	成	周女儿	Female	Tiger	1926	5
Fujin Xiang 1st Bao	288	8	1	The household head	董	治连	Male	Rabbit	1878	8
Fujin Xiang 1st Bao				The household head's wife	连	儿	Female	Horse	1929	5
Fujin Xiang 1st Bao	289	8	2	The household head	吴	怌心	Male	Goat	1882	10
Fujin Xiang 1st Bao				The household head's wife	赵	崴秀	Female	Ox	1900	3
Fujin Xiang 1st Bao	290	8	3	The household head	陈	万富	Male	Pig	1910	1

Date	The beginning time of inhabitation in this county	Educational level	Profession sector	Occupation	Marital status	Physical disability	Religion
9	Inhabited for generations						
27	Inhabited for generations						
30	Inhabited for generations						
17	Inhabited for generations	Illiteracy	Agriculture	Land-holding peasant	Married	No	No religion
15	Inhabited for generations	Illiteracy	Agriculture	Land-holding peasant	Married	No	No religion
15	Inhabited for generations	Illiteracy	Agriculture	Land-holding peasant	Married	No	No religion
24	Inhabited for generations	Illiteracy	Agriculture	Land-holding peasant	Married	No	No religion
16	Inhabited for generations						
31	Inhabited for generations	Higher primary school 5 years	Agriculture	Land-holding peasant	Unmarried	No	No religion
21	Inhabited for generations	Illiteracy	Agriculture	Land-holding peasant	Widowed	No	No religion
16	Inhabited for generations	Illiteracy	Agriculture	Land-holding peasant	Married	No	No religion
1	Inhabited for generations	Illiteracy	Others	Weaving	Married	No	No religion
23	Inhabited for generations	Illiteracy	Agriculture	Land-holding peasant	Married	No	No religion
29	Inhabited for generations	Illiteracy	Others	Land-holding peasant	Married	No	No religion
16	Inhabited for generations	Illiteracy	Agriculture	Land-holding peasant	Married	No	No religion
1	Inhabited for generations	Illiteracy	Others	Land-holding peasant	Married	No	No religion
23	Inhabited for generations	Illiteracy	Agriculture	Land-holding peasant	Married	No	No religion

Appendix: Overview of Microdata for Wudu County (*cont.*)

	The no. of household	Jia	Household	Relationship with the household head	Surname	First name	Sex	The Chinese zodiac	Birthdate Year	Birthdate Month
Fujin Xiang 1st Bao				The household head's wife	赵	菊娃	Female	Rabbit	1915	4
Fujin Xiang 1st Bao				Father	陈	志元	Male	Snake	1881	6
Fujin Xiang 1st Bao				Mother	王	秀娃	Female	Dragon	1891	9
Fujin Xiang 1st Bao				Daughter	陈	吹娃	Female	Dog	1946	8
Fujin Xiang 1st Bao				The third younger brother	陈	万双	Male	Goat	1919	10
Fujin Xiang 1st Bao				The fourth younger brother	陈	万贵	Male	Ox	1922	9
Fujin Xiang 1st Bao				The fifth younger brother	陈	万有	Male	Ox	1922	8
Fujin Xiang 1st Bao	291	8	4	The household head	陈	希花	Male	Pig	1909	11
Fujin Xiang 1st Bao				The household head's wife	杨	张秀	Female	Snake	1904	10
Fujin Xiang 1st Bao				Father	陈	玉德	Male		1881	9
Fujin Xiang 1st Bao				Mother	赵	谢娃	Female	Rat	1875	10
Fujin Xiang 1st Bao				Daughter	陈	来娃	Female	Rabbit	1929	4
Fujin Xiang 1st Bao				The second daughter	陈	崴来儿	Female	Rooster	1945	9
Fujin Xiang 1st Bao				Older brother's wife	赵	月儿	Female	Rabbit	1902	10
Fujin Xiang 1st Bao				Younger brother	陈	希科	Male	Dragon	1916	6

Date	The beginning time of inhabitation in this county	Educational level	Profession sector	Occupation	Marital status	Physical disability	Religion
1	Inhabited for generations	Illiteracy	Agriculture	Land-holding peasant	Married	No	No religion
18	Inhabited for generations	Illiteracy			Married		
14	Inhabited for generations	Illiteracy			Married		
27	Inhabited for generations						
16	Inhabited for generations	Illiteracy	Agriculture	Land-holding peasant	Widowed	No	No religion
29	Inhabited for generations	Illiteracy	Agriculture	Land-holding peasant	Widowed	No	No religion
11	Inhabited for generations	Illiteracy	Agriculture	Land-holding peasant	Unmarried	No	No religion
17	Inhabited for generations	Illiteracy	Agriculture	Land-holding peasant	Married	No	No religion
1	Inhabited for generations	Illiteracy	Agriculture	Land-holding peasant	Married	No	No religion
10	Inhabited for generations				Married	No	
16	Inhabited for generations				Married		
1	Inhabited for generations						
27	Inhabited for generations						
26	Inhabited for generations	Illiteracy	Agriculture	Land-holding peasant	Widowed	No	No religion
27	Inhabited for generations	Illiteracy	Agriculture	Land-holding peasant	Married	No	No religion

Appendix: Overview of Microdata for Wudu County (*cont.*)

	The no. of household	Jia	Household	Relationship with the household head	Surname	First name	Sex	The Chinese zodiac	Birthdate	
									Year	Month
Fujin Xiang 1st Bao				Younger brother's wife	马	唤哇	Female	Horse	1918	7
Fujin Xiang 1st Bao	292	8	5	The household head	马	张生	Male	Goat	1906	4
Fujin Xiang 1st Bao				The household head's wife	聂	住连	Female	Rabbit	1915	2
Fujin Xiang 1st Bao				Son	马	九玉	Male	Dog	1934	10
Fujin Xiang 1st Bao	293	8	6	The household head	岳	小照	Female	Dragon	1891	11
Fujin Xiang 1st Bao				Son	马	明贵	Male	Goat	1931	7
Fujin Xiang 1st Bao	294	8	7	The household head	马	富儿	Male	Rat	1887	1
Fujin Xiang 1st Bao				The household head's wife	李	秀儿	Female	Tiger	1901	8
Fujin Xiang 1st Bao				Son	马	明星	Male	Dog	1924	9
Fujin Xiang 1st Bao	295	8	8	The household head	崔	满库	Male	Monkey	1886	4
Fujin Xiang 1st Bao				The household head's wife	李	桃儿	Female	Dog	1885	11
Fujin Xiang 1st Bao				Younger brother's wife	李	银香	Female	Dragon	1891	6
Fujin Xiang 1st Bao				Nephew	崔	玉德	Male	Dragon	1928	12
Fujin Xiang 1st Bao				Nephew's wife	李	琴娃	Female	Rabbit	1915	11
Fujin Xiang 1st Bao				Nephew's son	崔	泰涵	Male	Tiger	1938	2
Fujin Xiang 1st Bao				The second grandson	崔	崴寒	Male	Dragon	1940	5

Date	The beginning time of inhabitation in this county	Educational level	Profession sector	Occupation	Marital status	Physical disability	Religion
1	Inhabited for generations	Illiteracy	Agriculture	Land-holding peasant	Married	No	No religion
3	Inhabited for generations	Illiteracy	Agriculture	Land-holding peasant	Married	No	No religion
27	Inhabited for generations	Illiteracy	Others	Land-holding peasant	Married	No	No religion
6	Inhabited for generations	Bao school 2 years					
16	Inhabited for generations	Illiteracy	Agriculture	Land-holding peasant	Widowed	No	No religion
10	Inhabited for generations	Illiteracy	Agriculture	Land-holding peasant	Unmarried	No	No religion
22	Inhabited for generations	Illiteracy	Agriculture	Land-holding peasant	Married	No	No religion
26	Inhabited for generations	Illiteracy	Others	Land-holding peasant	Married	No	No religion
13	Inhabited for generations	Bao school 2 years					
26	Inhabited for generations	Illiteracy	Agriculture	Land-holding peasant	Married	No	No religion
11	Inhabited for generations	Illiteracy	Agriculture	Land-holding peasant	Married	No	No religion
29	Inhabited for generations	Illiteracy	Agriculture	Land-holding peasant	Widowed	No	No religion
12	Inhabited for generations	Illiteracy	Agriculture	Land-holding peasant	Married	No	No religion
23	Inhabited for generations	Illiteracy	Agriculture	Land-holding peasant	Married	No	No religion
11	Inhabited for generations	Illiteracy	Agriculture				
18	Inhabited for generations						

Appendix: Overview of Microdata for Wudu County (*cont.*)

	The no. of household	Jia	Household	Relationship with the household head	Surname	First name	Sex	The Chinese zodiac	Birthdate Year	Month
Fujin Xiang 1st Bao				The third grandson	崔	崴炳	Male	Dog	1946	6
Fujin Xiang 1st Bao	296	8	9	The household head	马	银翠	Female	Tiger	1889	9
Fujin Xiang 1st Bao				Son's wife	罗	玄娃	Female	Ox	1913	3
Fujin Xiang 1st Bao				Nephew	马	星才	Male	Monkey	1932	10
Fujin Xiang 1st Bao				Nephew's wife	王	叩娃	Female	Rabbit	1927	1
Fujin Xiang 1st Bao				Family member	赵	万林	Male	Monkey	1883	10
Fujin Xiang 1st Bao				Family member	赵	永泰	Male	Ox	1909	7
Fujin Xiang 1st Bao	297	8	10	The household head	马	明德	Male	Goat	1894	4
Fujin Xiang 1st Bao				The household head's wife	赵	刘生	Female	Snake	1892	3
Fujin Xiang 1st Bao				Son	马	步仓	Male	Horse	1918	12
Fujin Xiang 1st Bao				Son's wife	王	玉娃	Female	Snake	1929	12
Fujin Xiang 1st Bao				The second son	马	步花	Male	Dog	1934	2
Fujin Xiang 1st Bao				Older brother	马	明福	Male	Pig	1886	8
Fujin Xiang 1st Bao				Older brother's wife	岳	何儿	Female	Pig	1886	8
Fujin Xiang 1st Bao				Nephew	马	其发	Male	Dragon	1928	3
Fujin Xiang 1st Bao				Nephew's wife	岳	桃秀	Female	Ox	1925	10
Fujin Xiang 1st Bao	298	8	11	The household head	王	国仓	Male	Ox	1900	9
Fujin Xiang 1st Bao				Son	王	永德	Male	Rabbit	1927	9

Date	The beginning time of inhabitation in this county	Educational level	Profession sector	Occupation	Marital status	Physical disability	Religion
4	Inhabited for generations						
30	Inhabited for generations	Illiteracy	Agriculture	Land-holding peasant	Married	No	No religion
29	Inhabited for generations	Illiteracy	Agriculture	Land-holding peasant	Married	No	No religion
1	Inhabited for generations	Illiteracy	Agriculture	Land-holding peasant	Married	No	No religion
22	Inhabited for generations	Illiteracy	Agriculture	Land-holding peasant	Married	No	No religion
20	Inhabited for generations	Illiteracy	Agriculture	Land-holding peasant	Married	No	No religion
1	Inhabited for generations	Illiteracy	Agriculture	Land-holding peasant	Married	No	No religion
26	Inhabited for generations	Illiteracy	Agriculture	Land-holding peasant	Married	No	No religion
5	Inhabited for generations	Illiteracy	Agriculture	Land-holding peasant	Married	No	No religion
27	Inhabited for generations	Illiteracy	Agriculture	Land-holding peasant	Married	No	No religion
14	Inhabited for generations	Illiteracy	Agriculture	Land-holding peasant	Married	No	No religion
15	Inhabited for generations	Bao school 2 years					
29	Inhabited for generations	Illiteracy	Agriculture	Land-holding peasant	Married	No	No religion
9	Inhabited for generations	Illiteracy	Agriculture	Land-holding peasant	Married	No	No religion
31	Inhabited for generations	Illiteracy	Agriculture	Land-holding peasant	Married	No	No religion
10	Inhabited for generations	Illiteracy	Agriculture	Land-holding peasant	Married	No	No religion
10	Inhabited for generations	Illiteracy	Agriculture	Land-holding peasant	Widowed	No	No religion
4	Inhabited for generations	Illiteracy	Agriculture	Land-holding peasant	Married	No	No religion

Appendix: Overview of Microdata for Wudu County (*cont.*)

	The no. of household	Jia	Household	Relationship with the household head	Surname	First name	Sex	The Chinese zodiac	Birthdate Year	Month
Fujin Xiang 1st Bao				Son's wife	岳	反桃	Female	Rabbit	1927	5
Fujin Xiang 1st Bao				The second son	王	仁仁	Male	Pig	1935	9
Fujin Xiang 1st Bao				The third son	王	义娃	Male	Goat	1943	3
Fujin Xiang 1st Bao				Daughter	王	处娃	Female	Goat	1931	12
Fujin Xiang 1st Bao	299	8	12	The household head	陈	明娃子	Female	Pig	1886	2
Fujin Xiang 1st Bao				Daughter	冯	儿	Female	Ox	1921	4
Fujin Xiang 1st Bao	300	8	13	The household head	陈	守云	Male	Snake	1904	8
Fujin Xiang 1st Bao				The household head's wife	浩	王女	Female	Snake	1917	6
Fujin Xiang 1st Bao				Mother	董	缠送	Female		1869	1
Fujin Xiang 1st Bao				Son	陈	赵顺	Male	Rat	1936	11
Fujin Xiang 1st Bao				The second son	陈	根才	Male	Rooster	1945	1
Fujin Xiang 1st Bao	301	8	14	The household head	李	枝德	Male	Snake	1892	6
Fujin Xiang 1st Bao				The household head's wife	岳	张秀	Female	Dragon	1903	4
Fujin Xiang 1st Bao				Son	李	四娃	Male	Ox	1937	3
Fujin Xiang 1st Bao				The second son	李	社娃	Male	Rabbit	1939	8
Fujin Xiang 1st Bao	302	8	15	The household head	李	友昇	Male	Tiger	1926	1
Fujin Xiang 1st Bao				The household head's wife	杨	彩儿	Female	Rooster	1921	11

Date	The beginning time of inhabitation in this county	Educational level	Profession sector	Occupation	Marital status	Physical disability	Religion	
25	Inhabited for generations	Illiteracy	Agriculture	Land-holding peasant	Married	No	No religion	
1	Inhabited for generations	Illiteracy						
6	Inhabited for generations							
1	Inhabited for generations	Illiteracy	Agriculture	Land-holding peasant	Unmarried	No	No religion	
25	Inhabited for generations	Illiteracy	Agriculture	Land-holding peasant	Widowed	No	No religion	
3	Inhabited for generations	Illiteracy			Unmarried			
8	Inhabited for generations	Illiteracy	Agriculture	Land-holding peasant	Married	No	No religion	
3	Inhabited for generations	Illiteracy	Agriculture	Land-holding peasant	Married	No	No religion	
9	Inhabited for generations							
30	Inhabited for generations	Bao school 1 year						
26	Inhabited for generations							
27	Inhabited for generations	Illiteracy	Agriculture	Land-holding peasant	Married	No	No religion	
2	Inhabited for generations	Illiteracy	Agriculture	Land-holding peasant	Married	No	No religion	Died on 20th July
11	Inhabited for generations	Bao school 2 years	Agriculture	Land-holding peasant	Unmarried	No	No religion	
25	Inhabited for generations							
26	Inhabited for generations	Higher primary school	Agriculture	Land-holding peasant	Married	No	No religion	
1	Inhabited for generations	Illiteracy	Agriculture	Land-holding peasant	Married	No	No religion	

Appendix: Overview of Microdata for Wudu County (*cont.*)

									Birthdate	
	The no. of household	Jia	Household	Relationship with the household head	Surname	First name	Sex	The Chinese zodiac	Year	Month
Fujin Xiang 1st Bao				Mother	杨	夂娃	Female	Pig	1898	4
Fujin Xiang 1st Bao				Older brother	李	友锡	Male	Ox	1925	5
Fujin Xiang 1st Bao				Older brother's wife	汪	蒋娃	Female	Rat	1924	6
Fujin Xiang 1st Bao				Niece	李	双宣	Female	Rooster	1945	9
Fujin Xiang 1st Bao				Younger brother	李	根成儿	Male	Tiger	1938	9
Fujin Xiang 1st Bao				Younger sister	李	丑儿	Female	Rat	1936	11
Fujin Xiang 1st Bao				The second younger sister	李	香成	Female	Rabbit	1939	12
Fujin Xiang 1st Bao				Family member	李	雄才	Male	Monkey	1907	1
Fujin Xiang 1st Bao	303	9	1	The household head	赵	永生	Male	Pig	1910	1
Fujin Xiang 1st Bao				The household head's wife	聂	歌犬	Female	Tiger	1914	3
Fujin Xiang 1st Bao				Son	赵	天社儿	Male	Rooster	1945	3
Fujin Xiang 1st Bao				Daughter	赵	社女子	Female	Rooster	1933	11
Fujin Xiang 1st Bao				The second daughter	赵	崴社	Female	Ox	1937	11
Fujin Xiang 1st Bao	304	9	2	The household head	赵	永奎	Male	Goat	1894	10
Fujin Xiang 1st Bao				The household head's wife	韩	润润	Female	Horse	1905	10
Fujin Xiang 1st Bao				Son	保	生儿	Male	Pig	1923	9
Fujin Xiang 1st Bao				The second son	赵	保才	Male	Tiger	1926	2

Date	The beginning time of inhabitation in this county	Educational level	Profession sector	Occupation	Marital status	Physical disability	Religion
1	Inhabited for generations	Illiteracy	Agriculture	Land-holding peasant	Married	No	No religion
1	Inhabited for generations	Illiteracy	Agriculture	Land-holding peasant	Married	No	No religion
5	Inhabited for generations	Illiteracy	Agriculture	Land-holding peasant	Married	No	No religion
6	Inhabited for generations						
29	Inhabited for generations	Bao school 1 year					
22	Inhabited for generations						
31	Inhabited for generations	Illiteracy					
1	Inhabited for generations	Illiteracy	Agriculture	Land-holding peasant	Married	No	No religion
22	Inhabited for generations	Illiteracy	Agriculture	Land-holding peasant	Married	No	No religion
1	Inhabited for generations	Illiteracy	Agriculture	Land-holding peasant	Married	No	No religion
10	Inhabited for generations						
15	Inhabited for generations	Illiteracy			Unmarried	No	No religion
21	Inhabited for generations	Illiteracy					
19	Inhabited for generations	Illiteracy	Agriculture	Land-holding peasant	Married	No	No religion
6	Inhabited for generations	Illiteracy	Others	Land-holding peasant	Married	No	No religion
5	Inhabited for generations	Illiteracy	Agriculture	Land-holding peasant	Unmarried	No	No religion
24	Inhabited for generations	Illiteracy	Agriculture	Land-holding peasant	Unmarried	No	No religion

Appendix: Overview of Microdata for Wudu County (*cont.*)

	The no. of household	Jia	Household	Relationship with the household head	Surname	First name	Sex	The Chinese zodiac	Birthdate	
									Year	Month
Fujin Xiang 1st Bao				The third son	赵	见运娃	Male	Rat	1936	7
Fujin Xiang 1st Bao				The fourth son	赵	桂才儿	Male	Dog	1946	1
Fujin Xiang 1st Bao				Daughter	赵	女女子	Female	Goat	1931	7
Fujin Xiang 1st Bao				The second daughter	赵	柱女儿	Female	Monkey	1944	2
Fujin Xiang 1st Bao	305	9	3	The household head	赵	蘇连	Female	Tiger	1901	11
Fujin Xiang 1st Bao				Son	赵	保娃	Male	Monkey	1920	1
Fujin Xiang 1st Bao				The second son	赵	崴保子	Male	Rat	1924	10
Fujin Xiang 1st Bao				Son's wife	韩	刘家	Female	Ox	1925	3
Fujin Xiang 1st Bao				Grandson	赵	安	Male	Pig	1947	3
Fujin Xiang 1st Bao				The third son	李	家拜	Male	Dog	1934	7
Fujin Xiang 1st Bao				The fourth son	赵	么么	Male	Pig	1947	1
Fujin Xiang 1st Bao				Daughter	赵	川娃	Female	Goat	1941	6
Fujin Xiang 1st Bao				The second daughter	赵	崴明子	Female	Dragon	1940	11
Fujin Xiang 1st Bao				The third daughter	赵	进足	Female	Dog	1934	1
Fujin Xiang 1st Bao				Family member	赵	六十三	Male	Horse	1893	6
Fujin Xiang 1st Bao	306	9	4	The household head	李	友朝	Male	Dragon	1916	1
Fujin Xiang 1st Bao				The household head's wife	赵	柱娃	Female	Horse	1918	9
Fujin Xiang 1st Bao				Father	李	枝有	Male	Horse	1881	11

Date	The beginning time of inhabitation in this county	Educational level	Profession sector	Occupation	Marital status	Physical disability	Religion
29	Inhabited for generations	Illiteracy	Agriculture	Land-holding peasant	Unmarried	No	No religion
2	Inhabited for generations						
26	Inhabited for generations	Illiteracy	Others	Land-holding peasant	Unmarried	No	No religion
3	Inhabited for generations						
12	Inhabited for generations	Illiteracy	Agriculture	Land-holding peasant	Married	No	No religion
12	Inhabited for generations	Illiteracy	Agriculture	Land-holding peasant	Unmarried	No	No religion
29	Inhabited for generations	Illiteracy	Agriculture	Land-holding peasant	Married	No	No religion
1	Inhabited for generations	Illiteracy	Agriculture	Land-holding peasant	Married	No	No religion
10	Inhabited for generations						
3	Inhabited for generations	Bao school 2 years					
22	Inhabited for generations						
14	Inhabited for generations	Illiteracy	Agriculture	Land-holding peasant	Unmarried	No	No religion
14	Inhabited for generations						
22	Inhabited for generations						
1	Inhabited for generations	Illiteracy	Agriculture	Land-holding peasant	Married	No	No religion
9	Inhabited for generations	Illiteracy	Agriculture	Land-holding peasant	Married	No	No religion
21	Inhabited for generations	Illiteracy	Others	Land-holding peasant	Married	No	No religion
22	Inhabited for generations	Illiteracy	Agriculture	Land-holding peasant	Married	No	No religion

Appendix: Overview of Microdata for Wudu County (*cont.*)

	The no. of household	Jia	Household	Relationship with the household head	Surname	First name	Sex	The Chinese zodiac	Birthdate Year	Month
Fujin Xiang 1st Bao				Mother	赵	润娃	Female	Monkey	1883	9
Fujin Xiang 1st Bao				Younger brother	李	友德	Male	Goat	1919	3
Fujin Xiang 1st Bao				Younger brother's wife	赵	菊香	Female	Pig	1923	9
Fujin Xiang 1st Bao				Niece	李	书香	Female	Pig	1946	2
Fujin Xiang 1st Bao	307	9	5	The household head	李	维泰	Male	Rat	1924	10
Fujin Xiang 1st Bao				The household head's wife	杨	香楼	Female	Tiger	1926	6
Fujin Xiang 1st Bao				Grandfather	李	枝高	Male	Ox	1888	10
Fujin Xiang 1st Bao				Grandmother	徐	王学子	Female	Horse	1881	5
Fujin Xiang 1st Bao				Mother	赵	贵连	Female	Rabbit	1902	2
Fujin Xiang 1st Bao				Uncle's wife	成	秀	Female	Dragon	1916	9
Fujin Xiang 1st Bao				Uncle's son	李	维智	Male	Rat	1936	4
Fujin Xiang 1st Bao				The second younger brother	李	继平	Male	Goat	1943	3
Fujin Xiang 1st Bao				Uncle's daughter	李	战中儿	Female	Tiger	1938	3
Fujin Xiang 1st Bao				The second younger sister	李	书秀	Female	Pig	1947	2
Fujin Xiang 1st Bao				Family member	李	友清	Male	Pig	1909	4
Fujin Xiang 1st Bao	308	9	6	The household head	李	友海	Male	Rabbit	1915	2

Date	The beginning time of inhabitation in this county	Educational level	Profession sector	Occupation	Marital status	Physical disability	Religion
2	Inhabited for generations	Illiteracy	Others	Land-holding peasant	Married	No	No religion
1	Inhabited for generations	Illiteracy	Agriculture	Land-holding peasant	Married	No	No religion
8	Inhabited for generations	Illiteracy	Others	Land-holding peasant	Married	No	No religion
16	Inhabited for generations						
24	Inhabited for generations	Illiteracy	Agriculture	Land-holding peasant	Married	No	No religion
11	Inhabited for generations	Illiteracy	Agriculture	Land-holding peasant	Married	No	No religion
30	Inhabited for generations	Illiteracy	Agriculture	Land-holding peasant	Married	No	No religion
22	Inhabited for generations	Illiteracy	Agriculture	Land-holding peasant	Married	No	No religion
27	Inhabited for generations	Illiteracy	Agriculture	Land-holding peasant	Widowed	No	No religion
24	Inhabited for generations	Illiteracy	Agriculture	Land-holding peasant	Married	No	No religion
4	Inhabited for generations	Bao school 4 years					
29	Inhabited for generations						
16	Inhabited for generations						
3	Inhabited for generations						
5	Inhabited for generations	Illiteracy	Agriculture	Land-holding peasant	Married	No	No religion
8	Inhabited for generations	Illiteracy	Agriculture	Land-holding peasant	Married	No	No religion

Appendix: Overview of Microdata for Wudu County (*cont.*)

									Birthdate	
	The no. of household	**Jia**	**Household**	**Relationship with the household head**	**Surname**	**First name**	**Sex**	**The Chinese zodiac**	**Year**	**Month**
Fujin Xiang 1st Bao				The household head's wife	足	连	Female	Snake	1917	3
Fujin Xiang 1st Bao				Mother	见	儿	Female	Monkey	1883	1
Fujin Xiang 1st Bao				Daughter	李	相哇	Female	Snake	1941	3
Fujin Xiang 1st Bao				The second daughter	李	崴相	Female	Rooster	1945	11
Fujin Xiang 1st Bao				Older brother	李	友兴	Male	Horse	1905	10
Fujin Xiang 1st Bao				Older brother's wife	马	转秀	Female	Pig	1909	6
Fujin Xiang 1st Bao				Nephew	李	继山	Male	Rabbit	1926	4
Fujin Xiang 1st Bao				Nephew's wife	陈	朝足娃	Female	Snake	1929	5
Fujin Xiang 1st Bao	309	9	7	The household head	李	支千	Male	Rooster	1884	5
Fujin Xiang 1st Bao				The household head's wife	兔	女	Female	Tiger	1889	11
Fujin Xiang 1st Bao				Son	李	友忠	Male	Horse	1923	11
Fujin Xiang 1st Bao				Grandson	李	祥贵	Male	Horse	1941	12
Fujin Xiang 1st Bao				The second son	李	友祥	Male	Horse	1930	3
Fujin Xiang 1st Bao				The third son	李	忠娃	Male	Rooster	1933	6
Fujin Xiang 1st Bao				The third son's wife	李	信女	Female	Dragon	1928	8
Fujin Xiang 1st Bao	310	9	8	The household head	李	支仓	Male	Rabbit	1890	12

Date	The beginning time of inhabitation in this county	Educational level	Profession sector	Occupation	Marital status	Physical disability	Religion
1	Inhabited for generations	Illiteracy	Agriculture	Land-holding peasant	Married	No	No religion
27	Inhabited for generations	Illiteracy	Agriculture	Land-holding peasant	Widowed	No	No religion
17	Inhabited for generations	Illiteracy					
27	Inhabited for generations						
1	Inhabited for generations				Married	Blind	No religion
7	Inhabited for generations	Illiteracy	Agriculture	Land-holding peasant	Married	No	No religion
24	Inhabited for generations	Illiteracy	Agriculture	Land-holding peasant	Married	No	No religion
9	Inhabited for generations	Illiteracy	Agriculture	Land-holding peasant	Married	No	No religion
17	Inhabited for generations	Illiteracy	Agriculture	Land-holding peasant	Married	No	No religion
16	Inhabited for generations	Illiteracy	Agriculture	Land-holding peasant	Married	No	No religion
21	Inhabited for generations	Illiteracy	Agriculture	Land-holding peasant	Widowed	No	No religion
6	Inhabited for generations						
21	Inhabited for generations	Primary school 5 years	Agriculture	Land-holding peasant	Unmarried	No	No religion
4	Inhabited for generations	Bao school 3 years	Agriculture	Land-holding peasant	Married	No	No religion
1	Inhabited for generations	Illiteracy	Agriculture	Land-holding peasant	Married	No	No religion
12	Inhabited for generations	Illiteracy	Agriculture	Land-holding peasant	Married	No	No religion

Appendix: Overview of Microdata for Wudu County (*cont.*)

	The no. of household	Jia	Household	Relationship with the household head	Surname	First name	Sex	The Chinese zodiac	Birthdate Year	Month
Fujin Xiang 1st Bao				The household head's wife	赵	跳娃	Female	Tiger	1901	12
Fujin Xiang 1st Bao				Mother	杨	月月	Female	Snake	1868	10
Fujin Xiang 1st Bao				Son	李	友明	Male	Ox	1925	9
Fujin Xiang 1st Bao				The second son	李	二丑	Male	Pig	1935	2
Fujin Xiang 1st Bao				Daughter	李	四娃	Female	Monkey	1932	2
Fujin Xiang 1st Bao				Uncle's son	李	枝银	Male	Dragon	1891	10
Fujin Xiang 1st Bao				Younger brother's wife	王	秋连	Female	Tiger	1901	8
Fujin Xiang 1st Bao				Nephew	李	友生	Male	Monkey	1920	10
Fujin Xiang 1st Bao				Nephew's wife	马	三谢	Female	Tiger	1926	10
Fujin Xiang 1st Bao				The second nephew	李	孟娃	Male	Horse	1930	3
Fujin Xiang 1st Bao				The third nephew	李	孟桂儿	Male	Rooster	1933	1
Fujin Xiang 1st Bao				The fourth nephew	李	四娃	Male	Tiger	1938	8
Fujin Xiang 1st Bao				Niece	李	能娃	Female	Snake	1941	9
Fujin Xiang 1st Bao	311	9	9	The household head	李	友堂	Male	Rat	1912	3
Fujin Xiang 1st Bao				The household head's wife	马	反娃	Female	Tiger	1915	5
Fujin Xiang 1st Bao				Father	李	支奎	Male	Goat	1882	3
Fujin Xiang 1st Bao				Son	李	赵生	Male	Dragon	1940	11

Date	The beginning time of inhabitation in this county	Educational level	Profession sector	Occupation	Marital status	Physical disability	Religion
1	Inhabited for generations	Illiteracy	Agriculture	Land-holding peasant	Married	No	No religion
21	Inhabited for generations						
7	Inhabited for generations	Illiteracy	Agriculture	Land-holding peasant	Married	No	No religion
16	Inhabited for generations	Bao school 2 years					
16	Inhabited for generations						
1	Inhabited for generations	Illiteracy	Agriculture	Land-holding peasant	Married	No	No religion
1	Inhabited for generations	Illiteracy	Agriculture	Land-holding peasant	Married	No	No religion
26	Inhabited for generations	Illiteracy	Agriculture	Land-holding peasant	Married	No	No religion
1	Inhabited for generations	Illiteracy	Agriculture	Land-holding peasant	Married	No	No religion
24	Inhabited for generations						
30	Inhabited for generations						
15	Inhabited for generations						
8	Inhabited for generations						
17	Inhabited for generations	Sishu 3 years	Agriculture	Land-holding peasant	Married	No	No religion
1	Inhabited for generations	Illiteracy	Others	Land-holding peasant	Married	No	No religion
4	Inhabited for generations	Illiteracy	Agriculture	Land-holding peasant	Widowed	No	No religion
6	Inhabited for generations						

Appendix: Overview of Microdata for Wudu County (*cont.*)

	The no. of household	Jia	Household	Relationship with the household head	Surname	First name	Sex	The Chinese zodiac	Birthdate	
									Year	Month
Fujin Xiang 1st Bao				Daughter	李	足女	Female	Rooster	1933	1
Fujin Xiang 1st Bao	312	9	10	The household head	李	枝清	Male	Ox	1888	1
Fujin Xiang 1st Bao				The household head's wife	徐	见女	Female	Rooster	1906	1
Fujin Xiang 1st Bao				Son	李	友林	Male	Dragon	1916	6
Fujin Xiang 1st Bao				The second son	李	友木	Male	Rooster	1933	1
Fujin Xiang 1st Bao				Younger brother's wife	赵	近旦	Female	Dragon	1928	1
Fujin Xiang 1st Bao				Daughter	李	冬女	Female	Tiger	1936	12
Fujin Xiang 1st Bao	313	9	11	The household head	李	维生	Male	Pig	1910	11
Fujin Xiang 1st Bao				The household head's wife	马	双娃	Female	Ox	1913	12
Fujin Xiang 1st Bao				Father	段	福仁	Male	Snake	1880	5
Fujin Xiang 1st Bao				Mother	成	长儿	Female	Ox	1888	10
Fujin Xiang 1st Bao				Daughter	李	中女儿	Female	Rooster	1933	4
Fujin Xiang 1st Bao				Younger brother	李	维隆	Male	Rooster	1921	10
Fujin Xiang 1st Bao				Grandfather	李	枝连	Male	Rat	1885	10
Fujin Xiang 1st Bao				Uncle	李	友章	Male	Goat	1906	7
Fujin Xiang 1st Bao				Uncle's wife	王	吾娃	Female	Dragon	1903	12
Fujin Xiang 1st Bao				Uncle's daughter	王	爱娃	Female	Horse	1942	1

Date	The beginning time of inhabitation in this county	Educational level	Profession sector	Occupation	Marital status	Physical disability	Religion
25	Inhabited for generations	Illiteracy					
1	Inhabited for generations	Sishu 3 years	Agriculture	Land-holding peasant	Married	No	No religion
21	Inhabited for generations	Illiteracy	Agriculture	Land-holding peasant	Married	No	No religion
1	Inhabited for generations	Higher primary school graduated	Agriculture	Land-holding peasant	Unmarried	No	No religion
7	Inhabited for generations	Junior middle school 5 years			Married	No	No religion
7	Inhabited for generations	Illiteracy	Agriculture	Land-holding peasant	Married	No	No religion
20	Inhabited for generations	Illiteracy					
18	Inhabited for generations	Illiteracy	Agriculture	Land-holding peasant	Married	No	No religion
8	Inhabited for generations	Illiteracy	Agriculture	Land-holding peasant	Married	No	No religion
3	Inhabited for generations	Illiteracy	Agriculture	Land-holding peasant	Married	No	No religion
25	Inhabited for generations	Illiteracy	Agriculture	Land-holding peasant	Married	No	No religion
5	Inhabited for generations	Illiteracy					
3	Inhabited for generations	Illiteracy	Agriculture	Land-holding peasant	Widowed	No	No religion
18	Inhabited for generations	Illiteracy	Agriculture	Land-holding peasant	Widowed	No	No religion
20	Inhabited for generations	Illiteracy	Agriculture	Land-holding peasant	Married	No	No religion
20	Inhabited for generations	Illiteracy	Agriculture	Land-holding peasant	Married	No	No religion
4	Inhabited for generations						

Appendix: Overview of Microdata for Wudu County (*cont.*)

	The no. of household	Jia	Household	Relationship with the household head	Surname	First name	Sex	The Chinese zodiac	Birthdate	
									Year	Month
Fujin Xiang 1st Bao	314	9	12	The household head	陈	义山	Male	Dragon	1891	10
Fujin Xiang 1st Bao				The household head's wife	赵	转儿	Female	Dog	1897	3
Fujin Xiang 1st Bao				Son	陈	国新	Male	Dog	1922	10
Fujin Xiang 1st Bao				Son's wife	董	双六	Female	Rooster	1921	10
Fujin Xiang 1st Bao				The second son	陈	国纪	Male	Rabbit	1927	10
Fujin Xiang 1st Bao				The second son's wife	李	旦娃	Female	Ox	1925	5
Fujin Xiang 1st Bao				Granddaughter	陈	必娃	Female	Monkey	1944	8
Fujin Xiang 1st Bao				Daughter	陈	住菊儿	Female	Rabbit	1939	12
Fujin Xiang 1st Bao	315	9	13	The household head	陈	义朝	Male	Horse	1905	4
Fujin Xiang 1st Bao				The household head's wife	史	福秀	Female	Snake	1904	7
Fujin Xiang 1st Bao				Father	陈	如贤	Male	Monkey	1883	3
Fujin Xiang 1st Bao				Mother	赵	利哇	Female	Monkey	1883	8
Fujin Xiang 1st Bao				Son	陈	六指儿	Male	Pig	1947	1
Fujin Xiang 1st Bao				Daughter	陈	菊娃	Female	Horse	1930	7
Fujin Xiang 1st Bao				Cohabitant	周	雲	Male	Horse	1918	5
Fujin Xiang 1st Bao				The second daughter	陈	春英娃	Female	Dragon	1940	7
Fujin Xiang 1st Bao	316	9	14	The household head	陈	如玉	Male	Monkey	1883	5

Date	The beginning time of inhabitation in this county	Educational level	Profession sector	Occupation	Marital status	Physical disability	Religion
12	Inhabited for generations	Illiteracy	Agriculture	Land-holding peasant	Married	No	No religion
1	Inhabited for generations	Illiteracy	Agriculture	Land-holding peasant	Married	No	No religion
7	Inhabited for generations	Illiteracy	Agriculture	Land-holding peasant	Married	No	No religion
1	Inhabited for generations	Illiteracy	Agriculture	Land-holding peasant	Married	No	No religion
1	Inhabited for generations	Bao school 4 years	Agriculture	Land-holding peasant	Married	No	No religion
1	Inhabited for generations	Illiteracy	Agriculture	Land-holding peasant	Married	No	No religion
1	Inhabited for generations						
17	Inhabited for generations						
1	Inhabited for generations	Illiteracy	Agriculture	Land-holding peasant	Married	No	No religion
18	Inhabited for generations	Illiteracy	Agriculture	Land-holding peasant	Married	No	No religion
21	Inhabited for generations				Married		
11	Inhabited for generations				Married		
25	Inhabited for generations						
15	Inhabited for generations	Illiteracy			Unmarried		
1	Inhabited for generations	Illiteracy			Unmarried		
23	Inhabited for generations						
1	Inhabited for generations	Illiteracy	Agriculture	Land-holding peasant	Married	Blind	No religion

Appendix: Overview of Microdata for Wudu County (*cont.*)

	The no. of household	Jia	Household	Relationship with the household head	Surname	First name	Sex	The Chinese zodiac	Birthdate	
									Year	Month
Fujin Xiang 1st Bao				The household head's wife	徐	拜儿	Female	Rooster	1884	7
Fujin Xiang 1st Bao				Daughter	陈	女子	Female	Dog	1922	5
Fujin Xiang 1st Bao				Granddaughter	陈	早桃儿	Female	Rooster	1945	7
Fujin Xiang 1st Bao				The second granddaughter	陈	秀梅	Female	Pig	1947	4
Fujin Xiang 1st Bao				Nephew	陈	义还	Male	Rabbit	1890	12
Fujin Xiang 1st Bao				Nephew's wife	陈	四秀	Female	Goat	1906	5
Fujin Xiang 1st Bao				Nephew's son	陈	正儿	Male	Ox	1925	9
Fujin Xiang 1st Bao				Grandson's wife	李	三菊	Female	Dragon	1928	7
Fujin Xiang 1st Bao				The second grandson	陈	裉称	Male	Monkey	1932	4
Fujin Xiang 1st Bao				Granddaughter	陈	玉娃	Female	Ox	1937	5
Fujin Xiang 1st Bao				Nephew's daughter	陈	满娃	Female	Horse	1942	7
Fujin Xiang 1st Bao				The third Granddaughter	陈	毛庭	Female	Pig	1947	6
Fujin Xiang 1st Bao	317	9	15	The household head	陈	义才	Male	Monkey	1983	10
Fujin Xiang 1st Bao				The household head's wife	王	秀儿	Female	Rat	1887	4
Fujin Xiang 1st Bao				Son	陈	国安	Male	Horse	1905	10
Fujin Xiang 1st Bao				Son's wife	王	萧家	Female	Rabbit	1902	5

Date	The beginning time of inhabitation in this county	Educational level	Profession sector	Occupation	Marital status	Physical disability	Religion	
7	Inhabited for generations	Illiteracy	Agriculture	Land-holding peasant	Married	No	No religion	
1	Inhabited for generations	Illiteracy	Agriculture	Land-holding peasant	Widowed	No	No religion	
2	Inhabited for generations							
14	Inhabited for generations							
15	Inhabited for generations	Illiteracy	Agriculture	Land-holding peasant	Married	No	No religion	
24	Inhabited for generations	Illiteracy	Agriculture	Land-holding peasant	Married	No	No religion	
10	Inhabited for generations	Illiteracy	Agriculture	Land-holding peasant	Married	No	No religion	Joined the army in 1943
11	Inhabited for generations	Illiteracy	Agriculture	Land-holding peasant	Married	No	No religion	
14	Inhabited for generations	Illiteracy						
26	Inhabited for generations							
9	Inhabited for generations							
1	Inhabited for generations							
27	Inhabited for generations	Illiteracy	Agriculture	Land-holding peasant	Married	No	No religion	
25	Inhabited for generations	Illiteracy	Agriculture	Land-holding peasant	Married	No	No religion	
28	Inhabited for generations	Illiteracy	Agriculture	Land-holding peasant	Married	No	No religion	
1	Inhabited for generations	Illiteracy	Agriculture	Land-holding peasant	Married	No	No religion	

Appendix: Overview of Microdata for Wudu County (*cont.*)

	The no. of household	Jia	Household	Relationship with the household head	Surname	First name	Sex	The Chinese zodiac	Birthdate Year	Month
Fujin Xiang 1st Bao				Grandson	陈	华娃	Male	Goat	1943	5
Fujin Xiang 1st Bao				The second son	陈	国宝	Male	Dog	1922	2
Fujin Xiang 1st Bao				The third son	陈	国山	Male	Rat	1924	5
Fujin Xiang 1st Bao				The third son's wife	王	三旦	Female	Rabbit	1927	6
Fujin Xiang 1st Bao	318	10	1	The household head	陈	义全	Male	Tiger	1879	10
Fujin Xiang 1st Bao				The household head's wife	赵	双羊	Female	Goat	1881	6
Fujin Xiang 1st Bao				Granddaughter	陈	所娃	Female	Rat	1924	3
Fujin Xiang 1st Bao				Great grandson	陈	谢臣	Male	Dog	1946	1
Fujin Xiang 1st Bao				Great Granddaughter	陈	义女	Female	Monkey	1944	4
Fujin Xiang 1st Bao				Family member	陈	彦山	Male	Rat	1912	6
Fujin Xiang 1st Bao	319	10	2	The household head	陈	国海	Male	Snake	1904	3
Fujin Xiang 1st Bao				The household head's wife	罗	谢娃	Female	Rat	1912	6
Fujin Xiang 1st Bao				Father	陈	义贤	Male	Snake	1879	3
Fujin Xiang 1st Bao				Mother	张	香女	Female	Rooster	1884	12
Fujin Xiang 1st Bao				Son	陈	守祥	Male	Dog	1922	2
Fujin Xiang 1st Bao				Son's wife	韩	东香	Female	Pig	1923	5
Fujin Xiang 1st Bao				Grandson	陈	早来儿	Male	Snake	1942	5

Date	The beginning time of inhabitation in this county	Educational level	Profession sector	Occupation	Marital status	Physical disability	Religion
1	Inhabited for generations						
21	Inhabited for generations	Illiteracy	Agriculture	Land-holding peasant	Unmarried	No	No religion
19	Inhabited for generations	Illiteracy	Agriculture	Land-holding peasant	Married	No	No religion
1	Inhabited for generations	Illiteracy	Agriculture	Land-holding peasant	Married	No	No religion
27	Inhabited for generations	Illiteracy	Agriculture	Land-holding peasant	Married	No	No religion
22	Inhabited for generations	Illiteracy	Agriculture	Land-holding peasant	Married	No	No religion
16	Inhabited for generations	Illiteracy	Agriculture	Land-holding peasant	Unmarried	No	No religion
22	Inhabited for generations						
22	Inhabited for generations						
24	Inhabited for generations	Illiteracy	Agriculture	Land-holding peasant	Married	No	No religion
23	Inhabited for generations	Illiteracy	Agriculture	Land-holding peasant	Married	No	No religion
5	Inhabited for generations	Illiteracy	Agriculture	Land-holding peasant	Married	No	No religion
20	Inhabited for generations	Illiteracy	Unemployed	Land-holding peasant	Married	No	No religion
17	Inhabited for generations	Illiteracy	Unemployed	Land-holding peasant	Married	No	No religion
19	Inhabited for generations	Illiteracy	Agriculture	Land-holding peasant	Married	No	No religion
31	Inhabited for generations	Illiteracy	Agriculture	Land-holding peasant	Married	No	No religion
2	Inhabited for generations						

Appendix: Overview of Microdata for Wudu County (*cont.*)

									Birthdate	
	The no. of house-hold	**Jia**	**House-hold**	**Relation-ship with the house-hold head**	**Surname**	**First name**	**Sex**	**The Chinese zodiac**	**Year**	**Month**
Fujin Xiang 1st Bao				Grand-daughter	陈	早女	Female	Dog	1946	8
Fujin Xiang 1st Bao				Daughter	陈	三菜娃	Female	Rooster	1933	1
Fujin Xiang 1st Bao				The second daughter	陈	五女	Female	Tiger	1938	6
Fujin Xiang 1st Bao				Uncle's wife	谈	林女	Female	Goat	1894	1
Fujin Xiang 1st Bao				Younger sister	陈	祼居娃	Female	Dragon	1916	2
Fujin Xiang 1st Bao				Nephew	陈	元道	Male	Rabbit	1939	7
Fujin Xiang 1st Bao				The second nephew	陈	聘元	Male	Snake	1941	10
Fujin Xiang 1st Bao				The third nephew	陈	崴元	Male	Rooster	1945	2
Fujin Xiang 1st Bao				Niece	陈	篩女	Female	Rat	1936	1
Fujin Xiang 1st Bao				Family member	国	正	Male	Tiger	1914	8
Fujin Xiang 1st Bao	320	10	3	The house-hold head	王	万清	Male	Pig	1912	10
Fujin Xiang 1st Bao				The house-hold head's wife	张	三路	Female	Rabbit	1915	12
Fujin Xiang 1st Bao				Mother	赵	见儿	Female	Dragon	1891	9
Fujin Xiang 1st Bao				Son	王	兹贵	Male	Dog	1946	8
Fujin Xiang 1st Bao				Daughter	王	吾香儿	Female	Dragon	1940	2
Fujin Xiang 1st Bao				Younger brother	王	万万	Male	Monkey	1932	2
Fujin Xiang 1st Bao	321	10	4	The house-hold head	陈	国青	Male	Horse	1918	10
Fujin Xiang 1st Bao				The house-hold head's wife	李	么娃	Female	Rooster	1921	7

Date	The beginning time of inhabitation in this county	Educational level	Profession sector	Occupation	Marital status	Physical disability	Religion
9	Inhabited for generations						
22	Inhabited for generations						
6	Inhabited for generations						
21	Inhabited for generations	Illiteracy	Agriculture	Land-holding peasant	Widowed	No	No religion
1	Inhabited for generations	Illiteracy	Agriculture	Land-holding peasant	Married	No	No religion
28	Inhabited for generations						
12	Inhabited for generations						
30	Inhabited for generations						
9	Inhabited for generations						
3	Inhabited for generations	Illiteracy	Agriculture	Land-holding peasant	Married	No	No religion
10	Inhabited for generations	Illiteracy	Agriculture	Land-holding peasant	Married	No	No religion
1	Inhabited for generations	Illiteracy	Agriculture	Land-holding peasant	Married	No	No religion
5	Inhabited for generations	Illiteracy	Agriculture	Land-holding peasant	Widowed	No	No religion
12	Inhabited for generations						
1	Inhabited for generations						
4	Inhabited for generations	Bao school 1 year			Unmarried		
5	Inhabited for generations	Illiteracy	Agriculture	Land-holding peasant	Married	No	No religion
1	Inhabited for generations	Illiteracy	Agriculture	Land-holding peasant	Married	No	No religion

Appendix: Overview of Microdata for Wudu County (*cont.*)

	The no. of household	Jia	Household	Relationship with the household head	Surname	First name	Sex	The Chinese zodiac	Birthdate Year	Month
Fujin Xiang 1st Bao				Mother	崔	林娃	Female	Monkey	1895	4
Fujin Xiang 1st Bao				Younger brother	陈	国发	Male	Dog	1922	1
Fujin Xiang 1st Bao				The third younger brother	陈	袓生	Male	Monkey	1932	6
Fujin Xiang 1st Bao	322	10	5	The household head	陈	义兴	Male	Monkey	1901	7
Fujin Xiang 1st Bao				The household head's wife	史	如娃	Female	Monkey	1912	5
Fujin Xiang 1st Bao				Father	陈	如元	Male	Horse	1881	3
Fujin Xiang 1st Bao				Son	陈	郭生儿	Male	Pig	1935	4
Fujin Xiang 1st Bao				Younger brother	陈	义荣	Male	Goat	1919	6
Fujin Xiang 1st Bao				Younger brother's wife	张	毛娃	Female	Ox	1925	9
Fujin Xiang 1st Bao				Uncle's son	陈	义转	Male	Monkey	1895	12
Fujin Xiang 1st Bao				Nephew's wife	王	崔娃	Female	Rooster	1908	1
Fujin Xiang 1st Bao				Nephew's son	陈	寸娃	Male	Monkey	1932	12
Fujin Xiang 1st Bao				Nephew's second son	陈	雷生儿	Male	Goat	1943	3
Fujin Xiang 1st Bao				The second granddaughter	朝	玉娃	Female	Dragon	1940	3
Fujin Xiang 1st Bao	323	10	6	The household head	韩	进富	Male	Rabbit	1915	2
Fujin Xiang 1st Bao				The household head's wife	亩	崴丑	Female	Rabbit	1915	3

Date	The beginning time of inhabitation in this county	Educational level	Profession sector	Occupation	Marital status	Physical disability	Religion
1	Inhabited for generations	Illiteracy	Agriculture	Land-holding peasant	Widowed	No	No religion
1	Inhabited for generations	Illiteracy	Agriculture	Land-holding peasant	Unmarried	No	No religion
1	Inhabited for generations	Illiteracy	Agriculture	Land-holding peasant	Unmarried	No	No religion
1	Inhabited for generations	Illiteracy	Agriculture	Land-holding peasant	Married	No	No religion
5	Inhabited for generations	Illiteracy	Agriculture	Land-holding peasant	Married	No	No religion
3	Inhabited for generations	Illiteracy	Agriculture	Land-holding peasant	Widowed	No	No religion
1	Inhabited for generations						
1	Inhabited for generations	Illiteracy	Agriculture	Land-holding peasant	Married	No	No religion
1	Inhabited for generations	Illiteracy	Agriculture	Land-holding peasant	Married	No	No religion
1	Inhabited for generations	Illiteracy	Agriculture	Land-holding peasant	Widowed	No	No religion
9	Inhabited for generations	Illiteracy	Agriculture	Land-holding peasant	Widowed	No	No religion
5	Inhabited for generations						
1	Inhabited for generations						
16							
17	Inhabited for generations	Illiteracy	Agriculture	Land-holding peasant	Married	No	No religion
17	Inhabited for generations	Illiteracy	Agriculture	Land-holding peasant	Married	No	No religion

Appendix: Overview of Microdata for Wudu County (*cont.*)

	The no. of household	Jia	Household	Relationship with the household head	Surname	First name	Sex	The Chinese zodiac	Birthdate Year	Birthdate Month
Fujin Xiang 1st Bao				Father	韩	仲仙	Male	Snake	1892	10
Fujin Xiang 1st Bao				Son	韩	金才	Male	Dog	1946	11
Fujin Xiang 1st Bao				Younger brother's wife	赵	焕朝	Female	Rooster	1921	7
Fujin Xiang 1st Bao				Niece	韩	香娃	Female	Rooster	1945	1
Fujin Xiang 1st Bao				Nephew	韩	生生	Male	Horse	1918	7
Fujin Xiang 1st Bao				Nephew's wife	马	祥子	Female	Horse	1918	10
Fujin Xiang 1st Bao				Nephew	韩	高正	Male	Dog	1946	7
Fujin Xiang 1st Bao				Niece	马	扣儿	Female	Dragon	1940	5
Fujin Xiang 1st Bao	324	10	7	The household head	韩	仲荣	Male	Monkey	1907	2
Fujin Xiang 1st Bao				The household head's wife	马	祥娃	Female	Dragon	1916	4
Fujin Xiang 1st Bao				Daughter	韩	马女	Female	Dog	1934	4
Fujin Xiang 1st Bao				The second daughter	韩	进女	Female	Rooster	1945	10
Fujin Xiang 1st Bao				The third daughter	韩	苏英	Female	Dog	1946	3
Fujin Xiang 1st Bao				Younger brother	韩	仲保	Male	Ox	1913	7
Fujin Xiang 1st Bao				Younger brother's wife	鲁	月连	Female	Rooster	1921	4
Fujin Xiang 1st Bao				Niece	韩	杜娃	Female	Tiger	1938	2
Fujin Xiang 1st Bao				The second niece	韩	英娃	Female	Dog	1946	2

Date	The beginning time of inhabitation in this county	Educational level	Profession sector	Occupation	Marital status	Physical disability	Religion
18	Inhabited for generations	Illiteracy	Agriculture	Land-holding peasant	Widowed	No	No religion
1	Inhabited for generations						
10	Inhabited for generations	Illiteracy	Agriculture	Land-holding peasant	Widowed	No	No religion
21	Inhabited for generations						
18	Inhabited for generations	Illiteracy	Agriculture	Land-holding peasant	Married	No	No religion
13	Inhabited for generations	Illiteracy	Agriculture	Land-holding peasant	Married	No	No religion
3	Inhabited for generations						
5	Inhabited for generations						
5	Inhabited for generations	Illiteracy	Agriculture	Land-holding peasant	Married	No	No religion
14	Inhabited for generations	Illiteracy	Agriculture	Land-holding peasant	Married	No	No religion
18	Inhabited for generations	Illiteracy	Agriculture		Unmarried		
16	Inhabited for generations						
4	Inhabited for generations						
5	Inhabited for generations	Illiteracy	Agriculture	Land-holding peasant	Married	No	No religion
10	Inhabited for generations	Illiteracy	Agriculture	Land-holding peasant	Married	No	No religion
4	Inhabited for generations						
14	Inhabited for generations						

Appendix: Overview of Microdata for Wudu County (*cont.*)

	The no. of household	Jia	Household	Relationship with the household head	Surname	First name	Sex	The Chinese zodiac	Birthdate	
									Year	Month
Fujin Xiang 1st Bao	325	10	8	The household head	韩	仲禄	Male	Ox	1913	10
Fujin Xiang 1st Bao				The household head's wife	李	菊娃	Female	Ox	1913	8
Fujin Xiang 1st Bao				Daughter	韩	才女	Female	Goat	1931	8
Fujin Xiang 1st Bao				The second daughter	韩	苏香	Female	Pig	1934	5
Fujin Xiang 1st Bao				The third daughter	韩	讨女	Female	Horse	1942	8
Fujin Xiang 1st Bao				Cohabitant	韩	家	Female	Ox	1925	2
Fujin Xiang 1st Bao				Daughter	韩	同女	Female	Dog	1946	2
Fujin Xiang 1st Bao				Family member	赵	三娃	Male	Goat	1919	9
Fujin Xiang 1st Bao	326	10	9	The household head	韩	仲玉	Male	Dragon	1903	9
Fujin Xiang 1st Bao				The household head's wife	马	香儿	Female	Tiger	1901	8
Fujin Xiang 1st Bao				Son	韩	招存	Male	Pig	1935	9
Fujin Xiang 1st Bao				The second son	韩	崴存	Male	Tiger	1938	2
Fujin Xiang 1st Bao				Daughter	韩	董家	Female	Monkey	1932	11
Fujin Xiang 1st Bao				The second daughter	韩	进女	Female	Snake	1941	11
Fujin Xiang 1st Bao	327	10	10	The household head	韩	进福	Male	Dragon	1916	3
Fujin Xiang 1st Bao				The household head's wife	赵	四娃	Female	Snake	1917	4
Fujin Xiang 1st Bao				Father	韩	仲林	Male	Goat	1894	10

Date	The beginning time of inhabitation in this county	Educational level	Profession sector	Occupation	Marital status	Physical disability	Religion
16	Inhabited for generations	Illiteracy	Agriculture	Land-holding peasant	Married	No	No religion
19	Inhabited for generations	Illiteracy	Agriculture	Land-holding peasant	Married	No	No religion
19	Inhabited for generations	Illiteracy					
5	Inhabited for generations						
13	Inhabited for generations						
10	Inhabited for generations	Illiteracy	Agriculture	Land-holding peasant	Married	No	No religion
18	Inhabited for generations						
29	Inhabited for generations	Illiteracy	Agriculture	Land-holding peasant	Married	No	No religion
7	Inhabited for generations	Illiteracy	Agriculture	Land-holding peasant	Married	No	No religion
8	Inhabited for generations	Illiteracy	Agriculture	Land-holding peasant	Married	No	No religion
3	Inhabited for generations	Bao school 3 years					
25	Inhabited for generations	Illiteracy					
3	Inhabited for generations	Illiteracy			Unmarried		
23	Inhabited for generations	Illiteracy					
28	Inhabited for generations	Illiteracy	Agriculture	Land-holding peasant	Married	No	No religion
1	Inhabited for generations	Illiteracy	Others	Land-holding peasant	Married	No	No religion
17	Inhabited for generations	Illiteracy	Agriculture	Land-holding peasant	Married	No	No religion

Appendix: Overview of Microdata for Wudu County (*cont.*)

	The no. of household	Jia	Household	Relationship with the household head	Surname	First name	Sex	The Chinese zodiac	Birthdate Year	Month
Fujin Xiang 1st Bao				Mother	赵	科女	Female	Horse	1893	8
Fujin Xiang 1st Bao				Son	韩	新来	Male	Tiger	1938	7
Fujin Xiang 1st Bao				Daughter	韩	新女	Female	Snake	1941	12
Fujin Xiang 1st Bao				The second daughter	韩	兹女儿	Female	Rooster	1945	8
Fujin Xiang 1st Bao				The third daughter	韩	毛娃	Female	Pig	1947	4
Fujin Xiang 1st Bao	328	10	11	The household head	赵	学有	Male	Pig	1898	3
Fujin Xiang 1st Bao				The household head's wife	史	客娃	Female	Tiger	1901	2
Fujin Xiang 1st Bao				Daughter	赵	苏家女	Female	Rooster	1933	2
Fujin Xiang 1st Bao				Younger brother	赵	学生	Male	Goat	1906	5
Fujin Xiang 1st Bao				Younger brother's wife	史	双玄儿	Female	Dragon	1916	4
Fujin Xiang 1st Bao				Niece	赵	此家	Female	Rat	1936	8
Fujin Xiang 1st Bao				The second niece	崴	此儿	Female	Snake	1941	12
Fujin Xiang 1st Bao				The third niece	稀	奇儿	Female	Snake	1941	12
Fujin Xiang 1st Bao				Nephew	送	成成	Male	Pig	1947	2
Fujin Xiang 1st Bao				Uncle's son	赵	学山	Male	Monkey	1895	9
Fujin Xiang 1st Bao	329	10	12	The household head	赵	学玉	Male	Dog	1909	8
Fujin Xiang 1st Bao				The household head's wife	李	连娃	Female	Tiger	1901	12

Date	The beginning time of inhabitation in this county	Educational level	Profession sector	Occupation	Marital status	Physical disability	Religion
1	Inhabited for generations	Illiteracy	Others	Land-holding peasant	Married	No	No religion
27	Inhabited for generations	Bao school 3 years					
8	Inhabited for generations						
12	Inhabited for generations						
25	Inhabited for generations						
23	Inhabited for generations	Illiteracy	Agricul-ture	Land-holding peasant	Married	No	No religion
2	Inhabited for generations	Illiteracy	Agricul-ture	Land-holding peasant	Married	No	No religion
13	Inhabited for generations	Illiteracy	Agricul-ture	Land-holding peasant			
15	Inhabited for generations	Illiteracy	Agricul-ture	Land-holding peasant	Married	No	No religion
12	Inhabited for generations	Illiteracy	Agricul-ture	Land-holding peasant	Married	No	No religion
8	Inhabited for generations	Illiteracy					
6	Inhabited for generations	Illiteracy					
6	Inhabited for generations	Illiteracy					
25	Inhabited for generations						
2	Inhabited for generations	Illiteracy	Agricul-ture	Land-holding peasant	Widowed	No	No religion
29	Inhabited for generations	Illiteracy	Agricul-ture	Land-holding peasant	Married	No	No religion
30	Inhabited for generations	Illiteracy	Agricul-ture	Land-holding peasant	Married	No	No religion

Appendix: Overview of Microdata for Wudu County (*cont.*)

	The no. of household	Jia	Household	Relationship with the household head	Surname	First name	Sex	The Chinese zodiac	Birthdate Year	Month
Fujin Xiang 1st Bao				Son	杨	生娃	Male	Rat	1924	11
Fujin Xiang 1st Bao				Son's wife	赵	书香	Female	Dog	1922	8
Fujin Xiang 1st Bao				The second son	赵	元成	Male	Goat	1931	2
Fujin Xiang 1st Bao				Younger brother	赵	学明	Male	Dragon	1916	5
Fujin Xiang 1st Bao				Younger brother's wife	崔	祥娃	Female	Goat	1919	11
Fujin Xiang 1st Bao				Nephew	赵	谢娃	Male	Rooster	1945	12
Fujin Xiang 1st Bao				The second nephew	赵	毛娃	Male	Pig	1947	4
Fujin Xiang 1st Bao	330	10	13	The household head	付	登才	Male	Monkey	1895	3
Fujin Xiang 1st Bao				The household head's wife	李	跳香	Female	Horse	1893	9
Fujin Xiang 1st Bao				Mother	徐	镇赛	Female	Pig	1874	6
Fujin Xiang 1st Bao				Son	付	君四儿	Male	Rabbit	1927	7
Fujin Xiang 1st Bao				Son's wife	陈	亥娃	Female	Pig	1923	9
Fujin Xiang 1st Bao	331	10	14	The household head	付	登云	Male	Monkey	1907	3
Fujin Xiang 1st Bao				The household head's wife	张	润娃	Female	Monkey	1920	8
Fujin Xiang 1st Bao				Son	付	杨兹	Male	Dragon	1940	11
Fujin Xiang 1st Bao				The second son	陈	娃	Male	Monkey	1944	5
Fujin Xiang 1st Bao				The third son	付	张信	Male	Pig	1947	2

Date	The beginning time of inhabitation in this county	Educational level	Profession sector	Occupation	Marital status	Physical disability	Religion
1	Inhabited for generations	Illiteracy	Agriculture	Land-holding peasant	Married	No	No religion
28	Inhabited for generations	Illiteracy	Agriculture	Land-holding peasant	Married	No	No religion
23	Inhabited for generations	Lower primary school 5 years					
3	Inhabited for generations	Illiteracy	Agriculture	Land-holding peasant	Married	No	No religion
25	Inhabited for generations	Illiteracy	Agriculture	Land-holding peasant	Married	No	No religion
28	Inhabited for generations						
16	Inhabited for generations						
1	Inhabited for generations	Illiteracy	Agriculture	Land-holding peasant	Married	No	No religion
1	Inhabited for generations	Illiteracy	Agriculture	Land-holding peasant	Married	No	No religion
4	Inhabited for generations	Illiteracy	Unemployed				
29	Inhabited for generations	Illiteracy	Agriculture	Land-holding peasant	Married	No	No religion
15	Inhabited for generations	Illiteracy	Agriculture	Land-holding peasant	Married	No	No religion
22	Inhabited for generations	Illiteracy	Agriculture	Land-holding peasant	Married	No	No religion
25	Inhabited for generations	Illiteracy	Agriculture	Land-holding peasant	Married	No	No religion
3	Inhabited for generations						
4	Inhabited for generations						
8	Inhabited for generations						

Appendix: Overview of Microdata for Wudu County (*cont.*)

	The no. of house-hold	Jia	House-hold	Relation-ship with the house-hold head	Surname	First name	Sex	The Chinese zodiac	Birthdate Year	Month
Fujin Xiang 1st Bao				Younger brother	付	登山	Male	Dog	1909	7
Fujin Xiang 1st Bao				Younger brother's wife	崔	安娃	Female	Ox	1913	8
Fujin Xiang 1st Bao				Nephew	付	仲堂	Male	Dog	1934	12
Fujin Xiang 1st Bao				The second nephew	付	仲林	Male	Rat	1936	6
Fujin Xiang 1st Bao				Niece	马	女	Female	Rabbit	1939	9
Fujin Xiang 1st Bao	332	10	15	The house-hold head	陈	登云	Male	Dragon	1903	3
Fujin Xiang 1st Bao				The house-hold head's wife	赵	官连	Female	Snake	1904	3
Fujin Xiang 1st Bao				Uncle	赵	学贤	Male	Monkey	1871	3
Fujin Xiang 1st Bao				Son	陈	松柏	Male	Rooster	1945	9
Fujin Xiang 1st Bao				Daughter	陈	唤器儿	Female	Tiger	1938	8
Fujin Xiang 1st Bao				Family member	史	桃秀儿	Female	Tiger	1901	3
Fujin Xiang 1st Bao	333	11	1	The house-hold head	韩	进锡	Male	Monkey	1907	1
Fujin Xiang 1st Bao				The house-hold head's wife	李	登娃	Female	Monkey	1907	2
Fujin Xiang 1st Bao				Father	韩	仲学	Male	Rooster	1884	12
Fujin Xiang 1st Bao				Mother	陈	连儿	Female	Rooster	1884	6
Fujin Xiang 1st Bao				Daughter	生	陈女	Female	Dog	1934	11

Date	The beginning time of inhabitation in this county	Educational level	Profession sector	Occupation	Marital status	Physical disability	Religion
18	Inhabited for generations	Illiteracy	Agriculture	Land-holding peasant	Married	No	No religion
6	Inhabited for generations	Illiteracy	Agriculture	Land-holding peasant	Married	No	No religion
31	Inhabited for generations	Lower primary school 5 years					
21	Inhabited for generations	Lower primary school 4 years					
1	Inhabited for generations						
2	Inhabited for generations	Illiteracy	Agriculture	Land-holding peasant	Married	No	No religion
20	Inhabited for generations	Illiteracy	Agriculture	Land-holding peasant	Married	No	No religion
5	Inhabited for generations						
12	Inhabited for generations						
24	Inhabited for generations						
24	Inhabited for generations	Illiteracy	Agriculture	Land-holding peasant	Married	No	No religion
14	Inhabited for generations	Illiteracy	Agriculture	Land-holding peasant	Married	No	No religion
21	Inhabited for generations	Illiteracy	Agriculture	Land-holding peasant	Married	No	No religion
2	Inhabited for generations	Illiteracy	Agriculture	Land-holding peasant	Married	No	No religion
2	Inhabited for generations	Illiteracy	Agriculture	Land-holding peasant	Married	No	No religion
11	Inhabited for generations						

Appendix: Overview of Microdata for Wudu County (*cont.*)

	The no. of household	Jia	Household	Relationship with the household head	Surname	First name	Sex	The Chinese zodiac	Birthdate Year	Birthdate Month
Fujin Xiang 1st Bao				Younger brother	韩	进花	Male	Rooster	1921	4
Fujin Xiang 1st Bao				Younger brother's wife	李	俊娃	Female	Monkey	1920	11
Fujin Xiang 1st Bao				Niece	韩	贵女	Female	Tiger	1938	7
Fujin Xiang 1st Bao				Adopted son	王	玉根	Male	Goat	1931	7
Fujin Xiang 1st Bao				Cohabitant	韩	李相	Male	Horse	1918	7
Fujin Xiang 1st Bao	334	11	2	The household head	韩	文财	Male	Snake	1904	1
Fujin Xiang 1st Bao				The household head's wife	杨	六娃	Female	Goat	1906	3
Fujin Xiang 1st Bao				Son	韩	富才	Male	Dog	1934	6
Fujin Xiang 1st Bao				Daughter	韩	兹女儿	Female	Rooster	1935	8
Fujin Xiang 1st Bao				Older brother's wife	王	秋家	Female	Tiger	1889	4
Fujin Xiang 1st Bao				Nephew's wife	王	翠女	Female	Dog	1880	4
Fujin Xiang 1st Bao				Nephew's son	胡	家女	Female	Horse	1942	11
Fujin Xiang 1st Bao				The second nephew	韩	双宣	Male	Rabbit	1915	11
Fujin Xiang 1st Bao				Nephew's son	韩	孩贵保	Male	Monkey	1944	8
Fujin Xiang 1st Bao				Nephew	韩	进娃	Male	Dragon	1928	12
Fujin Xiang 1st Bao				Nephew's wife	王	水娃	Female	Dragon	1916	3
Fujin Xiang 1st Bao				The fourth nephew	五	朝儿	Male	Horse	1930	11
Fujin Xiang 1st Bao				Uncle's son's wife	张	丁家	Female	Rooster	1896	6

Date	The beginning time of inhabitation in this county	Educational level	Profession sector	Occupation	Marital status	Physical disability	Religion
14	Inhabited for generations	Illiteracy	Agriculture	Land-holding peasant	Married	No	No religion
11	Inhabited for generations	Illiteracy	Agriculture	Land-holding peasant	Married	No	No religion
26	Inhabited for generations						
13	Inhabited for generations	Primary school 6 years					
1	Inhabited for generations	Illiteracy	Agriculture	Land-holding peasant	Unmarried	No	No religion
19	Inhabited for generations	Illiteracy	Agriculture	Land-holding peasant	Married	No	No religion
21	Inhabited for generations	Illiteracy	Agriculture	Land-holding peasant	Married	No	No religion
13	Inhabited for generations	Illiteracy					
7	Inhabited for generations	Illiteracy	Agriculture	Land-holding peasant			
17	Inhabited for generations	Illiteracy	Agriculture	Land-holding peasant	Widowed	No	No religion
19	Inhabited for generations	Illiteracy	Agriculture	Land-holding peasant	Widowed	No	No religion
19	Inhabited for generations						
11	Inhabited for generations	Illiteracy	Agriculture	Land-holding peasant	Widowed	No	No religion
23	Inhabited for generations						
10	Inhabited for generations	Illiteracy	Agriculture	Land-holding peasant	Unmarried	No	No religion
1	Inhabited for generations	Illiteracy	Agriculture	Land-holding peasant	Married	No	No religion
1	Inhabited for generations	Illiteracy	Agriculture	Land-holding peasant	Unmarried	No	No religion
30	Inhabited for generations	Illiteracy	Agriculture	Land-holding peasant	Widowed	No	No religion

Appendix: Overview of Microdata for Wudu County (*cont.*)

									Birthdate	
	The no. of household	Jia	Household	Relationship with the household head	Surname	First name	Sex	The Chinese zodiac	Year	Month
Fujin Xiang 1st Bao				Nephew	考	才儿	Male	Rat	1924	10
Fujin Xiang 1st Bao				Nephew's wife	聂	胖娃	Female	Pig	1923	7
Fujin Xiang 1st Bao				Nephew	王	财儿	Male	Tiger	1926	8
Fujin Xiang 1st Bao				Nephew's wife	赵	润月	Female	Horse	1930	3
Fujin Xiang 1st Bao				Niece	韩	堂女	Female	Dog	1922	2
Fujin Xiang 1st Bao	335	11	3	The household head	韩	仲臣	Male	Rat	1912	8
Fujin Xiang 1st Bao				The household head's wife	李	贤娃	Female	Ox	1913	7
Fujin Xiang 1st Bao				Father	汉	文向	Male	Monkey	1905	7
Fujin Xiang 1st Bao				Daughter	韩	双女子	Female	Horse	1930	11
Fujin Xiang 1st Bao				Younger brother	韩	仲发	Male	Horse	1930	11
Fujin Xiang 1st Bao				Younger brother's wife	董	贵娃	Female	Dragon	1928	6
Fujin Xiang 1st Bao				Family member	陈	进财	Male	Goat	1931	9
Fujin Xiang 1st Bao	336	11	4	The household head	韩	进铭	Male	Rabbit	1915	13
Fujin Xiang 1st Bao				The household head's wife	陈	称女	Female	Pig	1910	2
Fujin Xiang 1st Bao				Daughter	韩	贵英	Female	Ox	1937	4
Fujin Xiang 1st Bao				Younger brother	韩	进山	Male	Rat	1924	11
Fujin Xiang 1st Bao				Younger brother's wife	赵	秋娃	Female	Ox	1925	9
Fujin Xiang 1st Bao				Niece	韩	早临	Female	Monkey	1944	1

Date	The beginning time of inhabitation in this county	Educational level	Profession sector	Occupation	Marital status	Physical disability	Religion
5	Inhabited for generations	Illiteracy	Agriculture	Land-holding peasant	Married	No	No religion
21	Inhabited for generations	Illiteracy	Agriculture	Land-holding peasant	Married	No	No religion
14	Inhabited for generations	Illiteracy	Agriculture	Land-holding peasant	Married	No	No religion
17	Inhabited for generations	Illiteracy	Agriculture	Land-holding peasant	Married	No	No religion
21	Inhabited for generations	Illiteracy	Agriculture	Land-holding peasant	Unmarried	No	No religion
19	Inhabited for generations	Illiteracy	Agriculture	Land-holding peasant	Married	No	No religion
9	Inhabited for generations	Illiteracy	Agriculture	Land-holding peasant	Married	No	No religion
2	Inhabited for generations				Widowed		
17	Inhabited for generations	Illiteracy	Agriculture	Land-holding peasant	Married	No	No religion
22	Inhabited for generations	Illiteracy	Agriculture	Land-holding peasant	Married	No	No religion
1	Inhabited for generations	Illiteracy	Agriculture	Land-holding peasant	Married	No	No religion
26	Inhabited for generations	Illiteracy	Agriculture	Land-holding peasant	Married	No	No religion
10	Inhabited for generations	Illiteracy	Agriculture	Land-holding peasant	Married	No	No religion
20	Inhabited for generations	Illiteracy	Agriculture	Land-holding peasant	Married	No	No religion
11	Inhabited for generations						
20	Inhabited for generations	Illiteracy	Agriculture	Land-holding peasant	Married	No	No religion
23	Inhabited for generations	Illiteracy	Agriculture	Land-holding peasant	Married	No	No religion
2	Inhabited for generations						

Appendix: Overview of Microdata for Wudu County (*cont.*)

	The no. of household	Jia	Household	Relationship with the household head	Surname	First name	Sex	The Chinese zodiac	Birthdate	
									Year	Month
Fujin Xiang 1st Bao				Uncle's son's wife	李	东姐	Female	Monkey	1883	2
Fujin Xiang 1st Bao				Nephew	路	根儿	Male	Goat	1919	12
Fujin Xiang 1st Bao				Nephew's son	傅	成儿	Male	Horse	1942	12
Fujin Xiang 1st Bao				Cohabitant	成	永银	Male	Monkey	1907	3
Fujin Xiang 1st Bao	337	11	5	The household head	韩	进荣	Male	Horse	1918	9
Fujin Xiang 1st Bao				The household head's wife	段	知娃	Female	Tiger	1926	9
Fujin Xiang 1st Bao				Father	韩	仲富	Male	Dragon	1891	2
Fujin Xiang 1st Bao				Daughter	韩	蝉女	Female	Snake	1941	3
Fujin Xiang 1st Bao				Uncle	韩	仲全	Male	Rooster	1884	3
Fujin Xiang 1st Bao				Uncle's wife	史	谁双	Female	Goat	1894	8
Fujin Xiang 1st Bao				Uncle's son	韩	进宝	Male	Rooster	1921	6
Fujin Xiang 1st Bao				The second uncle	韩	仲家	Male	Pig	1886	11
Fujin Xiang 1st Bao				Older sister	玉	娃子	Female	Snake	1917	11
Fujin Xiang 1st Bao				Nephew	韩	元生	Male	Pig	1947	3
Fujin Xiang 1st Bao				Family member	韩	进德	Male	Snake	1917	12
Fujin Xiang 1st Bao				Cohabitant	韩	苏毛儿	Male	Ox	1925	12
Fujin Xiang 1st Bao	338	11	6	The household head	赵	学万	Male	Rooster	1908	4
Fujin Xiang 1st Bao				The household head's wife	董	招娃	Female	Dragon	1916	5
Fujin Xiang 1st Bao				Mother	贾	称姐	Female	Dragon	1877	8

Date	The beginning time of inhabitation in this county	Educational level	Profession sector	Occupation	Marital status	Physical disability	Religion	
1	Inhabited for generations	Illiteracy	Agriculture	Land-holding peasant	Married	No	No religion	
6	Inhabited for generations	Illiteracy	Agriculture	Land-holding peasant	Married	No	No religion	
30	Inhabited for generations							
1	Inhabited for generations	Illiteracy	Agriculture	Land-holding peasant	Widowed	No	No religion	
29	Inhabited for generations	Sishu 3 years	Agriculture	Land-holding peasant	Married	No	No religion	
29	Inhabited for generations	Illiteracy	Agriculture	Land-holding peasant	Married	No	No religion	
1	Inhabited for generations	Illiteracy	Agriculture	Land-holding peasant	Widowed	No	No religion	
2	Inhabited for generations							
24	Inhabited for generations	Illiteracy	Agriculture	Land-holding peasant	Married	No	No religion	
1	Inhabited for generations	Illiteracy	Agriculture	Land-holding peasant	Married	No	No religion	
10								
24	Inhabited for generations	Illiteracy	Agriculture	Land-holding peasant	Widowed	No	No religion	
26	Inhabited for generations	Illiteracy	Agriculture	Land-holding peasant	Married	No	No religion	
19	Inhabited for generations							
19	Inhabited for generations	Illiteracy	Agriculture	Land-holding peasant	Married	No	No religion	
6	Inhabited for generations	Illiteracy	Agriculture	Land-holding peasant	Unmarried	No	No religion	Joined the army in 1947
30	Inhabited for generations	Illiteracy	Agriculture	Land-holding peasant	Married	No	No religion	
15	Inhabited for generations	Illiteracy	Agriculture	Land-holding peasant	Married	No	No religion	
14	Inhabited for generations	Illiteracy	Agriculture	Land-holding peasant				

Appendix: Overview of Microdata for Wudu County (*cont.*)

	The no. of household	Jia	Household	Relationship with the household head	Surname	First name	Sex	The Chinese zodiac	Birthdate	
									Year	Month
Fujin Xiang 1st Bao				Son	赵	喜儿	Male	Tiger	1938	2
Fujin Xiang 1st Bao				The second son	赵	满才	Male	Monkey	1944	11
Fujin Xiang 1st Bao				Older brother	赵	学贵	Male	Horse	1905	7
Fujin Xiang 1st Bao				Older brother's wife	李	葳连	Female	Dog	1909	1
Fujin Xiang 1st Bao				Cohabitant	姚	仲云	Male	Dog	1885	11
Fujin Xiang 1st Bao				Cohabitant's wife	王	萧佐	Female	Dog	1885	10
Fujin Xiang 1st Bao	339	11	7	The household head	赵	思福	Male	Rabbit	1915	2
Fujin Xiang 1st Bao				The household head's wife	赵	连娃	Female	Rat	1912	11
Fujin Xiang 1st Bao				Son	赵	保儿	Male	Ox	1937	6
Fujin Xiang 1st Bao				Daughter	赵	必家	Female	Pig	1934	8
Fujin Xiang 1st Bao				The second daughter	赵	来女	Female	Snake	1941	7
Fujin Xiang 1st Bao				The third daughter	赵	九月	Female	Monkey	1944	11
Fujin Xiang 1st Bao				The fourth daughter	赵	正月	Female	Pig	1947	1
Fujin Xiang 1st Bao				Uncle's wife	赵	裉娃	Female	Rat	1899	3
Fujin Xiang 1st Bao				Younger sister	赵	三送子	Female	Horse	1930	1
Fujin Xiang 1st Bao				The second younger sister	赵	家	Female	Ox	1937	11
Fujin Xiang 1st Bao	340	11	8	The household head	赵	学强	Male	Rat	1912	5
Fujin Xiang 1st Bao				The household head's wife	贾	兔娃	Female	Rabbit	1927	5

Date	The beginning time of inhabitation in this county	Educational level	Profession sector	Occupation	Marital status	Physical disability	Religion
6	Inhabited for generations					Blind	
2	Inhabited for generations						
12	Inhabited for generations	Illiteracy	Agriculture	Land-holding peasant	Married	No	No religion
14	Inhabited for generations	Illiteracy	Agriculture	Land-holding peasant	Married	No	No religion
9	Inhabited for generations	Illiteracy	Agriculture	Land-holding peasant	Married	No	No religion
14	Inhabited for generations	Illiteracy	Agriculture	Land-holding peasant	Married	No	No religion
23	Inhabited for generations	Illiteracy	Agriculture	Land-holding peasant	Married	No	No religion
23	Inhabited for generations	Illiteracy	Agriculture	Land-holding peasant	Married	No	No religion
29	Inhabited for generations						
2	Inhabited for generations						
6	Inhabited for generations						
20	Inhabited for generations						
26	Inhabited for generations						
21	Inhabited for generations	Illiteracy	Agriculture	Land-holding peasant	Widowed	No	No religion
23	Inhabited for generations	Illiteracy	Agriculture	Land-holding peasant	Unmarried	No	No religion
19	Inhabited for generations						
28	Inhabited for generations	Illiteracy	Agriculture	Land-holding peasant	Married	No	No religion
1	Inhabited for generations	Illiteracy	Agriculture	Land-holding peasant	Married	No	No religion

Appendix: Overview of Microdata for Wudu County (*cont.*)

									Birthdate	
	The no. of household	Jia	Household	Relationship with the household head	Surname	First name	Sex	The Chinese zodiac	Year	Month
Fujin Xiang 1st Bao				Father	赵	希选	Male	Dragon	1879	7
Fujin Xiang 1st Bao				Mother	李	招儿	Female	Monkey	1883	6
Fujin Xiang 1st Bao				Daughter	赵	琴娃	Female	Rooster	1945	6
Fujin Xiang 1st Bao				Older brother	赵	崴旦	Male	Rooster	1908	11
Fujin Xiang 1st Bao				Older brother's wife	赵	兴娃	Female	Tiger	1914	5
Fujin Xiang 1st Bao				Younger sister	赵	秋英	Female	Goat	1931	9
Fujin Xiang 1st Bao				Cohabitant	袁	花金	Male	Ox	1925	5
Fujin Xiang 1st Bao				Cohabitant's wife	赵	送女	Female	Tiger	1926	9
Fujin Xiang 1st Bao	341	11	9	The household head	赵	元炳	Male	Snake	1917	2
Fujin Xiang 1st Bao				The household head's wife	徐	仓娃	Female	Dragon	1916	3
Fujin Xiang 1st Bao				Mother	李	明儿	Female	Snake	1880	8
Fujin Xiang 1st Bao				Daughter	赵	石女	Female	Dog	1946	4
Fujin Xiang 1st Bao				Niece	崴	喜儿	Female	Monkey	1907	6
Fujin Xiang 1st Bao				Nephew's son	赵	早儿	Male	Monkey	1932	9
Fujin Xiang 1st Bao				The second grandson	赵	補儿	Male	Pig	1935	3
Fujin Xiang 1st Bao				Family member	赵	思玉	Male	Rooster	1914	11
Fujin Xiang 1st Bao				Cohabitant	马	杰	Male	Horse	1918	2
Fujin Xiang 1st Bao				Cohabitant's wife	赵	锁女子	Female	Goat	1919	1

Date	The beginning time of inhabitation in this county	Educational level	Profession sector	Occupation	Marital status	Physical disability	Religion
15	Inhabited for generations	Illiteracy	Agriculture	Land-holding peasant	Married	No	No religion
13	Inhabited for generations	Illiteracy	Agriculture	Land-holding peasant	Married	No	No religion
26	Inhabited for generations						
30	Inhabited for generations	Illiteracy	Agriculture	Land-holding peasant	Married	No	No religion
1	Inhabited for generations	Illiteracy	Agriculture	Land-holding peasant	Married	No	No religion
3	Inhabited for generations						
1	Inhabited for generations	Illiteracy	Agriculture	Land-holding peasant	Married	No	No religion
4	Inhabited for generations	Illiteracy	Agriculture	Peasant	Married	No	No religion
3	Inhabited for generations	Illiteracy	Agriculture	Land-holding peasant	Married	No	No religion
1	Inhabited for generations	Illiteracy	Agriculture	Land-holding peasant	Married	No	No religion
22	Inhabited for generations	Illiteracy	Agriculture	Land-holding peasant	Widowed	No	No religion
2	Inhabited for generations						
1	Inhabited for generations	Illiteracy	Agriculture	Land-holding peasant	Widowed	No	No religion
1	Inhabited for generations						
16	Inhabited for generations						
4	Inhabited for generations	Illiteracy	Agriculture	Land-holding peasant	Unmarried	No	No religion
19	Inhabited for generations	Illiteracy	Agriculture	Land-holding peasant	Married	No	No religion
16	Inhabited for generations	Illiteracy	Agriculture	Land-holding peasant	Married	No	No religion

Appendix: Overview of Microdata for Wudu County (*cont.*)

	The no. of household	Jia	Household	Relationship with the household head	Surname	First name	Sex	The Chinese zodiac	Birthdate	
									Year	Month
Fujin Xiang 1st Bao				Cohabitant's daughter	赵	拆娃	Female	Goat	1919	1
Fujin Xiang 1st Bao				Cohabitant's second daughter	赵	谷卖	Female	Rooster	1945	10
Fujin Xiang 1st Bao	342	11	10	The household head	赵	元虎	Male	Rat	1899	2
Fujin Xiang 1st Bao				The household head's wife	唐	还娃	Female	Monkey	1895	11
Fujin Xiang 1st Bao				Mother	赵	氏	Female	Tiger	1865	5
Fujin Xiang 1st Bao				Son	赵	思顺	Male	Pig	1923	8
Fujin Xiang 1st Bao				Son's wife	陈	发亥	Female	Pig	1923	1
Fujin Xiang 1st Bao				Granddaughter	赵	智娃	Female	Rooster	1945	2
Fujin Xiang 1st Bao				Daughter	赵	菊连儿	Female	Pig	1935	7
Fujin Xiang 1st Bao	343	11	11	The household head	赵	韩娃	Male	Horse	1905	2
Fujin Xiang 1st Bao				The household head's wife	李	裉女	Female	Snake	1904	4
Fujin Xiang 1st Bao				Son	赵	思全	Male	Rooster	1934	1
Fujin Xiang 1st Bao				Younger brother's wife	赵	三女子	Female	Dragon	1903	5
Fujin Xiang 1st Bao				Nephew	赵	思财	Male	Tiger	1926	4
Fujin Xiang 1st Bao				Nephew's wife	徐	叩明	Female	Tiger	1926	7
Fujin Xiang 1st Bao				The second nephew	赵	聚才	Male	Snake	1941	4
Fujin Xiang 1st Bao				Nephew	赵	思林	Male	Rabbit	1915	1

Date	The beginning time of inhabitation in this county	Educational level	Profession sector	Occupation	Marital status	Physical disability	Religion
16	Inhabited for generations	Illiteracy	Agriculture	Land-holding peasant	Married	No	No religion
25	Inhabited for generations						
13	Inhabited for generations	Illiteracy	Agriculture	Land-holding peasant	Married	No	No religion
17	Inhabited for generations	Illiteracy	Agriculture	Land-holding peasant	Married	No	No religion
13	Inhabited for generations	Illiteracy	Agriculture	Land-holding peasant	Widowed	No	No religion
2	Inhabited for generations	Illiteracy	Agriculture	Land-holding peasant	Married	No	No religion
30	Inhabited for generations	Illiteracy	Agriculture	Land-holding peasant	Married	No	No religion
15	Inhabited for generations						
27	Inhabited for generations						
17	Inhabited for generations	Illiteracy	Agriculture	Land-holding peasant	Married	No	No religion
21	Inhabited for generations	Illiteracy	Agriculture	Land-holding peasant	Married	No	No religion
16	Inhabited for generations	Illiteracy	Agriculture	Land-holding peasant	Married	No	No religion
1	Inhabited for generations	Illiteracy	Agriculture	Land-holding peasant	Married	No	No religion
11	Inhabited for generations	Illiteracy	Agriculture	Land-holding peasant	Married	No	No religion
4	Inhabited for generations	Illiteracy	Agriculture	Land-holding peasant	Married	No	No religion
1	Inhabited for generations						
1	Inhabited for generations	Illiteracy	Agriculture	Land-holding peasant	Married	No	No religion

Appendix: Overview of Microdata for Wudu County (*cont.*)

	The no. of household	Jia	Household	Relationship with the household head	Surname	First name	Sex	The Chinese zodiac	Birthdate	
									Year	Month
Fujin Xiang 1st Bao				Nephew's wife	雷	贱娃	Female	Rabbit	1915	1
Fujin Xiang 1st Bao				Nephew's son	赵	五斤	Male	Tiger	1938	9
Fujin Xiang 1st Bao				Nephew's second son	赵	欸娃	Male	Horse	1943	6
Fujin Xiang 1st Bao	344	11	12	The household head	赵	元部	Male	Tiger	1901	7
Fujin Xiang 1st Bao				The household head's wife	岳	菜娃	Female	Ox	1900	7
Fujin Xiang 1st Bao				Mother	赵	苏姐	Female	Rat	1875	10
Fujin Xiang 1st Bao				Daughter	赵	不稀器	Female	Rat	1924	4
Fujin Xiang 1st Bao				The second daughter	赵	反娃子	Female	Goat	1931	3
Fujin Xiang 1st Bao				Uncle	赵	学洪	Male	Snake	1892	5
Fujin Xiang 1st Bao				Uncle's daughter	赵	月娃	Female	Rooster	1921	9
Fujin Xiang 1st Bao	345	11	13	The household head	刘	玉宝	Male	Dragon	1903	8
Fujin Xiang 1st Bao				The household head's wife	石	崴连	Female	Goat	1906	10
Fujin Xiang 1st Bao				Son	刘	五兹娃	Male	Rabbit	1927	7
Fujin Xiang 1st Bao				Nephew	刘	膀娃	Male	Ox	1925	1
Fujin Xiang 1st Bao	346	11	14	The household head	韩	仲山	Male	Tiger	1901	3
Fujin Xiang 1st Bao				The household head's wife	李	跳秀	Female	Rooster	1908	7
Fujin Xiang 1st Bao				Daughter	赵	娃	Female	Tiger	1926	9
Fujin Xiang 1st Bao				Family member	刘	羊毛	Male	Rabbit	1927	11
Fujin Xiang 1st Bao	347	11	15	The household head	韩	仲洪	Male	Snake	1917	4

Date	The beginning time of inhabitation in this county	Educational level	Profession sector	Occupation	Marital status	Physical disability	Religion
24	Inhabited for generations	Illiteracy	Agriculture	Land-holding peasant	Married	No	No religion
24	Inhabited for generations						
3	Inhabited for generations						
7	Inhabited for generations	Illiteracy	Agriculture	Land-holding peasant	Married	No	No religion
17	Inhabited for generations	Illiteracy	Agriculture	Land-holding peasant	Married	No	No religion
11	Inhabited for generations	Illiteracy	Unemployed	Land-holding peasant	Widowed	No	No religion
8	Inhabited for generations						
23	Inhabited for generations						
25	Inhabited for generations	Illiteracy	Agriculture	Land-holding peasant	Widowed	No	No religion
1	Inhabited for generations	Illiteracy	Agriculture	Land-holding peasant	Unmarried	No	No religion
24	Inhabited for generations	Illiteracy	Agriculture	Land-holding peasant	Married	No	No religion
25	Inhabited for generations	Illiteracy	Agriculture	Land-holding peasant	Married	No	No religion
1	Inhabited for generations	Illiteracy	Agriculture	Land-holding peasant	Unmarried	No	No religion
28	Inhabited for generations	Illiteracy	Agriculture	Land-holding peasant	Unmarried	No	No religion
10	Inhabited for generations	Illiteracy	Agriculture	Land-holding peasant	Married	No	No religion
2	Inhabited for generations	Illiteracy	Agriculture	Land-holding peasant	Married	No	No religion
14	Inhabited for generations	Illiteracy	Agriculture	Land-holding peasant	Married	No	No religion
7	Inhabited for generations	Illiteracy	Agriculture	Land-holding peasant	Married	No	No religion
29	Inhabited for generations	Illiteracy	Agriculture	Land-holding peasant	Married	No	No religion

Appendix: Overview of Microdata for Wudu County (*cont.*)

	The no. of household	Jia	Household	Relationship with the household head	Surname	First name	Sex	The Chinese zodiac	Birthdate Year	Month
Fujin Xiang 1st Bao				The household head's wife	徐	能娃	Female	Ox	1925	10
Fujin Xiang 1st Bao				Father	韩	文贵	Male	Rooster	1871	4
Fujin Xiang 1st Bao				Mother	赵	路姐	Female	Rat	1887	5
Fujin Xiang 1st Bao				Daughter	韩	董娃	Female	Goat	1943	1
Fujin Xiang 1st Bao				The second younger brother	韩	仲连	Male	Pig	1923	8
Fujin Xiang 1st Bao				The third younger brother	韩	进学儿	Male	Tiger	1926	4
Fujin Xiang 1st Bao				Younger sister	韩	應唤子	Female	Goat	1931	4
Fujin Xiang 1st Bao	348	12	1	The household head	韩	仲彦	Male	Rabbit	1927	6
Fujin Xiang 1st Bao				The household head's wife	成	周女	Female	Horse	1930	4
Fujin Xiang 1st Bao				Grandfather	韩	登武	Male	Rat	1875	7
Fujin Xiang 1st Bao				Mother	韩	春娃	Female	Horse	1905	11
Fujin Xiang 1st Bao				Younger brother	韩	仲明	Male	Monkey	1932	4
Fujin Xiang 1st Bao				Younger brother's wife	冯	等娃	Female	Snake	1929	7
Fujin Xiang 1st Bao				Younger sister	韩	金香	Female	Ox	1937	2
Fujin Xiang 1st Bao				Grandfather's brother's wife	李	银福	Female	Snake	1880	9
Fujin Xiang 1st Bao				Uncle	韩	文义	Male	Snake	1904	6

Date	The beginning time of inhabitation in this county	Educational level	Profession sector	Occupation	Marital status	Physical disability	Religion
11	Inhabited for generations	Illiteracy	Agriculture	Land-holding peasant	Married	No	No religion
7	Inhabited for generations						
27	Inhabited for generations	Illiteracy	Agriculture	Land-holding peasant	Married	No	No religion
11	Inhabited for generations						
1	Inhabited for generations	Illiteracy	Agriculture	Land-holding peasant	Unmarried	No	No religion
8	Inhabited for generations	Illiteracy	Agriculture	Land-holding peasant	Unmarried	No	No religion
7	Inhabited for generations	Illiteracy					
21	Inhabited for generations	Bao school 3 years	Agriculture	Land-holding peasant	Married	No	No religion
1	Inhabited for generations	Illiteracy	Agriculture	Land-holding peasant	Married	No	No religion
7	Inhabited for generations						
25	Inhabited for generations	Illiteracy	Agriculture	Land-holding peasant	Widowed	No	No religion
6	Inhabited for generations	Bao school 2 years					
1	Inhabited for generations	Illiteracy	Agriculture	Land-holding peasant	Married	No	No religion
15	Inhabited for generations						
13	Inhabited for generations						
6	Inhabited for generations	Illiteracy	Agriculture	Land-holding peasant	Married	No	No religion

Appendix: Overview of Microdata for Wudu County (*cont.*)

									Birthdate	
	The no. of household	**Jia**	**Household**	**Relationship with the household head**	**Surname**	**First name**	**Sex**	**The Chinese zodiac**	**Year**	**Month**
Fujin Xiang 1st Bao				Uncle's wife	巩	芝明	Female	Monkey		
Fujin Xiang 1st Bao				Uncle's son	韩	仲虎	Male	Rat	1936	6
Fujin Xiang 1st Bao				The second uncle	韩	文举	Male	Dragon	1916	3
Fujin Xiang 1st Bao				Family member	王	芝英	Female	Pig	1909	5
Fujin Xiang 1st Bao	349	12	2	The household head	刘	义成	Male	Pig	1910	12
Fujin Xiang 1st Bao				The household head's wife	张	跳选	Female	Dog	1909	1
Fujin Xiang 1st Bao				Uncle	刘	德禄	Male	Dog	1874	8
Fujin Xiang 1st Bao				Daughter	刘	双宣女	Female	Rat	1936	9
Fujin Xiang 1st Bao				The second daughter	刘	董香女	Female	Tiger	1938	1
Fujin Xiang 1st Bao				The third daughter	刘	香女	Female	Monkey	1943	1
Fujin Xiang 1st Bao				Cohabitant	梁	万花	Male	Ox	1925	5
Fujin Xiang 1st Bao				Cohabitant's wife	朝	女	Female	Dog	1909	9
Fujin Xiang 1st Bao				Cohabitant's son	刘	伶儿	Male	Pig	1947	4
Fujin Xiang 1st Bao				Cohabitant's daughter	刘	双双	Female	Rooster	1945	11
Fujin Xiang 1st Bao	350	12	3	The household head	刘	义德	Male	Rooster	1921	1
Fujin Xiang 1st Bao				The household head's wife	韩	富女	Female	Monkey	1920	12
Fujin Xiang 1st Bao				Mother	韩	润娃	Female	Pig	1886	9
Fujin Xiang 1st Bao				Son	刘	义娃	Male	Pig	1946	1

Date	The beginning time of inhabitation in this county	Educational level	Profession sector	Occupation	Marital status	Physical disability	Religion
	Inhabited for generations	Illiteracy	Agriculture	Land-holding peasant	Married	No	No religion
25	Inhabited for generations	Bao school 5 years					
21	Inhabited for generations	Illiteracy	Agriculture	Land-holding peasant	Married	No	No religion
4	Inhabited for generations	Illiteracy	Agriculture	Land-holding peasant	Married	No	No religion
12	Inhabited for generations	Illiteracy	Agriculture	Land-holding peasant	Married	No	No religion
4	Inhabited for generations	Illiteracy	Agriculture	Land-holding peasant	Married	No	No religion
18	Inhabited for generations						
8	Inhabited for generations						
25	Inhabited for generations						
25	Inhabited for generations						
16	Inhabited for generations	Illiteracy	Agriculture	Land-holding peasant	Married	No	No religion
1	Inhabited for generations	Illiteracy	Agriculture	Land-holding peasant	Married	No	No religion
1	Inhabited for generations						
7	Inhabited for generations						
2	Inhabited for generations	Illiteracy	Agriculture	Land-holding peasant	Married	No	No religion
24	Inhabited for generations	Illiteracy	Agriculture	Land-holding peasant	Married	No	No religion
21	Inhabited for generations	Illiteracy	Agriculture	Land-holding peasant	Widowed	No	No religion
10	Inhabited for generations						

Appendix: Overview of Microdata for Wudu County (*cont.*)

	The no. of household	Jia	Household	Relationship with the household head	Surname	First name	Sex	The Chinese zodiac	Birthdate	
									Year	Month
Fujin Xiang 1st Bao				Younger sister	桂	花女	Female	Ox	1937	10
Fujin Xiang 1st Bao				Cohabitant	李	秋姐	Female	Rabbit	1890	12
Fujin Xiang 1st Bao				Cohabitant's son	韩	兆儿	Male	Horse	1930	1
Fujin Xiang 1st Bao	351	12	4	The household head	刘	义财	Male	Dog	1897	3
Fujin Xiang 1st Bao				The household head's wife	李	跳连	Female	Pig	1898	1
Fujin Xiang 1st Bao				Mother	冯	吉女	Female	Horse	1869	9
Fujin Xiang 1st Bao				Son	刘	国华	Male	Rooster	1921	8
Fujin Xiang 1st Bao				Son's wife	赵	二娃	Female	Monkey	1920	10
Fujin Xiang 1st Bao				Grandson	刘	节节	Male	Pig	1947	4
Fujin Xiang 1st Bao				Daughter	刘	富娃	Female	Monkey	1932	10
Fujin Xiang 1st Bao				Older brother	刘	义弟	Male	Goat	1894	6
Fujin Xiang 1st Bao				Older brother's wife	冯	四娃	Female	Snake	1892	11
Fujin Xiang 1st Bao				Nephew	刘	国富	Male	Ox	1925	2
Fujin Xiang 1st Bao				Niece	刘	川娃	Female	Goat	1931	1
Fujin Xiang 1st Bao				Adopted daughter	赵	扣儿	Female	Goat	1931	3
Fujin Xiang 1st Bao	352	12	5	The household head	赵	学兴	Male	Dragon	1903	1
Fujin Xiang 1st Bao				The household head's wife	马	双秀	Female	Ox	1913	2
Fujin Xiang 1st Bao				Son	赵	元明	Male	Dog	1934	9

Date	The beginning time of inhabitation in this county	Educational level	Profession sector	Occupation	Marital status	Physical disability	Religion
14	Inhabited for generations						
12	Inhabited for generations	Illiteracy	Agricul-ture	Land-holding peasant	Widowed	No	No religion
28	Inhabited for generations	Illiteracy	Agricul-ture	Land-holding peasant	Unmarried	No	No religion
3	Inhabited for generations	Illiteracy	Agricul-ture	Land-holding peasant	Married	No	No religion
11	Inhabited for generations	Illiteracy	Agricul-ture	Land-holding peasant	Married	No	No religion
11	Inhabited for generations						
17	Inhabited for generations	Illiteracy	Agricul-ture	Land-holding peasant	Married	No	No religion
1	Inhabited for generations	Illiteracy	Agricul-ture	Land-holding peasant	Married	No	No religion
28	Inhabited for generations						
17	Inhabited for generations						
12	Inhabited for generations	Illiteracy	Agricul-ture	Land-holding peasant	Married	No	No religion
17	Inhabited for generations	Illiteracy	Agricul-ture	Land-holding peasant	Married	No	No religion
13	Inhabited for generations	Illiteracy	Agricul-ture	Land-holding peasant	Unmarried	No	No religion
17	Inhabited for generations	Illiteracy	Agricul-ture	Land-holding peasant	Unmarried	No	No religion
1	Inhabited for generations	Illiteracy	Agricul-ture	Land-holding peasant	Unmarried	No	No religion
19	Inhabited for generations	Illiteracy	Agricul-ture	Land-holding peasant	Married	No	No religion
4	Inhabited for generations	Illiteracy	Agricul-ture	Land-holding peasant	Married	No	No religion
16	Inhabited for generations	Bao school 4 years					

Appendix: Overview of Microdata for Wudu County (*cont.*)

	The no. of household	Jia	Household	Relationship with the household head	Surname	First name	Sex	The Chinese zodiac	Birthdate Year	Month
Fujin Xiang 1st Bao	353	12	6	The household head	韩	仲生	Male	Rooster	1908	10
Fujin Xiang 1st Bao				The household head's wife	陈	艰娃	Female	Pig	1910	12
Fujin Xiang 1st Bao				Son	韩	琴儿	Male	Rabbit	1939	4
Fujin Xiang 1st Bao				The second son	韩	全生	Male	Rooster	1945	9
Fujin Xiang 1st Bao				The third son	韩	称全	Male	Pig	1947	5
Fujin Xiang 1st Bao				Daughter	韩	琴娃	Female	Rooster	1933	12
Fujin Xiang 1st Bao	354	12	7	The household head	刘	国发	Male	Ox	1913	3
Fujin Xiang 1st Bao				The household head's wife	冯	秋娃	Female	Ox	1913	12
Fujin Xiang 1st Bao				Father	刘	义田	Male	Ox	1888	5
Fujin Xiang 1st Bao				Son	刘	世明	Male	Rooster	1933	1
Fujin Xiang 1st Bao				The second son	刘	尚娃	Male	Tiger	1938	2
Fujin Xiang 1st Bao				Daughter	刘	苏娃	Female	Pig	1947	2
Fujin Xiang 1st Bao	355	12	8	The household head	赵	田科	Male	Ox	1888	11
Fujin Xiang 1st Bao				The household head's wife	周	来儿子	Female	Rabbit	1890	12
Fujin Xiang 1st Bao				Son	赵	兴隆	Male	Horse	1918	11
Fujin Xiang 1st Bao				Son's wife	赵	耕娃	Female	Dog	1922	12
Fujin Xiang 1st Bao				Granddaughter	赵	梧儿	Female	Dragon	1941	2
Fujin Xiang 1st Bao				The second son	赵	兴明	Male	Monkey	1932	1
Fujin Xiang 1st Bao	356	12	9	The household head	赵	兴仁	Male	Dog	1897	6

Date	The beginning time of inhabitation in this county	Educational level	Profession sector	Occupation	Marital status	Physical disability	Religion
17	Inhabited for generations	Illiteracy	Agriculture	Land-holding peasant	Married	No	No religion
30	Inhabited for generations	Illiteracy	Agriculture	Land-holding peasant	Married	No	No religion
2	Inhabited for generations	Bao school 1 year					
23	Inhabited for generations						
8	Inhabited for generations						
26	Inhabited for generations	Illiteracy					
22	Inhabited for generations	Illiteracy	Agriculture	Land-holding peasant	Married	No	No religion
12	Inhabited for generations	Illiteracy	Agriculture	Land-holding peasant	Married	No	No religion
22	Inhabited for generations	Illiteracy	Agriculture	Land-holding peasant	Widowed	No	No religion
4	Inhabited for generations	Bao school 4 years					
26	Inhabited for generations	Illiteracy					
7	Inhabited for generations						
21	Inhabited for generations	Illiteracy	Agriculture	Land-holding peasant	Married	No	No religion
17	Inhabited for generations	Illiteracy	Agriculture	Land-holding peasant	Married	No	No religion
17	Inhabited for generations	Illiteracy	Agriculture	Land-holding peasant	Married	No	No religion
4	Inhabited for generations	Illiteracy	Agriculture	Land-holding peasant	Married	No	No religion
9	Inhabited for generations	Illiteracy					
12	Inhabited for generations	Illiteracy	Agriculture	Land-holding peasant	Unmarried	No	No religion
23	Inhabited for generations	Illiteracy	Agriculture	Land-holding peasant	Married	No	No religion

Appendix: Overview of Microdata for Wudu County (*cont.*)

	The no. of household	Jia	Household	Relationship with the household head	Surname	First name	Sex	The Chinese zodiac	Birthdate	
									Year	Month
Fujin Xiang 1st Bao				The household head's wife	谈	董娃	Female	Tiger	1901	2
Fujin Xiang 1st Bao				Son	谢	称娃	Male	Goat	1931	12
Fujin Xiang 1st Bao				The second son	赵	刘全	Male	Tiger	1938	10
Fujin Xiang 1st Bao				The third son	赵	桂娃	Male	Monkey	1944	3
Fujin Xiang 1st Bao				Daughter	赵	省娃	Female	Dog	1946	5
Fujin Xiang 1st Bao	357	12	10	The household head	赵	田福	Male	Dog	1885	11
Fujin Xiang 1st Bao				The household head's wife	王	仓哇	Female	Pig	1886	9
Fujin Xiang 1st Bao				Son	赵	兴怀	Male	Rabbit	1927	5
Fujin Xiang 1st Bao				Son's wife	张	跳香	Female	Pig	1923	6
Fujin Xiang 1st Bao				Daughter	崴	玉女	Female	Goat	1931	8
Fujin Xiang 1st Bao				Cohabitant	茹	青花	Male	Tiger	1914	3
Fujin Xiang 1st Bao				Cohabitant's wife	李	田娃	Female	Monkey	1920	2
Fujin Xiang 1st Bao				Cohabitant's son	韩	娃	Male	Pig	1947	2
Fujin Xiang 1st Bao	358	12	11	The household head	赵	明福	Male	Horse	1905	4
Fujin Xiang 1st Bao				Father	赵	兴有	Male	Rabbit	1878	11
Fujin Xiang 1st Bao				Mother	谈	润羊	Female	Goat	1882	3
Fujin Xiang 1st Bao				Uncle's wife	马	羊娃	Female	Snake	1879	5
Fujin Xiang 1st Bao				Son	郭	生娃	Male	Rabbit	1927	4
Fujin Xiang 1st Bao				Son's wife	李	贵姓	Female	Tiger	1926	3

Date	The beginning time of inhabitation in this county	Educational level	Profession sector	Occupation	Marital status	Physical disability	Religion
12	Inhabited for generations	Illiteracy	Agriculture	Land-holding peasant	Married	No	No religion
5	Inhabited for generations	Illiteracy	Agriculture	Land-holding peasant	Unmarried	No	No religion
21	Inhabited for generations	Illiteracy	Agriculture	Land-holding peasant	Unmarried	No	No religion
22	Inhabited for generations						
7	Inhabited for generations						
5	Inhabited for generations	Illiteracy	Agriculture	Land-holding peasant	Married	No	No religion
11	Inhabited for generations	Illiteracy	Agriculture	Land-holding peasant	Married	No	No religion
2	Inhabited for generations	Illiteracy	Agriculture	Land-holding peasant	Married	No	No religion
10	Inhabited for generations	Illiteracy	Agriculture	Land-holding peasant	Married	No	No religion
12	Inhabited for generations	Illiteracy	Agriculture	Land-holding peasant	Unmarried	No	No religion
2	Inhabited for generations	Illiteracy	Agriculture	Land-holding peasant	Married	No	No religion
30	Inhabited for generations	Illiteracy	Agriculture	Land-holding peasant	Married	No	No religion
26	Inhabited for generations						
21	Inhabited for generations	Sishu 3 years	Agriculture	Land-holding peasant	Widowed	No	No religion
11	Inhabited for generations	Illiteracy	Agriculture	Land-holding peasant	Married	No	No religion
18	Inhabited for generations	Illiteracy	Agriculture	Land-holding peasant	Married	No	No religion
15	Inhabited for generations	Illiteracy	Agriculture	Land-holding peasant	Widowed	No	No religion
18	Inhabited for generations	Illiteracy	Agriculture	Land-holding peasant	Married	No	No religion
29	Inhabited for generations	Illiteracy	Agriculture	Land-holding peasant	Married	No	No religion

Appendix: Overview of Microdata for Wudu County (*cont.*)

	The no. of household	Jia	Household	Relationship with the household head	Surname	First name	Sex	The Chinese zodiac	Birthdate Year	Month
Fujin Xiang 1st Bao				Grandson	林	森	Male	Dog	1946	10
Fujin Xiang 1st Bao				The second son	赵	正娃	Male	Snake	1929	9
Fujin Xiang 1st Bao				The third son	赵	全全	Male	Pig	1934	4
Fujin Xiang 1st Bao				Daughter	赵	书秀	Female	Snake	1941	2
Fujin Xiang 1st Bao	359	12	12	The household head	赵	田德	Male	Dog	1885	1
Fujin Xiang 1st Bao				The household head's wife	李	翠连	Female	Dog	1885	7
Fujin Xiang 1st Bao				Son	赵	兴昌	Male	Ox	1925	12
Fujin Xiang 1st Bao				Son's wife	赵	存香	Female	Pig	1923	8
Fujin Xiang 1st Bao				Granddaughter	赵	四斤	Female	Rooster	1945	12
Fujin Xiang 1st Bao				Younger brother	赵	田赐	Male	Tiger	1889	3
Fujin Xiang 1st Bao				Younger brother's wife	史	近娃	Female	Goat	1894	3
Fujin Xiang 1st Bao				Niece	赵	大双儿	Female	Goat	1931	6
Fujin Xiang 1st Bao				The second niece	赵	崴双儿	Female	Rooster	1933	12
Fujin Xiang 1st Bao	360	12	13	The household head	赵	田禄	Male	Rat	1899	5
Fujin Xiang 1st Bao				Son	赵	兴义	Male	Horse	1918	9
Fujin Xiang 1st Bao				Son's wife	周	跳娃	Female	Dragon	1914	8
Fujin Xiang 1st Bao				Grandson	赵	成保	Male	Dragon	1940	5
Fujin Xiang 1st Bao				Granddaughter	赵	緒娃	Female	Ox	1937	9

Date	The beginning time of inhabitation in this county	Educational level	Profession sector	Occupation	Marital status	Physical disability	Religion	
3	Inhabited for generations	Illiteracy						
4	Inhabited for generations	Illiteracy	Agriculture	Land-holding peasant	Unmarried	No	No religion	Joined the army in 1946
23	Inhabited for generations	Illiteracy	Agriculture	Land-holding peasant		No	No religion	
12	Inhabited for generations	Illiteracy	Agriculture	Land-holding peasant		No	No religion	
12	Inhabited for generations	Illiteracy	Agriculture	Land-holding peasant	Married	No	No religion	
8	Inhabited for generations	Illiteracy	Agriculture	Land-holding peasant	Married	No	No religion	
21	Inhabited for generations	Illiteracy	Agriculture	Land-holding peasant	Married	No	No religion	
19	Inhabited for generations	Illiteracy	Agriculture	Land-holding peasant	Married	No	No religion	
2	Inhabited for generations							
16	Inhabited for generations	Illiteracy	Agriculture	Land-holding peasant	Married	No	No religion	
4	Inhabited for generations	Illiteracy	Agriculture	Land-holding peasant	Married	No	No religion	
18	Inhabited for generations	Illiteracy	Agriculture	Land-holding peasant	Unmarried	No	No religion	
18	Inhabited for generations	Illiteracy	Agriculture	Land-holding peasant	Unmarried	No	No religion	
18	Inhabited for generations	Illiteracy	Agriculture	Land-holding peasant	Widowed	No	No religion	
2	Inhabited for generations	Illiteracy	Agriculture	Land-holding peasant	Married	No	No religion	
13	Inhabited for generations	Illiteracy	Agriculture	Land-holding peasant	Married	No	No religion	
25	Inhabited for generations	Illiteracy	Agriculture	Land-holding peasant	Unmarried	No	No religion	
27	Inhabited for generations	Illiteracy	Agriculture	Land-holding peasant		No		

Appendix: Overview of Microdata for Wudu County (*cont.*)

	The no. of household	Jia	Household	Relationship with the household head	Surname	First name	Sex	The Chinese zodiac	Birthdate Year	Month
Fujin Xiang 1st Bao				The second Grand-daughter	赵	贵娃	Female	Rat	1936	1
Fujin Xiang 1st Bao	361	12	14	The household head	赵	田海	Male	Goat	1882	7
Fujin Xiang 1st Bao				The household head's wife	韩	成姐子	Female	Snake	1880	10
Fujin Xiang 1st Bao				Son	赵	兴治	Male	Dog	1909	7
Fujin Xiang 1st Bao				Son's wife	陈	翠女	Female	Monkey	1907	3
Fujin Xiang 1st Bao				Grand-daughter	赵	祥娃	Female	Goat	1931	9
Fujin Xiang 1st Bao				The second Grand-daughter	赵	贵英	Female	Pig	1935	12
Fujin Xiang 1st Bao				Grandson	赵	贵忠	Male	Horse	1942	1
Fujin Xiang 1st Bao				The second son	赵	兴周	Male	Snake	1917	12
Fujin Xiang 1st Bao				The second son's wife	张	六女	Female	Pig	1923	7
Fujin Xiang 1st Bao				The third son	赵	兴山	Male	Dog	1922	4
Fujin Xiang 1st Bao				The third son's wife	李	雪哇	Female	Ox	1925	3
Fujin Xiang 1st Bao				Grand-daughter	赵	忠英	Female	Pig	1947	5
Fujin Xiang 1st Bao	362	12	15	The household head	王	义堂	Male	Dog	1897	4
Fujin Xiang 1st Bao				The household head's wife	谈	寸玉	Female	Rat	1899	9
Fujin Xiang 1st Bao				Son	王	孟娃	Male	Pig	1934	6
Fujin Xiang 1st Bao				Younger brother	王	义珍	Male	Horse	1905	1
Fujin Xiang 1st Bao				Younger brother's wife	李	柱女	Female	Dog	1909	5

Date	The beginning time of inhabitation in this county	Educational level	Profession sector	Occupation	Marital status	Physical disability	Religion
2	Inhabited for generations	Illiteracy	Agriculture	Land-holding peasant		No	
18	Inhabited for generations	Illiteracy	Agriculture	Land-holding peasant	Married	No	No religion
10	Inhabited for generations	Illiteracy	Agriculture	Land-holding peasant	Married	No	No religion
9	Inhabited for generations	Illiteracy	Agriculture	Land-holding peasant	Married	Unable to work	No religion
19	Inhabited for generations	Illiteracy	Agriculture	Land-holding peasant	Married	No	No religion
19	Inhabited for generations	Illiteracy	Agriculture	Land-holding peasant	Unmarried	No	No religion
21	Inhabited for generations	Illiteracy	Agriculture	Land-holding peasant	Unmarried	No	No religion
13	Inhabited for generations						
21	Inhabited for generations	Illiteracy	Agriculture	Land-holding peasant	Married	No	No religion
15	Inhabited for generations	Illiteracy	Agriculture	Land-holding peasant	Married	No	No religion
19	Inhabited for generations	Illiteracy	Agriculture	Land-holding peasant	Married	No	No religion
13	Inhabited for generations	Illiteracy	Agriculture	Land-holding peasant	Married	No	No religion
13	Inhabited for generations						
19	Inhabited for generations	Illiteracy	Agriculture	Land-holding peasant	Married	Leg broken	No religion
18	Inhabited for generations	Illiteracy	Agriculture	Land-holding peasant	Married	No	No religion
28	Inhabited for generations	Illiteracy	Agriculture				
6	Inhabited for generations	Illiteracy	Agriculture	Land-holding peasant	Married	No	No religion
23	Inhabited for generations	Illiteracy	Agriculture	Land-holding peasant	Married	No	No religion

Appendix: Overview of Microdata for Wudu County (*cont.*)

	The no. of household	Jia	Household	Relationship with the household head	Surname	First name	Sex	The Chinese zodiac	Birthdate Year	Month
Fujin Xiang 1st Bao				Niece	王	三女	Female	Dog	1934	10
Fujin Xiang 1st Bao				The second niece	王	必家	Female	Dragon	1940	7
Fujin Xiang 1st Bao				The third younger brother	王	义隆	Male	Ox	1913	2
Fujin Xiang 1st Bao				Younger brother's wife	张	崴香	Female	Ox	1913	12
Fujin Xiang 1st Bao				Nephew	王	仁娃	Male	Dog	1934	10
Fujin Xiang 1st Bao				The second nephew	王	拜儿	Male	Dragon	1940	10
Fujin Xiang 1st Bao	363	13	1	The household head	王	义清	Male	Goat	1906	8
Fujin Xiang 1st Bao				Mother	赵	比姐	Female	孙	1882	9
Fujin Xiang 1st Bao				Son	王	生生	Male	Dog	1934	6
Fujin Xiang 1st Bao				Younger sister	王	菊娃	Female	Horse	1918	9
Fujin Xiang 1st Bao				Older brother's wife	谈	四娃	Female	Snake	1904	2
Fujin Xiang 1st Bao				Nephew	崴	坡娃	Male	Goat	1931	10
Fujin Xiang 1st Bao				The second nephew	王	三换	Male	Ox	1937	6
Fujin Xiang 1st Bao				The third nephew	王	转生	Male	Snake	1941	6
Fujin Xiang 1st Bao				Daughter	王	舞女	Female	Dog	1946	2
Fujin Xiang 1st Bao	364	13	2	The household head	王	义福	Male	Monkey	1920	12
Fujin Xiang 1st Bao				The household head's wife	赵	春明	Female	Monkey	1920	9
Fujin Xiang 1st Bao				Mother	史	得秀	Female	Monkey	1895	5

Date	The beginning time of inhabitation in this county	Educational level	Profession sector	Occupation	Marital status	Physical disability	Religion	
15	Inhabited for generations	Illiteracy						
17	Inhabited for generations	Illiteracy						
21	Inhabited for generations	Illiteracy	Agriculture	Land-holding peasant	Married	No	No religion	Died on 23rd September 1947
16	Inhabited for generations	Illiteracy	Agriculture	Land-holding peasant	Married	No	No religion	
1	Inhabited for generations	Illiteracy						
10	Inhabited for generations							
16	Inhabited for generations	Illiteracy	Agriculture	Land-holding peasant	Widowed	Left leg injured	No religion	
11	Inhabited for generations	Illiteracy	Agriculture	Land-holding peasant	Widowed	No	No religion	
12	Inhabited for generations	Illiteracy	Agriculture	Land-holding peasant	Unmarried	No	No religion	
14	Inhabited for generations	Illiteracy	Agriculture	Land-holding peasant	Unmarried	Legs injured	No religion	
14	Inhabited for generations	Illiteracy	Agriculture	Land-holding peasant	Widowed	Blind	No religion	
1	Inhabited for generations	Illiteracy	Agriculture	Land-holding peasant	Unmarried	No	No religion	
25	Inhabited for generations	Illiteracy	Agriculture	Land-holding peasant	Unmarried	No	No religion	
14	Inhabited for generations	Illiteracy	Agriculture	Land-holding peasant		No		
23	Inhabited for generations							
27	Inhabited for generations	Illiteracy	Agriculture	Land-holding peasant	Married	No	No religion	
29	Inhabited for generations	Illiteracy	Agriculture	Land-holding peasant	Married	No	No religion	
22	Inhabited for generations	Illiteracy	Agriculture	Land-holding peasant	Widowed	No	No religion	

Appendix: Overview of Microdata for Wudu County (*cont.*)

									Birthdate	
	The no. of household	Jia	Household	Relationship with the household head	Surname	First name	Sex	The Chinese zodiac	Year	Month
Fujin Xiang 1st Bao				Son	王	富儿	Male	Dragon	1940	1
Fujin Xiang 1st Bao				Daughter	王	緒女	Female	Dog	1946	6
Fujin Xiang 1st Bao				Younger brother	王	慈女	Male	Dragon	1940	2
Fujin Xiang 1st Bao				Younger sister	王	慈娃	Female	Rat	1946	2
Fujin Xiang 1st Bao				Younger brother's wife	陈	润祥	Female	Dragon	1928	3
Fujin Xiang 1st Bao				Uncle's son	王	义朝	Male	Rabbit	1902	3
Fujin Xiang 1st Bao				Uncle's son's wife	赵	香娃	Female	Horse	1905	4
Fujin Xiang 1st Bao				Nephew	王	芝奎	Male	Rat	1924	2
Fujin Xiang 1st Bao	365	13	4	The household head	赵	兴海	Male	Goat	1895	6
Fujin Xiang 1st Bao				The household head's wife	谈	早玉	Female	Monkey	1895	1
Fujin Xiang 1st Bao				Son	赵	朝娃	Male	Snake	1917	7
Fujin Xiang 1st Bao				Son's wife	马	芦娃	Female	Rat	1924	4
Fujin Xiang 1st Bao				Grandson	赵	羊娃	Male	Dog	1946	11
Fujin Xiang 1st Bao				The second son	赵	禄娃	Male	Ox	1925	12
Fujin Xiang 1st Bao				Daughter	赵	双环	Female	Monkey	1932	11
Fujin Xiang 1st Bao				The second daughter	赵	住生	Female	Pig	1935	4
Fujin Xiang 1st Bao				The third daughter	赵	柱哇	Female	Rabbit	1939	7
Fujin Xiang 1st Bao	366	13	5	The household head	赵	兴万	Male	Pig	1898	1

Date	The beginning time of inhabitation in this county	Educational level	Profession sector	Occupation	Marital status	Physical disability	Religion	
31	Inhabited for generations	Illiteracy	Agriculture	Land-holding peasant		No	No religion	
26	Inhabited for generations	Illiteracy	Agriculture	Land-holding peasant		No		
11	Inhabited for generations	Illiteracy	Agriculture	Land-holding peasant		No		
14	Inhabited for generations	Illiteracy	Agriculture	Land-holding peasant		No		
1	Inhabited for generations	Illiteracy	Agriculture	Land-holding peasant	Married	No	No religion	
6	Inhabited for generations	Illiteracy	Agriculture	Land-holding peasant	Married	No	No religion	
28	Inhabited for generations	Illiteracy	Agriculture	Land-holding peasant	Married	No	No religion	
5	Inhabited for generations	Illiteracy	Agriculture	Land-holding peasant	Married	No	No religion	
29	Inhabited for generations	Illiteracy	Agriculture	Land-holding peasant	Married	No	No religion	
23	Inhabited for generations	Illiteracy	Agriculture	Land-holding peasant	Married	No	No religion	
8	Inhabited for generations	Illiteracy	Agriculture	Land-holding peasant	Married	No	No religion	Joined the army in May 1942
1	Inhabited for generations	Illiteracy	Agriculture	Land-holding peasant	Married	No	No religion	
3	Inhabited for generations							
6	Inhabited for generations	Illiteracy	Agriculture	Land-holding peasant	Unmarried	No	No religion	
7	Inhabited for generations	Illiteracy	Agriculture	Land-holding peasant	Unmarried	No	No religion	
1	Inhabited for generations		Agriculture	Land-holding peasant		No	No religion	
8	Inhabited for generations							
21	Inhabited for generations	Illiteracy	Agriculture	Land-holding peasant	Married	No	No religion	

Appendix: Overview of Microdata for Wudu County (*cont.*)

	The no. of household	Jia	Household	Relationship with the household head	Surname	First name	Sex	The Chinese zodiac	Birthdate Year	Month
Fujin Xiang 1st Bao				The household head's wife	赵	灶送	Female	Rabbit	1902	10
Fujin Xiang 1st Bao				Son	赵	高才	Male	Snake	1929	4
Fujin Xiang 1st Bao				Son's wife	韩	家女	Female	Tiger	1926	5
Fujin Xiang 1st Bao				The second son	赵	有才	Male	Goat	1931	10
Fujin Xiang 1st Bao				The second son's wife	张	全女	Female	Goat	1931	6
Fujin Xiang 1st Bao				The third son	赵	有生	Male	Dragon	1940	11
Fujin Xiang 1st Bao				Daughter	赵	么哇	Female	Dog	1934	10
Fujin Xiang 1st Bao				The second daughter	赵	有女	Female	Goat	1943	2
Fujin Xiang 1st Bao	367	13	6	The household head	赵	兴满	Male	Ox	1888	12
Fujin Xiang 1st Bao				Son	赵	明兴	Male	Dog	1922	12
Fujin Xiang 1st Bao				Son's wife	贾	全秀	Female	Tiger	1926	1
Fujin Xiang 1st Bao				Uncle's wife	董	土歌	Female	Goat	1882	3
Fujin Xiang 1st Bao				Nephew	赵	兴贤	Male	Goat	1894	7
Fujin Xiang 1st Bao				Nephew's wife	张	秀儿	Female	Rabbit	1902	9
Fujin Xiang 1st Bao				Niece	崴	赵娃	Female	Rabbit	1927	3
Fujin Xiang 1st Bao	368	13	6	The household head	赵	兴元	Male	Dragon	1891	8
Fujin Xiang 1st Bao				The household head's wife	谈	喜哇	Female	Rooster	1896	9
Fujin Xiang 1st Bao				Mother	赵	凤姐	Female	Goat	1872	2

Date	The beginning time of inhabitation in this county	Educational level	Profession sector	Occupation	Marital status	Physical disability	Religion	
28	Inhabited for generations	Illiteracy	Agriculture	Land-holding peasant	Married	No	No religion	
16	Inhabited for generations	Illiteracy	Agriculture	Land-holding peasant	Married	No	No religion	Joined the army in July 1947
31	Inhabited for generations	Illiteracy	Agriculture	Land-holding peasant	Married	No	No religion	
1	Inhabited for generations	Illiteracy	Agriculture	Land-holding peasant	Married	No	No religion	
26	Inhabited for generations	Illiteracy	Agriculture	Land-holding peasant	Married	No	No religion	
11	Inhabited for generations	Illiteracy	Agriculture	Land-holding peasant	Married	No	No religion	
18	Inhabited for generations	Illiteracy	Agriculture	Land-holding peasant	Married	No	No religion	
13	Inhabited for generations							
4	Inhabited for generations	Illiteracy	Agriculture	Land-holding peasant	Widowed	No	No religion	
6	Inhabited for generations	Illiteracy	Agriculture	Land-holding peasant	Married	No	No religion	
21	Inhabited for generations	Illiteracy	Agriculture	Land-holding peasant	Married	No	No religion	
15	Inhabited for generations							
28	Inhabited for generations	Illiteracy	Agriculture	Land-holding peasant	Married	No	No religion	
2	Inhabited for generations	Illiteracy	Agriculture	Land-holding peasant	Married	No	No religion	
15	Inhabited for generations	Illiteracy	Agriculture	Land-holding peasant	Married	No	No religion	
21	Inhabited for generations	Illiteracy	Agriculture	Land-holding peasant	Married	No	No religion	
11	Inhabited for generations	Illiteracy	Agriculture	Land-holding peasant	Married	No	No religion	
20	Inhabited for generations			Land-holding peasant				

Appendix: Overview of Microdata for Wudu County (*cont.*)

	The no. of household	Jia	Household	Relationship with the household head	Surname	First name	Sex	The Chinese zodiac	Birthdate	
									Year	Month
Fujin Xiang 1st Bao				Son	赵	明德	Male	Rooster	1911	9
Fujin Xiang 1st Bao				Son's wife	谈	香娃	Female	Dragon	1916	3
Fujin Xiang 1st Bao				Grandson	赵	保林	Male	Rabbit	1939	5
Fujin Xiang 1st Bao				The second grandson	赵	中中	Male	Rooster	1945	10
Fujin Xiang 1st Bao				The third grandson	赵	吉林	Male	Dog	1946	8
Fujin Xiang 1st Bao				The second son	赵	明忠	Male	Tiger	1926	2
Fujin Xiang 1st Bao				The second son's wife	王	芦娃	Female	Horse	1930	9
Fujin Xiang 1st Bao				The third son	赵	明青	Male	Dragon	1928	12
Fujin Xiang 1st Bao				The third son's wife	陈	菊儿	Female	Ox	1925	3
Fujin Xiang 1st Bao				Younger brother's wife	史	坡女子	Female	Horse	1905	11
Fujin Xiang 1st Bao				Nephew	赵	惠生	Male	Tiger	1938	5
Fujin Xiang 1st Bao				The second nephew	赵	对生	Male	Horse	1942	8
Fujin Xiang 1st Bao				Niece	崴	秋儿	Female	Dog	1934	6
Fujin Xiang 1st Bao				The second niece	崴	见儿	Female	Rooster	1945	7
Fujin Xiang 1st Bao	369	13	7	The household head	陈	国贤	Male	Ox	1913	11
Fujin Xiang 1st Bao				The household head's wife	王	柏香儿	Female	Monkey	1907	12
Fujin Xiang 1st Bao				Father	陈	义福	Male	Dragon	1879	10
Fujin Xiang 1st Bao				Mother	王	喜儿	Female	Dragon	1879	10

Date	The beginning time of inhabitation in this county	Educational level	Profession sector	Occupation	Marital status	Physical disability	Religion
24	Inhabited for generations	Illiteracy	Agriculture	Land-holding peasant	Married	No	No religion
23	Inhabited for generations	Illiteracy	Agriculture	Land-holding peasant	Married	No	No religion
22	Inhabited for generations	Bao school 2 years	Agriculture				
9	Inhabited for generations						
1	Inhabited for generations						
19	Inhabited for generations	Illiteracy	Agriculture	Land-holding peasant	Married	No	No religion
29	Inhabited for generations	Illiteracy	Agriculture	Land-holding peasant	Married	No	No religion
13	Inhabited for generations	Illiteracy	Agriculture	Land-holding peasant	Married	No	No religion
30	Inhabited for generations	Illiteracy	Agriculture	Land-holding peasant	Married	No	No religion
17	Inhabited for generations	Illiteracy	Agriculture	Land-holding peasant	Widowed	No	No religion
3	Inhabited for generations	Illiteracy	Agriculture	Land-holding peasant			
30	Inhabited for generations	Illiteracy	Agriculture	Land-holding peasant			
16	Inhabited for generations	Illiteracy	Agriculture				
20	Inhabited for generations						
15	Inhabited for generations	Illiteracy	Agriculture	Land-holding peasant	Married	No	No religion
4	Inhabited for generations	Illiteracy	Agriculture	Land-holding peasant	Married	No	No religion
15	Inhabited for generations	Illiteracy	Agriculture	Land-holding peasant	Married	No	No religion
1	Inhabited for generations	Illiteracy	Agriculture	Land-holding peasant	Married	No	No religion

Appendix: Overview of Microdata for Wudu County (*cont.*)

	The no. of household	Jia	Household	Relationship with the household head	Surname	First name	Sex	The Chinese zodiac	Birthdate	
									Year	Month
Fujin Xiang 1st Bao				Son	陈	守花	Male	Ox	1934	12
Fujin Xiang 1st Bao				Son's wife	王	富哇	Female	Snake	1929	11
Fujin Xiang 1st Bao				Grandson	张	林	Male	Pig	1947	5
Fujin Xiang 1st Bao				Younger brother	陈	国三	Male	Dragon	1916	2
Fujin Xiang 1st Bao				Younger brother's wife	王	鱼娃	Female	Ox	1925	10
Fujin Xiang 1st Bao				Niece	陈	中娃	Female	Monkey	1944	9
Fujin Xiang 1st Bao				The second niece	陈	润润	Female	Pig	1947	2
Fujin Xiang 1st Bao				Cohabitant	陈	义德	Male	Dragon	1903	4
Fujin Xiang 1st Bao				Cohabitant's mother	张	女女子	Female	Dog	1873	5
Fujin Xiang 1st Bao				Cohabitant's daughter	陈	腮娃	Female	Snake	1929	8
Fujin Xiang 1st Bao				Cohabitant's younger brother's wife	张	崔娃	Female	Rat	1912	9
Fujin Xiang 1st Bao				Cohabitant's nephew	陈	丙生	Male	Rat	1936	12
Fujin Xiang 1st Bao				Cohabitant's second nephew	陈	里佪大	Male	Monkey	1944	12
Fujin Xiang 1st Bao	370	13	8	The household head	陈	守银	Male	Rat	1912	9

Date	The beginning time of inhabitation in this county	Educational level	Profession sector	Occupation	Marital status	Physical disability	Religion
18	Inhabited for generations	Illiteracy	Agriculture	Land-holding peasant	Married	No	No religion
1	Inhabited for generations	Illiteracy	Agriculture	Land-holding peasant	Married	No	No religion
18	Inhabited for generations						
11	Inhabited for generations	Sishu 3 years	Agriculture	Land-holding peasant	Married	No	No religion
6	Inhabited for generations	Illiteracy	Agriculture	Land-holding peasant	Married	No	No religion
17	Inhabited for generations						
28	Inhabited for generations						
10	Inhabited for generations	Illiteracy	Agriculture	Land-holding peasant	Widowed	No	No religion
11	Inhabited for generations	Illiteracy					
8	Inhabited for generations	Illiteracy	Agriculture	Land-holding peasant	Unmarried	No	No religion
14	Inhabited for generations	Illiteracy	Agriculture	Land-holding peasant	Widowed	No	No religion
31	Inhabited for generations	Illiteracy	Agriculture	Land-holding peasant		No	No religion
13	Inhabited for generations						
3	Inhabited for generations	Illiteracy	Agriculture	Land-holding peasant	Married	No	No religion

Appendix: Overview of Microdata for Wudu County (*cont.*)

	The no. of household	Jia	Household	Relationship with the household head	Surname	First name	Sex	The Chinese zodiac	Birthdate Year	Month
Fujin Xiang 1st Bao				The household head's wife	王	全女	Female	Dog	1922	10
Fujin Xiang 1st Bao				Mother	石	转女子	Female	Rat	1873	2
Fujin Xiang 1st Bao	371	13	9	The household head	陈	国柱	Male	Ox	1925	1
Fujin Xiang 1st Bao				The household head's wife	冯	毛娃	Female	Ox	1925	12
Fujin Xiang 1st Bao				Father	陈	义堂	Male	Rooster	1884	7
Fujin Xiang 1st Bao				Son	陈	海生	Male	Monkey	1947	4
Fujin Xiang 1st Bao				Uncle's wife	耕	有	Female	Dog	1873	4
Fujin Xiang 1st Bao				Uncle's son	陈	国斌	Male	Horse	1893	2
Fujin Xiang 1st Bao				Older brother's wife	马	禄娃	Female	Pig	1891	4
Fujin Xiang 1st Bao				Nephew	陈	怌生	Male	Monkey	1932	3
Fujin Xiang 1st Bao				Niece	陈	怌女	Female	Horse	1942	4
Fujin Xiang 1st Bao	372	13	10	The household head	陈	国均	Male	Ox	1913	12
Fujin Xiang 1st Bao				The household head's wife	赵	崴见	Female	Horse	1918	1
Fujin Xiang 1st Bao				Son	陈	举平	Male	Goat	1943	4
Fujin Xiang 1st Bao				The second son	陈	冯拜	Male	Dog	1946	2
Fujin Xiang 1st Bao	373	13	11	The household head	陈	菊双儿	Female	Monkey	1920	10
Fujin Xiang 1st Bao				Mother	王	唤儿	Female	Rooster	1896	10

Date	The beginning time of inhabitation in this county	Educational level	Profession sector	Occupation	Marital status	Physical disability	Religion
3	Inhabited for generations	Illiteracy	Agriculture	Land-holding peasant	Married	No	No religion
5	Inhabited for generations						
3	Inhabited for generations	Sishu 2 years	Agriculture	Land-holding peasant	Married	No	No religion
4	Inhabited for generations	Illiteracy	Agriculture	Land-holding peasant	Married	No	No religion
2	Inhabited for generations	Sishu 4 years	Agriculture	Land-holding peasant	Widowed	No	No religion
26	Inhabited for generations						
25	Inhabited for generations	Illiteracy	Agriculture	Land-holding peasant	Widowed	No	No religion
14	Inhabited for generations	Illiteracy	Agriculture	Land-holding peasant	Married	No	No religion
18	Inhabited for generations	Illiteracy	Agriculture	Land-holding peasant	Married	No	No religion
10	Inhabited for generations	Illiteracy	Agriculture	Land-holding peasant	Unmarried	No	No religion
6	Inhabited for generations						
15	Inhabited for generations	Illiteracy	Agriculture	Land-holding peasant	Married	No	No religion
17	Inhabited for generations	Illiteracy	Agriculture	Land-holding peasant	Married	No	No religion
13	Inhabited for generations						
27	Inhabited for generations						
29	Inhabited for generations	Illiteracy	Agriculture	Land-holding peasant	Married	No	No religion
13	Inhabited for generations	Illiteracy	Agriculture	Land-holding peasant	Widowed	No	No religion

Appendix: Overview of Microdata for Wudu County (*cont.*)

	The no. of household	Jia	Household	Relationship with the household head	Surname	First name	Sex	The Chinese zodiac	Birthdate	
									Year	Month
Fujin Xiang 1st Bao				Son	陈	谢哇	Male	Tiger	1938	1
Fujin Xiang 1st Bao				Family member	陈	双全	Male	Tiger	1914	9
Fujin Xiang 1st Bao				Uncle	陈	国仓	Male	Rat	1899	8
Fujin Xiang 1st Bao				Uncle's wife	聂	四生	Female	Dragon	1916	5
Fujin Xiang 1st Bao				Nephew	陈	乔娃	Male	Monkey	1932	8
Fujin Xiang 1st Bao				Uncle's son	陈	守家	Male	Rabbit	1902	5
Fujin Xiang 1st Bao				Nephew	陈	慈来儿	Male	Snake	1941	7
Fujin Xiang 1st Bao	374	13	12	The household head	陈	守义	Male	Monkey	1920	3
Fujin Xiang 1st Bao				The household head's wife	李	宣娃	Female	Goat	1919	7
Fujin Xiang 1st Bao				Father	陈	国章	Male	Rat	1887	2
Fujin Xiang 1st Bao				Older brother	陈	守业	Male	Rooster	1908	9
Fujin Xiang 1st Bao				Older brother's wife	马	菊娃	Female	Monkey	1907	11
Fujin Xiang 1st Bao				Nephew	陈	贵隆儿	Male	Dragon	1918	2
Fujin Xiang 1st Bao				Nephew's wife	岳	贵娃	Female	Tiger	1926	3
Fujin Xiang 1st Bao				The second nephew	陈	隆林	Male	Tiger	1938	6
Fujin Xiang 1st Bao				Nephew's daughter	陈	隐隐	Female	Rooster	1945	1
Fujin Xiang 1st Bao				Younger brother	陈	兹姑儿	Male	Ox	1937	12
Fujin Xiang 1st Bao				The second younger brother	陈	二斤	Male	Snake	1941	7

Date	The beginning time of inhabitation in this county	Educational level	Profession sector	Occupation	Marital status	Physical disability	Religion
10	Inhabited for generations	Illiteracy	Agriculture				
1	Inhabited for generations	Illiteracy	Agriculture	Land-holding peasant	Married	No	No religion
5	Inhabited for generations	Illiteracy	Agriculture	Land-holding peasant	Widowed	No	No religion
6	Inhabited for generations	Illiteracy	Agriculture	Land-holding peasant	Widowed	No	No religion
30	Inhabited for generations	Illiteracy	Agriculture	Land-holding peasant			
4	Inhabited for generations	Illiteracy	Agriculture	Land-holding peasant	Widowed	No	No religion
6	Inhabited for generations	Illiteracy					
16	Inhabited for generations	Sishu 3 years	Agriculture	Land-holding peasant	Married	No	No religion
20	Inhabited for generations	Illiteracy	Agriculture	Land-holding peasant	Married	No	No religion
7	Inhabited for generations	Illiteracy	Agriculture	Land-holding peasant	Widowed	No	No religion
10	Inhabited for generations	Illiteracy	Agriculture	Land-holding peasant	Married	No	No religion
12	Inhabited for generations	Illiteracy	Agriculture	Land-holding peasant	Married	No	No religion
19	Inhabited for generations	Illiteracy	Agriculture	Land-holding peasant	Married	No	No religion
27	Inhabited for generations	Illiteracy	Agriculture	Land-holding peasant	Married	No	No religion
7	Inhabited for generations	Bao school 4 years					
12	Inhabited for generations						
26	Inhabited for generations	Illiteracy	Agriculture				
1	Inhabited for generations						

Appendix: Overview of Microdata for Wudu County (*cont.*)

	The no. of household	Jia	Household	Relationship with the household head	Surname	First name	Sex	The Chinese zodiac	Birthdate	
									Year	Month
Fujin Xiang 1st Bao				Uncle	陈	国贞	Male	Snake	1892	9
Fujin Xiang 1st Bao				Uncle's wife	敬	崴吾	Female	Dragon	1891	12
Fujin Xiang 1st Bao				Uncle's son	陈	仁甲儿	Male	Dog	1922	12
Fujin Xiang 1st Bao				Younger brother's wife	马	苏明	Female	Rat	1924	1
Fujin Xiang 1st Bao				Niece	陈	贵林	Female	Snake	1941	12
Fujin Xiang 1st Bao	375	13	13	The household head	陈	守科	Male	Ox	1925	12
Fujin Xiang 1st Bao				The household head's wife	张	润香	Female	Rabbit	1927	11
Fujin Xiang 1st Bao				Younger sister	崴	香儿	Female	Ox	1937	11
Fujin Xiang 1st Bao				Uncle's son	陈	守山	Male	Rooster	1908	1
Fujin Xiang 1st Bao				Older brother's wife	马	登娃	Female	Monkey	1907	3
Fujin Xiang 1st Bao				Niece	陈	马女	Female	Dragon	1940	5
Fujin Xiang 1st Bao				Nephew	陈	满才	Male	Goat	1943	2
Fujin Xiang 1st Bao	376	13	14	The household head	陈	永香	Male	Tiger	1926	12
Fujin Xiang 1st Bao				The household head's wife	赵	梅娃	Female	Tiger	1926	4
Fujin Xiang 1st Bao				Grandfather	陈	国礼	Male	Ox	1875	6
Fujin Xiang 1st Bao				Son	陈	忠？儿	Male	Pig	1947	4
Fujin Xiang 1st Bao				Uncle's son	陈	永琴	Male	Rooster	1921	3

Date	The beginning time of inhabitation in this county	Educational level	Profession sector	Occupation	Marital status	Physical disability	Religion	
19	Inhabited for generations	Illiteracy	Agriculture	Land-holding peasant	Married	No	No religion	
11	Inhabited for generations	Illiteracy	Agriculture	Land-holding peasant	Married	No	No religion	
14	Inhabited for generations	Illiteracy	Agriculture	Land-holding peasant	Married	No	No religion	
13	Inhabited for generations	Illiteracy	Agriculture	Land-holding peasant	Married	No	No religion	
20	Inhabited for generations							
8	Inhabited for generations	Illiteracy	Agriculture	Land-holding peasant	Married	No	No religion	
6	Inhabited for generations	Illiteracy	Agriculture	Land-holding peasant	Married	No	No religion	
13	Inhabited for generations	Illiteracy	Agriculture	Land-holding peasant	Unmarried	No	No religion	
26	Inhabited for generations	Illiteracy	Agriculture	Land-holding peasant	Married	No	No religion	
28	Inhabited for generations	Illiteracy	Agriculture	Land-holding peasant	Married	No	No religion	
26	Inhabited for generations							
11	Inhabited for generations							
17	Inhabited for generations	Illiteracy	Agriculture	Land-holding peasant	Married	No	No religion	Died in July 1947
26	Inhabited for generations	Illiteracy	Agriculture	Land-holding peasant	Married	No	No religion	
3	Inhabited for generations	Illiteracy	Agriculture	Land-holding peasant	Widowed	No	No religion	
5	Inhabited for generations							
30	Inhabited for generations	Illiteracy	Agriculture	Land-holding peasant	Married	No	No religion	

Appendix: Overview of Microdata for Wudu County (*cont.*)

	The no. of household	Jia	Household	Relationship with the household head	Surname	First name	Sex	The Chinese zodiac	Birthdate Year	Month
Fujin Xiang 1st Bao				Older brother's wife	张	框儿	Female	Monkey	1920	12
Fujin Xiang 1st Bao				Uncle's wife	赵	润连	Female	Rabbit	1902	2
Fujin Xiang 1st Bao				Nephew	陈	俱生	Male	Rooster	1945	3
Fujin Xiang 1st Bao				The second nephew	陈	玉生	Male	Pig	1947	3
Fujin Xiang 1st Bao				Uncle's son	陈	田娃	Male	Horse	1942	7
Fujin Xiang 1st Bao	377	13	15	The household head	陈	守隆	Male	Horse	1905	2
Fujin Xiang 1st Bao				The household head's wife	张	苏家	Female	Tiger	1901	2
Fujin Xiang 1st Bao				Father	陈	国虎	Male	Dragon	1881	1
Fujin Xiang 1st Bao				Son	陈	忙人	Male	Tiger	1938	6
Fujin Xiang 1st Bao				The second son	陈	小毛子	Male	Dragon	1940	6
Fujin Xiang 1st Bao				Daughter	陈	罗儿	Female	Dog	1946	3
Fujin Xiang 1st Bao	378	14	1	The household head	陈	国玉	Male	Rabbit	1902	11
Fujin Xiang 1st Bao				The household head's wife	马	安娃	Female	Dragon	1919	1
Fujin Xiang 1st Bao				Son	陈	同娃	Male	Goat	1919	5
Fujin Xiang 1st Bao				Son's wife	马	玉娃	Female	Pig	1923	1
Fujin Xiang 1st Bao				Daughter	陈	周娃	Female	Rooster	1945	6
Fujin Xiang 1st Bao				Younger brother	陈	国福	Male	Dragon	1903	5
Fujin Xiang 1st Bao				Younger brother's wife	史	刘柏	Female	Snake	1904	9

Date	The beginning time of inhabitation in this county	Educational level	Profession sector	Occupation	Marital status	Physical disability	Religion
8	Inhabited for generations	Illiteracy	Agriculture	Land-holding peasant	Married	No	No religion
20	Inhabited for generations	Illiteracy	Agriculture	Land-holding peasant	Widowed	No	No religion
15	Inhabited for generations						
16	Inhabited for generations						
24	Inhabited for generations						
16	Inhabited for generations	Illiteracy	Agriculture	Land-holding peasant	Married	No	No religion
29	Inhabited for generations	Illiteracy	Agriculture	Land-holding peasant	Married	No	No religion
16	Inhabited for generations	Illiteracy	Agriculture	Land-holding peasant	Widowed	No	No religion
12	Inhabited for generations	Illiteracy	Agriculture	Land-holding peasant			
27	Inhabited for generations						
23	Inhabited for generations						
27	Inhabited for generations	Illiteracy	Agriculture	Land-holding peasant	Married	No	No religion
28	Inhabited for generations	Illiteracy	Agriculture	Land-holding peasant	Married	No	No religion
24	Inhabited for generations	Illiteracy	Agriculture	Land-holding peasant	Married	No	No religion
10	Inhabited for generations	Illiteracy	Agriculture	Land-holding peasant	Married	No	No religion
6	Inhabited for generations						
24	Inhabited for generations	Illiteracy	Agriculture	Land-holding peasant	Married	No	No religion
4	Inhabited for generations	Illiteracy	Agriculture	Land-holding peasant	Married	No	No religion

Appendix: Overview of Microdata for Wudu County (*cont.*)

									Birthdate	
	The no. of house-hold	Jia	House-hold	Relation-ship with the house-hold head	Surname	First name	Sex	The Chinese zodiac	Year	Month
Fujin Xiang 1st Bao				Nephew	陈	双全儿	Male	Monkey	1920	3
Fujin Xiang 1st Bao				Nephew's wife	史	润娃	Female	Dragon	1928	7
Fujin Xiang 1st Bao	379	14	2	The house-hold head	陈	守彦	Male	Tiger	1914	3
Fujin Xiang 1st Bao				The house-hold head's wife	赵	苏信	Female	Snake	1917	1
Fujin Xiang 1st Bao				Father	陈	国顺	Male	Goat	1881	12
Fujin Xiang 1st Bao				Son	陈	印娃	Male	Rabbit	1939	5
Fujin Xiang 1st Bao				The second son	陈	祿印	Male	Dog	1941	1
Fujin Xiang 1st Bao				Daughter	陈	印女	Female	Snake	1941	7
Fujin Xiang 1st Bao				Uncle's son	陈	全儿	Male	Rat	1899	11
Fujin Xiang 1st Bao				Uncle's wife	刘	尕老	Female	Pig	1886	3
Fujin Xiang 1st Bao				Family member	义	田	Male	Rabbit	1902	2
Fujin Xiang 1st Bao				Niece	陈	花香	Female	Rooster	1933	8
Fujin Xiang 1st Bao	380	14	3	The house-hold head	陈	国建	Male	Rooster	1908	5
Fujin Xiang 1st Bao				The house-hold head's wife	史	早女子	Female	Rabbit	1902	2
Fujin Xiang 1st Bao				Son	陈	守福	Male	Rooster	1933	4
Fujin Xiang 1st Bao				The second son	陈	么儿	Male	Tiger	1937	4
Fujin Xiang 1st Bao				The second daughter	陈	柱香	Female	Dog	1934	11
Fujin Xiang 1st Bao				Uncle's son	陈	国生	Male	Pig	1912	6

Date	The beginning time of inhabitation in this county	Educational level	Profession sector	Occupation	Marital status	Physical disability	Religion
13	Inhabited for generations	Illiteracy	Agriculture	Land-holding peasant	Married	No	No religion
14	Inhabited for generations	Illiteracy	Agriculture	Land-holding peasant	Married	No	No religion
28	Inhabited for generations	Illiteracy	Agriculture	Land-holding peasant	Married	No	No religion
12	Inhabited for generations	Illiteracy	Agriculture	Land-holding peasant	Married	No	No religion
20	Inhabited for generations	Illiteracy	Agriculture	Land-holding peasant	Widowed	No	No religion
10	Inhabited for generations	Bao school 4 years					
12	Inhabited for generations						
10	Inhabited for generations	Illiteracy					
28	Inhabited for generations	Illiteracy	Agriculture	Land-holding peasant	Widowed	No	No religion
12	Inhabited for generations	Illiteracy	Agriculture	Land-holding peasant	Widowed	No	No religion
12	Inhabited for generations	Illiteracy	Agriculture	Land-holding peasant	Widowed	No	No religion
7	Inhabited for generations	Illiteracy					
5	Inhabited for generations	Illiteracy	Agriculture	Land-holding peasant	Married	No	No religion
1	Inhabited for generations	Illiteracy	Agriculture	Land-holding peasant	Married	No	No religion
14	Inhabited for generations	Bao school 4 years	Agriculture	Land-holding peasant	Unmarried	No	No religion
28	Inhabited for generations	Illiteracy					
28	Inhabited for generations	Illiteracy					
28	Inhabited for generations	Illiteracy	Agriculture	Land-holding peasant	Married	No	No religion

Appendix: Overview of Microdata for Wudu County (*cont.*)

	The no. of household	Jia	Household	Relationship with the household head	Surname	First name	Sex	The Chinese zodiac	Birthdate Year	Month
Fujin Xiang 1st Bao				The household head's wife	马	九儿	Female	Horse	1918	9
Fujin Xiang 1st Bao				Father	陈	义雄	Male	Pig	1886	8
Fujin Xiang 1st Bao				Son	陈	过房	Male	Tiger	1938	12
Fujin Xiang 1st Bao				The second son	陈	二房	Male	Goat	1943	1
Fujin Xiang 1st Bao				The third son	陈	谢林	Male	Rooster	1945	5
Fujin Xiang 1st Bao				Daughter	陈	双娃	Female	Pig	1935	3
Fujin Xiang 1st Bao	381	14	4	The household head	陈	国邦	Male	Rat	1912	7
Fujin Xiang 1st Bao				The household head's wife	石	条双儿	Female	Ox	1925	2
Fujin Xiang 1st Bao				Father	陈	义禄	Male	Horse	1879	8
Fujin Xiang 1st Bao				Mother	董	换儿	Female	Tiger	1889	10
Fujin Xiang 1st Bao				Son	陈	杨忠	Male	Horse	1942	11
Fujin Xiang 1st Bao				Uncle's son	陈	国兴	Male	Rooster	1908	10
Fujin Xiang 1st Bao				The household head's wife	冯	秋林儿	Female	Dragon	1906	11
Fujin Xiang 1st Bao				Son	陈	王姓娃	Male	Rat	1936	1
Fujin Xiang 1st Bao				The second son	陈	保保	Male	Rabbit	1936	10
Fujin Xiang 1st Bao				Daughter	陈	东儿	Female	Dragon	1940	1
Fujin Xiang 1st Bao				The second daughter	陈	香娃	Female	Goat	1943	12
Fujin Xiang 1st Bao	382	14	5	The household head	冯	玉成	Male	Dragon	1916	3

Date	The beginning time of inhabitation in this county	Educational level	Profession sector	Occupation	Marital status	Physical disability	Religion
18	Inhabited for generations	Illiteracy	Agriculture	Land-holding peasant	Married	No	No religion
1	Inhabited for generations	Illiteracy	Agriculture	Land-holding peasant	Widowed	No	No religion
10	Inhabited for generations	Bao school 2 years					
6	Inhabited for generations						
1	Inhabited for generations						
12	Inhabited for generations	Illiteracy	Agriculture		Unmarried	No	No religion
8	Inhabited for generations	Sishu 3 years	Agriculture	Land-holding peasant	Married	No	No religion
26	Inhabited for generations	Illiteracy	Agriculture	Land-holding peasant	Married	No	No religion
12	Inhabited for generations	Illiteracy	Agriculture	Land-holding peasant	Married	No	No religion
25	Inhabited for generations	Illiteracy	Agriculture	Land-holding peasant	Married	No	No religion
12	Inhabited for generations						
15	Inhabited for generations	Illiteracy	Agriculture	Land-holding peasant	Married	No	No religion
5	Inhabited for generations	Illiteracy	Agriculture	Land-holding peasant	Married	No	No religion
27	Inhabited for generations	Bao school 4 years	Agriculture	Land-holding peasant	Unmarried	No	No religion
23	Inhabited for generations	Illiteracy	Agriculture	Land-holding peasant		No	No religion
21	Inhabited for generations						
21	Inhabited for generations						
23	Inhabited for generations	Illiteracy	Agriculture	Land-holding peasant	Married	No	No religion

Appendix: Overview of Microdata for Wudu County (*cont.*)

	The no. of household	Jia	Household	Relationship with the household head	Surname	First name	Sex	The Chinese zodiac	Birthdate	
									Year	Month
Fujin Xiang 1st Bao				The household head's wife	马	苏香	Female	Dog	1922	10
Fujin Xiang 1st Bao				Father	冯	义林	Male	Ox	1888	1
Fujin Xiang 1st Bao				Mother	陈	知娃	Female	Monkey	1883	9
Fujin Xiang 1st Bao				Son	冯	杨才	Male	Monkey	1944	12
Fujin Xiang 1st Bao				Daughter	冯	爱娃	Female	Snake	1941	9
Fujin Xiang 1st Bao	383	14	6	The household head	冯	玉喜	Male	Tiger	1889	4
Fujin Xiang 1st Bao				The household head's wife	史	慈娃	Female	Pig	1886	10
Fujin Xiang 1st Bao				Son	冯	崴喜	Male	Monkey	1920	10
Fujin Xiang 1st Bao				Son's wife	陈	仓生	Female	Rabbit	1922	4
Fujin Xiang 1st Bao				Granddaughter	冯	九玉儿	Female	Dog	1946	5
Fujin Xiang 1st Bao	384	14	7	The household head	冯	义全	Male	Ox	1900	4
Fujin Xiang 1st Bao				The household head's wife	史	崴秀儿	Female	Rabbit	1902	4
Fujin Xiang 1st Bao				Mother	冯	宋儿	Female	Rabbit	1878	2
Fujin Xiang 1st Bao				Son	冯	崴朝哇	Male	Goat	1919	5
Fujin Xiang 1st Bao				Son's wife	岳	医娃	Female	Rabbit	1927	12
Fujin Xiang 1st Bao				Older brother	冯	义库	Male	Tiger	1889	6
Fujin Xiang 1st Bao				Older brother's wife	陈	年香儿	Female	Dog	1897	1

Date	The beginning time of inhabitation in this county	Educational level	Profession sector	Occupation	Marital status	Physical disability	Religion
3	Inhabited for generations	Illiteracy	Agriculture	Land-holding peasant	Married	No	No religion
7	Inhabited for generations	Illiteracy	Agriculture	Land-holding peasant	Married	No	No religion
3	Inhabited for generations	Illiteracy	Agriculture	Land-holding peasant	Married	No	No religion
10	Inhabited for generations						
29	Inhabited for generations						
20	Inhabited for generations	Illiteracy	Agriculture	Land-holding peasant	Married	No	No religion
28	Inhabited for generations	Illiteracy	Agriculture	Land-holding peasant	Married	No	No religion
25	Inhabited for generations	Illiteracy	Agriculture	Land-holding peasant	Married	No	No religion
17	Inhabited for generations	Illiteracy	Agriculture	Land-holding peasant	Married	No	No religion
24	Inhabited for generations						
7	Inhabited for generations	Illiteracy	Agriculture	Land-holding peasant	Married	No	No religion
4	Inhabited for generations	Illiteracy	Agriculture	Land-holding peasant	Married	No	No religion
20	Inhabited for generations						
19	Inhabited for generations	Illiteracy	Agriculture	Land-holding peasant	Married	No	No religion
1	Inhabited for generations	Illiteracy	Agriculture	Land-holding peasant	Married	No	No religion
30	Inhabited for generations	Illiteracy	Agriculture	Land-holding peasant	Married	No	No religion
22	Inhabited for generations	Illiteracy	Agriculture	Land-holding peasant	Married	No	No religion

Appendix: Overview of Microdata for Wudu County (*cont.*)

	The no. of household	Jia	Household	Relationship with the household head	Surname	First name	Sex	The Chinese zodiac	Birthdate	
									Year	Month
Fujin Xiang 1st Bao				Nephew	冯	申娃	Male	Pig	1935	1
Fujin Xiang 1st Bao				The second nephew	进	三儿	Male	Tiger	1938	12
Fujin Xiang 1st Bao	385	14	8	The household head	陈	国泰	Male	Horse	1879	12
Fujin Xiang 1st Bao				The household head's wife	王	见称	Female	Snake	1892	10
Fujin Xiang 1st Bao				Son	陈	守义	Male	Snake	1917	3
Fujin Xiang 1st Bao				Son's wife	史	住生	Female	Snake	1917	5
Fujin Xiang 1st Bao				Grandson	陈	忠信	Male	Ox	1937	9
Fujin Xiang 1st Bao				Younger brother	陈	国玉	Male	Tiger	1889	2
Fujin Xiang 1st Bao				Younger brother's wife	史	德女	Female	Dragon	1891	12
Fujin Xiang 1st Bao				Nephew	陈	守德	Male	Goat	1919	3
Fujin Xiang 1st Bao				Nephew's wife	赵	召娃	Female	Dragon	1916	9
Fujin Xiang 1st Bao				Nephew's son	陈	中林	Male	Horse	1942	11
Fujin Xiang 1st Bao				Nephew's wife	杜	连儿	Female	Pig	1924	4
Fujin Xiang 1st Bao				Nephew's son	陈	中平	Male	Rat	1936	4
Fujin Xiang 1st Bao	386	14	9	The household head	冯	玉隆	Male	Snake	1917	2
Fujin Xiang 1st Bao				The household head's wife	岳	琴娃	Female	Snake	1917	4
Fujin Xiang 1st Bao				Father	冯	义德	Male	Pig	1897	1

Date	The beginning time of inhabitation in this county	Educational level	Profession sector	Occupation	Marital status	Physical disability	Religion
28	Inhabited for generations	Illiteracy	Agriculture	Land-holding peasant			
28	Inhabited for generations	Bao school 2 years					
4	Inhabited for generations	Sishu 3 years			Married	Blind	
3	Inhabited for generations	Illiteracy	Agriculture	Land-holding peasant	Married	No	No religion
17	Inhabited for generations	Sishu 5 years	Agriculture	Land-holding peasant	Married	No	No religion
13	Inhabited for generations	Illiteracy	Agriculture	Land-holding peasant	Married	No	No religion
29	Inhabited for generations						
3	Inhabited for generations	Sishu 5 years	Agriculture	Land-holding peasant	Married	No	No religion
11	Inhabited for generations	Illiteracy	Agriculture	Land-holding peasant	Married	No	No religion
9	Inhabited for generations	Sishu 3 years	Agriculture	Land-holding peasant	Married	No	No religion
6	Inhabited for generations	Illiteracy	Agriculture	Land-holding peasant	Married	No	No religion
7	Inhabited for generations						
22	Inhabited for generations	Illiteracy	Agriculture	Land-holding peasant	Widowed	No	No religion
16	Inhabited for generations	Bao school 3 years					
26	Inhabited for generations	Illiteracy	Agriculture	Land-holding peasant	Married	No	No religion
11	Inhabited for generations	Illiteracy	Agriculture	Land-holding peasant	Married	No	No religion
26	Inhabited for generations	Illiteracy	Agriculture	Land-holding peasant	Married	No	No religion

Appendix: Overview of Microdata for Wudu County (*cont.*)

	The no. of household	Jia	Household	Relationship with the household head	Surname	First name	Sex	The Chinese zodiac	Birthdate Year	Month
Fujin Xiang 1st Bao				Mother	史	崴房	Female	Pig	1897	1
Fujin Xiang 1st Bao				Son	冯	聚足	Male	Snake	1941	1
Fujin Xiang 1st Bao				Daughter	冯	周家	Female	Rabbit	1939	6
Fujin Xiang 1st Bao				Younger brother	冯	仁娃	Male	Goat	1931	11
Fujin Xiang 1st Bao				Younger sister	冯	白鸡	Female	Rabbit	1939	1
Fujin Xiang 1st Bao				The second younger sister	冯	麻鸡	Female	Monkey	1941	2
Fujin Xiang 1st Bao	387	14	10	The household head	冯	世清	Male	Dog	1922	2
Fujin Xiang 1st Bao				The household head's wife	赵	元英	Female	Pig	1923	6
Fujin Xiang 1st Bao				Father	冯	玉山	Male	Rabbit	1890	7
Fujin Xiang 1st Bao				Daughter	冯	菊桃	Female	Horse	1943	9
Fujin Xiang 1st Bao	388	14	11	The household head	冯	义仓	Male	Horse	1879	9
Fujin Xiang 1st Bao				The household head's wife	张	丑儿	Female	Tiger	1901	10
Fujin Xiang 1st Bao				Son	冯	柱娃	Male	Goat	1919	2
Fujin Xiang 1st Bao				Son's wife	润	秋儿	Female	Ox	1925	3
Fujin Xiang 1st Bao				Grandson	冯	马娃	Male	Goat	1943	7
Fujin Xiang 1st Bao				Daughter	冯	义女	Female	Rooster	1945	6
Fujin Xiang 1st Bao	389	14	12	The household head	冯	玉哥	Male	Dragon	1903	11

Date	The beginning time of inhabitation in this county	Educational level	Profession sector	Occupation	Marital status	Physical disability	Religion
26	Inhabited for generations	Illiteracy	Agriculture	Land-holding peasant	Married	No	No religion
9	Inhabited for generations	Illiteracy					
28	Inhabited for generations	Illiteracy					
4	Inhabited for generations	Illiteracy	Agriculture	Land-holding peasant	Unmarried	No	No religion
8	Inhabited for generations	Illiteracy					
13	Inhabited for generations						
2	Inhabited for generations	Junior middle school 5 years	Agriculture	Land-holding peasant	Married	No	No religion
2	Inhabited for generations	Illiteracy	Agriculture	Land-holding peasant	Married	No	No religion
15	Inhabited for generations	Sishu 3 years	Agriculture	Land-holding peasant	Widowed	No	No religion
1	Inhabited for generations						
25	Inhabited for generations	Illiteracy	Agriculture	Land-holding peasant	Married	No	No religion
4	Inhabited for generations	Illiteracy	Agriculture	Land-holding peasant	Married	No	No religion
2	Inhabited for generations	Illiteracy	Agriculture	Land-holding peasant	Married	No	No religion
23	Inhabited for generations	Illiteracy	Agriculture	Land-holding peasant	Married	No	No religion
2	Inhabited for generations						
2	Inhabited for generations						
24	Inhabited for generations	Illiteracy	Agriculture	Land-holding peasant	Married	No	No religion

Appendix: Overview of Microdata for Wudu County (*cont.*)

	The no. of household	Jia	Household	Relationship with the household head	Surname	First name	Sex	The Chinese zodiac	Birthdate Year	Month
Fujin Xiang 1st Bao				The household head's wife	赵	跳娃	Female	Dog	1909	2
Fujin Xiang 1st Bao				Son	冯	佪大	Male	Rooster	1945	12
Fujin Xiang 1st Bao				Daughter	冯	换换	Female	Rat	1936	12
Fujin Xiang 1st Bao				The second daughter	冯	香哇	Female	Dragon	1940	6
Fujin Xiang 1st Bao	390	14	13	The household head	冯	玉田	Male	Rabbit	1902	2
Fujin Xiang 1st Bao				The household head's wife	李	柱生娃	Female	Horse	1905	8
Fujin Xiang 1st Bao				Son	冯	世虎	Male	Monkey	1920	7
Fujin Xiang 1st Bao				Son's wife	赵	崴全	Female	Dragon	1928	11
Fujin Xiang 1st Bao	391	14	14	The household head	王	枝虎	Male	Rat	1912	11
Fujin Xiang 1st Bao				The household head's wife	李	龙菊	Female	Rooster	1921	2
Fujin Xiang 1st Bao				Father	王	玉林	Male	Rabbit	1891	8
Fujin Xiang 1st Bao				Mother	马	崴秀	Female	Dog	1897	1
Fujin Xiang 1st Bao				Grandson	王	三女	Male	Goat	1943	12
Fujin Xiang 1st Bao				The second grandson	崴	三斤	Male	Rooster	1945	2
Fujin Xiang 1st Bao	392	14	15	The household head	赵	思荣	Male	Pig	1898	11
Fujin Xiang 1st Bao				The household head's wife	赵	跳儿	Female	Pig	1898	11
Fujin Xiang 1st Bao				Father	赵	元奎	Male	Ox	1876	9

Date	The beginning time of inhabitation in this county	Educational level	Profession sector	Occupation	Marital status	Physical disability	Religion
28	Inhabited for generations	Illiteracy	Agriculture	Land-holding peasant	Married	No	No religion
18	Inhabited for generations						
31	Inhabited for generations	Illiteracy					
18	Inhabited for generations						
4	Inhabited for generations	Illiteracy	Agriculture	Land-holding peasant	Married	No	No religion
15	Inhabited for generations	Illiteracy	Agriculture	Land-holding peasant	Married	No	No religion
21	Inhabited for generations	Illiteracy	Agriculture	Land-holding peasant	Married	No	No religion
29	Inhabited for generations	Illiteracy	Agriculture	Land-holding peasant	Married	No	No religion
29	Inhabited for generations	Illiteracy	Agriculture	Land-holding peasant	Married	No	No religion
23	Inhabited for generations	Illiteracy	Agriculture	Land-holding peasant	Married	No	No religion
15	Inhabited for generations	Illiteracy	Agriculture	Land-holding peasant	Married	No	No religion
29	Inhabited for generations	Illiteracy	Agriculture	Land-holding peasant	Married	No	No religion
6	Inhabited for generations						
15	Inhabited for generations						
1	Inhabited for generations	Sishu 3 years	Agriculture	Land-holding peasant	Married	No	No religion
18	Inhabited for generations	Illiteracy	Agriculture	Land-holding peasant	Married	No	No religion
9	Inhabited for generations				Married		

Appendix: Overview of Microdata for Wudu County (*cont.*)

	The no. of household	Jia	Household	Relationship with the household head	Surname	First name	Sex	The Chinese zodiac	Birthdate Year	Month
Fujin Xiang 1st Bao				Mother	陈	党官	Female	Tiger	1877	5
Fujin Xiang 1st Bao				Son	赵	子贞	Male	Goat	1919	12
Fujin Xiang 1st Bao				Son's wife	李	祥娃	Female	Monkey	1920	11
Fujin Xiang 1st Bao				Grandson	赵	柏哇	Male	Snake	1941	2
Fujin Xiang 1st Bao				The second grandson	赵	东儿	Male	Horse	1942	11
Fujin Xiang 1st Bao				Grand-daughter	赵	良女	Female	Monkey	1945	12
Fujin Xiang 1st Bao				The second Grand-daughter	赵	二女	Female	Dog	1946	1
Fujin Xiang 1st Bao				Younger brother	赵	思花	Male	Horse	1918	2
Fujin Xiang 1st Bao				Younger brother's wife	李	佐哇	Female	Tiger	1926	10
Fujin Xiang 1st Bao				The second nephew	赵	早儿	Male	Pig	1935	5
Fujin Xiang 1st Bao	393	15	1	The household head	赵	志德	Male	Tiger	1901	11
Fujin Xiang 1st Bao				The household head's wife	社	香	Female	Dragon	1903	9
Fujin Xiang 1st Bao				Son	赵	宋保	Male	Dog	1934	5
Fujin Xiang 1st Bao				Daughter	赵	存儿	Female	Tiger	1938	1
Fujin Xiang 1st Bao				The second daughter	赵	存长儿	Female	Horse	1942	7
Fujin Xiang 1st Bao	394	15	2	The household head	赵	思隆	Male	Dragon	1916	9
Fujin Xiang 1st Bao				The household head's wife	刘	二赛	Female	Monkey	1920	1

Date	The beginning time of inhabitation in this county	Educational level	Profession sector	Occupation	Marital status	Physical disability	Religion
8	Inhabited for generations				Married		
21	Inhabited for generations	Illiteracy	Agriculture	Land-holding peasant	Married	No	No religion
27	Inhabited for generations	Illiteracy	Agriculture	Land-holding peasant	Married	No	No religion
1	Inhabited for generations						
9	Inhabited for generations						
30	Inhabited for generations						
30	Inhabited for generations						
17	Inhabited for generations	Illiteracy	Agriculture	Land-holding peasant	Married	No	No religion
27	Inhabited for generations	Illiteracy	Agriculture	Land-holding peasant	Married	No	No religion
23	Inhabited for generations	Illiteracy					
19	Inhabited for generations	Illiteracy	Agriculture	Land-holding peasant	Married	No	No religion
30	Inhabited for generations	Illiteracy	Agriculture	Land-holding peasant	Married	No	No religion
24	Inhabited for generations	Bao school 2 years					
1	Inhabited for generations	Illiteracy					
5	Inhabited for generations						
5	Inhabited for generations	Illiteracy	Agriculture	Land-holding peasant	Married	No	No religion
3	Inhabited for generations	Illiteracy	Agriculture	Land-holding peasant	Married	No	No religion

Appendix: Overview of Microdata for Wudu County (*cont.*)

	The no. of house-hold	Jia	House-hold	Relation-ship with the house-hold head	Surname	First name	Sex	The Chinese zodiac	Birthdate	
									Year	Month
Fujin Xiang 1st Bao				Father	赵	元生	Male	Snake	1892	10
Fujin Xiang 1st Bao				Mother	李	官兆	Female	Goat	1894	10
Fujin Xiang 1st Bao				Son	赵	甲子哇	Male	Monkey	1944	1
Fujin Xiang 1st Bao				The second son	赵	存养	Male	Monkey	1946	11
Fujin Xiang 1st Bao	395	15	3	The house-hold head	赵	思孔	Male	Monkey	1907	9
Fujin Xiang 1st Bao				The house-hold head's wife	潘	海秀	Female	Ox	1913	2
Fujin Xiang 1st Bao				Son	赵	国儿	Male	Tiger	1938	6
Fujin Xiang 1st Bao				The second son	赵	保玉	Male	Rabbit	1939	1
Fujin Xiang 1st Bao				The third son	赵	园玉	Male	Goat	1943	12
Fujin Xiang 1st Bao				Uncle	赵	元合	Male	Dog	1909	9
Fujin Xiang 1st Bao				Younger brother	赵	长受	Male	Ox	1937	11
Fujin Xiang 1st Bao				Uncle's son	赵	思福	Male	Rooster	1884	3
Fujin Xiang 1st Bao				Older broth-er's wife	杨	翠香	Female	Dog	1897	4
Fujin Xiang 1st Bao				Nephew	赵	杨哇	Male	Horse	1905	5
Fujin Xiang 1st Bao				Nephew's wife	贾	六女	Female	Horse	1918	2
Fujin Xiang 1st Bao				Nephew's son	赵	乔长	Male	Goat	1943	10
Fujin Xiang 1st Bao	396	15	4	The house-hold head	赵	元忠	Male	Rat	1899	3
Fujin Xiang 1st Bao				The house-hold head's wife	徐	社社	Female	Snake	1904	6
Fujin Xiang 1st Bao				Son	赵	思银	Male	Rat	1924	2

Date	The beginning time of inhabitation in this county	Educational level	Profession sector	Occupation	Marital status	Physical disability	Religion
13	Inhabited for generations	Illiteracy	Agriculture	Land-holding peasant	Married	No	No religion
19	Inhabited for generations	Illiteracy	Agriculture	Land-holding peasant	Married	No	No religion
11	Inhabited for generations	Illiteracy	Agriculture	Land-holding peasant	Married	No	No religion
19	Inhabited for generations						
18	Inhabited for generations	Illiteracy	Agriculture	Land-holding peasant	Married	No	No religion
25	Inhabited for generations	Illiteracy	Agriculture	Land-holding peasant	Married	No	No religion
12	Inhabited for generations	Illiteracy					
22	Inhabited for generations	Illiteracy					
26	Inhabited for generations						
21	Inhabited for generations	Illiteracy	Agriculture	Land-holding peasant	Widowed	No	No religion
1	Inhabited for generations	Illiteracy	Agriculture	Land-holding peasant			
1	Inhabited for generations	Illiteracy	Agriculture	Land-holding peasant	Married	No	No religion
15	Inhabited for generations						
1	Inhabited for generations	Illiteracy	Agriculture	Land-holding peasant	Married	No	No religion
24	Inhabited for generations	Illiteracy	Agriculture	Land-holding peasant	Married	No	No religion
20	Inhabited for generations						
8	Inhabited for generations	Illiteracy	Agriculture	Land-holding peasant	Married	No	No religion
12	Inhabited for generations	Illiteracy	Agriculture	Land-holding peasant	Married	No	No religion
2	Inhabited for generations	Illiteracy	Agriculture	Land-holding peasant	Married	No	No religion

Appendix: Overview of Microdata for Wudu County (*cont.*)

	The no. of household	Jia	Household	Relationship with the household head	Surname	First name	Sex	The Chinese zodiac	Birthdate	
									Year	Month
Fujin Xiang 1st Bao				Son's wife	陈	连秀	Female	Pig	1923	12
Fujin Xiang 1st Bao				Grandson	赵	昊成	Male	Dog	1946	2
Fujin Xiang 1st Bao				Granddaughter	赵	周芦	Female	Rooster	1945	3
Fujin Xiang 1st Bao				Nephew	赵	思成	Male	Tiger	1901	4
Fujin Xiang 1st Bao				Nephew's wife	史	双柱	Female	Tiger	1901	10
Fujin Xiang 1st Bao				Nephew's son	赵	昊家	Male	Dragon	1928	1
Fujin Xiang 1st Bao				Grandson's wife	陈	书香	Female	Tiger	1926	3
Fujin Xiang 1st Bao				The second grandson	赵	三斤	Male	Rooster	1933	5
Fujin Xiang 1st Bao				The third grandson	赵	崴三斤	Male	Pig	1935	4
Fujin Xiang 1st Bao				The fourth grandson	赵	四学儿	Male	Tiger	1918	11
Fujin Xiang 1st Bao				Granddaughter	赵	怌女子	Female	Rooster	1945	4
Fujin Xiang 1st Bao	397	15	5	The household head	赵	思武	Male	Pig	1885	12
Fujin Xiang 1st Bao				The household head's wife	李	月儿	Female	Ox	1900	7
Fujin Xiang 1st Bao				Son	赵	全哇	Male	Dog	1922	10
Fujin Xiang 1st Bao				The second son	赵	寒哇	Male	Rabbit	1927	2
Fujin Xiang 1st Bao				The third son	赵	三全儿	Male	Ox	1937	2
Fujin Xiang 1st Bao				Daughter	赵	寒女	Female	Monkey	1932	8
Fujin Xiang 1st Bao				The second daughter	赵	元女	Female	Goat	1943	6
Fujin Xiang 1st Bao	398	15	6	The household head	赵	元兴	Male	Goat	1906	9

Date	The beginning time of inhabitation in this county	Educational level	Profession sector	Occupation	Marital status	Physical disability	Religion
12	Inhabited for generations	Illiteracy	Agriculture	Land-holding peasant	Married	No	No religion
3	Inhabited for generations						
21	Inhabited for generations						
22	Inhabited for generations	Illiteracy	Agriculture	Land-holding peasant	Married	No	No religion
7	Inhabited for generations	Illiteracy	Agriculture	Land-holding peasant	Married	No	No religion
9	Inhabited for generations	Illiteracy	Agriculture	Land-holding peasant	Married	No	No religion
10	Inhabited for generations	Illiteracy	Agriculture	Land-holding peasant	Married	No	No religion
1	Inhabited for generations	Illiteracy					
31	Inhabited for generations	Illiteracy					
21	Inhabited for generations	Illiteracy					
13	Inhabited for generations						
1	Inhabited for generations	Illiteracy	Agriculture	Land-holding peasant	Married	Blind	No religion
3	Inhabited for generations	Illiteracy	Agriculture	Land-holding peasant	Married	No	No religion
11	Inhabited for generations	Illiteracy	Agriculture	Land-holding peasant	Widowed	No	No religion
12	Inhabited for generations	Illiteracy	Agriculture	Land-holding peasant	Unmarried	No	No religion
2	Inhabited for generations	Illiteracy					
22	Inhabited for generations	Illiteracy					
21	Inhabited for generations						
1	Inhabited for generations	Illiteracy	Agriculture	Land-holding peasant	Married	No	No religion

Appendix: Overview of Microdata for Wudu County (*cont.*)

	The no. of household	Jia	Household	Relationship with the household head	Surname	First name	Sex	The Chinese zodiac	Birthdate	
									Year	Month
Fujin Xiang 1st Bao				The household head's wife	石	邦女	Female	Rabbit	1915	4
Fujin Xiang 1st Bao				Son	赵	韩才	Male	Horse	1942	2
Fujin Xiang 1st Bao				Daughter	赵	應女	Female	Pig	1947	1
Fujin Xiang 1st Bao	399	15	7	The household head	赵	思玉	Male	Pig	1897	6
Fujin Xiang 1st Bao				The household head's wife	潘	付连	Female	Tiger	1901	2
Fujin Xiang 1st Bao				Son	赵	买哇	Male	Pig	1935	9
Fujin Xiang 1st Bao				The second son	赵	具哇	Male	Goat	1937	3
Fujin Xiang 1st Bao				The third son	赵	柱平儿	Male	Snake	1941	7
Fujin Xiang 1st Bao				Daughter	赵	卯香	Female	Rabbit	1939	3
Fujin Xiang 1st Bao	400	15	8	The household head	赵	志旺	Male	Dog	1909	12
Fujin Xiang 1st Bao				The household head's wife	徐	三女	Female	Snake	1917	2
Fujin Xiang 1st Bao				Father	赵	思明	Male	Dog	1897	2
Fujin Xiang 1st Bao				Mother	李	霏香	Female	Rooster	1896	4
Fujin Xiang 1st Bao				Grandfather	赵	元凤	Male	Rabbit	1866	9
Fujin Xiang 1st Bao				Daughter	赵	马家	Female	Tiger	1938	4
Fujin Xiang 1st Bao				The second daughter	赵	崴马家	Female	Snake	1941	9
Fujin Xiang 1st Bao				Son	赵	称成	Male	Pig	1947	1
Fujin Xiang 1st Bao	401	15	9	The household head	苏	凤荣	Male	Rat	1924	2

Date	The beginning time of inhabitation in this county	Educational level	Profession sector	Occupation	Marital status	Physical disability	Religion
18	Inhabited for generations	Illiteracy	Others		Married	No	No religion
2	Inhabited for generations						
30	Inhabited for generations						
6	Inhabited for generations	Illiteracy	Agriculture	Land-holding peasant	Married	No	No religion
4	Inhabited for generations	Illiteracy	Agriculture	Land-holding peasant	Married	No	No religion
22	Inhabited for generations	Illiteracy					
28	Inhabited for generations	Illiteracy					
11	Inhabited for generations	Illiteracy					
31	Inhabited for generations	Illiteracy					
9	Inhabited for generations	Illiteracy	Agriculture	Land-holding peasant	Married	No	No religion
1	Inhabited for generations	Illiteracy	Agriculture	Land-holding peasant	Married	No	No religion
23	Inhabited for generations	Illiteracy	Agriculture	Land-holding peasant	Married	No	No religion
4	Inhabited for generations	Illiteracy	Agriculture	Land-holding peasant	Married	No	No religion
13	Inhabited for generations	Illiteracy	Agriculture	Land-holding peasant	Widowed	No	No religion
16	Inhabited for generations	Illiteracy	Agriculture	Land-holding peasant		No	
10	Inhabited for generations	Illiteracy					
23	Inhabited for generations						
17	Inhabited for generations	Illiteracy	Agriculture	Land-holding peasant	Married	No	No religion

Appendix: Overview of Microdata for Wudu County (*cont.*)

									Birthdate	
	The no. of household	**Jia**	**Household**	**Relationship with the household head**	**Surname**	**First name**	**Sex**	**The Chinese zodiac**	**Year**	**Month**
Fujin Xiang 1st Bao				The household head's wife	谈	拜女	Female	Rooster	1921	1
Fujin Xiang 1st Bao				Father	苏	明贤	Male	Rabbit	1890	12
Fujin Xiang 1st Bao				Mother	李	顺结	Female	Goat	1894	4
Fujin Xiang 1st Bao				Daughter	苏	竹女子	Female	Horse	1942	12
Fujin Xiang 1st Bao				Younger brother	苏	凤清	Male	Snake	1929	2
Fujin Xiang 1st Bao				Uncle	苏	明堂	Male	Rooster	1884	7
Fujin Xiang 1st Bao				Uncle's wife	环	环	Female	Horse	1893	8
Fujin Xiang 1st Bao				Uncle's son	苏	凤银	Male	Monkey	1932	7
Fujin Xiang 1st Bao	402	15	10	The household head	苏	明山	Male	Goat	1906	11
Fujin Xiang 1st Bao				The household head's wife	李	统哇	Female	Snake	1917	10
Fujin Xiang 1st Bao				Father	苏	登董	Male	Goat	1882	2
Fujin Xiang 1st Bao				Son	苏	称全	Male	Rabbit	1927	1
Fujin Xiang 1st Bao				The second son	苏	三保	Male	Ox	1937	12
Fujin Xiang 1st Bao				Daughter	苏	玉女	Female	Goat	1943	9
Fujin Xiang 1st Bao				The second daughter	苏	崴女	Female	Monkey	1944	2
Fujin Xiang 1st Bao				Older brother	苏	明忠	Male	Pig	1898	11
Fujin Xiang 1st Bao				Older brother's wife	董	菜秀	Female	Monkey	1895	9

Date	The beginning time of inhabitation in this county	Educational level	Profession sector	Occupation	Marital status	Physical disability	Religion
16	Inhabited for generations	Illiteracy	Agriculture	Land-holding peasant	Married	No	No religion
2	Inhabited for generations	Illiteracy	Agriculture	Land-holding peasant	Married	No	No religion
13	Inhabited for generations	Illiteracy	Agriculture	Land-holding peasant	Married	No	No religion
1	Inhabited for generations	Illiteracy	Agriculture			No	No religion
14	Inhabited for generations	Higher primary school 1 year		Land-holding peasant		No	No religion
9	Inhabited for generations	Illiteracy	Agriculture	Land-holding peasant	Married	No	No religion
30	Inhabited for generations	Illiteracy	Agriculture	Land-holding peasant	Married	No	No religion
20	Inhabited for generations	Illiteracy	Agriculture	Land-holding peasant	Unmarried	No	No religion
17	Inhabited for generations	Illiteracy	Agriculture	Land-holding peasant	Married	No	No religion
24	Inhabited for generations	Illiteracy	Agriculture	Land-holding peasant	Married	No	No religion
15	Inhabited for generations	Illiteracy			Widowed		
25	Inhabited for generations	Illiteracy	Agriculture	Land-holding peasant	Widowed	No	No religion
24	Inhabited for generations	Illiteracy					
18	Inhabited for generations	Illiteracy					
26	Inhabited for generations						
13	Inhabited for generations	Illiteracy	Agriculture	Land-holding peasant	Married	No	No religion
18	Inhabited for generations	Illiteracy	Agriculture	Land-holding peasant	Married	No	No religion

Appendix: Overview of Microdata for Wudu County (*cont.*)

	The no. of household	Jia	Household	Relationship with the household head	Surname	First name	Sex	The Chinese zodiac	Birthdate Year	Month
Fujin Xiang 1st Bao				Niece	苏	董女	Female	Rabbit	1927	9
Fujin Xiang 1st Bao				Family member	老	王	Male	Rooster	1921	8
Fujin Xiang 1st Bao	403	15	11	The household head	李	万清	Male	Rabbit	1927	3
Fujin Xiang 1st Bao				The household head's wife	杨	桂香	Female	Rabbit	1927	10
Fujin Xiang 1st Bao				Mother	李	落落	Female	Dog	1909	9
Fujin Xiang 1st Bao				Grandmother	杨	善儿	Female	Pig	1898	9
Fujin Xiang 1st Bao				Son	李	成玉	Male	Dog	1946	1
Fujin Xiang 1st Bao				Family member	韩	新朝儿	Male	Ox	1913	6
Fujin Xiang 1st Bao	404	15	12	The household head	陈	守忠	Male	Rabbit	1915	5
Fujin Xiang 1st Bao				The household head's wife	赵	近儿	Female	Horse	1918	3
Fujin Xiang 1st Bao				Mother	赵	氏	Female	Monkey	1883	2
Fujin Xiang 1st Bao				Son	陈	永清	Male	Pig	1935	4
Fujin Xiang 1st Bao				Cohabitant	陈	俭哇	Male	Horse	1918	7
Fujin Xiang 1st Bao	405	15	13	The household head	陈	守义	Male	Pig	1895	4
Fujin Xiang 1st Bao				The household head's wife	杨	朱秀	Female	Snake	1904	12
Fujin Xiang 1st Bao				Son	陈	桂生子	Male	Goat	1943	7
Fujin Xiang 1st Bao				Daughter	陈	桂花	Female	Tiger	1938	6
Fujin Xiang 1st Bao				The second daughter	陈	星星	Female	Pig	1947	1

Date	The beginning time of inhabitation in this county	Educational level	Profession sector	Occupation	Marital status	Physical disability	Religion
27	Inhabited for generations	Illiteracy	Agriculture	Land-holding peasant	Married	No	No religion
13	Inhabited for generations	Illiteracy	Agriculture	Land-holding peasant	Married	No	No religion
29	Inhabited for generations	Illiteracy	Agriculture	Land-holding peasant	Married	No	No religion
4	Inhabited for generations	Illiteracy	Agriculture	Land-holding peasant	Married	No	No religion
12	Inhabited for generations	Illiteracy	Agriculture	Land-holding peasant	Married	No	No religion
21	Inhabited for generations	Illiteracy	Agriculture	Land-holding peasant	Married	No	No religion
17	Inhabited for generations						
1	Inhabited for generations	Illiteracy	Agriculture	Land-holding peasant	Married	No	No religion
8	Inhabited for generations	Sishu 3 years	Agriculture	Land-holding peasant	Married	No	No religion
13	Inhabited for generations	Illiteracy	Agriculture	Land-holding peasant	Married	No	No religion
15	Inhabited for generations	Illiteracy	Agriculture		Widowed		
4	Inhabited for generations	Bao school 4 years	Agriculture		Unmarried		
7	Inhabited for generations	Illiteracy	Agriculture		Unmarried	No	No religion
25	Inhabited for generations	Illiteracy	Agriculture	Land-holding peasant	Married	No	No religion
1	Inhabited for generations	Illiteracy	Agriculture	Land-holding peasant	Married	No	No religion
10	Inhabited for generations						
16	Inhabited for generations	Illiteracy					
23	Inhabited for generations						

Appendix: Overview of Microdata for Wudu County (*cont.*)

	The no. of household	Jia	Household	Relationship with the household head	Surname	First name	Sex	The Chinese zodiac	Birthdate	
									Year	Month
Fujin Xiang 1st Bao				Uncle	陈	思贤	Male	Monkey	1883	9
Fujin Xiang 1st Bao				Younger brother	陈	守吉	Male	Pig	1910	1
Fujin Xiang 1st Bao				Younger brother's wife	苏	香	Female	Goat	1919	10
Fujin Xiang 1st Bao	406	15	14	The household head	陈	思全	Male	Rooster	1896	4
Fujin Xiang 1st Bao				The household head's wife	苟	唤儿	Female	Horse		
Fujin Xiang 1st Bao				Son	陈	合成	Male	Rooster	1933	8
Fujin Xiang 1st Bao				Son's wife	罗	巧娃	Female	Goat	1931	10
Fujin Xiang 1st Bao				Daughter	送	香子	Female	Ox	1937	11
Fujin Xiang 1st Bao				Uncle's son	陈	思道	Male	Monkey	1907	3
Fujin Xiang 1st Bao				Uncle's son's wife	陈	召儿	Female	Dog	1909	8
Fujin Xiang 1st Bao				Nephew	陈	进生子	Male	Snake	1941	10
Fujin Xiang 1st Bao				The second nephew	陈	崴进娃	Male	Monkey	1944	8
Fujin Xiang 1st Bao				Niece	陈	移娃	Female	Ox	1937	11
Fujin Xiang 1st Bao				The second niece	陈	进女子	Female	Dog	1946	3
Fujin Xiang 1st Bao	407	15	15	The household head	苏	明益	Male	Tiger	1901	6
Fujin Xiang 1st Bao				The household head's wife	李	大姐	Female	Dog		
Fujin Xiang 1st Bao				Son	苏	杜娃	Male	Dragon	1940	1

Date	The beginning time of inhabitation in this county	Educational level	Profession sector	Occupation	Marital status	Physical disability	Religion
1	Inhabited for generations	Illiteracy	Agriculture	Land-holding peasant	Widowed		
2	Inhabited for generations	Illiteracy	Agriculture	Land-holding peasant	Married	No	No religion
7	Inhabited for generations	Illiteracy	Agriculture	Land-holding peasant	Married	No	No religion
3	Inhabited for generations	Illiteracy	Agriculture	Land-holding peasant	Married	No	No religion
	Inhabited for generations	Illiteracy	Agriculture	Land-holding peasant	Married	No	No religion
26	Inhabited for generations	Higher primary school 5 years	Agriculture	Land-holding peasant	Married	No	No religion
2	Inhabited for generations	Illiteracy	Agriculture	Land-holding peasant	Married	No	No religion
5	Inhabited for generations						
10	Inhabited for generations	Illiteracy	Agriculture	Land-holding peasant	Married	No	No religion
10	Inhabited for generations	Illiteracy	Agriculture	Land-holding peasant	Married	No	No religion
31	Inhabited for generations						
17	Inhabited for generations						
15	Inhabited for generations	Illiteracy	Agriculture	Land-holding peasant	Unmarried	No	No religion
7	Inhabited for generations	Illiteracy	Agriculture	Land-holding peasant	Unmarried	No	No religion
16	Inhabited for generations	Illiteracy	Agriculture	Land-holding peasant	Married	No	No religion
	Inhabited for generations	Illiteracy	Agriculture	Land-holding peasant	Married	No	No religion
25	Inhabited for generations						

Appendix: Overview of Microdata for Wudu County (*cont.*)

	The no. of household	Jia	Household	Relationship with the household head	Surname	First name	Sex	The Chinese zodiac	Birthdate Year	Month
Fujin Xiang 1st Bao				The second son	苏	崴杜儿	Male	Rooster	1945	10
Fujin Xiang 1st Bao				Younger brother	苏	明才	Male	Rat	1912	1
Fujin Xiang 1st Bao				Younger brother's wife	韩	琴娃	Female	Dragon	1916	8
Fujin Xiang 5th Bao	408	1	1	The household head	王	宝玉	Male	Goat	1919	6
Fujin Xiang 5th Bao				The household head's wife	谈	款娃	Female	Monkey	1920	6
Fujin Xiang 5th Bao				Mother	马	氏	Female	Horse	1905	10
Fujin Xiang 5th Bao				Grandmother	石	氏	Female	Dog	1875	9
Fujin Xiang 5th Bao				Daughter	王	刘英	Female	Dog	1947	9
Fujin Xiang 5th Bao				Younger brother	王	风明	Male	Ox	1937	6
Fujin Xiang 5th Bao				Uncle's father	王	治邦	Male	Dragon	1903	5
Fujin Xiang 5th Bao	409	1	2	The household head	王	海荣	Male	Pig	1898	9
Fujin Xiang 5th Bao				Son	王	高哇	Male	Dragon	1940	10
Fujin Xiang 5th Bao				Daughter	女	必家	Female	Rooster	1944	3
Fujin Xiang 5th Bao	410	1	3	The household head	张	仲成	Male	Dragon	1891	6
Fujin Xiang 5th Bao				Son	张	啟银	Male	Dog	1922	7
Fujin Xiang 5th Bao				Son's wife	李	金環	Female	Rabbit	1927	11
Fujin Xiang 5th Bao				The second son	张	啟富	Male	Tiger	1926	4
Fujin Xiang 5th Bao				The second son's wife	石	当娃	Female	Tiger	1926	2

Date	The beginning time of inhabitation in this county	Educational level	Profession sector	Occupation	Marital status	Physical disability	Religion
9	Inhabited for generations						
24	Inhabited for generations	Illiteracy	Agriculture	Land-holding peasant	Married	No	No religion
16	Inhabited for generations	Illiteracy	Agriculture	Land-holding peasant	Married	No	No religion
8	Inhabited for generations	Illiteracy	Agriculture				
21	Inhabited for generations	Illiteracy					
22	Inhabited for generations	Illiteracy					
28	Inhabited for generations						
1							
9							
21							
13	Inhabited for generations	Illiteracy	Agriculture	Land-holding peasant	Widowed	Pain on the lung	No religion
1	Inhabited for generations	Illiteracy					
8	Inhabited for generations						
23	Inhabited for generations	Sishu 3 years	Agriculture	Land-holding peasant	Widowed	No	No religion
25	Inhabited for generations	Illiteracy	Agriculture	Land-holding peasant	Married	No	No religion
3	Inhabited for generations	Illiteracy	Agriculture	Land-holding peasant	Married	No	No religion
19	Inhabited for generations	Illiteracy	Agriculture	Land-holding peasant	Married	No	No religion
14	Inhabited for generations	Illiteracy	Agriculture	Land-holding peasant	Married	No	No religion

Appendix: Overview of Microdata for Wudu County (*cont.*)

	The no. of household	Jia	Household	Relationship with the household head	Surname	First name	Sex	The Chinese zodiac	Birthdate	
									Year	Month
Fujin Xiang 5th Bao				Grandson	张	宝财	Male	Pig	1947	3
Fujin Xiang 5th Bao	411	1	4	The household head	张	有科	Male	Tiger	1914	10
Fujin Xiang 5th Bao				The household head's wife	赵	岁琴	Female	Tiger	1914	1
Fujin Xiang 5th Bao				Father	赵	成喜	Male	Goat	1882	9
Fujin Xiang 5th Bao				Mother	石	喜儿	Female	Ox	1888	10
Fujin Xiang 5th Bao				Son	张	宝时	Male	Tiger	1938	12
Fujin Xiang 5th Bao				Daughter	张	菊娃	Female	Monkey	1932	1
Fujin Xiang 5th Bao	412	1	5	The household head	张	仲富	Male	Horse	1893	8
Fujin Xiang 5th Bao				The household head's wife	赵	女子	Female	Rabbit	1902	3
Fujin Xiang 5th Bao				Son	张	平娃	Male	Dog	1922	10
Fujin Xiang 5th Bao				Son's wife	赵	菊哇	Female	Ox	1925	7
Fujin Xiang 5th Bao				The second son	张	明	Male	Rabbit	1939	11
Fujin Xiang 5th Bao				The third son	张	友儿	Male	Monkey	1944	4
Fujin Xiang 5th Bao	413	1	6	The household head	张	啟文	Male	Rabbit	1915	1
Fujin Xiang 5th Bao				The household head's wife	谈	当娃	Female	Goat	1919	8
Fujin Xiang 5th Bao				Son	张	学儿	Male	Tiger	1938	12
Fujin Xiang 5th Bao				The second son	双	牛儿	Male	Horse	1942	12
Fujin Xiang 5th Bao				Daughter	张	全牛	Female	Dog	1946	5

Date	The beginning time of inhabitation in this county	Educational level	Profession sector	Occupation	Marital status	Physical disability	Religion
13	Inhabited for generations						
25	Inhabited for generations	Illiteracy	Agriculture	Land-holding peasant	Married	No	No religion
30	Inhabited for generations	Illiteracy	Agriculture	Land-holding peasant	Married	No	No religion
7	Inhabited for generations	Illiteracy	Agriculture	Land-holding peasant	Married	No	No religion
19	Inhabited for generations	Illiteracy	Agriculture	Land-holding peasant	Married	No	No religion
3	Inhabited for generations	Bao school 1 year					
22	Inhabited for generations	Illiteracy					
22	Inhabited for generations	Illiteracy	Agriculture	Land-holding peasant	Married	No	No religion
16	Inhabited for generations	Illiteracy	Agriculture	Land-holding peasant	Married	No	No religion
6	Inhabited for generations	Illiteracy	Agriculture	Land-holding peasant	Married	No	No religion
1	Inhabited for generations	Illiteracy	Agriculture	Land-holding peasant	Married	No	No religion
1	Inhabited for generations	Bao school 1 year					
7	Inhabited for generations						
19	Inhabited for generations	Illiteracy	Agriculture	Land-holding peasant	Married	No	No religion
12	Inhabited for generations	Illiteracy	Agriculture	Land-holding peasant	Married	No	No religion
8	Inhabited for generations	Illiteracy					
28	Inhabited for generations	Illiteracy					
16	Inhabited for generations						

Appendix: Overview of Microdata for Wudu County (*cont.*)

	The no. of household	Jia	Household	Relationship with the household head	Surname	First name	Sex	The Chinese zodiac	Birthdate	
									Year	Month
Fujin Xiang 5th Bao	414	1	7	The household head	张	有贵	Male	Rat	1893	5
Fujin Xiang 5th Bao				The household head's wife	赵	雷女	Female	Dragon	1903	3
Fujin Xiang 5th Bao				Mother	张	秀儿	Female	Dragon	1878	9
Fujin Xiang 5th Bao				Son's wife	张	能娃	Female	Ox	1925	7
Fujin Xiang 5th Bao				Grandson	张	长才	Male	Pig	1947	3
Fujin Xiang 5th Bao				Granddaughter	张	良英	Female	Monkey	1944	6
Fujin Xiang 5th Bao				Younger brother	张	有库	Male	Pig	1910	9
Fujin Xiang 5th Bao				Younger brother's wife	石	进娃	Female	Tiger	1914	5
Fujin Xiang 5th Bao				The third younger brother	张	有富	Male	Rabbit	1915	10
Fujin Xiang 5th Bao				Nephew	张	星顺	Male	Rabbit	1939	10
Fujin Xiang 5th Bao	415	1	8	The household head	辈	进和	Male	Horse	1908	7
Fujin Xiang 5th Bao				The household head's wife	石	柱哇	Female	Horse	1916	10
Fujin Xiang 5th Bao				Daughter	石	亭哇	Female	Tiger	1937	8
Fujin Xiang 5th Bao	416	1	9	The household head	张	啟生	Male	Rooster	1921	3
Fujin Xiang 5th Bao				The household head's wife	石	银建	Female	Goat	1919	12
Fujin Xiang 5th Bao				Son	张	俊娃	Male	Dragon	1940	9
Fujin Xiang 5th Bao				Daughter	全	女儿	Female	Rooster	1945	11

Date	The beginning time of inhabitation in this county	Educational level	Profession sector	Occupation	Marital status	Physical disability	Religion
30	Inhabited for generations	Illiteracy	Agriculture	Land-holding peasant	Married	No	No religion
13	Inhabited for generations	Illiteracy	Agriculture	Land-holding peasant	Married	No	No religion
13	Inhabited for generations	Illiteracy	Agriculture	Land-holding peasant			
4	Inhabited for generations	Illiteracy	Agriculture	Land-holding peasant	Widowed	No	No religion
6	Inhabited for generations						
8	Inhabited for generations						
17	Inhabited for generations	Illiteracy	Agriculture	Land-holding peasant	Married	No	No religion
21	Inhabited for generations	Illiteracy	Agriculture	Land-holding peasant	Married	No	No religion
27	Inhabited for generations	Illiteracy	Agriculture	Land-holding peasant	Widowed	No	No religion
27	Inhabited for generations	Bao school 1 year					
6	Inhabited for generations	Sishu 3 years	Others		Married	No	No religion
24	Inhabited for generations	Illiteracy	Weaving		Married	No	No religion
20	Inhabited for generations	Illiteracy					
13	Inhabited for generations	Illiteracy	Agriculture	Land-holding peasant	Married	No	No religion
3	Inhabited for generations	Illiteracy	Agriculture	Land-holding peasant	Married	No	No religion
5	Inhabited for generations	Illiteracy					
16	Inhabited for generations						

Appendix: Overview of Microdata for Wudu County (*cont.*)

	The no. of household	Jia	Household	Relationship with the household head	Surname	First name	Sex	The Chinese zodiac	Birthdate	
									Year	Month
Fujin Xiang 5th Bao				The second daughter	张	女娃	Female	Pig	1947	6
Fujin Xiang 5th Bao	417	2	1	The household head	黄	登荣	Male	Dragon	1903	7
Fujin Xiang 5th Bao				The household head's wife	张	何儿	Female	Monkey	1907	10
Fujin Xiang 5th Bao				Son	黄	万娃	Male	Monkey	1920	3
Fujin Xiang 5th Bao				Son's wife	长	清儿	Female	Horse	1930	11
Fujin Xiang 5th Bao				The second son	黄	宝儿	Male	Tiger	1928	7
Fujin Xiang 5th Bao				Daughter	岁	清儿	Female	Rooster	1933	1
Fujin Xiang 5th Bao				Mother	黄	根娃	Female	Dog	1885	6
Fujin Xiang 5th Bao				Younger brother	黄	登林	Male	Rooster	1908	4
Fujin Xiang 5th Bao				Younger brother's wife	石	二女	Female	Goat	1906	3
Fujin Xiang 5th Bao				Nephew	黄	未生	Male	Ox	1937	6
Fujin Xiang 5th Bao				Niece	黄	猪英	Female	Monkey	1932	5
Fujin Xiang 5th Bao				The second niece	黄	玉挂	Female	Horse	1942	4
Fujin Xiang 5th Bao				The third niece	黄	柏家	Female	Rooster	1945	1
Fujin Xiang 5th Bao	418	2	2	The household head	黄	登富	Male	Snake	1917	6
Fujin Xiang 5th Bao				The household head's wife	赵	田娃	Female	Rooster	1921	10
Fujin Xiang 5th Bao				Father	黄	近堂	Male	Monkey	1895	11
Fujin Xiang 5th Bao				Mother	陈	秀儿	Female	Monkey	1895	4

Date	The beginning time of inhabitation in this county	Educational level	Profession sector	Occupation	Marital status	Physical disability	Religion
14	Inhabited for generations						
3	Inhabited for generations	Sishu 3 years	Agriculture	Land-holding peasant	Married	No	No religion
23	Inhabited for generations	Illiteracy	Agriculture	Land-holding peasant	Married	No	No religion
30	Inhabited for generations	Illiteracy	Agriculture	Land-holding peasant	Married	No	No religion
7	Inhabited for generations	Illiteracy	Agriculture	Land-holding peasant	Married	No	No religion
1	Inhabited for generations	Bao school 1 year					
11	Inhabited for generations	Illiteracy	Agriculture				
13	Inhabited for generations	Illiteracy	Agriculture	Land-holding peasant	Married	No	No religion
30	Inhabited for generations	Illiteracy	Agriculture	Land-holding peasant	Married	No	No religion
29	Inhabited for generations	Illiteracy	Agriculture	Land-holding peasant	Married	No	No religion
13	Inhabited for generations	Bao school 2 years					
2	Inhabited for generations	Illiteracy	Agriculture	Land-holding peasant	Unmarried	No	No religion
13	Inhabited for generations						
8	Inhabited for generations						
18	Inhabited for generations	Illiteracy	Agriculture	Land-holding peasant	Married	No	No religion
7	Inhabited for generations	Illiteracy	Agriculture	Land-holding peasant	Married	No	No religion
5	Inhabited for generations	Sishu 5 years	Agriculture	Land-holding peasant	Married	No	No religion
2	Inhabited for generations	Illiteracy	Agriculture	Land-holding peasant	Married	No	No religion

Appendix: Overview of Microdata for Wudu County (*cont.*)

	The no. of household	Jia	Household	Relationship with the household head	Surname	First name	Sex	The Chinese zodiac	Birthdate	
									Year	Month
Fujin Xiang 5th Bao				Son	黄	银才	Male	Horse	1942	1
Fujin Xiang 5th Bao				The second son	黄	东才	Male	Dog	1946	1
Fujin Xiang 5th Bao				Daughter	黄	必义娃	Female	Rabbit	1939	7
Fujin Xiang 5th Bao				Older brother	黄	登万	Male	Goat	1906	3
Fujin Xiang 5th Bao				Older brother's wife	王	採娃	Female	Pig	1910	12
Fujin Xiang 5th Bao				Niece	黄	冬英	Female	Rooster	1933	12
Fujin Xiang 5th Bao				The second niece	黄	毛人子	Female	Ox	1937	12
Fujin Xiang 5th Bao				Nephew	黄	全德	Male	Snake	1941	6
Fujin Xiang 5th Bao				Nephew's son	黄	石朝娃	Male	Monkey	1920	6
Fujin Xiang 5th Bao				Nephew's son's wife	石	拜女	Female	Snake	1917	6
Fujin Xiang 5th Bao	419	2	3	The household head	黄	登朝	Male	Pig	1910	3
Fujin Xiang 5th Bao				The household head's wife	石	必娃	Female	Tiger	1914	12
Fujin Xiang 5th Bao				Father	黄	进全	Male	Pig	1886	8
Fujin Xiang 5th Bao				Mother	罗	秀儿	Female	Monkey	1895	6
Fujin Xiang 5th Bao				Son	黄	吉利	Male	Dragon	1940	4
Fujin Xiang 5th Bao				Younger brother	黄	万才	Male	Goat	1931	9
Fujin Xiang 5th Bao				Younger brother's wife	石	田女	Female	Tiger	1926	6

Date	The beginning time of inhabitation in this county	Educational level	Profession sector	Occupation	Marital status	Physical disability	Religion
28	Inhabited for generations						
5	Inhabited for generations						
18	Inhabited for generations	Illiteracy					
1	Inhabited for generations	Illiteracy	Agriculture	Land-holding peasant	Married	No	No religion
15	Inhabited for generations	Illiteracy	Agriculture	Land-holding peasant	Married	No	No religion
30	Inhabited for generations	Illiteracy					
21	Inhabited for generations	Illiteracy					
24	Inhabited for generations	Bao school 1 year					
22	Inhabited for generations	Illiteracy	Agriculture	Land-holding peasant	Married	No	No religion
29	Inhabited for generations	Illiteracy	Agriculture	Land-holding peasant	Married	No	No religion
14	Inhabited for generations	Illiteracy	Agriculture	Land-holding peasant	Married	No	No religion
27	Inhabited for generations	Illiteracy	Agriculture	Land-holding peasant	Married	No	No religion
25	Inhabited for generations	Sishu 3 years	Agriculture	Land-holding peasant	Married	No	No religion
16	Inhabited for generations	Illiteracy	Weaving	Land-holding peasant	Married	No	No religion
21	Inhabited for generations						
3	Inhabited for generations	Illiteracy	Agriculture	Land-holding peasant	Married	No	No religion
14	Inhabited for generations	Illiteracy	Agriculture	Land-holding peasant	Married	No	No religion

Appendix: Overview of Microdata for Wudu County (*cont.*)

	The no. of household	Jia	Household	Relationship with the household head	Surname	First name	Sex	The Chinese zodiac	Birthdate Year	Month
Fujin Xiang 5th Bao				The third younger brother	黄	三牛	Male	Dragon	1940	10
Fujin Xiang 5th Bao	420	2	4	The household head	黄	万益	Male	Ox	1913	7
Fujin Xiang 5th Bao				The household head's wife	石	赛娃	Female	Rabbit	1915	12
Fujin Xiang 5th Bao				Father	黄	登崔	Male	Dragon	1878	9
Fujin Xiang 5th Bao				Mother	谈	顺香	Female	Snake	1979	9
Fujin Xiang 5th Bao				Daughter	黄	石称	Female	Dragon	1940	7
Fujin Xiang 5th Bao				The second daughter	德	赐哇	Female	Rooster	1945	1
Fujin Xiang 5th Bao				Older brother	黄	万高	Male	Snake	1904	7
Fujin Xiang 5th Bao				Nephew	黄	旺娃	Male	Pig	1935	2
Fujin Xiang 5th Bao				Niece	黄	赐娃	Female	Goat	1931	5
Fujin Xiang 5th Bao				The second niece	黄	石家女	Female	Ox	1937	7
Fujin Xiang 5th Bao	421	2	5	The household head	黄	万清	Male	Dog	1922	3
Fujin Xiang 5th Bao				The household head's wife	石	早女	Female	Tiger	1926	3
Fujin Xiang 5th Bao				Daughter	黄	苏英	Female	Monkey	1944	2
Fujin Xiang 5th Bao	422	2	6	The household head	黄	登耀	Male	Rat	1912	4
Fujin Xiang 5th Bao				The household head's wife	黄	琴娃	Female	Rooster	1921	11
Fujin Xiang 5th Bao				Father	黄	进德	Male	Monkey	1883	2

Date	The beginning time of inhabitation in this county	Educational level	Profession sector	Occupation	Marital status	Physical disability	Religion
7	Inhabited for generations	Illiteracy					
23	Inhabited for generations	Illiteracy	Agriculture	Land-holding peasant	Married	No	No religion
12	Inhabited for generations	Illiteracy	Agriculture	Land-holding peasant	Married	No	No religion
15	Inhabited for generations	Illiteracy	Agriculture	Land-holding peasant	Married	No	No religion
15	Inhabited for generations	Illiteracy	Agriculture	Land-holding peasant	Married	No	No religion
24	Inhabited for generations	Illiteracy					
19	Inhabited for generations						
2	Inhabited for generations	Illiteracy	Agriculture	Land-holding peasant	Widowed	No	No religion
19	Inhabited for generations	Illiteracy	Agriculture				
21	Inhabited for generations	Illiteracy	Agriculture	Land-holding peasant	Unmarried	No	No religion
20	Inhabited for generations	Illiteracy					
9	Inhabited for generations	Sishu 3 years	Agriculture	Land-holding peasant	Married	No	No religion
12	Inhabited for generations	Illiteracy	Agriculture	Land-holding peasant	Married	No	No religion
22	Inhabited for generations						
1	Inhabited for generations	Illiteracy	Agriculture	Land-holding peasant	Married	No	No religion
2	Inhabited for generations	Illiteracy	Weaving	Land-holding peasant	Married	No	No religion
19	Inhabited for generations	Illiteracy	Agriculture	Land-holding peasant	Married	No	No religion

Appendix: Overview of Microdata for Wudu County (*cont.*)

	The no. of household	Jia	Household	Relationship with the household head	Surname	First name	Sex	The Chinese zodiac	Birthdate	
									Year	Month
Fujin Xiang 5th Bao				Mother	赵	崴女子	Female	Rabbit	1890	10
Fujin Xiang 5th Bao				Son	黄	玉哇子	Male	Rat	1936	9
Fujin Xiang 5th Bao				The second son	黄	正娃	Male	Tiger	1939	8
Fujin Xiang 5th Bao				Younger brother	黄	歪娃	Male	Ox	1925	11
Fujin Xiang 5th Bao				Younger brother's wife	菊	香	Female	Horse	1930	1
Fujin Xiang 5th Bao	423	2	7	The household head	黄	登味	Male	Horse	1905	2
Fujin Xiang 5th Bao				The household head's wife	石	女儿子	Female	Goat	1906	1
Fujin Xiang 5th Bao				Father	黄	进山	Male	Dog	1885	10
Fujin Xiang 5th Bao				Son	黄	德儿	Male	Tiger	1938	4
Fujin Xiang 5th Bao				Daughter	黄	必娃	Female	Rooster	1933	8
Fujin Xiang 5th Bao				The second daughter	黄	旦旦	Female	Pig	1935	9
Fujin Xiang 5th Bao	424	2	8	The household head	黄	万进	Male	Dog	1921	1
Fujin Xiang 5th Bao				The household head's wife	石	长女	Female	Rabbit	1927	5
Fujin Xiang 5th Bao				Mother	石	唤儿	Female	Monkey	1895	10
Fujin Xiang 5th Bao				Younger brother	黄	忠娃	Male	Ox	1937	11
Fujin Xiang 5th Bao				Younger sister	黄	贤哇	Female	Dog	1934	6
Fujin Xiang 5th Bao	425	2	9	The household head	黄	登武	Male	Horse	1905	5

Date	The beginning time of inhabitation in this county	Educational level	Profession sector	Occupation	Marital status	Physical disability	Religion
19	Inhabited for generations	Illiteracy	Weaving	Land-holding peasant	Married	No	No religion
27	Inhabited for generations	Bao school 2 years					
27	Inhabited for generations	Bao school 1 year					
10	Inhabited for generations	Illiteracy	Agricul-ture	Land-holding peasant	Married	No	No religion
21	Inhabited for generations	Illiteracy	Agricul-ture	Land-holding peasant	Married	No	No religion
13	Inhabited for generations	Sishu 3 years	Agricul-ture	Land-holding peasant	Married	No	No religion
1	Inhabited for generations	Illiteracy	Agricul-ture	Land-holding peasant	Married	No	No religion
30	Inhabited for generations	Illiteracy	Agricul-ture	Land-holding peasant	Widowed	No	No religion
16	Inhabited for generations	Bao school 1 year					
8	Inhabited for generations	Illiteracy					
12	Inhabited for generations	Illiteracy					
30	Inhabited for generations	Sishu 3 years	Agricul-ture	Land-holding peasant	Married	No	No religion
1	Inhabited for generations	Illiteracy	Agricul-ture	Land-holding peasant	Married	No	No religion
14	Inhabited for generations	Illiteracy	Agricul-ture	Land-holding peasant	Widowed	No	No religion
6	Inhabited for generations	Bao school 2 years					
26	Inhabited for generations	Illiteracy					
6	Inhabited for generations	Illiteracy	Agricul-ture	Land-holding peasant	Married	No	No religion

Appendix: Overview of Microdata for Wudu County (*cont.*)

	The no. of household	Jia	Household	Relationship with the household head	Surname	First name	Sex	The Chinese zodiac	Birthdate	
									Year	Month
Fujin Xiang 5th Bao				The household head's wife	石	书儿	Female	Monkey	1907	11
Fujin Xiang 5th Bao				Mother	赵	氏	Female	Ox	1871	5
Fujin Xiang 5th Bao				Son	黄	王有	Male	Monkey	1932	10
Fujin Xiang 5th Bao				The second son	黄	王清	Male	Pig	1935	12
Fujin Xiang 5th Bao				The third son	黄	祥儿	Male	Tiger	1938	5
Fujin Xiang 5th Bao				Daughter	黄	有女	Female	Dragon	1940	11
Fujin Xiang 5th Bao	426	2	10	The household head	黄	登山	Male	Rooster	1896	3
Fujin Xiang 5th Bao				The household head's wife	石	银娃	Female	Ox	1908	2
Fujin Xiang 5th Bao				Daughter	黄	卯娃	Female	Ox	1937	11
Fujin Xiang 5th Bao				The second daughter	黄	旺娃	Female	Dragon	1940	11
Fujin Xiang 5th Bao	427	2	11	The household head	黄	登治	Male	Pig	1898	9
Fujin Xiang 5th Bao				Son	黄	万科	Male	Rat	1924	9
Fujin Xiang 5th Bao				Son's wife	石	廷娃	Female	Dragon	1928	11
Fujin Xiang 5th Bao				The second son	黄	东冬	Male	Dragon	1940	11
Fujin Xiang 5th Bao	428	2	12	The household head	黄	登义	Male	Ox	1925	7
Fujin Xiang 5th Bao				The household head's wife	石	文女	Female	Goat	1931	7
Fujin Xiang 5th Bao				Mother	陈	秀儿	Female	Dog	1885	2
Fujin Xiang 5th Bao				Younger brother	黄	团娃	Male	Rabbit	1927	4

Date	The beginning time of inhabitation in this county	Educational level	Profession sector	Occupation	Marital status	Physical disability	Religion
18	Inhabited for generations	Illiteracy	Agriculture	Land-holding peasant	Married	No	No religion
5	Inhabited for generations						
18	Inhabited for generations	Bao school 4 years					
15	Inhabited for generations	Bao school 2 years					
18	Inhabited for generations	Bao school 2 years					
8	Inhabited for generations						
1	Inhabited for generations	Illiteracy	Agriculture	Land-holding peasant	Married	No	No religion
1	Inhabited for generations	Illiteracy	Agriculture	Land-holding peasant	Married	No	No religion
17	Inhabited for generations	Illiteracy					
17	Inhabited for generations	Illiteracy					
4	Inhabited for generations	Illiteracy	Weaving	Land-holding peasant	Widowed	No	No religion
3	Inhabited for generations	Illiteracy	Agriculture	Land-holding peasant	Married	No	No religion
27	Inhabited for generations	Illiteracy	Agriculture	Land-holding peasant	Married	No	No religion
15	Inhabited for generations	Bao school 1 year					
6	Inhabited for generations	Illiteracy	Agriculture	Land-holding peasant	Married	No	No religion
24	Inhabited for generations	Illiteracy	Agriculture	Land-holding peasant	Married	No	No religion
17	Inhabited for generations	Illiteracy	Weaving	Land-holding peasant	Widowed	No	No religion
28	Inhabited for generations	Illiteracy	Agriculture	Land-holding peasant	Married	No	No religion

Appendix: Overview of Microdata for Wudu County (*cont.*)

	The no. of household	Jia	Household	Relationship with the household head	Surname	First name	Sex	The Chinese zodiac	Birthdate	
									Year	Month
Fujin Xiang 5th Bao				Younger brother's wife	石	双玉	Female	Rabbit	1927	2
Fujin Xiang 5th Bao				Uncle's wife	李	陈秀	Female	Pig	1886	12
Fujin Xiang 5th Bao				Niece	黄	清香	Female	Pig	1947	1
Fujin Xiang 5th Bao				Uncle's son	黄	登生	Male	Ox	1925	9
Fujin Xiang 5th Bao				Uncle's second son	黄	张生	Female	Monkey	1932	3
Fujin Xiang 5th Bao	429	3	1	The household head	石	银奇	Male	Rooster	1908	7
Fujin Xiang 5th Bao				The household head's wife	赵	九连儿	Female	Snake	1917	9
Fujin Xiang 5th Bao				Son	石	马裉儿	Male	Ox	1937	7
Fujin Xiang 5th Bao				Daughter	石	染娃	Female	Rooster	1933	10
Fujin Xiang 5th Bao				Uncle	石	仲念	Male	Rabbit	1915	5
Fujin Xiang 5th Bao				Uncle's wife	李	润娃	Female	Rooster	1908	4
Fujin Xiang 5th Bao				Uncle's son	石	银学	Male	Rat	1936	8
Fujin Xiang 5th Bao				Uncle's daughter	石	毛儿子	Female	Goat	1943	5
Fujin Xiang 5th Bao				Uncle's second son	石	亭娃	Male	Pig	1917	1
Fujin Xiang 5th Bao	430	3	2	The household head	石	宝转	Male	Snake	1904	7
Fujin Xiang 5th Bao				The household head's wife	毛	旦娃	Female	Goat	1906	11
Fujin Xiang 5th Bao				Father	石	怌山	Male	Monkey	1883	10
Fujin Xiang 5th Bao				Mother	赵	喜秀儿	Female	Rooster	1884	6

Date	The beginning time of inhabitation in this county	Educational level	Profession sector	Occupation	Marital status	Physical disability	Religion
4	Inhabited for generations	Illiteracy	Agriculture	Land-holding peasant	Married	No	No religion
13	Inhabited for generations	Illiteracy	Weaving	Land-holding peasant	Widowed	No	No religion
24	Inhabited for generations						
16	Inhabited for generations	Illiteracy	Agriculture	Land-holding peasant	Divorced	No	No religion
9	Inhabited for generations	Illiteracy	Agriculture	Land-holding peasant	Unmarried	No	No religion
21	Inhabited for generations	Illiteracy	Agriculture	Land-holding peasant	Married	No	No religion
19	Inhabited for generations	Illiteracy	Agriculture	Land-holding peasant	Married	No	No religion
7	Inhabited for generations	Bao school 1 year					
2	Inhabited for generations	Illiteracy					
19	Inhabited for generations	Illiteracy	Agriculture	Land-holding peasant	Married	No	No religion
4	Inhabited for generations	Illiteracy	Agriculture	Land-holding peasant	Married	No	No religion
22	Inhabited for generations	Bao school 2 years					
18	Inhabited for generations	Illiteracy					
1	Inhabited for generations	Illiteracy					
31	Inhabited for generations	Illiteracy	Agriculture	Land-holding peasant	Married	No	No religion
3	Inhabited for generations	Illiteracy	Agriculture	Land-holding peasant	Married	No	No religion
13	Inhabited for generations	Illiteracy	Agriculture	Land-holding peasant	Married	No	No religion
27	Inhabited for generations	Illiteracy	Agriculture	Land-holding peasant	Married	No	No religion

Appendix: Overview of Microdata for Wudu County (*cont.*)

	The no. of household	Jia	Household	Relationship with the household head	Surname	First name	Sex	The Chinese zodiac	Birthdate Year	Month
Fujin Xiang 5th Bao				Son	石	德儿	Male	Rooster	1921	8
Fujin Xiang 5th Bao				Son's wife	石	亭哇	Female	Rat	1924	7
Fujin Xiang 5th Bao				Granddaughter	石	卢香儿	Female	Dog	1946	5
Fujin Xiang 5th Bao				Adopted son	石	召才儿	Male	Rooster	1931	10
Fujin Xiang 5th Bao				Daughter	石	芦儿	Female	Monkey	1932	6
Fujin Xiang 5th Bao	431	3	3	The household head	石	李进娃	Male	Rat	1899	6
Fujin Xiang 5th Bao				Son	石	仲观	Male	Rooster	1921	11
Fujin Xiang 5th Bao				Son's wife	德	林娃	Female	Rat	1924	8
Fujin Xiang 5th Bao				Grandson	石	冬英	Female	Dog	1946	8
Fujin Xiang 5th Bao				The second son	石	茶还儿	Male	Dog	1934	6
Fujin Xiang 5th Bao				The third son	石	发儿	Male	Rabbit	1939	3
Fujin Xiang 5th Bao				Nephew	石	仲具	Male	Goat	1906	5
Fujin Xiang 5th Bao				Nephew's wife	石	知娃	Female	Monkey	1910	11
Fujin Xiang 5th Bao				Nephew's son	石	富娃	Male	Horse	1930	8
Fujin Xiang 5th Bao				Nephew's daughter	石	听娃	Female	Monkey	1932	6
Fujin Xiang 5th Bao	432	3	4	The household head	石	银库	Male	Snake	1892	11
Fujin Xiang 5th Bao				The household head's wife	李	玉节子	Female	Rat	1887	9
Fujin Xiang 5th Bao				Father	石	仲礼	Male	Monkey	1881	10
Fujin Xiang 5th Bao				Son	石	东娃	Male	Rabbit	1915	1

Date	The beginning time of inhabitation in this county	Educational level	Profession sector	Occupation	Marital status	Physical disability	Religion	
18	Inhabited for generations	Illiteracy	Agriculture	Land-holding peasant	Married	No	No religion	
23	Inhabited for generations	Illiteracy	Agriculture	Land-holding peasant	Married	No	No religion	
9	Inhabited for generations							
12	Inhabited for generations	Illiteracy	Agriculture	Land-holding peasant	Unmarried	No	No religion	
28	Inhabited for generations	Illiteracy	Agriculture	Land-holding peasant	Unmarried	No	No religion	
8	Inhabited for generations	Illiteracy	Agriculture	Land-holding peasant	Widowed	No	No religion	
1	Inhabited for generations	Illiteracy	Agriculture	Land-holding peasant	Married	No	No religion	Joined the army
4	Inhabited for generations	Illiteracy	Agriculture	Land-holding peasant	Married	No	No religion	
8	Inhabited for generations							
17	Inhabited for generations	Illiteracy	Agriculture					
7	Inhabited for generations							
25	Inhabited for generations	Illiteracy	Agriculture	Land-holding peasant	Married	No	No religion	
19	Inhabited for generations	Illiteracy	Agriculture	Land-holding peasant	Married	No	No religion	
5	Inhabited for generations	Illiteracy	Agriculture	Land-holding peasant	Unmarried	No	No religion	
29	Inhabited for generations	Illiteracy	Agriculture	Land-holding peasant	Unmarried	No	No religion	
30	Inhabited for generations	Sishu 3 years	Agriculture	Land-holding peasant	Married	No	No religion	
29	Inhabited for generations	Illiteracy	Agriculture	Land-holding peasant	Married	No	No religion	
9	Inhabited for generations	Illiteracy	Agriculture	Land-holding peasant	Widowed	No	No religion	
31	Inhabited for generations	Illiteracy	Agriculture	Land-holding peasant	Married	No	No religion	

Appendix: Overview of Microdata for Wudu County (*cont.*)

	The no. of household	Jia	Household	Relationship with the household head	Surname	First name	Sex	The Chinese zodiac	Birthdate	
									Year	Month
Fujin Xiang 5th Bao				Son's wife	润	女子	Female	Monkey	1920	10
Fujin Xiang 5th Bao				Grandson	石	满娃	Male	Snake	1941	4
Fujin Xiang 5th Bao	433	3	5	The household head	石	永转	Male	Horse	1905	2
Fujin Xiang 5th Bao				The household head's wife	张	崔娃	Female	Rooster	1908	2
Fujin Xiang 5th Bao				Son	石	生生	Male	Horse	1930	2
Fujin Xiang 5th Bao				The second son	石	全生娃	Male		1933	8
Fujin Xiang 5th Bao				Daughter	石	满娃	Female	Tiger	1938	8
Fujin Xiang 5th Bao				The second daughter	石	马家女	Female	Horse	1942	4
Fujin Xiang 5th Bao	434	3	6	The household head	石	登應	Male	Rat	1912	8
Fujin Xiang 5th Bao				The household head's wife	石	禄儿	Female	Rooster	1908	11
Fujin Xiang 5th Bao				Son	石	银堂	Male	Rooster	1933	1
Fujin Xiang 5th Bao				Daughter	石	女娃	Female	Tiger	1938	5
Fujin Xiang 5th Bao				The second daughter	石	毛儿子	Female	Horse	1942	8
Fujin Xiang 5th Bao				Nephew	石	银斌	Male	Rooster	1921	6
Fujin Xiang 5th Bao				Nephew's wife	李	王家	Female	Tiger	1926	6
Fujin Xiang 5th Bao				Nephew's son	石	凤明	Male	Pig	1947	5
Fujin Xiang 5th Bao				Uncle's daughter	石	哑娃	Female	Tiger	1926	7
Fujin Xiang 5th Bao	435	3	7	The household head	石	银宽	Male	Tiger	1889	1

Date	The beginning time of inhabitation in this county	Educational level	Profession sector	Occupation	Marital status	Physical disability	Religion
14	Inhabited for generations	Illiteracy	Agriculture	Land-holding peasant	Married	No	No religion
19	Inhabited for generations						
20	Inhabited for generations	Illiteracy	Agriculture	Land-holding peasant	Married	No	No religion
7	Inhabited for generations	Illiteracy	Agriculture	Land-holding peasant	Married	No	No religion
11	Inhabited for generations	Illiteracy	Agriculture	Land-holding peasant	Unmarried	No	No religion
12	Inhabited for generations	Illiteracy					
12	Inhabited for generations	Illiteracy	Agriculture	Land-holding peasant	Unmarried	No	No religion
29	Inhabited for generations						
31	Inhabited for generations	Sishu 2 years	Agriculture	Land-holding peasant	Married	No	No religion
25	Inhabited for generations	Illiteracy	Agriculture	Land-holding peasant	Married	No	No religion
28	Inhabited for generations	Bao school 3 years	Agriculture				
3	Inhabited for generations	Illiteracy					
6	Inhabited for generations						
10	Inhabited for generations	Illiteracy	Agriculture	Land-holding peasant	Married	No	No religion
14	Inhabited for generations	Illiteracy	Agriculture	Land-holding peasant	Married	No	No religion
5	Inhabited for generations						
24	Inhabited for generations	Illiteracy	Agriculture	Land-holding peasant	Unmarried	Mental disease	No religion
4	Inhabited for generations	Illiteracy	Agriculture	Land-holding peasant	Married	No	No religion

Appendix: Overview of Microdata for Wudu County (*cont.*)

	The no. of household	Jia	Household	Relationship with the household head	Surname	First name	Sex	The Chinese zodiac	Birthdate	
									Year	Month
Fujin Xiang 5th Bao				The household head's wife	李	凉儿	Female	Snake	1892	7
Fujin Xiang 5th Bao				Father	石	仲金	Male	Dog	1873	10
Fujin Xiang 5th Bao				Son	石	永明	Male	Pig	1933	1
Fujin Xiang 5th Bao				Son's wife	石	钉儿	Female	Tiger	1926	11
Fujin Xiang 5th Bao				Grandson	石	三生	Male	Dog	1946	4
Fujin Xiang 5th Bao				Older brother	石	银顺	Male	Dragon	1891	11
Fujin Xiang 5th Bao				Older brother's wife	石	唤儿	Female	Tiger	1901	6
Fujin Xiang 5th Bao				Niece	林	林子	Female	Horse	1941	3
Fujin Xiang 5th Bao	436	3	8	The household head	石	银花	Male	Tiger	1889	6
Fujin Xiang 5th Bao				The household head's wife	谈	裉三儿	Female	Rat	1899	3
Fujin Xiang 5th Bao				Son	石	永怌	Male	Goat	1919	11
Fujin Xiang 5th Bao				Daughter	石	毛旦女	Female	Pig	1937	5
Fujin Xiang 5th Bao				The second daughter	石	菊娃	Female	Monkey	1935	2
Fujin Xiang 5th Bao				Nephew	石	永清	Male	Monkey	1907	12
Fujin Xiang 5th Bao				The second nephew	石	永亭	Male	Dragon	1915	3
Fujin Xiang 5th Bao				The second nephew's wife	石	见娃	Female	Horse	1918	1
Fujin Xiang 5th Bao				Grandson	石	全儿	Male	Horse	1942	5
Fujin Xiang 5th Bao				Nephew's daughter	石	明儿	Female	Rooster	1945	5

Date	The beginning time of inhabitation in this county	Educational level	Profession sector	Occupation	Marital status	Physical disability	Religion
8	Inhabited for generations	Illiteracy	Agriculture	Land-holding peasant	Married	No	No religion
3	Inhabited for generations	Illiteracy	Agriculture	Land-holding peasant	Widowed	No	No religion
26	Inhabited for generations	Illiteracy	Agriculture	Land-holding peasant	Married	No	No religion
19	Inhabited for generations	Illiteracy	Agriculture	Land-holding peasant	Married	No	No religion
19	Inhabited for generations						
21	Inhabited for generations	Illiteracy	Agriculture	Land-holding peasant	Married	No	No religion
13	Inhabited for generations	Illiteracy	Agriculture	Land-holding peasant	Married	No	No religion
18	Inhabited for generations	Illiteracy					
24	Inhabited for generations	Illiteracy	Agriculture	Land-holding peasant	Married	No	No religion
4	Inhabited for generations	Illiteracy	Agriculture	Land-holding peasant	Married	No	No religion
25	Inhabited for generations	Illiteracy	Agriculture	Land-holding peasant	Unmarried	No	No religion
15	Inhabited for generations	Illiteracy					
4	Inhabited for generations	Illiteracy	Agriculture	Land-holding peasant	Unmarried	No	No religion
18	Inhabited for generations	Illiteracy	Agriculture	Land-holding peasant	Divorced	No	No religion
1	Inhabited for generations	Illiteracy	Agriculture	Land-holding peasant	Married	No	No religion
18	Inhabited for generations	Illiteracy	Agriculture	Land-holding peasant	Married	No	No religion
13	Inhabited for generations						
19	Inhabited for generations						

Appendix: Overview of Microdata for Wudu County (*cont.*)

									Birthdate	
	The no. of household	Jia	Household	Relationship with the household head	Surname	First name	Sex	The Chinese zodiac	Year	Month
Fujin Xiang 5th Bao	437	3	9	The household head	石	永朝	Male	Snake	1892	7
Fujin Xiang 5th Bao				Son	石	生林	Male	Dog	1922	9
Fujin Xiang 5th Bao				Son's wife	赵	反哇	Female	Dragon	1928	12
Fujin Xiang 5th Bao				The second son	石	进学儿	Male	Rat	1924	6
Fujin Xiang 5th Bao	438	3	10	The household head	石	银周	Male	Monkey	1907	2
Fujin Xiang 5th Bao				The household head's wife	李	乐乐子	Female	Goat	1919	12
Fujin Xiang 5th Bao				Son	石	琚應	Male	Monkey	1941	11
Fujin Xiang 5th Bao				The second son	石	琚生	Male	Dog	1946	10
Fujin Xiang 5th Bao				Daughter	石	菊香	Female	Rabbit	1939	1
Fujin Xiang 5th Bao				Younger brother	石	银现	Male	Rooster	1919	2
Fujin Xiang 5th Bao				Younger brother's wife	谈	马哇	Female	Horse	1918	3
Fujin Xiang 5th Bao				Niece	石	全英	Female	Tiger	1938	4
Fujin Xiang 5th Bao				Uncle's son	石	银全	Male		1921	
Fujin Xiang 5th Bao				Uncle's son's wife	谈	巧娃	Female	Horse	1918	6
Fujin Xiang 5th Bao	439	3	11	The household head	石	银林	Male	Dragon	1903	11
Fujin Xiang 5th Bao				Son	石	永德	Male	Rooster	1908	1
Fujin Xiang 5th Bao				Grandson	讨	口子	Male	Rat	1940	8
Fujin Xiang 5th Bao				Granddaughter	何	娃子	Female	Monkey	1944	4

Date	The beginning time of inhabitation in this county	Educational level	Profession sector	Occupation	Marital status	Physical disability	Religion
8	Inhabited for generations	Illiteracy	Agriculture	Land-holding peasant	Widowed	No	No religion
27	Inhabited for generations	Illiteracy	Agriculture	Land-holding peasant	Married	No	No religion
13	Inhabited for generations	Illiteracy	Agriculture	Land-holding peasant	Married	No	No religion
2	Inhabited for generations	Illiteracy	Agriculture	Land-holding peasant	Unmarried	No	No religion
17	Inhabited for generations	Sishu 2 years	Agriculture	Land-holding peasant	Married	No	No religion
6	Inhabited for generations	Illiteracy	Agriculture	Land-holding peasant	Married	No	No religion
30	Inhabited for generations	Illiteracy					
9	Inhabited for generations						
12	Inhabited for generations	Illiteracy					
6	Inhabited for generations	Illiteracy	Agriculture	Land-holding peasant	Married	No	No religion
15	Inhabited for generations	Illiteracy	Agriculture	Land-holding peasant	Married	No	No religion
4	Inhabited for generations	Illiteracy					
	Inhabited for generations	Illiteracy	Agriculture	Land-holding peasant	Married	No	No religion
5	Inhabited for generations	Illiteracy	Agriculture	Land-holding peasant	Married	No	No religion
28	Inhabited for generations	Illiteracy	Agriculture	Land-holding peasant	Married	No	No religion
4	Inhabited for generations	Illiteracy	Agriculture	Land-holding peasant	Married	No	No religion
18	Inhabited for generations	Illiteracy					
15	Inhabited for generations						

Appendix: Overview of Microdata for Wudu County (*cont.*)

	The no. of household	Jia	Household	Relationship with the household head	Surname	First name	Sex	The Chinese zodiac	Birthdate Year	Month
Fujin Xiang 5th Bao				Nephew's son	白	娃	Male	Monkey	1944	10
Fujin Xiang 5th Bao				Uncle's wife	赵	鸡鸣儿	Female	Tiger	1889	11
Fujin Xiang 5th Bao				Nephew	石	永洪	Male	Rooster	1921	8
Fujin Xiang 5th Bao				Nephew's wife	黄	照才哇	Female	Rooster	1921	3
Fujin Xiang 5th Bao				Nephew's daughter	石	万娃	Female	Dog	1946	4
Fujin Xiang 5th Bao	440	3	12	The household head	石	永琴	Male	Ox	1913	4
Fujin Xiang 5th Bao				The household head's wife	张	知娃	Female	Rat	1912	4
Fujin Xiang 5th Bao				Son	石	娃	Male	Rat	1936	11
Fujin Xiang 5th Bao				Younger brother	石	永贤	Male	Dog	1922	5
Fujin Xiang 5th Bao				Younger brother's wife	李	月英	Female	Rabbit	1927	4
Fujin Xiang 5th Bao				Nephew	石	妖精	Male	Rooster	1945	6
Fujin Xiang 5th Bao	441	3	13	The household head	石	永禄	Male	Tiger	1901	2
Fujin Xiang 5th Bao				The household head's wife	赵	双儿	Female	Rat	1899	12
Fujin Xiang 5th Bao				Father	石	银仓	Male	Dog	1872	2
Fujin Xiang 5th Bao				Mother	谈	李子	Female	Dragon	1879	6
Fujin Xiang 5th Bao				Son	石	大满娃	Male	Ox	1925	11
Fujin Xiang 5th Bao				The second son	石	崴满哇	Male	Dragon	1928	2
Fujin Xiang 5th Bao				The third son	石	相儿	Male	Snake	1940	7

Date	The beginning time of inhabitation in this county	Educational level	Profession sector	Occupation	Marital status	Physical disability	Religion
26	Inhabited for generations						
23	Inhabited for generations	Illiteracy	Agriculture	Land-holding peasant	Widowed	No	No religion
21	Inhabited for generations	Illiteracy	Agriculture	Land-holding peasant	Married	No	No religion
16	Inhabited for generations	Illiteracy	Agriculture	Land-holding peasant	Married	No	No religion
4	Inhabited for generations						
21	Inhabited for generations	Illiteracy	Agriculture	Land-holding peasant	Married	No	No religion
21	Inhabited for generations	Illiteracy	Agriculture	Land-holding peasant	Married	No	No religion
23	Inhabited for generations	Illiteracy					
14	Inhabited for generations	Illiteracy	Agriculture	Land-holding peasant	Married	No	No religion
15	Inhabited for generations	Illiteracy	Agriculture	Land-holding peasant	Married	No	No religion
17	Inhabited for generations						
26	Inhabited for generations	Sishu 3 years	Agriculture	Land-holding peasant	Married	No	No religion
6	Inhabited for generations	Illiteracy	Agriculture	Land-holding peasant	Married	No	No religion
23	Inhabited for generations	Illiteracy	Agriculture				
3	Inhabited for generations	Illiteracy					
4	Inhabited for generations	Illiteracy	Agriculture	Land-holding peasant	Unmarried	No	No religion
26	Inhabited for generations	Illiteracy	Agriculture	Land-holding peasant	Unmarried	No	No religion
8	Inhabited for generations	Illiteracy					

Appendix: Overview of Microdata for Wudu County (*cont.*)

	The no. of household	Jia	Household	Relationship with the household head	Surname	First name	Sex	The Chinese zodiac	Birthdate Year	Month
Fujin Xiang 5th Bao				Daughter	石	彬娃	Female	Monkey	1932	8
Fujin Xiang 5th Bao				The second daughter	石	兹娃	Female	Tiger	1938	2
Fujin Xiang 5th Bao				Younger brother	石	永全	Male	Tiger	1914	11
Fujin Xiang 5th Bao				The third younger brother	石	永富	Male	Goat	1919	5
Fujin Xiang 5th Bao				The third younger brother's wife	谈	菊娃	Female	Tiger	1926	3
Fujin Xiang 5th Bao				The third niece	石	和女子	Female	Pig	1947	3
Fujin Xiang 5th Bao	442	3	14	The household head	石	冬娃	Male	Horse	1905	11
Fujin Xiang 5th Bao				The household head's wife	谈	招哇	Female	Goat	1906	6
Fujin Xiang 5th Bao				Son	石	平娃	Male	Dragon	1928	12
Fujin Xiang 5th Bao				The second son	石	旺哇	Male	Tiger	1938	8
Fujin Xiang 5th Bao				Daughter	石	大巧娃	Female	Pig	1935	7
Fujin Xiang 5th Bao				The second daughter	石	崴巧娃	Female	Goat	1943	1
Fujin Xiang 5th Bao	443	4	1	The household head	石	仲芳	Male	Dragon	1903	5
Fujin Xiang 5th Bao				Son	石	昇儿	Male	Goat	1931	4
Fujin Xiang 5th Bao				Younger brother	石	仲唤	Male	Rabbit	1915	10
Fujin Xiang 5th Bao				Younger brother's wife	赵	庆娃	Female	Rabbit	1915	2
Fujin Xiang 5th Bao				Nephew	石	柏指子	Male	Pig	1947	2

Date	The beginning time of inhabitation in this county	Educational level	Profession sector	Occupation	Marital status	Physical disability	Religion
13	Inhabited for generations	Illiteracy	Agriculture	Land-holding peasant	Unmarried	No	No religion
22	Inhabited for generations						
28	Inhabited for generations	Illiteracy	Agriculture	Land-holding peasant	Unmarried	No	No religion
25	Inhabited for generations	Illiteracy	Agriculture	Land-holding peasant	Married	No	No religion
9	Inhabited for generations	Illiteracy	Agriculture	Land-holding peasant	Married	No	No religion
9	Inhabited for generations						
28	Inhabited for generations	Illiteracy	Agriculture	Land-holding peasant	Married	No	No religion
11	Inhabited for generations	Illiteracy	Agriculture	Land-holding peasant	Married	No	No religion
6	Inhabited for generations	Illiteracy	Agriculture	Land-holding peasant	Unmarried	No	No religion
21	Inhabited for generations	Illiteracy					
3	Inhabited for generations	Illiteracy	Agriculture	Land-holding peasant	Unmarried	No	No religion
6	Inhabited for generations						
27	Inhabited for generations	Illiteracy	Agriculture	Land-holding peasant	Widowed	No	No religion
19	Inhabited for generations	Illiteracy	Agriculture	Land-holding peasant	Unmarried	No	No religion
11	Inhabited for generations	Sishu 2 years	Agriculture	Land-holding peasant	Married	No	No religion
20	Inhabited for generations	Illiteracy	Agriculture	Land-holding peasant	Married	No	No religion
26	Inhabited for generations						

Appendix: Overview of Microdata for Wudu County (*cont.*)

	The no. of household	Jia	Household	Relationship with the household head	Surname	First name	Sex	The Chinese zodiac	Birthdate Year	Month
Fujin Xiang 5th Bao				Niece	石	侄娃	Female	Tiger	1938	8
Fujin Xiang 5th Bao				The second niece	石	书英	Female	Horse	1942	6
Fujin Xiang 5th Bao	444	4	2	The household head	石	元光武	Male	Monkey	1907	12
Fujin Xiang 5th Bao				The household head's wife	张	年哇	Female	Ox	1900	2
Fujin Xiang 5th Bao				Son	石	兹儿	Male	Rooster	1943	2
Fujin Xiang 5th Bao	445	4	3	The household head	石	银含	Male	Tiger	1914	2
Fujin Xiang 5th Bao				The household head's wife	李	反娃	Female	Pig	1910	2
Fujin Xiang 5th Bao				Father	石	四银儿	Male	Dog	1873	8
Fujin Xiang 5th Bao				Younger brother	石	富生儿	Male	Snake	1917	3
Fujin Xiang 5th Bao				Younger brother's wife	黄	岁娃	Female	Pig	1923	1
Fujin Xiang 5th Bao				Niece	石	益娃	Female	Dragon	1940	2
Fujin Xiang 5th Bao				Uncle's son	石	相哇	Male	Tiger	1926	12
Fujin Xiang 5th Bao				Uncle's son's wife	崴	女	Female	Rabbit	1927	7
Fujin Xiang 5th Bao				Niece	来	女	Female	Pig	1917	5
Fujin Xiang 5th Bao	446	4	4	The household head	石	银郭	Male	Rat	1904	4
Fujin Xiang 5th Bao				The household head's wife	东	听儿	Female	Rabbit	1902	12
Fujin Xiang 5th Bao				Son	石	执娃	Male	Snake	1939	8
Fujin Xiang 5th Bao				Daughter	石	平儿	Female		1936	5

Date	The beginning time of inhabitation in this county	Educational level	Profession sector	Occupation	Marital status	Physical disability	Religion
14	Inhabited for generations	Illiteracy					
27	Inhabited for generations						
8	Inhabited for generations	Illiteracy	Agriculture	Land-holding peasant	Married	No	No religion
7	Inhabited for generations	Illiteracy	Weaving	Land-holding peasant	Married	No	No religion
10	Inhabited for generations						
9	Inhabited for generations	Illiteracy	Agriculture	Land-holding peasant	Married		No religion
11	Inhabited for generations	Illiteracy	Agriculture	Land-holding peasant	Married	No	No religion
13	Inhabited for generations	Illiteracy	Agriculture	Land-holding peasant	Widowed	No	No religion
12	Inhabited for generations	Illiteracy	Agriculture	Land-holding peasant	Married	No	No religion
30	Inhabited for generations	Illiteracy	Agriculture	Land-holding peasant	Married	No	No religion
27	Inhabited for generations	Illiteracy					
11	Inhabited for generations	Illiteracy	Agriculture	Land-holding peasant	Married	No	No religion
3	Inhabited for generations	Illiteracy	Agriculture	Land-holding peasant	Married	No	No religion
28	Inhabited for generations						
26	Inhabited for generations	Illiteracy	Agriculture	Land-holding peasant	Married	No	No religion
28	Inhabited for generations	Illiteracy	Agriculture	Land-holding peasant	Married	No	No religion
11	Inhabited for generations	Illiteracy	Agriculture	Land-holding peasant	Unmarried	No	No religion
29	Inhabited for generations	Illiteracy					

Appendix: Overview of Microdata for Wudu County (*cont.*)

	The no. of household	Jia	Household	Relationship with the household head	Surname	First name	Sex	The Chinese zodiac	Birthdate Year	Month
Fujin Xiang 5th Bao				The second daughter	石	长命儿	Female	Tiger	1938	11
Fujin Xiang 5th Bao	447	4	5	The household head	石	仲元	Male	Snake	1904	8
Fujin Xiang 5th Bao				The household head's wife	赵	毛儿	Female	Rooster	1908	4
Fujin Xiang 5th Bao				Son	石	崴冯儿	Male	Rat	1936	10
Fujin Xiang 5th Bao	448	4	6	The household head	石	永德	Male	Pig	1898	12
Fujin Xiang 5th Bao				The household head's wife	李	银连	Female	Dragon	1903	5
Fujin Xiang 5th Bao				Son	石	杨兹	Male	Pig	1935	12
Fujin Xiang 5th Bao				The second son	石	进旦	Male	Dragon	1940	8
Fujin Xiang 5th Bao	449	4	7	The household head	石	银双	Male	Ox	1925	9
Fujin Xiang 5th Bao				The household head's wife	谈	书家	Female	Rat	1924	9
Fujin Xiang 5th Bao				Father	石	文仲	Male	Rat	1887	6
Fujin Xiang 5th Bao				Mother	马	五儿	Female	Dragon	1891	2
Fujin Xiang 5th Bao	450	4	8	The household head	石	仲否	Male	Rabbit	1902	2
Fujin Xiang 5th Bao				The household head's wife	赵	亭娃	Female	Horse	1905	5
Fujin Xiang 5th Bao				Mother	崔	银姐	Female	Tiger	1889	5
Fujin Xiang 5th Bao				Daughter	石	班班	Female	Dog	1934	3
Fujin Xiang 5th Bao	451	4	9	The household head	石	银旺	Male	Tiger	1914	4
Fujin Xiang 5th Bao				The household head's wife	党	娃	Female	Horse	1918	10

Date	The beginning time of inhabitation in this county	Educational level	Profession sector	Occupation	Marital status	Physical disability	Religion
21	Inhabited for generations	Illiteracy					
14	Inhabited for generations	Illiteracy	Agriculture	Land-holding peasant	Married	No	No religion
24	Inhabited for generations	Illiteracy	Weaving	Land-holding peasant	Married	No	No religion
24	Inhabited for generations	Bao school 2 years					
26	Inhabited for generations	Illiteracy	Agriculture	Land-holding peasant	Married	No	No religion
17	Inhabited for generations	Illiteracy	Weaving	Land-holding peasant	Married	No	No religion
13	Inhabited for generations	Illiteracy	Agriculture				
22	Inhabited for generations	Bao school 1 year					
17	Inhabited for generations	Illiteracy	Agriculture	Land-holding peasant	Married	No	No religion
15	Inhabited for generations	Illiteracy	Agriculture	Land-holding peasant	Married	No	No religion
6	Inhabited for generations	Illiteracy	Agriculture	Land-holding peasant	Married	No	No religion
9	Inhabited for generations	Illiteracy	Weaving	Land-holding peasant	Married	No	No religion
21	Inhabited for generations	Illiteracy	Agriculture	Land-holding peasant	Married	No	No religion
12	Inhabited for generations	Illiteracy	Agriculture	Land-holding peasant	Married	No	No religion
23	Inhabited for generations	Illiteracy	Agriculture	Land-holding peasant	Widowed	No	No religion
31	Inhabited for generations	Bao school 4 years	Agriculture				
6	Inhabited for generations	Illiteracy	Agriculture	Land-holding peasant	Married	No	No religion
4	Inhabited for generations	Illiteracy	Agriculture	Land-holding peasant	Married	No	No religion

Appendix: Overview of Microdata for Wudu County (*cont.*)

	The no. of household	Jia	Household	Relationship with the household head	Surname	First name	Sex	The Chinese zodiac	Birthdate	
									Year	Month
Fujin Xiang 5th Bao				Father	石	仲品	Male	Snake	1892	1
Fujin Xiang 5th Bao				Mother	马	生女	Female	Snake	1892	5
Fujin Xiang 5th Bao				Daughter	石	柏柏	Female	Dragon	1940	3
Fujin Xiang 5th Bao				The second daughter	石	梅梅	Female	Goat	1943	12
Fujin Xiang 5th Bao				The third daughter	石	龄龄	Female	Dog	1946	9
Fujin Xiang 5th Bao	452	4	10	The household head	石	永生	Male	Dragon	1916	3
Fujin Xiang 5th Bao				The household head's wife	崔	冬香	Female	Rabbit	1927	6
Fujin Xiang 5th Bao				Mother	赵	[illegible]儿	Female	Horse	1893	4
Fujin Xiang 5th Bao				Son	石	生旺	Male	Tiger	1938	10
Fujin Xiang 5th Bao				The second son	石	旺才儿	Male	Pig	1947	2
Fujin Xiang 5th Bao	453	4	11	The household head	石	仲全	Male	Ox	1913	12
Fujin Xiang 5th Bao				The household head's wife	李	喜女子	Female	Rat	1912	4
Fujin Xiang 5th Bao				Son	石	仓娃	Male	Rat	1937	6
Fujin Xiang 5th Bao	454	4	12	The household head	石	相儿	Male	Tiger	1901	3
Fujin Xiang 5th Bao				The household head's wife	杨	天花	Female	Monkey	1895	6
Fujin Xiang 5th Bao				Son	石	杨才	Male	Rabbit	1927	11
Fujin Xiang 5th Bao				The second son	石	银惠	Male	Rat	1936	6
Fujin Xiang 5th Bao				Daughter	石	才女子	Female	Horse	1930	2
Fujin Xiang 5th Bao	455	4	13	The household head	石	仲辉	Male	Rat	1887	2

Date	The beginning time of inhabitation in this county	Educational level	Profession sector	Occupation	Marital status	Physical disability	Religion
27	Inhabited for generations	Illiteracy	Agriculture	Land-holding peasant	Married	No	No religion
2	Inhabited for generations	Illiteracy	Agriculture	Land-holding peasant	Married	No	No religion
1	Inhabited for generations	Illiteracy					
18	Inhabited for generations						
30	Inhabited for generations						
26	Inhabited for generations	Illiteracy	Agriculture	Land-holding peasant	Married	No	No religion
12	Inhabited for generations	Illiteracy	Agriculture	Land-holding peasant	Married	No	No religion
10	Inhabited for generations	Illiteracy	Agriculture	Land-holding peasant	Widowed	No	No religion
21	Inhabited for generations	Bao school 1 year					
7	Inhabited for generations						
24	Inhabited for generations	Illiteracy	Agriculture	Land-holding peasant	Married	No	No religion
7	Inhabited for generations	Illiteracy	Agriculture	Land-holding peasant	Married	No	No religion
12	Inhabited for generations	Bao school 1 year					
8	Inhabited for generations	Illiteracy	Agriculture	Land-holding peasant	Married	No	No religion
7	Inhabited for generations	Illiteracy	Agriculture	Land-holding peasant	Married	No	No religion
9	Inhabited for generations	Illiteracy	Agriculture	Land-holding peasant	Unmarried	No	No religion
21	Inhabited for generations	Bao school 2 years					
28	Inhabited for generations	Illiteracy	Agriculture	Land-holding peasant	Unmarried	No	No religion
13	Inhabited for generations	Illiteracy	Agriculture	Land-holding peasant	Married	No	No religion

Appendix: Overview of Microdata for Wudu County (*cont.*)

	The no. of household	Jia	Household	Relationship with the household head	Surname	First name	Sex	The Chinese zodiac	Birthdate Year	Month
Fujin Xiang 5th Bao				The household head's wife	谈	崴廷	Female	Tiger	1889	3
Fujin Xiang 5th Bao	456	4	14	The household head	石	永忠	Male	Dragon	1916	10
Fujin Xiang 5th Bao				The household head's wife	石	双女	Female	Horse	1918	6
Fujin Xiang 5th Bao				Father	石	银炳	Male	Monkey	1895	6
Fujin Xiang 5th Bao				Mother	桂	女	Female	Monkey	1897	8
Fujin Xiang 5th Bao				Son	石	建娃	Male	Horse	1942	7
Fujin Xiang 5th Bao	457	4	15	The household head	石	仲珍	Male	Dog	1924	8
Fujin Xiang 5th Bao				The household head's wife	当	哇	Female	Dog	1924	8
Fujin Xiang 5th Bao				Father	石	宝	Male	Rat	1887	6
Fujin Xiang 5th Bao				Older brother's wife	马	春娃	Female	Rabbit	1915	12
Fujin Xiang 5th Bao				Niece	石	巧娃	Female	Dog	1946	6
Fujin Xiang 5th Bao				Younger brother	石	仲應	Male	Rooster	1933	1
Fujin Xiang 5th Bao	458	5	1	The household head	谈	篆儿	Female	Pig	1910	2
Fujin Xiang 5th Bao				Daughter	谈	明哇	Female	Dog	1934	6
Fujin Xiang 5th Bao				The second daughter	谈	进女	Female	Tiger	1938	1
Fujin Xiang 5th Bao				The third daughter	谈	苏英	Female	Rooster	1945	9
Fujin Xiang 5th Bao	459	5	2	The household head	石	宝高	Male	Tiger	1889	8
Fujin Xiang 5th Bao				The household head's wife	称	姐	Female	Tiger	1901	8

Date	The beginning time of inhabitation in this county	Educational level	Profession sector	Occupation	Marital status	Physical disability	Religion	
13	Inhabited for generations	Illiteracy	Agriculture	Land-holding peasant	Married	No	No religion	
12	Inhabited for generations	Illiteracy	Agriculture	Land-holding peasant	Married	No	No religion	
23	Inhabited for generations	Illiteracy	Weaving	Land-holding peasant	Married	No	No religion	
23	Inhabited for generations	Illiteracy	Agriculture	Land-holding peasant	Married	No	No religion	
15	Inhabited for generations	Illiteracy	Agriculture	Land-holding peasant	Married	No	No religion	
18	Inhabited for generations							
12	Inhabited for generations	Illiteracy	Agriculture	Land-holding peasant	Married	No	No religion	Serving in the army
12	Inhabited for generations	Illiteracy	Agriculture	Land-holding peasant	Married	No	No religion	
20	Inhabited for generations	Illiteracy	Agriculture	Land-holding peasant	Widowed	No	No religion	
25	Inhabited for generations	Illiteracy	Agriculture	Land-holding peasant	Widowed	No	No religion	
12	Inhabited for generations							
8	Inhabited for generations	Bao school 4 years	Agriculture					
8	Inhabited for generations	Illiteracy	Agriculture	Land-holding peasant	Widowed	No	No religion	
23	Inhabited for generations	Bao school 4 years	Agriculture	Land-holding peasant				
20	Inhabited for generations	Illiteracy						
7	Inhabited for generations							
2	Inhabited for generations	Illiteracy	Agriculture	Land-holding peasant	Married	No	No religion	
2	Inhabited for generations	Illiteracy	Agriculture	Land-holding peasant	Married	No	No religion	

Appendix: Overview of Microdata for Wudu County (*cont.*)

	The no. of household	Jia	Household	Relationship with the household head	Surname	First name	Sex	The Chinese zodiac	Birthdate	
									Year	Month
Fujin Xiang 5th Bao				Son	石	必学儿	Male	Monkey	1932	7
Fujin Xiang 5th Bao				Nephew's wife	谈	雪娃	Female	Rabbit	1902	9
Fujin Xiang 5th Bao				Nephew's son	石	杨学儿	Male	Rooster	1921	11
Fujin Xiang 5th Bao				Nephew's daughter	石	贱女子	Female	Horse	1930	2
Fujin Xiang 5th Bao				The second nephew's wife	石	崴当娃	Female	Rat	1899	4
Fujin Xiang 5th Bao				Nephew's son	石	岁佛儿	Male	Rabbit	1927	6
Fujin Xiang 5th Bao				Nephew's son's wife	赵	必娃	Female	Dragon	1928	8
Fujin Xiang 5th Bao				The second nephew's son's wife	石	崴长	Male	Dragon	1940	3
Fujin Xiang 5th Bao				Nephew	石	郭娃	Male	Horse	1918	5
Fujin Xiang 5th Bao	460	5	3	The household head	石	永连	Male	Tiger	1926	3
Fujin Xiang 5th Bao				The household head's wife	黄	见娃	Female	Dragon	1928	7
Fujin Xiang 5th Bao				Father	石	银义	Male	Dog	1897	11
Fujin Xiang 5th Bao				Mother	赵	同儿子	Female	Monkey	1895	1
Fujin Xiang 5th Bao				Younger brother	石	崴学哇	Male	Monkey	1932	4
Fujin Xiang 5th Bao				The third younger brother	石	崴旦旦	Male	Snake	1941	6
Fujin Xiang 5th Bao				Younger sister	石	无信儿	Female	Rat	1936	1
Fujin Xiang 5th Bao	461	5	4	The household head	石	银海	Male	Rat	1912	6
Fujin Xiang 5th Bao				The household head's wife	马	反娃	Female	Rat	1912	6

Date	The beginning time of inhabitation in this county	Educational level	Profession sector	Occupation	Marital status	Physical disability	Religion
27	Inhabited for generations	Bao school 3 years	Agriculture	Land-holding peasant	Unmarried	No	No religion
28	Inhabited for generations	Illiteracy	Agriculture	Land-holding peasant	Widowed	No	No religion
14	Inhabited for generations	Illiteracy	Agriculture	Land-holding peasant	Unmarried	No	No religion
7	Inhabited for generations	Illiteracy	Agriculture	Land-holding peasant	Unmarried	No	No religion
24	Inhabited for generations	Illiteracy	Agriculture	Land-holding peasant	Widowed	No	No religion
27	Inhabited for generations	Illiteracy	Agriculture	Land-holding peasant	Married	No	No religion
15	Inhabited for generations	Illiteracy	Agriculture	Land-holding peasant	Married	No	No religion
10	Inhabited for generations	Illiteracy					
10	Inhabited for generations	Illiteracy	Agriculture	Land-holding peasant	Widowed	No	No religion
3	Inhabited for generations	Illiteracy	Agriculture	Land-holding peasant	Married	No	No religion
5	Inhabited for generations	Illiteracy	Agriculture	Land-holding peasant	Married	No	No religion
9	Inhabited for generations	Illiteracy	Agriculture	Land-holding peasant	Married	No	No religion
5	Inhabited for generations	Illiteracy	Agriculture	Land-holding peasant	Married	No	No religion
7	Inhabited for generations	Bao school 2 years	Agriculture	Land-holding peasant	Unmarried	No	No religion
7	Inhabited for generations	Bao school 1 year					
5	Inhabited for generations	Illiteracy					
15	Inhabited for generations	Illiteracy	Agriculture	Land-holding peasant	Married	No	No religion
15	Inhabited for generations	Illiteracy	Agriculture	Land-holding peasant	Married	No	No religion

Appendix: Overview of Microdata for Wudu County (*cont.*)

	The no. of household	Jia	Household	Relationship with the household head	Surname	First name	Sex	The Chinese zodiac	Birthdate	
									Year	Month
Fujin Xiang 5th Bao				Daughter	石	柳明儿	Female	Rat	1936	8
Fujin Xiang 5th Bao				The second daughter	石	明六娃	Female	Snake	1941	8
Fujin Xiang 5th Bao				The third daughter	石	崔家	Female	Goat	1943	8
Fujin Xiang 5th Bao				The fourth daughter	石	毛子女	Female	Pig	1947	1
Fujin Xiang 5th Bao				Uncle's son	石	荣山	Male	Goat	1894	2
Fujin Xiang 5th Bao	462	5	5	The household head	石	仲举	Male	Rabbit	1915	8
Fujin Xiang 5th Bao				The household head's wife	赵	东娃	Female	Rat	1912	2
Fujin Xiang 5th Bao				Son	石	马义	Male	Dog	1934	6
Fujin Xiang 5th Bao				The second son	石	刘明	Male	Goat	1943	4
Fujin Xiang 5th Bao				Daughter	石	义女	Female	Rat	1936	6
Fujin Xiang 5th Bao				The second daughter	石	进女	Female	Horse	1942	9
Fujin Xiang 5th Bao				Younger brother	赵	何全	Male	Horse	1918	3
Fujin Xiang 5th Bao				Younger brother's wife	谈	岁娃	Female	Tiger	1926	4
Fujin Xiang 5th Bao	463	5	6	The household head	石	仲学	Male	Pig	1926	11
Fujin Xiang 5th Bao				The household head's wife	萧	书	Female	Pig	1923	2
Fujin Xiang 5th Bao				Grandfather	石	怌发	Male	Tiger	1885	6

Date	The beginning time of inhabitation in this county	Educational level	Profession sector	Occupation	Marital status	Physical disability	Religion
27	Inhabited for generations	Illiteracy					
24	Inhabited for generations	Illiteracy					
25	Inhabited for generations						
25	Inhabited for generations						
3	Inhabited for generations	Sishu 2 years	Agriculture	Land-holding peasant	Widowed	No	No religion
28	Inhabited for generations	Illiteracy	Agriculture	Land-holding peasant	Married	No	No religion
19	Inhabited for generations	Illiteracy	Agriculture	Land-holding peasant	Married	No	No religion
17	Inhabited for generations	Lower primary school 4 years					
7	Inhabited for generations						
19	Inhabited for generations	Illiteracy					
17	Inhabited for generations						
12	Inhabited for generations	Illiteracy	Agriculture	Land-holding peasant	Married	No	No religion
20	Inhabited for generations	Illiteracy	Agriculture	Land-holding peasant	Married	No	No religion
19	Inhabited for generations	Sishu 3 years	Agriculture	Land-holding peasant	Married	No	No religion
24	Inhabited for generations	Illiteracy	Agriculture	Land-holding peasant	Married	No	No religion
20	Inhabited for generations						

Appendix: Overview of Microdata for Wudu County (*cont.*)

	The no. of household	Jia	Household	Relationship with the household head	Surname	First name	Sex	The Chinese zodiac	Birthdate	
									Year	Month
Fujin Xiang 5th Bao				Father	石	宝文	Male	Rooster	1896	3
Fujin Xiang 5th Bao				Uncle's wife	赵	拜女	Female	Pig	1886	8
Fujin Xiang 5th Bao				Younger brother	石	仲生	Male	Dragon	1928	8
Fujin Xiang 5th Bao				Younger brother's wife	石	成娃	Female	Rat	1924	5
Fujin Xiang 5th Bao				Grand-mother	赵	氏	Female	Goat	1870	5
Fujin Xiang 5th Bao				Uncle	石	宝田	Male	Dragon	1931	10
Fujin Xiang 5th Bao				Uncle's son	石	仲绪	Male	Rooster	1921	6
Fujin Xiang 5th Bao	464	5	7	The household head	石	永库	Male	Rabbit	1915	8
Fujin Xiang 5th Bao				The household head's wife	石	双廷	Female	Rabbit	1915	2
Fujin Xiang 5th Bao				Grand-mother	赵	氏	Female	Ox	1864	2
Fujin Xiang 5th Bao				Son	石	生有	Male	Rat	1936	6
Fujin Xiang 5th Bao				Older brother	石	永仓	Male	Dog	1910	1
Fujin Xiang 5th Bao				Older brother's wife	五	进	Female	Rabbit	1927	3
Fujin Xiang 5th Bao				Uncle's son	石	永山	Male	Snake	1917	6
Fujin Xiang 5th Bao				Uncle's son's wife	张	平足	Female	Rat	1924	9
Fujin Xiang 5th Bao				Nephew	鸡	儿子	Male	Dog	1946	5
Fujin Xiang 5th Bao	465	5	8	The household head	石	银生	Male	Horse	1905	10

Date	The beginning time of inhabitation in this county	Educational level	Profession sector	Occupation	Marital status	Physical disability	Religion
3	Inhabited for generations	Illiteracy	Agriculture	Land-holding peasant	Widowed	No	No religion
15	Inhabited for generations	Illiteracy	Agriculture	Land-holding peasant	Married	No	No religion
15	Inhabited for generations	Illiteracy	Agriculture	Land-holding peasant	Married	No	No religion
18	Inhabited for generations	Illiteracy	Agriculture	Land-holding peasant	Married	No	No religion
7	Inhabited for generations						
21	Inhabited for generations	Illiteracy	Agriculture	Land-holding peasant	Widowed	No	No religion
20	Inhabited for generations	Illiteracy	Agriculture	Land-holding peasant	Unmarried	No	No religion
11	Inhabited for generations	Illiteracy	Agriculture	Land-holding peasant	Married	No	No religion
18	Inhabited for generations	Illiteracy	Agriculture	Land-holding peasant	Married	No	No religion
5	Inhabited for generations	Illiteracy					
23	Inhabited for generations	Bao school 1 year					
4	Inhabited for generations	Illiteracy	Agriculture	Land-holding peasant	Married	No	No religion
28	Inhabited for generations	Illiteracy	Agriculture	Land-holding peasant	Married	No	No religion
12	Inhabited for generations	Illiteracy	Agriculture	Land-holding peasant	Married	No	No religion
11	Inhabited for generations	Illiteracy	Agriculture	Land-holding peasant	Married	No	No religion
25	Inhabited for generations						
18	Inhabited for generations	Illiteracy	Agriculture	Land-holding peasant	Married	No	No religion

Appendix: Overview of Microdata for Wudu County (*cont.*)

	The no. of household	Jia	Household	Relationship with the household head	Surname	First name	Sex	The Chinese zodiac	Birthdate Year	Month
Fujin Xiang 5th Bao				The household head's wife	谈	节女	Female	Snake	1904	6
Fujin Xiang 5th Bao				Son	石	永孔	Male	Rooster	1933	2
Fujin Xiang 5th Bao				Daughter	石	苏英	Female	Goat	1943	12
Fujin Xiang 5th Bao	466	5	9	The household head	石	银耀	Male	Rabbit	1902	8
Fujin Xiang 5th Bao				Son	石	根儿	Male	Horse	1930	8
Fujin Xiang 5th Bao				The second son	石	马姓儿	Male	Dog	1946	3
Fujin Xiang 5th Bao				Daughter	石	代英	Female	Rooster	1933	5
Fujin Xiang 5th Bao				The second daughter	石	胡女	Female	Tiger	1938	7
Fujin Xiang 5th Bao				Younger brother	石	柱娃	Male	Horse	1918	6
Fujin Xiang 5th Bao				Younger brother's wife	当	娃	Female	Dragon	1916	2
Fujin Xiang 5th Bao				Nephew	石	丑儿	Male	Snake	1941	4
Fujin Xiang 5th Bao				Niece	石	丑女子	Female	Pig	1947	6
Fujin Xiang 5th Bao	467	5	10	The household head	石	仲鑫	Male	Horse	1894	1
Fujin Xiang 5th Bao				The household head's wife	赵	氏	Female	Rat	1887	3
Fujin Xiang 5th Bao				Son	石	银章	Male	Pig	1923	1
Fujin Xiang 5th Bao				Son's wife	石	社娃	Female	Ox	1925	2
Fujin Xiang 5th Bao				Granddaughter	石	早明娃	Female	Dog	1946	9
Fujin Xiang 5th Bao	468	5	11	The household head	石	永昇	Male	Horse	1905	10

Date	The beginning time of inhabitation in this county	Educational level	Profession sector	Occupation	Marital status	Physical disability	Religion
10	Inhabited for generations	Illiteracy	Agriculture	Land-holding peasant	Married	No	No religion
25	Inhabited for generations	Illiteracy					
20	Inhabited for generations						
28	Inhabited for generations	Sishu 5 years	Agriculture	Land-holding peasant	Widowed	No	No religion
14	Inhabited for generations	Illiteracy	Agriculture	Land-holding peasant	Married	No	No religion
14	Inhabited for generations						
8	Inhabited for generations	Illiteracy	Agriculture				
30	Inhabited for generations	Illiteracy					
23	Inhabited for generations	Illiteracy	Agriculture	Land-holding peasant	Married	No	No religion
16	Inhabited for generations	Illiteracy	Agriculture	Land-holding peasant	Married	No	No religion
9	Inhabited for generations	Illiteracy					
26	Inhabited for generations						
1	Inhabited for generations	Sishu 2 years	Agriculture	Land-holding peasant	Married	No	No religion
14	Inhabited for generations	Illiteracy	Agriculture	Land-holding peasant	Married	No	No religion
20	Inhabited for generations	Illiteracy	Agriculture	Land-holding peasant	Married	No	No religion
13	Inhabited for generations	Illiteracy	Agriculture	Land-holding peasant	Married	No	No religion
25	Inhabited for generations						
9	Inhabited for generations	Illiteracy	Agriculture	Land-holding peasant	Married	No	No religion

Appendix: Overview of Microdata for Wudu County (*cont.*)

	The no. of household	Jia	Household	Relationship with the household head	Surname	First name	Sex	The Chinese zodiac	Birthdate Year	Month
Fujin Xiang 5th Bao				The household head's wife	李	秋儿	Female	Horse	1918	7
Fujin Xiang 5th Bao				Father	石	银德	Male	Horse	1881	3
Fujin Xiang 5th Bao				Daughter	石	鹤娃	Female	Rabbit	1939	8
Fujin Xiang 5th Bao				The second daughter	石	苏梅	Female	Horse	1942	1
Fujin Xiang 5th Bao				Uncle's wife	赵	氏	Female	Dragon	1891	10
Fujin Xiang 5th Bao				Uncle's son	石	永梅	Male	Ox	1913	12
Fujin Xiang 5th Bao				Younger brother's wife	谈	岁宝	Female	Goat	1919	6
Fujin Xiang 5th Bao				Nephew	石	人仁	Male	Goat	1943	10
Fujin Xiang 5th Bao	469	5	12	The household head	石	银连	Male	Monkey	1920	6
Fujin Xiang 5th Bao				The household head's wife	谈	扣娃	Female	Horse	1918	8
Fujin Xiang 5th Bao				Mother	黄	氏	Female	Horse	1893	8
Fujin Xiang 5th Bao				Son	石	高旺儿	Male	Rooster	1945	7
Fujin Xiang 5th Bao				Younger brother	石	银贤	Male	Ox	1925	1
Fujin Xiang 5th Bao				Younger brother's wife	石	郭娃	Female	Ox	1925	9
Fujin Xiang 5th Bao				The second younger brother	石	银否	Male	Pig	1935	12
Fujin Xiang 5th Bao				Younger sister	石	裉娃	Female	Snake	1941	4
Fujin Xiang 5th Bao	470	5	13	The household head	石	银拜	Male	Rabbit	1902	5

Date	The beginning time of inhabitation in this county	Educational level	Profession sector	Occupation	Marital status	Physical disability	Religion
25	Inhabited for generations	Illiteracy	Agriculture	Land-holding peasant	Married	No	No religion
1	Inhabited for generations	Illiteracy	Agriculture	Land-holding peasant	Widowed	No	No religion
5	Inhabited for generations						
10	Inhabited for generations						
21	Inhabited for generations	Illiteracy	Agriculture	Land-holding peasant	Widowed	No	No religion
27	Inhabited for generations	Illiteracy	Agriculture	Land-holding peasant	Married	No	No religion
21	Inhabited for generations	Illiteracy	Agriculture	Land-holding peasant	Married	Blind	No religion
29	Inhabited for generations						
16	Inhabited for generations	Illiteracy	Agriculture	Land-holding peasant	Married	No	No religion
11	Inhabited for generations	Illiteracy	Agriculture	Land-holding peasant	Married	No	No religion
11	Inhabited for generations	Illiteracy	Agriculture	Land-holding peasant	Widowed	No	No religion
15	Inhabited for generations						
13	Inhabited for generations	Sishu 2 years	Agriculture	Land-holding peasant	Married	No	No religion
12	Inhabited for generations	Illiteracy	Agriculture	Land-holding peasant	Married	No	No religion
28	Inhabited for generations	Lower primary school 4 years					
26	Inhabited for generations						
26	Inhabited for generations	Illiteracy	Agriculture	Land-holding peasant	Married	No	No religion

Appendix: Overview of Microdata for Wudu County (*cont.*)

									Birthdate	
	The no. of household	Jia	Household	Relationship with the household head	Surname	First name	Sex	The Chinese zodiac	Year	Month
Fujin Xiang 5th Bao				The household head's wife	石	秋儿	Female	Horse	1905	8
Fujin Xiang 5th Bao				Son	石	永近	Male	Goat	1931	12
Fujin Xiang 5th Bao				The second son	石	权娃	Male	Goat	1943	3
Fujin Xiang 5th Bao				Daughter	石	义娃	Female	Monkey	1932	8
Fujin Xiang 5th Bao	471	5	14	The household head	赵	思林	Male	Rat	1924	4
Fujin Xiang 5th Bao				Father	赵	元银	Male	Pig	1898	11
Fujin Xiang 5th Bao				Mother	张	荔儿	Female	Pig	1898	4
Fujin Xiang 5th Bao				Younger brother	王	代娃	Male	Tiger	1939	1
Fujin Xiang 5th Bao				The second Grandmother	谈	毛娃	Female	Goat	1882	11
Fujin Xiang 5th Bao	472	6	1	The household head	赵	元才	Male	Ox	1913	10
Fujin Xiang 5th Bao				The household head's wife	扣	娃	Female	Rabbit	1915	7
Fujin Xiang 5th Bao				Son	赵	马信儿	Male	Rooster	1945	7
Fujin Xiang 5th Bao				Daughter	赵	崴柏儿	Female	Tiger	1938	2
Fujin Xiang 5th Bao				Adopted daughter	见	娃	Female	Horse	1942	11
Fujin Xiang 5th Bao	473	6	2	The household head	赵	元金	Male	Rabbit	1915	11
Fujin Xiang 5th Bao				The household head's wife	李	龙哇	Female	Dragon	1916	9
Fujin Xiang 5th Bao				Daughter	赵	兵女子	Female	Rabbit	1939	4

Date	The beginning time of inhabitation in this county	Educational level	Profession sector	Occupation	Marital status	Physical disability	Religion
31	Inhabited for generations	Illiteracy	Agriculture	Land-holding peasant	Married	No	No religion
13	Inhabited for generations	Bao school 4 years					
6	Inhabited for generations						
16	Inhabited for generations	Illiteracy	Agriculture	Land-holding peasant	Unmarried	No	No religion
5	Inhabited for generations	Illiteracy	Agriculture	Land-holding peasant	Unmarried	No	No religion
13	Inhabited for generations	Illiteracy	Agriculture	Land-holding peasant	Married	No	No religion
19	Inhabited for generations	Illiteracy	Agriculture	Land-holding peasant	Married	No	No religion
9	Inhabited for generations	Bao school 1 year					
1	Inhabited for generations						
7	Inhabited for generations	Illiteracy	Agriculture	Land-holding peasant	Married	No	No religion
24	Inhabited for generations	Illiteracy	Agriculture	Land-holding peasant	Married	No	No religion
4	Inhabited for generations						
5	Inhabited for generations	Illiteracy					
25	Inhabited for generations						
21	Inhabited for generations	Illiteracy	Agriculture	Land-holding peasant	Married	No	No religion
29	Inhabited for generations	Illiteracy	Agriculture	Land-holding peasant	Married	No	No religion
27	Inhabited for generations	Illiteracy					

Appendix: Overview of Microdata for Wudu County (*cont.*)

	The no. of household	Jia	Household	Relationship with the household head	Surname	First name	Sex	The Chinese zodiac	Birthdate	
									Year	Month
Fujin Xiang 5th Bao				The second daughter	赵	等路儿	Female	Snake	1941	10
Fujin Xiang 5th Bao				The third daughter	赵	英英	Female	Rooster	1945	2
Fujin Xiang 5th Bao	474	6	3	The household head	赵	玉琴	Male	Tiger	1914	3
Fujin Xiang 5th Bao				The household head's wife	石	琴哇	Female	Tiger	1914	1
Fujin Xiang 5th Bao				Father	赵	文奎	Male	Rooster	1884	8
Fujin Xiang 5th Bao				Stepmother	篆	登	Female	Snake	1904	2
Fujin Xiang 5th Bao				Daughter	赵	明明	Female	Goat	1933	2
Fujin Xiang 5th Bao				The second daughter	赵	崴明明	Female	Rooster	1938	1
Fujin Xiang 5th Bao				The third daughter	赵	讨口子	Female	Dog	1945	3
Fujin Xiang 5th Bao				Older brother	赵	玉德	Male	Monkey	1907	12
Fujin Xiang 5th Bao				Older brother's wife	毛	旦	Female	Rat	1903	1
Fujin Xiang 5th Bao				Nephew	赵	永忠	Male	Ox	1936	1
Fujin Xiang 5th Bao				Nephew's wife	谈	早廷	Female	Ox	1926	3
Fujin Xiang 5th Bao				Nephew's son	郭	哇	Male	Pig	1947	5
Fujin Xiang 5th Bao				Niece	莫	明哇	Female	Horse	1942	7
Fujin Xiang 5th Bao				The second niece	赵	女子	Female	Rooster	1945	11
Fujin Xiang 5th Bao	475	6	4	The household head	石	玉赵	Male	Snake	1892	11
Fujin Xiang 5th Bao				The household head's wife	路	顺	Female	Rooster	1884	9

Date	The beginning time of inhabitation in this county	Educational level	Profession sector	Occupation	Marital status	Physical disability	Religion
20	Inhabited for generations	Illiteracy					
24	Inhabited for generations						
27	Inhabited for generations	Illiteracy	Agriculture	Land-holding peasant	Married	No	No religion
30	Inhabited for generations	Illiteracy	Agriculture	Land-holding peasant	Married	No	No religion
14	Inhabited for generations	Illiteracy	Agriculture	Land-holding peasant	Married	No	No religion
11	Inhabited for generations	Illiteracy	Agriculture	Land-holding peasant	Married	No	No religion
7	Inhabited for generations	Illiteracy	Agriculture	Land-holding peasant			
20	Inhabited for generations	Illiteracy					
25	Inhabited for generations	Illiteracy					
25	Inhabited for generations	Illiteracy	Agriculture	Land-holding peasant	Married	No	No religion
14	Inhabited for generations	Illiteracy	Agriculture	Land-holding peasant	Married	No	No religion
23	Inhabited for generations	Illiteracy	Agriculture	Land-holding peasant	Married	No	No religion
24	Inhabited for generations	Illiteracy	Agriculture	Land-holding peasant	Married	No	No religion
14	Inhabited for generations	Illiteracy	Agriculture	Land-holding peasant			
15	Inhabited for generations						
24	Inhabited for generations						
17	Inhabited for generations	Illiteracy	Agriculture	Land-holding peasant	Married	No	No religion
20	Inhabited for generations	Illiteracy	Agriculture	Land-holding peasant	Married	No	No religion

Appendix: Overview of Microdata for Wudu County (*cont.*)

									Birthdate	
	The no. of household	**Jia**	**Household**	**Relationship with the household head**	**Surname**	**First name**	**Sex**	**The Chinese zodiac**	**Year**	**Month**
Fujin Xiang 5th Bao				Son	石	学仲	Male	Dog	1910	1
Fujin Xiang 5th Bao				Son's wife	石	女子	Female	Pig	1909	3
Fujin Xiang 5th Bao				Granddaughter	石	苏因	Female	Goat	1943	2
Fujin Xiang 5th Bao				The second son	石	文郭	Male	Dragon	1928	6
Fujin Xiang 5th Bao	476	6	5	The household head	崔	连举	Male	Tiger	1914	8
Fujin Xiang 5th Bao				The household head's wife	黄	岁知	Female	Rooster	1921	10
Fujin Xiang 5th Bao				Mother	赵	罗儿	Female	Snake	1892	10
Fujin Xiang 5th Bao				Daughter	崔	黑娃	Female	Rooster	1945	5
Fujin Xiang 5th Bao				Older brother	崔	信王儿	Male	Dragon	1916	5
Fujin Xiang 5th Bao				Older brother's wife	陈	岳哇	Female	Horse	1918	6
Fujin Xiang 5th Bao				Niece	崔	当女子	Female	Dog	1946	5
Fujin Xiang 5th Bao	477	6	6	The household head	赵	文太	Male	Horse	1905	1
Fujin Xiang 5th Bao				The household head's wife	石	金娃	Female	Tiger	1901	10
Fujin Xiang 5th Bao				Son	赵	代娃	Male	Ox	1925	10
Fujin Xiang 5th Bao				The second son	赵	玉银	Male	Pig	1935	6
Fujin Xiang 5th Bao				Daughter	赵	四女	Female	Rabbit	1939	4
Fujin Xiang 5th Bao	478	6	7	The household head	赵	文潘	Male	Pig	1898	7
Fujin Xiang 5th Bao				Mother	崴	缠	Female	Tiger	1889	8

Date	The beginning time of inhabitation in this county	Educational level	Profession sector	Occupation	Marital status	Physical disability	Religion
28	Inhabited for generations	Illiteracy	Agriculture	Land-holding peasant	Married	No	No religion
22	Inhabited for generations	Illiteracy	Agriculture	Land-holding peasant	Married	No	No religion
5	Inhabited for generations						
6	Inhabited for generations	Bao school 4 years	Agriculture	Land-holding peasant	Unmarried	No	No religion
26	Inhabited for generations	Illiteracy	Agriculture	Land-holding peasant	Married	No	No religion
17	Inhabited for generations	Illiteracy	Agriculture	Land-holding peasant	Married	No	No religion
4	Inhabited for generations	Illiteracy	Agriculture	Land-holding peasant	Married	No	No religion
23	Inhabited for generations						
13	Inhabited for generations	Illiteracy	Agriculture	Land-holding peasant	Married	No	No religion
19	Inhabited for generations	Illiteracy	Agriculture	Land-holding peasant	Married	No	No religion
1	Inhabited for generations						
25	Inhabited for generations	Illiteracy	Agriculture	Land-holding peasant	Married	No	No religion
1	Inhabited for generations	Illiteracy	Agriculture	Land-holding peasant	Married	No	No religion
23	Inhabited for generations	Illiteracy	Agriculture	Land-holding peasant	Unmarried	No	No religion
28	Inhabited for generations	Bao school 3 years					
1	Inhabited for generations						
8	Inhabited for generations	Illiteracy	Agriculture	Land-holding peasant	Widowed	No	No religion
15	Inhabited for generations	Illiteracy	Agriculture	Land-holding peasant	Widowed	No	No religion

Appendix: Overview of Microdata for Wudu County (*cont.*)

	The no. of household	Jia	Household	Relationship with the household head	Surname	First name	Sex	The Chinese zodiac	Birthdate	
									Year	Month
Fujin Xiang 5th Bao				Son	赵	春官	Male	Snake	1941	4
Fujin Xiang 5th Bao				Daughter	赵	未生	Female	Ox	1936	9
Fujin Xiang 5th Bao				Younger brother	张	代娃	Male	Rat	1909	1
Fujin Xiang 5th Bao				Younger brother's wife	谈	禄娃	Female	Rat	1912	6
Fujin Xiang 5th Bao				Nephew	赵	谈称称	Male	Dragon	1940	4
Fujin Xiang 5th Bao				The second nephew	赵	石虎儿	Male	Horse	1942	6
Fujin Xiang 5th Bao				Niece	赵	谈家女	Female	Dog	1946	6
Fujin Xiang 5th Bao				The second younger brother's wife	罗	旦旦	Female	Rat	1924	7
Fujin Xiang 5th Bao	479	6	8	The household head	石	文概	Male	Monkey	1907	8
Fujin Xiang 5th Bao				The household head's wife	张	双拜	Female	Rat	1912	7
Fujin Xiang 5th Bao				Mother	谈	氏	Female	Rat	1887	6
Fujin Xiang 5th Bao				Son	石	张才	Male	Rooster	1934	6
Fujin Xiang 5th Bao				Daughter	石	雪娃	Female	Tiger	1938	10
Fujin Xiang 5th Bao				The second daughter	安	明儿	Female	Dog	1946	10
Fujin Xiang 5th Bao				Uncle	杨	占春	Male	Rabbit	1890	12
Fujin Xiang 5th Bao				Uncle's son	石	回来娃	Male	Rat	1936	3
Fujin Xiang 5th Bao	480	6	9	The household head	石	银祥	Male	Dragon	1916	9

Date	The beginning time of inhabitation in this county	Educational level	Profession sector	Occupation	Marital status	Physical disability	Religion
6	Inhabited for generations						
26	Inhabited for generations						
1	Inhabited for generations	Sishu 3 years	Agriculture	Land-holding peasant	Married	No	No religion
1	Inhabited for generations	Illiteracy	Agriculture	Land-holding peasant	Married	No	No religion
27	Inhabited for generations	Illiteracy					
29	Inhabited for generations						
29	Inhabited for generations						
31	Inhabited for generations	Illiteracy	Agriculture	Land-holding peasant	Widowed	No	No religion
17	Inhabited for generations	Illiteracy	Agriculture	Land-holding peasant	Married	No	No religion
1	Inhabited for generations	Illiteracy	Agriculture	Land-holding peasant	Married	No	No religion
10	Inhabited for generations	Illiteracy	Agriculture	Land-holding peasant	Widowed	No	No religion
17	Inhabited for generations	Bao school 1 year	Agriculture				
14	Inhabited for generations	Illiteracy					
14	Inhabited for generations						
3	Inhabited for generations	Illiteracy	Agriculture	Land-holding peasant	Widowed	No	No religion
6	Inhabited for generations	Illiteracy					
27	Inhabited for generations	Illiteracy	Agriculture	Land-holding peasant	Married	No	No religion

Appendix: Overview of Microdata for Wudu County (*cont.*)

	The no. of household	Jia	Household	Relationship with the household head	Surname	First name	Sex	The Chinese zodiac	Birthdate	
									Year	Month
Fujin Xiang 5th Bao				The household head's wife	清	娃	Female	Rabbit	1915	6
Fujin Xiang 5th Bao				Mother	崔	根连	Female	Snake	1892	7
Fujin Xiang 5th Bao				Son	石	專富	Male	Horse	1942	7
Fujin Xiang 5th Bao				Daughter	石	專福	Female	Ox	1937	6
Fujin Xiang 5th Bao				The second daughter	石	勤娃	Female	Rabbit	1939	4
Fujin Xiang 5th Bao				The third daughter	石	进学儿	Female	Goat	1943	4
Fujin Xiang 5th Bao				Younger brother	石	银荣	Male	Ox	1925	8
Fujin Xiang 5th Bao				Younger brother's wife	柘	儿	Female	Horse	1930	10
Fujin Xiang 5th Bao	481	6	10	The household head	石	银清	Male	Goat	1906	4
Fujin Xiang 5th Bao				The household head's wife	金	女	Female	Horse	1905	1
Fujin Xiang 5th Bao				Son	早	儿子	Male	Rooster	1933	10
Fujin Xiang 5th Bao				Daughter	双	清娃	Female	Rooster	1933	10
Fujin Xiang 5th Bao	482	6	11	The household head	石	仲益	Male	Rat	1875	10
Fujin Xiang 5th Bao				The household head's wife	赵	知娃	Female	Rat	1887	7
Fujin Xiang 5th Bao				Son	石	银才	Male	Horse	1905	12
Fujin Xiang 5th Bao				Son's wife	谈	东春	Female	Horse	1905	4
Fujin Xiang 5th Bao				Grandson	张	信儿	Male	Tiger	1937	3

Date	The beginning time of inhabitation in this county	Educational level	Profession sector	Occupation	Marital status	Physical disability	Religion
23	Inhabited for generations	Illiteracy	Agriculture	Land-holding peasant	Married	No	No religion
13	Inhabited for generations	Illiteracy	Agriculture	Land-holding peasant	Widowed	No	No religion
23	Inhabited for generations						
25	Inhabited for generations	Illiteracy					
9	Inhabited for generations	Illiteracy					
15	Inhabited for generations						
24	Inhabited for generations	Illiteracy	Agriculture	Land-holding peasant	Married	No	No religion
29	Inhabited for generations	Illiteracy	Agriculture	Land-holding peasant	Married	No	No religion
3	Inhabited for generations	Illiteracy	Agriculture	Land-holding peasant	Married	No	No religion
21	Inhabited for generations	Illiteracy	Weaving	Land-holding peasant	Married	No	No religion
28	Inhabited for generations	Illiteracy	Agriculture	Land-holding peasant	Married	No	No religion
28	Inhabited for generations	Illiteracy	Agriculture	Land-holding peasant	Married	No	No religion
1	Inhabited for generations	Illiteracy	Agriculture	Land-holding peasant	Married	No	No religion
18	Inhabited for generations	Illiteracy	Agriculture	Land-holding peasant	Married	No	No religion
9	Inhabited for generations	Illiteracy	Agriculture	Land-holding peasant	Married	No	No religion
12	Inhabited for generations	Illiteracy	Agriculture	Land-holding peasant	Married	No	No religion
2	Inhabited for generations	Illiteracy					

Appendix: Overview of Microdata for Wudu County (*cont.*)

	The no. of household	Jia	Household	Relationship with the household head	Surname	First name	Sex	The Chinese zodiac	Birthdate	
									Year	Month
Fujin Xiang 5th Bao	483	6	12	The household head	石	仲贤	Male	Rabbit	1890	5
Fujin Xiang 5th Bao				The household head's wife	崴	女	Female	Rooster	1884	5
Fujin Xiang 5th Bao				Son	石	银均	Male	Rabbit	1915	10
Fujin Xiang 5th Bao				Son's wife	石	必哇	Female	Rabbit	1915	10
Fujin Xiang 5th Bao				Grandson	石	具才	Male	Dog	1946	11
Fujin Xiang 5th Bao				Granddaughter	石	相相	Female	Rabbit	1939	12
Fujin Xiang 5th Bao				The second granddaughter	石	二娃	Female	Snake	1941	5
Fujin Xiang 5th Bao				Younger brother	石	仲会	Male	Rooster	1896	12
Fujin Xiang 5th Bao				Nephew	石	银利	Female	Rooster	1921	11
Fujin Xiang 5th Bao				Nephew's wife	杨	大桃	Female	Ox	1926	1
Fujin Xiang 5th Bao				Niece	毕	明娃	Female	Pig	1935	7
Fujin Xiang 5th Bao	484	6	13	The household head	石	仲库	Male	Dragon	1891	6
Fujin Xiang 5th Bao				The household head's wife	肖	儿	Female	Snake	1905	6
Fujin Xiang 5th Bao				Son	石	银琴	Male	Goat	1919	7
Fujin Xiang 5th Bao				Son's wife	石	具廷	Female	Goat	1925	9
Fujin Xiang 5th Bao				Grandson	石	专乘儿	Male	Dog	1946	12
Fujin Xiang 5th Bao	485	6	14	The household head	石	文科	Male	Dog	1909	3

Date	The beginning time of inhabitation in this county	Educational level	Profession sector	Occupation	Marital status	Physical disability	Religion
17	Inhabited for generations	Sishu 3 years	Agriculture	Land-holding peasant	Married	No	No religion
23	Inhabited for generations	Illiteracy	Agriculture	Land-holding peasant	Married	No	No religion
13	Inhabited for generations	Illiteracy	Agriculture	Land-holding peasant	Married	No	No religion
13	Inhabited for generations	Illiteracy	Agriculture	Land-holding peasant	Married	No	No religion
28	Inhabited for generations						
13	Inhabited for generations	Illiteracy					
27	Inhabited for generations	Illiteracy					
5	Inhabited for generations	Sishu 3 years	Agriculture	Land-holding peasant	Widowed	No	No religion
3	Inhabited for generations	Sishu 3 years	Agriculture	Land-holding peasant	Married	No	No religion
18	Inhabited for generations	Illiteracy	Agriculture	Land-holding peasant	Married	No	No religion
1	Inhabited for generations	Illiteracy	Agriculture	Land-holding peasant			
20	Inhabited for generations	Illiteracy	Agriculture	Land-holding peasant	Married	No	No religion
22	Inhabited for generations	Illiteracy	Agriculture	Land-holding peasant	Married	No	No religion
22	Inhabited for generations	Illiteracy	Agriculture	Land-holding peasant	Married	No	No religion
15	Inhabited for generations	Illiteracy	Agriculture	Land-holding peasant	Married	No	No religion
7	Inhabited for generations						
22	Inhabited for generations	Sishu 2 years	Agriculture	Land-holding peasant	Married	No	No religion

Appendix: Overview of Microdata for Wudu County (*cont.*)

	The no. of household	Jia	Household	Relationship with the household head	Surname	First name	Sex	The Chinese zodiac	Birthdate	
									Year	Month
Fujin Xiang 5th Bao				The household head's wife	双	儿	Female	Horse	1905	6
Fujin Xiang 5th Bao				Son	石	含斌	Male	Monkey	1932	12
Fujin Xiang 5th Bao				Daughter	董	清	Female	Snake	1941	12
Fujin Xiang 5th Bao				Uncle's wife	石	氏	Female	Pig	1886	9
Fujin Xiang 5th Bao				Uncle's son	石	文清	Male	Rooster	1921	3
Fujin Xiang 5th Bao				Uncle's son's wife	称	称	Female	Goat	1919	5
Fujin Xiang 5th Bao	486	6	15	The household head	石	文杜	Male	Rabbit	1915	10
Fujin Xiang 5th Bao				The household head's wife	马	才娃	Female	Dog	1909	7
Fujin Xiang 5th Bao				Son	石	含怌	Male	Pig	1935	6
Fujin Xiang 5th Bao				Younger brother	石	文生	Male	Horse	1918	8
Fujin Xiang 5th Bao				Younger brother's wife	肖	专儿	Female	Horse	1918	10
Fujin Xiang 5th Bao				Nephew	石	正怌	Male	Rabbit	1939	9
Fujin Xiang 5th Bao				Niece	石	进女子	Female	Pig	1947	1
Fujin Xiang 5th Bao	487	7	1	The household head	赵	文贵	Male	Ox	1900	8
Fujin Xiang 5th Bao				The household head's wife	石	专女	Female	Horse	1905	2
Fujin Xiang 5th Bao				Son	赵	玉成	Male	Rooster	1932	7
Fujin Xiang 5th Bao				The second son	赵	韩生	Male	Dragon	1940	2
Fujin Xiang 5th Bao				Daughter	赵	成女	Female	Pig	1935	12

Date	The beginning time of inhabitation in this county	Educational level	Profession sector	Occupation	Marital status	Physical disability	Religion
22	Inhabited for generations	Illiteracy	Agriculture	Land-holding peasant	Married	No	No religion
11	Inhabited for generations	Bao school 4 years	Agriculture	Land-holding peasant	Married	No	No religion
22	Inhabited for generations						
19	Inhabited for generations	Illiteracy	Agriculture	Land-holding peasant	Widowed	No	No religion
15	Inhabited for generations	Illiteracy	Agriculture	Land-holding peasant	Married	No	No religion
10	Inhabited for generations	Illiteracy	Agriculture	Land-holding peasant	Married	No	No religion
9	Inhabited for generations	Illiteracy	Agriculture	Land-holding peasant	Married	No	No religion
18	Inhabited for generations	Illiteracy	Agriculture	Land-holding peasant	Married	No	No religion
8	Inhabited for generations	Bao school 4 years					
20	Inhabited for generations	Illiteracy	Agriculture	Land-holding peasant	Married	No	No religion
5	Inhabited for generations	Illiteracy	Agriculture	Land-holding peasant	Married	No	No religion
13	Inhabited for generations	Illiteracy					
28	Inhabited for generations						
20	Inhabited for generations	Sishu 2 years	Agriculture	Land-holding peasant	Married	No	No religion
5	Inhabited for generations	Illiteracy	Agriculture	Land-holding peasant	Married	No	No religion
13	Inhabited for generations	Illiteracy	Agriculture	Land-holding peasant	Unmarried	No	No religion
8	Inhabited for generations	Illiteracy					
28	Inhabited for generations	Illiteracy					

Appendix: Overview of Microdata for Wudu County (*cont.*)

	The no. of household	Jia	Household	Relationship with the household head	Surname	First name	Sex	The Chinese zodiac	Birthdate	
									Year	Month
Fujin Xiang 5th Bao				Younger brother	赵	文全	Male	Pig	1913	2
Fujin Xiang 5th Bao				Younger brother's wife	李	岁龙	Female	Dragon	1918	11
Fujin Xiang 5th Bao				Nephew	赵	高生	Male	Ox	1937	4
Fujin Xiang 5th Bao				The second nephew	赵	怀生	Male	Dragon	1940	10
Fujin Xiang 5th Bao				Uncle's third son	赵	成生	Male	Dog	1946	5
Fujin Xiang 5th Bao				Uncle's son	赵	文福	Male	Pig	1898	11
Fujin Xiang 5th Bao				Nephew	赵	玉旺	Male	Dragon	1916	7
Fujin Xiang 5th Bao				Nephew's wife	石	查娃	Female	Dragon	1916	4
Fujin Xiang 5th Bao				Nephew's daughter	赵	印女	Female	Rabbit	1939	2
Fujin Xiang 5th Bao	488	7	2	The household head	赵	文科	Male	Ox	1913	11
Fujin Xiang 5th Bao				The household head's wife	石	讲女	Female	Dragon	1916	7
Fujin Xiang 5th Bao				Mother	谈	家女	Female	Rooster	1884	3
Fujin Xiang 5th Bao				Son	赵	韩良儿	Male	Rabbit	1928	1
Fujin Xiang 5th Bao				Older brother	赵	文发	Male	Monkey	1907	1
Fujin Xiang 5th Bao				Older brother's wife	石	楼娃	Female	Pig	1910	10
Fujin Xiang 5th Bao				Nephew	徐	才	Male	Snake	1941	4
Fujin Xiang 5th Bao				Niece	必	明娃	Female	Dragon	1939	8
Fujin Xiang 5th Bao	489	7	3	The household head	赵	文斌	Male	Goat	1919	1

Date	The beginning time of inhabitation in this county	Educational level	Profession sector	Occupation	Marital status	Physical disability	Religion
2	Inhabited for generations	Illiteracy	Agriculture	Land-holding peasant	Married	No	No religion
4	Inhabited for generations	Illiteracy	Agriculture	Land-holding peasant	Married	No	No religion
23	Inhabited for generations	Bao school 3 years					
9	Inhabited for generations	Illiteracy					
9	Inhabited for generations						
13	Inhabited for generations	Sishu 4 years	Agriculture	Land-holding peasant	Married	No	No religion
10	Inhabited for generations	Illiteracy	Agriculture	Land-holding peasant	Married	No	No religion
12	Inhabited for generations	Illiteracy	Agriculture	Land-holding peasant	Married	No	No religion
29	Inhabited for generations	Illiteracy					
8	Inhabited for generations	Illiteracy	Agriculture	Land-holding peasant	Married	No	No religion
4	Inhabited for generations	Illiteracy	Agriculture	Land-holding peasant	Married	No	No religion
18	Inhabited for generations	Illiteracy	Agriculture	Land-holding peasant	Widowed	No	No religion
22	Inhabited for generations	Illiteracy	Agriculture	Land-holding peasant	Unmarried	No	No religion
1	Inhabited for generations	Illiteracy	Agriculture	Land-holding peasant	Married	No	No religion
18	Inhabited for generations	Illiteracy	Agriculture	Land-holding peasant	Married	No	No religion
21	Inhabited for generations	Illiteracy					
18	Inhabited for generations	Illiteracy					
25	Inhabited for generations	Illiteracy	Agriculture	Land-holding peasant	Married	No	No religion

Appendix: Overview of Microdata for Wudu County (*cont.*)

	The no. of household	Jia	Household	Relationship with the household head	Surname	First name	Sex	The Chinese zodiac	Birthdate	
									Year	Month
Fujin Xiang 5th Bao				The household head's wife	李	早早	Female	Tiger	1926	5
Fujin Xiang 5th Bao				Father	赵	金学	Male	Dragon	1943	11
Fujin Xiang 5th Bao				Mother	李	拜女	Female	Horse	1881	9
Fujin Xiang 5th Bao				Son	赵	應平	Male	Pig	1947	2
Fujin Xiang 5th Bao	490	7	4	The household head	赵	文孔	Male	Goat	1906	5
Fujin Xiang 5th Bao				The household head's wife	陈	岁路	Female	Dog	1909	2
Fujin Xiang 5th Bao				Son	赵	六斤	Male	Dog	1943	6
Fujin Xiang 5th Bao				The second son	赵	长旺儿	Male	Dog	1946	5
Fujin Xiang 5th Bao	491	7	5	The household head	赵	文润	Male	Horse	1905	3
Fujin Xiang 5th Bao				The household head's wife	柱	哇	Female	Horse	1905	3
Fujin Xiang 5th Bao				Son	赵	讲哇	Male	Dragon	1916	11
Fujin Xiang 5th Bao				Son's wife	有	宿	Female	Rooster	1921	12
Fujin Xiang 5th Bao				Granddaughter	赵	富女	Female	Goat	1943	11
Fujin Xiang 5th Bao				Daughter	崔	顺哇	Female	Rooster	1933	9
Fujin Xiang 5th Bao				The second daughter	赵	四女子	Female	Tiger	1938	9
Fujin Xiang 5th Bao				Uncle's son's wife	慈	女	Female	Dog	1897	2
Fujin Xiang 5th Bao	492	7	6	The household head	赵	玉哇子	Male	Rat	1912	5
Fujin Xiang 5th Bao				Younger brother	赵	天赐儿	Male	Snake	1941	3

Date	The beginning time of inhabitation in this county	Educational level	Profession sector	Occupation	Marital status	Physical disability	Religion
22	Inhabited for generations	Illiteracy	Agriculture	Land-holding peasant	Married	No	No religion
20	Inhabited for generations	Illiteracy	Agriculture	Land-holding peasant	Married	No	No religion
13	Inhabited for generations	Illiteracy	Agriculture	Land-holding peasant	Married	No	No religion
6	Inhabited for generations						
1	Inhabited for generations	Illiteracy	Agriculture	Land-holding peasant	Married	No	No religion
22	Inhabited for generations	Illiteracy	Agriculture	Land-holding peasant	Married	No	No religion
26	Inhabited for generations						
1	Inhabited for generations						
25	Inhabited for generations	Illiteracy	Agriculture	Land-holding peasant	Married	No	No religion
25	Inhabited for generations	Illiteracy	Agriculture	Land-holding peasant	Married	No	No religion
13	Inhabited for generations	Illiteracy	Agriculture	Land-holding peasant	Married	No	No religion
26	Inhabited for generations	Illiteracy	Agriculture	Land-holding peasant	Married	No	No religion
15	Inhabited for generations						
4	Inhabited for generations	Illiteracy	Agriculture	Land-holding peasant	Unmarried	No	No religion
30	Inhabited for generations	Illiteracy					
11	Inhabited for generations	Illiteracy	Agriculture	Land-holding peasant	Widowed	No	No religion
28	Inhabited for generations	Illiteracy	Agriculture	Land-holding peasant	Widowed	No	No religion
28	Inhabited for generations	Illiteracy					

Appendix: Overview of Microdata for Wudu County (*cont.*)

	The no. of household	Jia	Household	Relationship with the household head	Surname	First name	Sex	The Chinese zodiac	Birthdate	
									Year	Month
Fujin Xiang 5th Bao				Younger sister	赵	松哇	Female	Dog	1935	1
Fujin Xiang 5th Bao				The second younger sister	赵	崴女	Female	Tiger	1938	9
Fujin Xiang 5th Bao				Nephew	赵	才哇	Male	Rabbit	1940	1
Fujin Xiang 5th Bao	493	7	7	The household head	赵	文仓	Male	Horse	1917	7
Fujin Xiang 5th Bao				The household head's wife	石	崴贵	Female	Ox	1900	3
Fujin Xiang 5th Bao				Father	赵	金具	Male	Dragon	1879	8
Fujin Xiang 5th Bao				Son	赵	春哇	Male	Rabbit	1939	10
Fujin Xiang 5th Bao				Uncle	赵	金满	Male	Goat	1882	6
Fujin Xiang 5th Bao				Uncle's wife	鲁	刘女	Female	Horse	1881	5
Fujin Xiang 5th Bao				Uncle's son	赵	文库	Male	Dragon	1916	6
Fujin Xiang 5th Bao				Cohabitant	玉	才	Male	Horse	1918	11
Fujin Xiang 5th Bao				Cohabitant's wife	崴	转	Female	Horse	1918	11
Fujin Xiang 5th Bao				Cohabitant's daughter	赵	东女	Female	Goat	1933	6
Fujin Xiang 5th Bao	494	7	8	The household head	赵	文德	Male	Goat	1919	12
Fujin Xiang 5th Bao				The household head's wife	石	亭娃	Female	Goat	1919	10
Fujin Xiang 5th Bao				Mother	谈	冯娃	Female	Tiger	1889	2
Fujin Xiang 5th Bao				Son	赵	玉昇	Male	Ox	1937	10

Date	The beginning time of inhabitation in this county	Educational level	Profession sector	Occupation	Marital status	Physical disability	Religion
29	Inhabited for generations	Illiteracy	Agriculture	Land-holding peasant			
8	Inhabited for generations	Illiteracy					
2	Inhabited for generations	Illiteracy					
5	Inhabited for generations	Illiteracy	Agriculture	Land-holding peasant	Married	No	No religion
30	Inhabited for generations	Illiteracy	Agriculture	Land-holding peasant	Married	No	No religion
10	Inhabited for generations						
17	Inhabited for generations	Illiteracy					
5	Inhabited for generations	Illiteracy	Agriculture	Land-holding peasant	Married	No	No religion
9	Inhabited for generations	Illiteracy	Agriculture	Land-holding peasant	Married	No	No religion
4	Inhabited for generations	Illiteracy	Agriculture	Land-holding peasant	Unmarried	No	No religion
5	Inhabited for generations	Illiteracy	Agriculture	Land-holding peasant	Married	No	No religion
13	Inhabited for generations	Illiteracy	Agriculture	Land-holding peasant	Married	No	No religion
5	Inhabited for generations	Illiteracy	Agriculture	Land-holding peasant			
31	Inhabited for generations	Illiteracy	Agriculture	Land-holding peasant	Married	No	No religion
3	Inhabited for generations	Illiteracy	Agriculture	Land-holding peasant	Married	No	No religion
19	Inhabited for generations	Illiteracy	Agriculture	Land-holding peasant	Widowed	No	No religion
24	Inhabited for generations	Bao school 3 years					

Appendix: Overview of Microdata for Wudu County (*cont.*)

	The no. of house-hold	Jia	House-hold	Relation-ship with the house-hold head	Surname	First name	Sex	The Chinese zodiac	Birthdate	
									Year	Month
Fujin Xiang 5th Bao				The second son	赵	成足儿	Male	Dog	1946	6
Fujin Xiang 5th Bao				Younger brother	赵	岁才娃	Male	Rooster	1921	12
Fujin Xiang 5th Bao				The third younger brother	赵	高娃	Male	Ox	1925	8
Fujin Xiang 5th Bao				The fourth younger brother	赵	岁高娃	Male	Rabbit	1927	12
Fujin Xiang 5th Bao				Uncle's son	赵	文富	Male	Dragon	1891	7
Fujin Xiang 5th Bao				Uncle's son's wife	石	拓生	Female	Pig	1898	7
Fujin Xiang 5th Bao				Nephew	赵	三学娃	Male	Dragon	1928	6
Fujin Xiang 5th Bao	495	7	9	The house-hold head	赵	文讲	Male	Rabbit	1927	4
Fujin Xiang 5th Bao				The house-hold head's wife	成	岁廷	Female	Rabbit	1927	12
Fujin Xiang 5th Bao				Father	赵	金朝	Male	Horse	1881	9
Fujin Xiang 5th Bao				Mother	杨	寿银	Female	Tiger	1889	1
Fujin Xiang 5th Bao				Nephew	赵	玉春	Male	Rat	1936	7
Fujin Xiang 5th Bao	496	7	10	The house-hold head	赵	文发	Male	Dog	1923	1
Fujin Xiang 5th Bao				The house-hold head's wife	石	梅花	Female	Snake	1917	10
Fujin Xiang 5th Bao				Younger brother	赵	住儿	Male	Dog	1931	10
Fujin Xiang 5th Bao	497	7	11	The house-hold head	赵	文元	Male	Dog	1904	10
Fujin Xiang 5th Bao				The house-hold head's wife	崔	住秀	Female	Dog	1904	10

Date	The beginning time of inhabitation in this county	Educational level	Profession sector	Occupation	Marital status	Physical disability	Religion	
2	Inhabited for generations							
4	Inhabited for generations	Illiteracy	Agriculture	Land-holding peasant	Unmarried	No	No religion	
10	Inhabited for generations	Illiteracy	Agriculture	Land-holding peasant	Unmarried	No	No religion	Joined the army in 1943
31	Inhabited for generations	Illiteracy	Agriculture	Land-holding peasant	Unmarried	No	No religion	
30	Inhabited for generations	Illiteracy	Agriculture	Land-holding peasant	Married	No	No religion	
8	Inhabited for generations	Illiteracy	Agriculture	Land-holding peasant	Married	No	No religion	
15	Inhabited for generations	Illiteracy	Agriculture	Land-holding peasant	Unmarried	No	No religion	
21	Inhabited for generations	Illiteracy	Agriculture	Land-holding peasant	Married	No	No religion	
5	Inhabited for generations	Illiteracy	Agriculture	Land-holding peasant	Married	No	No religion	
10	Inhabited for generations	Illiteracy	Agriculture	Land-holding peasant	Married	No	No religion	
25	Inhabited for generations	Illiteracy	Agriculture	Land-holding peasant	Married	No	No religion	
12	Inhabited for generations	Bao school 3 years						
1	Inhabited for generations	Illiteracy	Agriculture	Land-holding peasant	Married	No	No religion	
28	Inhabited for generations	Illiteracy	Agriculture	Land-holding peasant	Married	No	No religion	
16	Inhabited for generations							
13	Inhabited for generations	Illiteracy	Agriculture	Land-holding peasant	Married	No	No religion	
13	Inhabited for generations	Illiteracy	Agriculture	Land-holding peasant	Married	No	No religion	

Appendix: Overview of Microdata for Wudu County (*cont.*)

	The no. of household	Jia	Household	Relationship with the household head	Surname	First name	Sex	The Chinese zodiac	Birthdate	
									Year	Month
Fujin Xiang 5th Bao				Son	谢	才娃	Male	Monkey	1932	12
Fujin Xiang 5th Bao				The second son	赵	长才娃	Male	Pig	1936	1
Fujin Xiang 5th Bao				The third son	赵	成才	Male	Pig	1947	5
Fujin Xiang 5th Bao				Daughter	赵	棌女	Female	Dragon	1940	11
Fujin Xiang 5th Bao				The second daughter	赵	名女	Female	Goat	1943	6
Fujin Xiang 5th Bao	498	7	12	The household head	赵	玉林	Male	Horse	1905	12
Fujin Xiang 5th Bao				The household head's wife	石	贤哇	Female	Dog	1909	5
Fujin Xiang 5th Bao				Father	赵	文建	Male	Rabbit	1878	7
Fujin Xiang 5th Bao				Son	赵	杨生	Male	Rabbit	1939	11
Fujin Xiang 5th Bao				Uncle	赵	文益	Male	Rabbit	1902	7
Fujin Xiang 5th Bao				Younger brother's wife	石	桂香	Female	Dog	1922	11
Fujin Xiang 5th Bao				Nephew	赵	款娃	Male	Dragon	1940	12
Fujin Xiang 5th Bao	499	7	13	The household head	赵	崴马娃	Male	Tiger	1926	1
Fujin Xiang 5th Bao				The household head's wife	石	五柱娃	Female	Tiger	1927	2
Fujin Xiang 5th Bao				Mother	陈	全姐	Female	Dragon	1891	11
Fujin Xiang 5th Bao				Grandfather's brother	赵	金泰	Male	Rooster	1884	12
Fujin Xiang 5th Bao				Uncle	赵	文芳	Male	Rooster	1908	5

Date	The beginning time of inhabitation in this county	Educational level	Profession sector	Occupation	Marital status	Physical disability	Religion
9	Inhabited for generations	Illiteracy	Agriculture	Land-holding peasant			
5	Inhabited for generations	Bao school 2 years					
14	Inhabited for generations						
24	Inhabited for generations	Illiteracy					
23	Inhabited for generations						
18	Inhabited for generations	Illiteracy	Agriculture	Land-holding peasant	Married	No	No religion
20	Inhabited for generations	Illiteracy	Agriculture	Land-holding peasant	Married	No	No religion
1	Inhabited for generations	Illiteracy	Agriculture	Land-holding peasant	Widowed	No	No religion
28	Inhabited for generations	Illiteracy					
27	Inhabited for generations	Illiteracy	Agriculture	Land-holding peasant	Widowed	No	No religion
3	Inhabited for generations	Illiteracy	Agriculture	Land-holding peasant	Widowed	No	No religion
16	Inhabited for generations	Illiteracy					
23	Inhabited for generations	Illiteracy	Agriculture	Land-holding peasant	Married	No	No religion
26	Inhabited for generations	Illiteracy	Agriculture	Land-holding peasant	Married	No	No religion
19	Inhabited for generations	Illiteracy	Agriculture	Land-holding peasant	Widowed	No	No religion
25	Inhabited for generations	Illiteracy	Agriculture	Land-holding peasant	Widowed	No	No religion
30	Inhabited for generations	Illiteracy	Agriculture	Land-holding peasant	Married	No	No religion

Appendix: Overview of Microdata for Wudu County (*cont.*)

	The no. of household	Jia	Household	Relationship with the household head	Surname	First name	Sex	The Chinese zodiac	Birthdate	
									Year	Month
Fujin Xiang 5th Bao				Uncle's wife	黄	扮秀	Female	Snake	1917	11
Fujin Xiang 5th Bao	500	8	1	The household head	石	会兴	Male	Rooster	1908	1
Fujin Xiang 5th Bao				The household head's wife	赵	廷哇	Female	Dragon	1916	9
Fujin Xiang 5th Bao				Mother	石	称儿	Female	Monkey	1883	5
Fujin Xiang 5th Bao				Son	石	春牛儿	Male	Goat	1943	12
Fujin Xiang 5th Bao				The second son	石	长娃	Male	Pig	1947	2
Fujin Xiang 5th Bao	501	8	2	The household head	石	含忠	Male	Horse	1905	5
Fujin Xiang 5th Bao				The household head's wife	崔	称儿	Female	Pig	1898	5
Fujin Xiang 5th Bao				Son	石	埋娃	Male	Ox	1937	4
Fujin Xiang 5th Bao				The second son	石	进学	Male	Rabbit	1939	6
Fujin Xiang 5th Bao				Uncle's son	石	含银	Male	Rat	1899	8
Fujin Xiang 5th Bao				Uncle's son's wife	赵	岁耀	Female	Snake	1892	4
Fujin Xiang 5th Bao				Nephew	石	六娃	Male	Rabbit	1927	6
Fujin Xiang 5th Bao				Nephew's wife	赵	岁菊	Female	Rabbit	1927	4
Fujin Xiang 5th Bao	502	8	3	The household head	石	耀生	Male	Rooster	1908	10
Fujin Xiang 5th Bao				The household head's wife	石	肖娃	Female	Goat	1906	7
Fujin Xiang 5th Bao				Mother	赵	氏	Female	Snake	1880	5
Fujin Xiang 5th Bao	503	8	4	The household head	石	含生	Male	Ox	1913	7

Date	The beginning time of inhabitation in this county	Educational level	Profession sector	Occupation	Marital status	Physical disability	Religion
20	Inhabited for generations	Illiteracy	Agriculture	Land-holding peasant	Married	No	No religion
29	Inhabited for generations	Illiteracy	Agriculture	Land-holding peasant	Married	No	No religion
12	Inhabited for generations	Illiteracy	Agriculture	Land-holding peasant	Married	No	No religion
1	Inhabited for generations	Illiteracy	Agriculture	Land-holding peasant	Widowed	No	No religion
9	Inhabited for generations						
17	Inhabited for generations						
25	Inhabited for generations	Illiteracy	Agriculture	Land-holding peasant	Married	No	No religion
23	Inhabited for generations	Illiteracy	Agriculture	Land-holding peasant	Married	No	No religion
1	Inhabited for generations	Illiteracy					
21	Inhabited for generations	Illiteracy					
25	Inhabited for generations	Illiteracy	Agriculture	Land-holding peasant	Married	No	No religion
1	Inhabited for generations	Illiteracy	Agriculture	Land-holding peasant	Married	No	No religion
14	Inhabited for generations	Illiteracy	Agriculture	Land-holding peasant	Married	No	No religion
14	Inhabited for generations	Illiteracy	Agriculture	Land-holding peasant	Married	No	No religion
26	Inhabited for generations	Illiteracy	Agriculture	Land-holding peasant	Married	No	No religion
22	Inhabited for generations	Illiteracy	Agriculture	Land-holding peasant	Married	No	No religion
8	Inhabited for generations	Illiteracy	Agriculture	Land-holding peasant	Widowed	No	No religion
4	Inhabited for generations	Illiteracy	Agriculture	Land-holding peasant	Married	No	No religion

Appendix: Overview of Microdata for Wudu County (*cont.*)

	The no. of household	Jia	Household	Relationship with the household head	Surname	First name	Sex	The Chinese zodiac	Birthdate Year	Month
Fujin Xiang 5th Bao				The household head's wife	李	信娃	Female	Dragon	1903	8
Fujin Xiang 5th Bao				Younger brother	石	仁才儿	Male	Tiger	1914	9
Fujin Xiang 5th Bao				Uncle's son's wife	石	节儿	Female	Dog	1897	6
Fujin Xiang 5th Bao				Nephew	石	柱生儿	Male	Ox	1913	6
Fujin Xiang 5th Bao				Nephew's wife	石	宣娃	Female	Dog	1919	3
Fujin Xiang 5th Bao				Nephew's son	石	高家	Male	Dog	1944	4
Fujin Xiang 5th Bao				Uncle's second son	石	会进	Male	Rabbit	1927	12
Fujin Xiang 5th Bao	504	8	5	The household head	石	含林	Male	Horse	1905	10
Fujin Xiang 5th Bao				The household head's wife	李	菊哇	Female	Ox	1913	7
Fujin Xiang 5th Bao				Father	石	文赵	Male	Monkey	1883	12
Fujin Xiang 5th Bao				Mother	石	赵娃	Female	Dragon	1880	5
Fujin Xiang 5th Bao				Son	石	会福	Male	Ox	1937	6
Fujin Xiang 5th Bao				The second son	石	鹤娃	Male	Horse	1942	10
Fujin Xiang 5th Bao				Daughter	五	女子	Female	Monkey	1942	7
Fujin Xiang 5th Bao	505	8	6	The household head	石	含均	Male	Dog	1885	6
Fujin Xiang 5th Bao				Younger brother	石	进娃	Male	Horse	1918	6
Fujin Xiang 5th Bao				Younger brother's wife	赵	早哇	Female	Dog	1922	9
Fujin Xiang 5th Bao				Niece	石	路哇	Female	Dog	1946	4

Date	The beginning time of inhabitation in this county	Educational level	Profession sector	Occupation	Marital status	Physical disability	Religion
18	Inhabited for generations	Illiteracy	Agriculture	Land-holding peasant	Married	No	No religion
17	Inhabited for generations	Illiteracy	Agriculture	Land-holding peasant	Widowed	No	No religion
5	Inhabited for generations	Illiteracy	Agriculture	Land-holding peasant	Widowed	No	No religion
8	Inhabited for generations	Illiteracy	Agriculture	Land-holding peasant	Married	No	No religion
14	Inhabited for generations	Illiteracy	Agriculture	Land-holding peasant	Married	No	No religion
14	Inhabited for generations						
17	Inhabited for generations	Sishu 2 years	Agriculture	Land-holding peasant	Widowed	No	No religion
5	Inhabited for generations	Sishu 3 years	Agriculture	Land-holding peasant	Married	No	No religion
8	Inhabited for generations	Illiteracy	Agriculture	Land-holding peasant	Married	No	No religion
1	Inhabited for generations	Illiteracy	Agriculture	Land-holding peasant	Married	No	No religion
8	Inhabited for generations	Illiteracy	Agriculture	Land-holding peasant	Married	No	No religion
21	Inhabited for generations	Bao school 4 years					
14	Inhabited for generations						
28	Inhabited for generations						
14	Inhabited for generations	Illiteracy	Agriculture	Land-holding peasant	Widowed	No	No religion
14	Inhabited for generations	Illiteracy	Agriculture	Land-holding peasant	Married	No	No religion
4	Inhabited for generations	Illiteracy	Agriculture	Land-holding peasant	Married	No	No religion
16	Inhabited for generations						

Appendix: Overview of Microdata for Wudu County (*cont.*)

	The no. of household	Jia	Household	Relationship with the household head	Surname	First name	Sex	The Chinese zodiac	Birthdate	
									Year	Month
Fujin Xiang 5th Bao	506	8	7	The household head	石	王儿	Male	Dog	1922	11
Fujin Xiang 5th Bao				Father	石	含仁	Male	Horse	1881	11
Fujin Xiang 5th Bao				Mother	石	翠哇	Female	Dog	1885	7
Fujin Xiang 5th Bao	507	8	8	The household head	石	进朝儿	Male	Dragon	1903	10
Fujin Xiang 5th Bao				Son	石	正娃	Male	Tiger	1926	8
Fujin Xiang 5th Bao				The second son	石	均有	Male	Pig	1935	10
Fujin Xiang 5th Bao				Grandfather's brother	石	文希	Male	Ox	1913	6
Fujin Xiang 5th Bao				Grandfather's second brother	石	文昇	Male	Horse	1928	5
Fujin Xiang 5th Bao	508	8	9	The household head	石	含敞	Male	Monkey	1907	10
Fujin Xiang 5th Bao				The household head's wife	王	春娃	Female	Ox	1913	7
Fujin Xiang 5th Bao				Mother	石	岁桂	Female	Rabbit	1878	8
Fujin Xiang 5th Bao				Son	陈	学儿	Male	Dragon	1928	10
Fujin Xiang 5th Bao				The second son	路	顺儿	Male	Dog	1947	1
Fujin Xiang 5th Bao				Older brother	石	含有	Male	Dog	1899	3
Fujin Xiang 5th Bao				Older brother's wife	赵	女子	Female	Horse	1905	9
Fujin Xiang 5th Bao				Nephew	刘	生儿	Male	Horse	1930	11
Fujin Xiang 5th Bao				Uncle	石	文双	Male	Dragon	1892	1

Date	The beginning time of inhabitation in this county	Educational level	Profession sector	Occupation	Marital status	Physical disability	Religion	
3	Inhabited for generations	Illiteracy	Agriculture	Land-holding peasant	Unmarried	No	No religion	
3	Inhabited for generations	Illiteracy	Agriculture	Land-holding peasant	Married	No	No religion	
29	Inhabited for generations	Illiteracy	Agriculture	Land-holding peasant	Married	Blind	No religion	
13	Inhabited for generations	Illiteracy	Agriculture	Land-holding peasant	Widowed	No	No religion	
22	Inhabited for generations	Illiteracy	Agriculture	Land-holding peasant	Unmarried	No	No religion	Joined the army in 1947
25	Inhabited for generations	Illiteracy	Agriculture	Land-holding peasant			No religion	
22	Inhabited for generations	Illiteracy	Agriculture	Land-holding peasant	Unmarried	Leg injured	No religion	
29	Inhabited for generations	Illiteracy	Agriculture	Land-holding peasant	Unmarried	Only one hand	No religion	
5	Inhabited for generations	Illiteracy	Agriculture	Land-holding peasant	Married	No	No religion	
8	Inhabited for generations	Illiteracy	Agriculture	Land-holding peasant	Married	No	No religion	
10	Inhabited for generations							
25	Inhabited for generations	Illiteracy	Agriculture	Land-holding peasant	Unmarried	No	No religion	
11	Inhabited for generations							
26	Inhabited for generations	Illiteracy	Agriculture	Land-holding peasant	Married	No	No religion	
13	Inhabited for generations	Illiteracy	Agriculture	Land-holding peasant	Married	No	No religion	
18	Inhabited for generations	Illiteracy	Agriculture	Land-holding peasant	Unmarried	No	No religion	
22	Inhabited for generations	Illiteracy	Agriculture	Land-holding peasant	Married	No	No religion	

Appendix: Overview of Microdata for Wudu County (*cont.*)

	The no. of household	Jia	Household	Relationship with the household head	Surname	First name	Sex	The Chinese zodiac	Birthdate Year	Month
Fujin Xiang 5th Bao				Uncle's wife	陈	秀儿	Female	Snake	1892	2
Fujin Xiang 5th Bao				Uncle's son	石	顺福儿	Male	Ox	1924	8
Fujin Xiang 5th Bao				Uncle's daughter	进	女子	Female	Goat	1931	5
Fujin Xiang 5th Bao	509	8	10	The household head	石	含才	Male	Horse	1918	12
Fujin Xiang 5th Bao				The household head's wife	王	银连	Female	Ox	1913	8
Fujin Xiang 5th Bao				Father	石	文高	Male	Monkey	1883	8
Fujin Xiang 5th Bao				Mother	杨	裉娃	Female	Rooster	1884	12
Fujin Xiang 5th Bao				Son	石	田娃子	Male	Rooster	1940	3
Fujin Xiang 5th Bao				Daughter	石	田女	Female	Monkey	1944	11
Fujin Xiang 5th Bao				Uncle	石	文富	Male	Dragon	1903	5
Fujin Xiang 5th Bao				Uncle's wife	赵	兹女	Female	Dragon	1903	7
Fujin Xiang 5th Bao				Uncle's son	石	正娃	Male	Ox	1937	8
Fujin Xiang 5th Bao				Uncle's second son	王	足儿	Male	Horse	1942	10
Fujin Xiang 5th Bao				Uncle's daughter	石	座女	Female	Dog	1946	4
Fujin Xiang 5th Bao	510	8	11	The household head	石	文元	Male	Snake	1892	9
Fujin Xiang 5th Bao				The household head's wife	张	崴东	Female	Dog	1909	10
Fujin Xiang 5th Bao				Son	石	房儿	Male	Monkey	1932	9
Fujin Xiang 5th Bao				Son's wife	赵	廷娃	Female	Snake	1929	3

Date	The beginning time of inhabitation in this county	Educational level	Profession sector	Occupation	Marital status	Physical disability	Religion	
17	Inhabited for generations	Illiteracy	Agriculture	Land-holding peasant	Married	No	No religion	
6	Inhabited for generations	Illiteracy	Agriculture	Land-holding peasant	Unmarried	No	No religion	
17	Inhabited for generations	Illiteracy	Agriculture					
5	Inhabited for generations	Illiteracy	Agriculture	Land-holding peasant	Married	No	No religion	
13	Inhabited for generations	Illiteracy	Agriculture	Land-holding peasant	Married	No	No religion	
1	Inhabited for generations	Illiteracy	Agriculture	Land-holding peasant	Married	No	No religion	
14	Inhabited for generations	Illiteracy	Agriculture	Land-holding peasant	Married	No	No religion	
19	Inhabited for generations	Illiteracy						
16	Inhabited for generations							
26	Inhabited for generations	Illiteracy	Agriculture	Land-holding peasant	Married	No	No religion	Dead
18	Inhabited for generations	Illiteracy	Agriculture	Land-holding peasant	Married	No	No religion	
17	Inhabited for generations	Illiteracy						
10	Inhabited for generations							
8	Inhabited for generations							
10	Inhabited for generations	Illiteracy	Agriculture	Land-holding peasant	Married	No	No religion	
12	Inhabited for generations	Illiteracy	Agriculture	Land-holding peasant	Married	No	No religion	
9	Inhabited for generations	Illiteracy	Agriculture	Land-holding peasant	Married	No	No religion	
20	Inhabited for generations	Illiteracy	Agriculture	Land-holding peasant	Married	No	No religion	

Appendix: Overview of Microdata for Wudu County (*cont.*)

	The no. of household	Jia	Household	Relationship with the household head	Surname	First name	Sex	The Chinese zodiac	Birthdate Year	Birthdate Month
Fujin Xiang 5th Bao				The second son	石	进学儿	Male	Pig	1935	9
Fujin Xiang 5th Bao				Daughter	赵	柏柱儿	Female	Rabbit	1939	8
Fujin Xiang 5th Bao				The second daughter	石	四柱儿	Female	Goat	1944	1
Fujin Xiang 5th Bao				Uncle	石	玉虎	Male	Pig	1874	8
Fujin Xiang 5th Bao	511	8	12	The household head	石	含贵	Male	Dragon	1891	4
Fujin Xiang 5th Bao				The household head's wife	杨	秀儿	Female	Dragon	1891	9
Fujin Xiang 5th Bao				Son	石	学儿	Male	Dragon	1919	8
Fujin Xiang 5th Bao				Son's wife	李	岁梅	Female	Ox	1925	1
Fujin Xiang 5th Bao				The second son	石	相娃	Male	Ox	1925	8
Fujin Xiang 5th Bao				The third son	石	会文	Male	Pig	1935	4
Fujin Xiang 5th Bao	512	8	13	The household head	石	含虎	Male	Tiger	1889	1
Fujin Xiang 5th Bao				The household head's wife	张	菜儿	Female	Tiger	1889	2
Fujin Xiang 5th Bao				Son	石	很娃	Male	Rat	1924	6
Fujin Xiang 5th Bao				Son's wife	崔	早女	Female	Tiger	1926	4
Fujin Xiang 5th Bao				Younger brother	石	含双	Male	Rooster	1896	11
Fujin Xiang 5th Bao				Nephew	石	桃娃	Male	Pig	1935	3
Fujin Xiang 5th Bao	513	8	14	The household head	石	含傲	Male	Goat	1894	2
Fujin Xiang 5th Bao				The household head's wife	赵	怌儿	Female	Goat	1894	12
Fujin Xiang 5th Bao				Son	石	双生娃	Male	Dragon	1928	6

Date	The beginning time of inhabitation in this county	Educational level	Profession sector	Occupation	Marital status	Physical disability	Religion
30	Inhabited for generations	Bao school 3 years					
29	Inhabited for generations	Illiteracy					
7	Inhabited for generations						
1	Inhabited for generations						
27	Inhabited for generations	Illiteracy	Agriculture	Land-holding peasant	Married	No	No religion
21	Inhabited for generations	Illiteracy	Agriculture	Land-holding peasant	Married	No	No religion
6	Inhabited for generations	Illiteracy	Agriculture	Land-holding peasant	Married	No	No religion
25	Inhabited for generations	Illiteracy	Agriculture	Land-holding peasant	Married	No	No religion
30	Inhabited for generations	Illiteracy	Agriculture	Land-holding peasant	Unmarried	No	No religion
23	Inhabited for generations	Bao school 4 years					
21	Inhabited for generations	Illiteracy	Agriculture	Land-holding peasant	Married	No	No religion
19	Inhabited for generations	Illiteracy	Agriculture	Land-holding peasant	Married	No	No religion
2	Inhabited for generations	Illiteracy	Agriculture	Land-holding peasant	Married	No	No religion
12	Inhabited for generations	Illiteracy	Agriculture	Land-holding peasant	Married	No	No religion
26	Inhabited for generations	Illiteracy	Agriculture	Land-holding peasant	Widowed	No	No religion
12	Inhabited for generations	Illiteracy	Agriculture	Land-holding peasant	Unmarried	No	No religion
25	Inhabited for generations	Illiteracy	Agriculture	Land-holding peasant	Married	No	No religion
1	Inhabited for generations	Illiteracy	Agriculture	Land-holding peasant	Married	No	No religion
29	Inhabited for generations	Illiteracy	Agriculture	Land-holding peasant	Unmarried	No	No religion

Appendix: Overview of Microdata for Wudu County (*cont.*)

	The no. of household	Jia	Household	Relationship with the household head	Surname	First name	Sex	The Chinese zodiac	Birthdate	
									Year	Month
Fujin Xiang 5th Bao				The second son	石	慈生哇	Male	Monkey	1932	5
Fujin Xiang 5th Bao				Daughter	石	柱女子	Female	Rabbit	1931	11
Fujin Xiang 5th Bao				The second daughter	石	岁柱儿	Female	Goat	1943	4
Fujin Xiang 5th Bao				Grandfather's brother's wife	谈	四香	Female	Rat	1875	9
Fujin Xiang 5th Bao				Uncle	石	有得	Male	Rooster	1884	7
Fujin Xiang 5th Bao				Uncle's son	石	讲娃	Male	Tiger	1926	2
Fujin Xiang 5th Bao	514	9	1	The household head	石	宝唤	Male	Ox	1900	8
Fujin Xiang 5th Bao				The household head's wife	石	频娃	Female	Ox	1913	9
Fujin Xiang 5th Bao				Father	石	文益	Male	Rooster	1884	9
Fujin Xiang 5th Bao				Mother	马	那玉	Female	Monkey	1896	1
Fujin Xiang 5th Bao				Son	石	陈学子	Male	Ox	1937	7
Fujin Xiang 5th Bao				The second son	石	学子	Male	Dragon	1940	12
Fujin Xiang 5th Bao				The third son	石	三学子	Male	Goat	1943	5
Fujin Xiang 5th Bao				Daughter	石	竹叶子	Female	Rooster	1933	12
Fujin Xiang 5th Bao	515	9	2	The household head	石	会成	Male	Dog	1909	4
Fujin Xiang 5th Bao				The household head's wife	李	林娃	Female	Goat	1906	11
Fujin Xiang 5th Bao				Father	石	含秀	Male	Goat	1882	3
Fujin Xiang 5th Bao				Mother	谈	董儿	Female	Ox	1888	12

Date	The beginning time of inhabitation in this county	Educational level	Profession sector	Occupation	Marital status	Physical disability	Religion
13	Inhabited for generations	Bao school 4 years	Agriculture	Land-holding peasant	Unmarried	No	No religion
23	Inhabited for generations	Illiteracy					
4	Inhabited for generations						
9	Inhabited for generations						
21	Inhabited for generations	Illiteracy	Agriculture	Land-holding peasant	Widowed	No	No religion
13	Inhabited for generations	Illiteracy	Agriculture	Land-holding peasant	Unmarried	No	No religion
28	Inhabited for generations	Illiteracy	Agriculture	Land-holding peasant	Married	No	No religion
30	Inhabited for generations	Illiteracy	Agriculture	Land-holding peasant	Married	No	No religion
10	Inhabited for generations	Illiteracy	Agriculture	Land-holding peasant	Married	No	No religion
3	Inhabited for generations	Illiteracy	Agriculture	Land-holding peasant	Married	No	No religion
19	Inhabited for generations	Bao school 2 years					
31	Inhabited for generations						
14	Inhabited for generations						
28	Inhabited for generations	Illiteracy	Agriculture	Land-holding peasant			
12	Inhabited for generations	Illiteracy	Agriculture	Land-holding peasant	Married	No	No religion
17	Inhabited for generations	Illiteracy	Agriculture	Land-holding peasant	Married	No	No religion
3	Inhabited for generations	Illiteracy	Agriculture	Land-holding peasant	Married	No	No religion
21	Inhabited for generations	Illiteracy	Agriculture	Land-holding peasant	Married	No	No religion

Appendix: Overview of Microdata for Wudu County (*cont.*)

									Birthdate	
	The no. of house-hold	Jia	House-hold	Relation-ship with the house-hold head	Surname	First name	Sex	The Chinese zodiac	Year	Month
Fujin Xiang 5th Bao				Son	石	两娃	Male	Rooster	1945	12
Fujin Xiang 5th Bao				Daughter	石	月女	Female	Horse	1942	4
Fujin Xiang 5th Bao				Younger brother	石	柱娃	Male	Tiger	1926	3
Fujin Xiang 5th Bao				Grand-father's brother's wife	成	根富	Female	Rat	1875	3
Fujin Xiang 5th Bao	516	9	3	The house-hold head	石	含云	Male	Tiger	1914	6
Fujin Xiang 5th Bao				The house-hold head's wife	四	女子	Female	Rat	1913	4
Fujin Xiang 5th Bao				Father	石	文崔	Male	Snake	1880	4
Fujin Xiang 5th Bao				Mother	崔	有姐	Female	Horse	1881	5
Fujin Xiang 5th Bao				Son	石	进旦	Male	Rat	1936	5
Fujin Xiang 5th Bao				The second son	石	代哇	Male	Snake	1941	3
Fujin Xiang 5th Bao				The third son	石	岁代哇	Male	Goat	1943	9
Fujin Xiang 5th Bao				Daughter	石	旦旦子	Female	Rooster	1933	5
Fujin Xiang 5th Bao				The second daughter	石	琴琴	Female	Dragon	1940	10
Fujin Xiang 5th Bao				Older brother	石	含明	Male	Goat	1906	9
Fujin Xiang 5th Bao				Older brother's wife	黄	篩娃	Female	Monkey	1920	3
Fujin Xiang 5th Bao				Nephew	佛	儿子	Male	Rabbit	1939	7
Fujin Xiang 5th Bao	517	9	4	The house-hold head	石	会斌	Male	Goat	1919	7
Fujin Xiang 5th Bao				The house-hold head's wife	黄	⍰娃	Female	Horse	1918	12

Date	The beginning time of inhabitation in this county	Educational level	Profession sector	Occupation	Marital status	Physical disability	Religion
5	Inhabited for generations						
17	Inhabited for generations						
15	Inhabited for generations	Illiteracy	Agriculture	Land-holding peasant	Unmarried	No	No religion
2	Inhabited for generations						
6	Inhabited for generations	Illiteracy	Agriculture	Land-holding peasant	Married	No	No religion
19	Inhabited for generations	Illiteracy	Agriculture	Land-holding peasant	Married	No	No religion
1	Inhabited for generations	Illiteracy	Agriculture	Land-holding peasant	Married	No	No religion
1	Inhabited for generations	Illiteracy	Agriculture	Land-holding peasant	Married	No	No religion
21	Inhabited for generations	Bao school 3 years					
5	Inhabited for generations						
11	Inhabited for generations						
6	Inhabited for generations	Illiteracy	Agriculture	Land-holding peasant			
1	Inhabited for generations	Illiteracy					
19	Inhabited for generations	Illiteracy	Agriculture	Land-holding peasant	Married	No	No religion
22	Inhabited for generations	Illiteracy	Agriculture	Land-holding peasant	Married	No	No religion
6	Inhabited for generations	Illiteracy					
2	Inhabited for generations	Illiteracy	Agriculture	Land-holding peasant	Married	No	No religion
3	Inhabited for generations	Illiteracy	Agriculture	Land-holding peasant	Married	No	No religion

Appendix: Overview of Microdata for Wudu County (*cont.*)

	The no. of household	Jia	Household	Relationship with the household head	Surname	First name	Sex	The Chinese zodiac	Birthdate Year	Month
Fujin Xiang 5th Bao				Father	石	含满	Male	Tiger	1889	4
Fujin Xiang 5th Bao				Son	石	根心	Male	Ox	1937	7
Fujin Xiang 5th Bao				Daughter	石	根未	Female	Rabbit	1939	9
Fujin Xiang 5th Bao				The second daughter	石	益娃	Female	Rooster	1945	5
Fujin Xiang 5th Bao	518	9	5	The household head	石	含宝	Male	Tiger	1914	6
Fujin Xiang 5th Bao				The household head's wife	马	拜娃	Female	Rooster	1908	10
Fujin Xiang 5th Bao				Mother	罗	赵娃	Female	Ox	1876	11
Fujin Xiang 5th Bao				Son	石	顺緒	Male	Rooster	1933	9
Fujin Xiang 5th Bao				The second son	石	三合乡	Male	Goat	1943	6
Fujin Xiang 5th Bao				Daughter	石	金緒	Female	Rabbit	1939	12
Fujin Xiang 5th Bao				Older brother	石	含山	Male	Rooster	1926	4
Fujin Xiang 5th Bao				Uncle's son	石	含才	Male	Monkey	1920	5
Fujin Xiang 5th Bao				Younger brother's wife	石	女子	Female	Goat	1919	2
Fujin Xiang 5th Bao				Niece	石	兵娃	Female	Snake	1941	12
Fujin Xiang 5th Bao				The second niece	石	谢谢	Female	Dog	1946	2
Fujin Xiang 5th Bao	519	9	6	The household head	石	会洪	Male	Rooster	1921	6
Fujin Xiang 5th Bao				Father	石	含虎	Male	Rooster	1899	6
Fujin Xiang 5th Bao				Mother	赵	苏儿	Female	Tiger	1901	6
Fujin Xiang 5th Bao				Younger brother	石	会彦	Male	Ox	1927	10

Date	The beginning time of inhabitation in this county	Educational level	Profession sector	Occupation	Marital status	Physical disability	Religion
19	Inhabited for generations	Illiteracy	Agriculture	Land-holding peasant	Widowed	No	No religion
18	Inhabited for generations	Bao school 3 years					
25	Inhabited for generations	Illiteracy					
11	Inhabited for generations						
3	Inhabited for generations	Illiteracy	Agriculture	Land-holding peasant	Married	No	No religion
22	Inhabited for generations	Illiteracy	Agriculture	Land-holding peasant	Married	No	No religion
1	Inhabited for generations						
20	Inhabited for generations	Illiteracy	Agriculture	Land-holding peasant			
13	Inhabited for generations						
11	Inhabited for generations	able to read					
8	Inhabited for generations	Illiteracy	Agriculture	Land-holding peasant	Widowed	No	No religion
22	Inhabited for generations	Illiteracy	Agriculture	Land-holding peasant	Married	No	No religion
12	Inhabited for generations	Illiteracy	Agriculture	Land-holding peasant	Married	No	No religion
18	Inhabited for generations	Illiteracy					
4	Inhabited for generations						
17	Inhabited for generations	Illiteracy	Agriculture	Land-holding peasant	Unmarried	No	No religion
17	Inhabited for generations	Illiteracy	Agriculture	Land-holding peasant	Married	Deaf	No religion
17	Inhabited for generations	Illiteracy	Agriculture	Land-holding peasant	Married	No	No religion
4	Inhabited for generations	Bao school 2 years					

Appendix: Overview of Microdata for Wudu County (*cont.*)

	The no. of household	Jia	Household	Relationship with the household head	Surname	First name	Sex	The Chinese zodiac	Birthdate	
									Year	Month
Fujin Xiang 5th Bao				The third younger brother	石	谢生	Male	Rabbit	1939	12
Fujin Xiang 5th Bao	520	9	7	The household head	石	含富	Male	Dragon	1903	7
Fujin Xiang 5th Bao				The household head's wife	谈	能娃	Female	Horse	1905	10
Fujin Xiang 5th Bao				Father	石	仲虎	Male	Dragon	1879	1
Fujin Xiang 5th Bao				Mother	谈	氏	Female	Horse	1881	12
Fujin Xiang 5th Bao				Son	石	会进	Male	Rooster	1933	11
Fujin Xiang 5th Bao				Daughter	石	进女子	Female	Tiger	1938	7
Fujin Xiang 5th Bao				Family member	石	成送年儿	Female	Tiger	1901	7
Fujin Xiang 5th Bao	521	9	8	The household head	石	会科	Male	Pig	1910	3
Fujin Xiang 5th Bao				The household head's wife	黄	知英	Female	Dragon	1928	4
Fujin Xiang 5th Bao				Father	石	舍珍	Male	Dog	1886	9
Fujin Xiang 5th Bao				Mother	谈	岁琴	Female	Tiger	1889	8
Fujin Xiang 5th Bao				Daughter	石	德女	Female	Rooster	1945	12
Fujin Xiang 5th Bao				Younger brother	石	麻儿	Male	Monkey	1920	8
Fujin Xiang 5th Bao				Younger brother's wife	谈	进娃	Female	Monkey	1920	8
Fujin Xiang 5th Bao				The third younger brother	石	吾生儿	Male	Ox	1925	3
Fujin Xiang 5th Bao				The third younger brother's wife	赵	送家	Female	Ox	1924	4
Fujin Xiang 5th Bao				Nephew	黄	宝娃	Male	Pig	1947	1

Date	The beginning time of inhabitation in this county	Educational level	Profession sector	Occupation	Marital status	Physical disability	Religion
13	Inhabited for generations	Illiteracy					
15	Inhabited for generations	Illiteracy	Agricul-ture	Land-holding peasant	Married	No	No religion
6	Inhabited for generations	Illiteracy	Agricul-ture	Land-holding peasant	Married	Blind	No religion
1	Inhabited for generations	Illiteracy	Agricul-ture	Land-holding peasant	Married	No	No religion
5	Inhabited for generations	Illiteracy	Agricul-ture	Land-holding peasant	Married	No	No religion
28	Inhabited for generations	Bao school 3 years					
19	Inhabited for generations	Illiteracy					
24	Inhabited for generations	Illiteracy	Agricul-ture	Land-holding peasant	Married	No	No religion
11	Inhabited for generations	Illiteracy	Agricul-ture	Land-holding peasant	Married	No	No religion
29	Inhabited for generations	Illiteracy	Agricul-ture	Land-holding peasant	Married	No	No religion
18	Inhabited for generations	Illiteracy	Agricul-ture	Land-holding peasant	Married	No	No religion
15	Inhabited for generations	Illiteracy	Agricul-ture	Land-holding peasant	Married	No	No religion
11	Inhabited for generations						
9	Inhabited for generations	Illiteracy	Agricul-ture	Land-holding peasant	Married	No	No religion
9	Inhabited for generations	Illiteracy	Agricul-ture	Land-holding peasant	Married	No	No religion
10	Inhabited for generations	Illiteracy	Agricul-ture	Land-holding peasant	Married	No	No religion
13	Inhabited for generations	Illiteracy	Agricul-ture	Land-holding peasant	Married	No	No religion
22	Inhabited for generations						

Appendix: Overview of Microdata for Wudu County (*cont.*)

	The no. of household	Jia	Household	Relationship with the household head	Surname	First name	Sex	The Chinese zodiac	Birthdate	
									Year	Month
Fujin Xiang 5th Bao	522	9	9	The household head	石	含昌	Male	Monkey	1895	6
Fujin Xiang 5th Bao				The household head's wife	崴	琴娃	Female	Horse	1905	5
Fujin Xiang 5th Bao				Son	石	会印	Male	Dragon	1928	1
Fujin Xiang 5th Bao				Son's wife	成	提女	Female	Tiger	1926	2
Fujin Xiang 5th Bao				Grandson	石	德赐儿	Male	Rooster	1945	9
Fujin Xiang 5th Bao				Older brother	石	含发	Male	Tiger	1889	3
Fujin Xiang 5th Bao				Nephew	石	张娃	Male	Tiger	1926	3
Fujin Xiang 5th Bao				Nephew's wife	王	伯女	Female	Tiger	1926	4
Fujin Xiang 5th Bao	523	9	10	The household head	石	含龙	Male	Rooster	1884	10
Fujin Xiang 5th Bao				The household head's wife	石	苟姐	Female	Dragon	1879	5
Fujin Xiang 5th Bao				Son	石	会德	Male	Dragon	1903	3
Fujin Xiang 5th Bao				Son's wife	黄	扣娃	Female	Dog	1909	7
Fujin Xiang 5th Bao				Grandson	石	长才	Male	Snake	1941	3
Fujin Xiang 5th Bao				Granddaughter	石	富英	Female	Goat	1943	10
Fujin Xiang 5th Bao				The second granddaughter	石	英女	Female	Dog	1946	3
Fujin Xiang 5th Bao	524	9	11	The household head	石	银尚	Male	Horse	1905	1
Fujin Xiang 5th Bao				The household head's wife	傅	生	Female	Goat	1906	4

Date	The beginning time of inhabitation in this county	Educational level	Profession sector	Occupation	Marital status	Physical disability	Religion
12	Inhabited for generations	Sishu 3 years	Agriculture	Land-holding peasant	Married	No	No religion
5	Inhabited for generations	Illiteracy	Agriculture	Land-holding peasant	Married	No	No religion
30	Inhabited for generations	Illiteracy	Agriculture	Land-holding peasant	Married	No	No religion
13	Inhabited for generations	Illiteracy	Agriculture	Land-holding peasant	Married	No	No religion
8	Inhabited for generations						
29	Inhabited for generations	Illiteracy	Agriculture	Land-holding peasant	Widowed	No	No religion
14	Inhabited for generations	Illiteracy	Agriculture	Land-holding peasant	Married	No	No religion
13	Inhabited for generations	Illiteracy	Agriculture	Land-holding peasant	Married	No	No religion
8	Inhabited for generations	Illiteracy	Agriculture	Land-holding peasant	Married	No	No religion
1	Inhabited for generations	Illiteracy	Agriculture	Land-holding peasant	Married	No	No religion
27	Inhabited for generations	Illiteracy	Agriculture	Land-holding peasant	Married	No	No religion
26	Inhabited for generations	Illiteracy	Agriculture	Land-holding peasant	Married	No	No religion
20	Inhabited for generations	Illiteracy					
8	Inhabited for generations						
7	Inhabited for generations						
9	Inhabited for generations	Illiteracy	Agriculture	Land-holding peasant	Married	No	No religion
9	Inhabited for generations	Illiteracy	Agriculture	Land-holding peasant	Married	No	No religion

Appendix: Overview of Microdata for Wudu County (*cont.*)

	The no. of household	Jia	Household	Relationship with the household head	Surname	First name	Sex	The Chinese zodiac	Birthdate	
									Year	Month
Fujin Xiang 5th Bao				Son	石	张生	Male	Rooster	1933	9
Fujin Xiang 5th Bao				The second son	石	柏生	Male	Pig	1935	3
Fujin Xiang 5th Bao				The third son	石	三孕儿	Male	Dog	1946	12
Fujin Xiang 5th Bao				Daughter	石	四女子	Female	Goat	1943	9
Fujin Xiang 5th Bao				The second younger brother	石	含金	Male	Monkey	1907	11
Fujin Xiang 5th Bao				Younger brother's wife	松	明	Female	Goat	1919	7
Fujin Xiang 5th Bao				Niece	马	廷儿	Female	Horse	1942	8
Fujin Xiang 5th Bao				The second niece	石	菜女子	Female	Dog	1946	12
Fujin Xiang 5th Bao				The third younger brother	王	送儿	Male	Dog	19110	11
Fujin Xiang 5th Bao				The third younger brother's wife	石	女儿	Female	Goat	1919	4
Fujin Xiang 5th Bao				Nephew	刘	才	Male	Dog	1946	8
Fujin Xiang 5th Bao				Niece	副	清儿	Female	Dragon	1940	12
Fujin Xiang 5th Bao				The second niece	符	英	Female	Goat	1943	6
Fujin Xiang 5th Bao	525	9	12	The household head	石	银忠	Male	Rat	1899	2
Fujin Xiang 5th Bao				The household head's wife	石	送哇	Female	Rooster	1908	5
Fujin Xiang 5th Bao				Father	石	仲清	Male	Horse	1869	6

Date	The beginning time of inhabitation in this county	Educational level	Profession sector	Occupation	Marital status	Physical disability	Religion
1	Inhabited for generations	Illiteracy	Agriculture	Land-holding peasant			
28	Inhabited for generations	Bao school 3 years					
30	Inhabited for generations						
11	Inhabited for generations						
17	Inhabited for generations	Illiteracy	Agriculture	Land-holding peasant	Married	No	No religion
12	Inhabited for generations	Illiteracy	Agriculture	Land-holding peasant	Married	No	No religion
25	Inhabited for generations						
5	Inhabited for generations						
9	Inhabited for generations	Illiteracy	Agriculture	Land-holding peasant	Married	No	No religion
29	Inhabited for generations	Illiteracy	Agriculture	Land-holding peasant	Married	No	No religion
20	Inhabited for generations						
1	Inhabited for generations	Illiteracy					
3	Inhabited for generations						
22	Inhabited for generations	Illiteracy	Agriculture	Land-holding peasant	Married	No	No religion
21	Inhabited for generations	Illiteracy	Agriculture	Land-holding peasant	Married	No	No religion
23	Inhabited for generations						

Appendix: Overview of Microdata for Wudu County (*cont.*)

	The no. of household	Jia	Household	Relationship with the household head	Surname	First name	Sex	The Chinese zodiac	Birthdate	
									Year	Month
Fujin Xiang 5th Bao				Son	石	平女儿	Male	Tiger	1938	11
Fujin Xiang 5th Bao				The second son	石	进生儿	Male	Dragon	1940	7
Fujin Xiang 5th Bao				Daughter	石	冬青儿	Female	Horse	1942	12
Fujin Xiang 5th Bao				The second daughter	石	柱女子	Female	Rooster	1945	4
Fujin Xiang 5th Bao				The third daughter	石	小平女	Female	Pig	1947	4
Fujin Xiang 5th Bao	526	9	13	The household head	石	银发	Male	Dog	1880	5
Fujin Xiang 5th Bao				The household head's wife	石	贵儿	Female	Ox	1888	12
Fujin Xiang 5th Bao				Son	石	王才儿	Male	Horse	1930	12
Fujin Xiang 5th Bao				The second son	石	郭才	Male	Monkey	1932	4
Fujin Xiang 5th Bao	527	9	14	The household head	石	含东	Male	Monkey	1883	9
Fujin Xiang 5th Bao				Son	石	怔儿	Male	Rooster	1937	4
Fujin Xiang 5th Bao				Nephew	石	会拜	Male	Tiger	1901	11
Fujin Xiang 5th Bao				Nephew's wife	陈	兹哇	Female		1890	2
Fujin Xiang 5th Bao				Nephew's daughter	石	能生	Female	Rooster	1933	7
Fujin Xiang 5th Bao				Nephew's second daughter	石	招家	Female	Dragon	1940	6
Fujin Xiang 5th Bao				Nephew's third daughter	石	赵女子	Female	Rooster	1935	11
Fujin Xiang 5th Bao	528	9	15	The household head	石	德义	Male	Pig	1910	8

Date	The beginning time of inhabitation in this county	Educational level	Profession sector	Occupation	Marital status	Physical disability	Religion
5	Inhabited for generations	Bao school 4 years					
9	Inhabited for generations	Illiteracy					
25	Inhabited for generations						
17	Inhabited for generations						
23	Inhabited for generations						
5	Inhabited for generations	Illiteracy	Agriculture	Land-holding peasant	Married	No	No religion
14	Inhabited for generations	Illiteracy	Agriculture	Land-holding peasant	Married	No	No religion
4	Inhabited for generations	Illiteracy	Agriculture	Land-holding peasant	Unmarried	No	No religion
28	Inhabited for generations	Illiteracy	Agriculture	Land-holding peasant	Unmarried	No	No religion
26	Inhabited for generations	Illiteracy	Agriculture	Land-holding peasant	Widowed	No	No religion
6	Inhabited for generations	Illiteracy	Agriculture	Land-holding peasant			
9	Inhabited for generations	Illiteracy	Agriculture	Land-holding peasant	Married	No	No religion
21	Inhabited for generations	Illiteracy	Agriculture	Land-holding peasant	Married	No	No religion
24	Inhabited for generations	Illiteracy	Agriculture	Land-holding peasant			
6	Inhabited for generations	Illiteracy					
1	Inhabited for generations						
30	Inhabited for generations	Illiteracy	Agriculture	Land-holding peasant	Married	No	No religion

Appendix: Overview of Microdata for Wudu County (*cont.*)

	The no. of household	Jia	Household	Relationship with the household head	Surname	First name	Sex	The Chinese zodiac	Birthdate	
									Year	Month
Fujin Xiang 5th Bao				The household head's wife	石	根女	Female	Snake	1917	9
Fujin Xiang 5th Bao				Father	石	虎儿	Male	Rat	1887	9
Fujin Xiang 5th Bao				Son	石	成全儿	Male	Pig	1935	3
Fujin Xiang 5th Bao				The second son	石	刘才	Male	Rabbit	1939	9
Fujin Xiang 5th Bao				Daughter	石	刘女子	Female	Rooster	1945	12
Fujin Xiang 5th Bao	529	10	1	The household head	石	学忠	Male	Monkey	1907	7
Fujin Xiang 5th Bao				The household head's wife	黄	当哇	Female	Snake	1917	4
Fujin Xiang 5th Bao				Son	石	谢哇	Male	Ox	1937	12
Fujin Xiang 5th Bao				The second son	石	长义	Male	Rabbit	1939	6
Fujin Xiang 5th Bao				The third son	石	马义	Male	Monkey	1944	9
Fujin Xiang 5th Bao				Daughter	石	庭哇	Female	Monkey	1932	9
Fujin Xiang 5th Bao				The second daughter	石	冬英	Female	Pig	1947	4
Fujin Xiang 5th Bao				Younger brother	石	学仓	Male	Rabbit	1915	6
Fujin Xiang 5th Bao				Younger brother's wife	同	哇	Female	Monkey	1920	9
Fujin Xiang 5th Bao				Nephew	进	才哇	Male	Snake	1941	12
Fujin Xiang 5th Bao				Niece	进	女	Female	Snake	1935	3
Fujin Xiang 5th Bao	530	10	2	The household head	石	含义	Male	Dog	1909	9

Date	The beginning time of inhabitation in this county	Educational level	Profession sector	Occupation	Marital status	Physical disability	Religion
22	Inhabited for generations	Illiteracy	Agriculture	Land-holding peasant	Married	No	No religion
1	Inhabited for generations	Illiteracy	Agriculture	Land-holding peasant	Widowed	No	No religion
10	Inhabited for generations	Illiteracy	Agriculture				
27	Inhabited for generations	Bao school 1 year					
14	Inhabited for generations						
24	Inhabited for generations	Illiteracy	Agriculture	Land-holding peasant	Married	No	No religion
1	Inhabited for generations	Illiteracy	Agriculture	Land-holding peasant	Married	No	No religion
5	Inhabited for generations	Bao school 1 year					
21	Inhabited for generations	Illiteracy					
12	Inhabited for generations						
10	Inhabited for generations						
10	Inhabited for generations						
30	Inhabited for generations	Illiteracy	Agriculture	Land-holding peasant			
21	Inhabited for generations						
7	Inhabited for generations						
25	Inhabited for generations						
12	Inhabited for generations	Illiteracy	Agriculture	Land-holding peasant	Married	No	No religion

Appendix: Overview of Microdata for Wudu County (*cont.*)

	The no. of household	Jia	Household	Relationship with the household head	Surname	First name	Sex	The Chinese zodiac	Birthdate Year	Month
Fujin Xiang 5th Bao				The household head's wife	谈	崔哇	Female	Rat	1912	4
Fujin Xiang 5th Bao				Mother	崔	氏	Female	Rat	1875	
Fujin Xiang 5th Bao	531	10	3	The household head	石	会才	Male	Dog	1897	10
Fujin Xiang 5th Bao				Father	石	毛儿	Male	Rat	1875	
Fujin Xiang 5th Bao				Mother	赵	氏	Female	Rat	1875	
Fujin Xiang 5th Bao				Younger brother	石	会全	Male	Rabbit	1915	2
Fujin Xiang 5th Bao				Younger brother's wife	赵	有儿	Female	Rabbit	1915	12
Fujin Xiang 5th Bao				Niece	石	等女	Female	Dragon	1940	7
Fujin Xiang 5th Bao	532	10	4	The household head	石	会有	Male	Goat	1904	5
Fujin Xiang 5th Bao				The household head's wife	杨	见儿	Female	Goat	1904	4
Fujin Xiang 5th Bao				Son	石	均朝	Male	Dragon	1916	4
Fujin Xiang 5th Bao				Son's wife	谈	青哇	Female	Tiger	1914	4
Fujin Xiang 5th Bao				Grandson	石	长哇	Male	Monkey	1944	3
Fujin Xiang 5th Bao				Granddaughter	石	长女	Female	Dog	1946	4
Fujin Xiang 5th Bao				The second son	石	崴禄子	Male	Horse	1918	3
Fujin Xiang 5th Bao				Uncle's son	石	会林	Male	Dog	1897	8
Fujin Xiang 5th Bao				Younger brother's wife	赵	崴琴	Female	Horse	1905	5
Fujin Xiang 5th Bao				Nephew's son	石	会儿	Male	Snake	1941	4

Date	The beginning time of inhabitation in this county	Educational level	Profession sector	Occupation	Marital status	Physical disability	Religion
27	Inhabited for generations	Illiteracy	Agriculture	Land-holding peasant	Married	No	No religion
	Inhabited for generations						
15	Inhabited for generations	Illiteracy	Agriculture	Land-holding peasant	Widowed	No	No religion
	Inhabited for generations						
	Inhabited for generations						
23	Inhabited for generations	Illiteracy	Agriculture	Land-holding peasant	Married	No	No religion
26	Inhabited for generations	Illiteracy	Agriculture	Land-holding peasant	Married	No	No religion
12	Inhabited for generations	Illiteracy					
25	Inhabited for generations	Illiteracy	Agriculture	Land-holding peasant	Married	No	No religion
27	Inhabited for generations	Illiteracy	Agriculture	Land-holding peasant	Married	No	No religion
4	Inhabited for generations	Illiteracy	Agriculture	Land-holding peasant	Married	No	No religion
29	Inhabited for generations	Illiteracy	Agriculture	Land-holding peasant	Married	No	No religion
5	Inhabited for generations						
11	Inhabited for generations						
30	Inhabited for generations	Illiteracy	Agriculture	Land-holding peasant	Unmarried	No	No religion
4	Inhabited for generations	Illiteracy	Agriculture	Land-holding peasant	Married	No	No religion
9	Inhabited for generations	Illiteracy	Agriculture	Land-holding peasant	Married	No	No religion
5	Inhabited for generations	Illiteracy					

Appendix: Overview of Microdata for Wudu County (*cont.*)

									Birthdate	
	The no. of household	Jia	Household	Relationship with the household head	Surname	First name	Sex	The Chinese zodiac	Year	Month
Fujin Xiang 5th Bao	533	10	5	The household head	石	学才	Male	Rat	1887	12
Fujin Xiang 5th Bao				The household head's wife	赵	羊女	Female	Rabbit	1902	5
Fujin Xiang 5th Bao				Son	石	会后	Male	Ox	1926	1
Fujin Xiang 5th Bao				Son's wife	赵	奈哇	Female	Ox	1925	7
Fujin Xiang 5th Bao				Grandson	德	赐儿	Male	Dog		
Fujin Xiang 5th Bao				The second son	张	栢	Male	Dog	1947	1
Fujin Xiang 5th Bao				Nephew	石	会法	Male	Rabbit	1890	1
Fujin Xiang 5th Bao				Nephew's wife	郭	秋儿	Female	Rat	1923	7
Fujin Xiang 5th Bao				Nephew's son	石	进生	Male	Rooster	1954	2
Fujin Xiang 5th Bao				Nephew's daughter	石	进应	Male	Ox	1937	5
Fujin Xiang 5th Bao				Nephew's daughter	石	进香	Female	Snake	1941	4
Fujin Xiang 5th Bao	534	10	6	The household head	石	转富	Male	Goat	1906	5
Fujin Xiang 5th Bao				The household head's wife	赵	马儿	Female	Horse	1905	3
Fujin Xiang 5th Bao				Father	石	学彦	Male	Snake	1880	12
Fujin Xiang 5th Bao				Mother	陈	氏	Female	Tiger	1876	1
Fujin Xiang 5th Bao				Son	兹	生儿	Male	Rabbit	1939	9
Fujin Xiang 5th Bao				The second son	石	玉娃	Male	Pig	1947	4
Fujin Xiang 5th Bao				Daughter	石	清英	Female	Horse	1942	3

Date	The beginning time of inhabitation in this county	Educational level	Profession sector	Occupation	Marital status	Physical disability	Religion
24	Inhabited for generations	Illiteracy	Agriculture	Land-holding peasant	Married	No	No religion
25	Inhabited for generations	Illiteracy	Agriculture	Land-holding peasant	Married	No	No religion
15	Inhabited for generations	Illiteracy	Agriculture	Land-holding peasant	Widowed	No	No religion
2	Inhabited for generations						
	Inhabited for generations						
2	Inhabited for generations						
1	Inhabited for generations	Illiteracy	Agriculture	Land-holding peasant	Married	No	No religion
26	Inhabited for generations	Illiteracy	Agriculture	Land-holding peasant	Married	No	No religion
9							
2							
28	Inhabited for generations	Illiteracy					
12	Inhabited for generations	Sishu 3 years	Agriculture	Land-holding peasant	Married	No	No religion
25	Inhabited for generations	Illiteracy	Agriculture	Land-holding peasant	Married	No	No religion
29	Inhabited for generations	Illiteracy	Agriculture	Land-holding peasant	Married	No	No religion
12	Inhabited for generations	Illiteracy	Agriculture	Land-holding peasant	Married	No	No religion
17	Inhabited for generations	Illiteracy					
8	Inhabited for generations						
24	Inhabited for generations						

Appendix: Overview of Microdata for Wudu County (*cont.*)

	The no. of household	Jia	Household	Relationship with the household head	Surname	First name	Sex	The Chinese zodiac	Birthdate	
									Year	Month
Fujin Xiang 5th Bao				Uncle's son	石	会会	Male	Goat	1894	4
Fujin Xiang 5th Bao				Older brother's wife	石	傍女	Female	Dog	1894	1
Fujin Xiang 5th Bao				Niece	代	女子	Female	Rooster	1933	9
Fujin Xiang 5th Bao	535	10	7	The household head	石	均义	Male	Dragon	1903	4
Fujin Xiang 5th Bao				The household head's wife	黄	润秀	Female	Rat	1899	11
Fujin Xiang 5th Bao				Son	石	成才儿	Male	Snake	1941	5
Fujin Xiang 5th Bao				Daughter	石	如意子	Female	Monkey	1942	4
Fujin Xiang 5th Bao				Younger brother	石	均才	Male	Dog	1909	6
Fujin Xiang 5th Bao				Nephew	石	义家	Male	Monkey	1944	4
Fujin Xiang 5th Bao	536	10	8	The household head	石	会玉	Male	Horse	1879	7
Fujin Xiang 5th Bao				The household head's wife	石	秀儿	Female	Horse	1879	5
Fujin Xiang 5th Bao				Mother	成	氏	Female	Ox	1884	6
Fujin Xiang 5th Bao				Son's wife	赵	宋三娃	Female	Monkey	1915	4
Fujin Xiang 5th Bao				Younger brother	石	会义	Male	Goat	1885	1
Fujin Xiang 5th Bao				Younger brother's wife	张	长儿	Female	Rooster	1896	5
Fujin Xiang 5th Bao				Nephew	石	琴学儿	Male	Rabbit	1915	7
Fujin Xiang 5th Bao				Nephew's wife	韩	赵廷	Female	Snake	1917	3

Date	The beginning time of inhabitation in this county	Educational level	Profession sector	Occupation	Marital status	Physical disability	Religion	
25	Inhabited for generations	Illiteracy	Agriculture	Land-holding peasant	Married	No	No religion	
24	Inhabited for generations	Illiteracy	Agriculture	Land-holding peasant	Married	No	No religion	
1	Inhabited for generations	Illiteracy						
28	Inhabited for generations	Illiteracy	Agriculture	Land-holding peasant	Married	No	No religion	
2	Inhabited for generations	Illiteracy	Agriculture	Land-holding peasant	Married	No	No religion	
7	Inhabited for generations	Illiteracy						
16	Inhabited for generations							
8	Inhabited for generations	Illiteracy	Agriculture	Land-holding peasant	Widowed	No	No religion	
21	Inhabited for generations							
12	Inhabited for generations	Illiteracy	Agriculture	Land-holding peasant	Married	No	No religion	
1	Inhabited for generations	Illiteracy	Agriculture	Land-holding peasant	Married	No	No religion	Dead
1	Inhabited for generations							
20	Inhabited for generations	Illiteracy	Agriculture	Land-holding peasant	Married	No	No religion	
22	Inhabited for generations	Illiteracy	Agriculture	Land-holding peasant	Married	No	No religion	
2	Inhabited for generations	Illiteracy	Agriculture	Land-holding peasant	Married	No	No religion	
23	Inhabited for generations	Illiteracy	Agriculture	Land-holding peasant	Married	No	No religion	
3	Inhabited for generations	Illiteracy	Agriculture	Land-holding peasant	Married	No	No religion	

Appendix: Overview of Microdata for Wudu County (*cont.*)

	The no. of household	Jia	Household	Relationship with the household head	Surname	First name	Sex	The Chinese zodiac	Birthdate Year	Month
Fujin Xiang 5th Bao				Nephew's son	石	柏子	Male	Goat	1943	11
Fujin Xiang 5th Bao				Granddaughter	石	女女儿	Female	Pig	1947	1
Fujin Xiang 5th Bao				The second nephew	石	徐牛娃	Male	Dragon	1940	4
Fujin Xiang 5th Bao				Niece	石	且娃	Female	Horse	1942	9
Fujin Xiang 5th Bao	537	10	9	The household head	石	慧德	Male	Dragon	1916	10
Fujin Xiang 5th Bao				The household head's wife	谈	四女	Female	Rat	1924	12
Fujin Xiang 5th Bao				Mother	谈	董姐	Female	Snake	1880	2
Fujin Xiang 5th Bao				Uncle's wife	谈	氏	Female	Pig	1874	10
Fujin Xiang 5th Bao				Uncle's son	石	拜省哇	Male	Rabbit	1915	2
Fujin Xiang 5th Bao				Younger brother's wife	李	冯金	Female	Rooster	1921	12
Fujin Xiang 5th Bao				Nephew	石	万赐儿	Male	Pig	1935	2
Fujin Xiang 5th Bao				The second nephew	石	刘成	Male	Rooster	1945	8
Fujin Xiang 5th Bao				Uncle's son	石	会元	Male	Monkey	1897	2
Fujin Xiang 5th Bao				Uncle's son's wife	李	送娃	Female	Dog	1909	2
Fujin Xiang 5th Bao				Nephew	刘	旺儿	Male	Rooster	1945	9
Fujin Xiang 5th Bao				Niece	旺	女子	Female	Dragon	1940	5
Fujin Xiang 5th Bao	538	10	10	The household head	石	会贤	Male	Monkey	1895	2
Fujin Xiang 5th Bao				The household head's wife	李	送娃	Female	Rat	1899	11

Date	The beginning time of inhabitation in this county	Educational level	Profession sector	Occupation	Marital status	Physical disability	Religion
27	Inhabited for generations						
30	Inhabited for generations						
24	Inhabited for generations						
21	Inhabited for generations						
27	Inhabited for generations	Illiteracy	Agriculture	Land-holding peasant	Married	No	No religion
4	Inhabited for generations	Illiteracy	Agriculture	Land-holding peasant	Married	No	No religion
5	Inhabited for generations	Illiteracy	Agriculture	Land-holding peasant		No	No religion
1	Inhabited for generations						
11	Inhabited for generations	Illiteracy	Agriculture	Land-holding peasant	Married	No	No religion
3	Inhabited for generations	Illiteracy	Agriculture	Land-holding peasant	Married	No	No religion
12	Inhabited for generations	Illiteracy	Agriculture	Land-holding peasant		No	No religion
21	Inhabited for generations						
10	Inhabited for generations	Illiteracy	Agriculture	Land-holding peasant	Married	No	No religion
10	Inhabited for generations	Illiteracy	Agriculture	Land-holding peasant	Married	No	No religion
11	Inhabited for generations						
17	Inhabited for generations	Illiteracy					
13	Inhabited for generations	Illiteracy	Agriculture	Land-holding peasant	Married	No	No religion
23	Inhabited for generations	Illiteracy	Agriculture	Land-holding peasant	Married	No	No religion

Appendix: Overview of Microdata for Wudu County (*cont.*)

	The no. of household	Jia	Household	Relationship with the household head	Surname	First name	Sex	The Chinese zodiac	Birthdate Year	Month
Fujin Xiang 5th Bao				Father	石	含德	Male	Tiger	1865	11
Fujin Xiang 5th Bao				Son	石	德时儿	Male	Tiger	1938	8
Fujin Xiang 5th Bao				Daughter	石	相娃	Female	Rooster	1935	4
Fujin Xiang 5th Bao				The second son	石	丑俊儿	Male	Goat	1942	8
Fujin Xiang 5th Bao	539	10	11	The household head	石	会田	Male	Pig	1885	1
Fujin Xiang 5th Bao				The household head's wife	史	必女	Female	Snake	1892	5
Fujin Xiang 5th Bao				Son	石	讨口子	Male	Ox	1925	8
Fujin Xiang 5th Bao				Son's wife	谈	娃儿	Female	Rabbit	1926	9
Fujin Xiang 5th Bao				Grandson	石	昇娃	Male	Monkey	1944	9
Fujin Xiang 5th Bao				The second son	石	埋娃	Male	Goat	1931	4
Fujin Xiang 5th Bao	540	10	12	The household head	石	学有	Male	Rooster	1896	11
Fujin Xiang 5th Bao				The household head's wife	杨	岁双	Female	Rat	1899	12
Fujin Xiang 5th Bao				Son	石	润昔娃	Male	Rooster	1933	7
Fujin Xiang 5th Bao				Son's wife	杨	二女	Female	Snake	1929	3
Fujin Xiang 5th Bao				The second son	石	长家	Male	Ox	1937	5
Fujin Xiang 5th Bao				Daughter	长	女	Female	Monkey	1944	5
Fujin Xiang 5th Bao	541	10	13	The household head	石	含文	Male	Ox	1913	2
Fujin Xiang 5th Bao				Mother	谈	氏	Female	Rooster	1872	3

Date	The beginning time of inhabitation in this county	Educational level	Profession sector	Occupation	Marital status	Physical disability	Religion
1	Inhabited for generations						
27	Inhabited for generations	Bao school 1 year					
23	Inhabited for generations	Illiteracy	Agriculture				
20	Inhabited for generations						
24	Inhabited for generations	Illiteracy	Agriculture	Land-holding peasant	Married	No	No religion
16	Inhabited for generations	Illiteracy	Agriculture	Land-holding peasant	Married	No	No religion
24	Inhabited for generations	Illiteracy	Agriculture	Land-holding peasant	Married	No	No religion
20	Inhabited for generations	Illiteracy	Agriculture	Land-holding peasant	Married	No	No religion
6	Inhabited for generations						
20	Inhabited for generations	Illiteracy	Agriculture	Land-holding peasant	Unmarried	No	No religion
4	Inhabited for generations	Illiteracy	Agriculture	Land-holding peasant	Married	No	No religion
24	Inhabited for generations	Illiteracy	Agriculture	Land-holding peasant	Married	No	No religion
31	Inhabited for generations	Illiteracy	Agriculture	Land-holding peasant	Married	No	No religion
20	Inhabited for generations	Illiteracy	Agriculture	Land-holding peasant	Married	No	No religion
22	Inhabited for generations	Bao school 2 years					
13	Inhabited for generations						
17	Inhabited for generations	Illiteracy	Agriculture	Land-holding peasant	Widowed	No	No religion
2	Inhabited for generations						

Appendix: Overview of Microdata for Wudu County (*cont.*)

	The no. of household	Jia	Household	Relationship with the household head	Surname	First name	Sex	The Chinese zodiac	Birthdate	
									Year	Month
Fujin Xiang 5th Bao				Older brother	石	含拜	Male	Dragon	1891	9
Fujin Xiang 5th Bao				Older brother's wife	刘	针姐	Female	Horse	1893	10
Fujin Xiang 5th Bao				Nephew	石	会银	Male	Rat	1912	2
Fujin Xiang 5th Bao				Nephew's wife	李	炉儿	Female	Ox	1913	11
Fujin Xiang 5th Bao				Grandson	石	平儿	Male	Horse	1942	6
Fujin Xiang 5th Bao				The second grandson	石	旺儿	Male	Monkey	1944	7
Fujin Xiang 5th Bao	542	10	14	The household head	石	会龙	Male	Pig	1912	6
Fujin Xiang 5th Bao				The household head's wife	谈	肖儿	Female	Ox	1913	12
Fujin Xiang 5th Bao				Father	石	含巧	Male	Monkey	1883	9
Fujin Xiang 5th Bao				Mother	李	氏	Female	Pig	1886	11
Fujin Xiang 5th Bao				Son	石	刘儿	Male	Tiger	1938	7
Fujin Xiang 5th Bao				Daughter	石	廷娃	Female	Horse	1942	4
Fujin Xiang 5th Bao				Older brother	石	会文	Male	Monkey	1907	10
Fujin Xiang 5th Bao				Older brother's wife	赵	困娃	Female	Dog	1909	12
Fujin Xiang 5th Bao				Nephew	石	很娃	Male	Rat	1938	2
Fujin Xiang 5th Bao				Niece	石	能娃	Female	Snake	1941	10
Fujin Xiang 5th Bao				The second niece	石	深女	Female	Goat	1943	8
Fujin Xiang 5th Bao				The third niece	石	能女子	Female	Pig	1947	1

Date	The beginning time of inhabitation in this county	Educational level	Profession sector	Occupation	Marital status	Physical disability	Religion
6	Inhabited for generations	Illiteracy	Agriculture	Land-holding peasant	Married	No	No religion
29	Inhabited for generations	Illiteracy	Agriculture	Land-holding peasant	Married	No	No religion
18	Inhabited for generations	Illiteracy	Agriculture	Land-holding peasant	Married	No	No religion
7	Inhabited for generations	Illiteracy	Agriculture	Land-holding peasant	Married	No	No religion
15	Inhabited for generations						
20	Inhabited for generations						
26	Inhabited for generations	Illiteracy	Agriculture	Land-holding peasant	Married	No	No religion
5	Inhabited for generations	Illiteracy	Agriculture	Land-holding peasant	Married	No	No religion
20	Inhabited for generations	Illiteracy	Agriculture	Land-holding peasant	Married	No	No religion
15	Inhabited for generations	Illiteracy	Agriculture	Land-holding peasant	Married	No	No religion
2	Inhabited for generations	Illiteracy					
17	Inhabited for generations	Illiteracy					
26	Inhabited for generations	Illiteracy	Agriculture	Land-holding peasant	Married	No	No religion
12	Inhabited for generations	Illiteracy	Agriculture	Land-holding peasant	Married	No	No religion
10	Inhabited for generations	Illiteracy					
10	Inhabited for generations						
12	Inhabited for generations						
23	Inhabited for generations						

Appendix: Overview of Microdata for Wudu County (*cont.*)

	The no. of household	Jia	Household	Relationship with the household head	Surname	First name	Sex	The Chinese zodiac	Birthdate Year	Month
Fujin Xiang 5th Bao	543	10	15	The household head	石	会永	Male	Rooster	1908	3
Fujin Xiang 5th Bao				The household head's wife	赵	进娃	Female	Horse	1905	4
Fujin Xiang 5th Bao				Mother	李	氏	Female	Ox	1876	5
Fujin Xiang 5th Bao				Son	石	张生儿	Male	Rabbit	1927	3
Fujin Xiang 5th Bao				Son's wife	成	柳娃	Female	Horse	1930	10
Fujin Xiang 5th Bao				The second son	成	娃	Male	Goat	1943	8
Fujin Xiang 5th Bao				Daughter	成	女	Female	Snake	1941	1
Fujin Xiang 5th Bao				Nephew	石	均秀	Male	Monkey	1920	12
Fujin Xiang 5th Bao				Nephew's wife	顺	义	Female	Monkey	1920	12
Fujin Xiang 5th Bao				Nephew's son	石	女才娃	Female	Rooster	1945	11
Fujin Xiang 5th Bao	544	11	1	The household head	石	会朝	Male	Rooster	1896	9
Fujin Xiang 5th Bao				The household head's wife	孟	贤哇	Female	Rabbit	1902	6
Fujin Xiang 5th Bao				Son	石	旺娃	Male	Ox	1931	1
Fujin Xiang 5th Bao				Younger brother	石	宝儿	Male	Ox	1901	1
Fujin Xiang 5th Bao				Grandfather's brother's wife	菜	儿	Female	Rabbit	1891	1
Fujin Xiang 5th Bao				Uncle	石	文建	Male	Rat	1899	5
Fujin Xiang 5th Bao				Uncle's wife	赵	女子	Female	Rat	1912	9

Date	The beginning time of inhabitation in this county	Educational level	Profession sector	Occupation	Marital status	Physical disability	Religion
25	Inhabited for generations	Illiteracy	Agriculture	Land-holding peasant	Married	No	No religion
10	Inhabited for generations	Illiteracy	Agriculture	Land-holding peasant	Married	No	No religion
2	Inhabited for generations						
15	Inhabited for generations	Illiteracy	Agriculture	Land-holding peasant	Married	No	No religion
30	Inhabited for generations	Illiteracy	Agriculture	Land-holding peasant	Married	No	No religion
20	Inhabited for generations						
26	Inhabited for generations	Illiteracy					
26	Inhabited for generations	Illiteracy	Agriculture	Land-holding peasant	Married	No	No religion
26	Inhabited for generations	Illiteracy	Agriculture	Land-holding peasant	Married	No	No religion
7	Inhabited for generations						
12	Inhabited for generations	Illiteracy	Agriculture	Land-holding peasant	Married	No	No religion
15	Inhabited for generations	Illiteracy	Agriculture	Land-holding peasant	Married	No	No religion
5	Inhabited for generations	Illiteracy	Agriculture	Land-holding peasant	Unmarried	No	No religion
1	Inhabited for generations	Illiteracy	Agriculture	Land-holding peasant	Unmarried	No	No religion
12	Inhabited for generations	Illiteracy	Agriculture	Land-holding peasant	Widowed	No	No religion
4	Inhabited for generations	Illiteracy	Agriculture	Land-holding peasant	Married	No	No religion
1	Inhabited for generations	Illiteracy	Agriculture	Land-holding peasant	Married	No	No religion

Appendix: Overview of Microdata for Wudu County (*cont.*)

	The no. of household	Jia	Household	Relationship with the household head	Surname	First name	Sex	The Chinese zodiac	Birthdate	
									Year	Month
Fujin Xiang 5th Bao				Older brother's wife	王	氏	Female	Pig	1874	5
Fujin Xiang 5th Bao	545	11	2	The household head	石	含海	Male	Rabbit	1890	10
Fujin Xiang 5th Bao				Father	石	文先	Male		1864	8
Fujin Xiang 5th Bao				Son	石	会银	Male	Monkey	1920	11
Fujin Xiang 5th Bao				Son's wife	黄	香音娃	Female	Monkey	1920	10
Fujin Xiang 5th Bao				Grandson	石	岁才	Male	Goat	1943	12
Fujin Xiang 5th Bao				Granddaughter	石	女	Female	Dog	1946	1
Fujin Xiang 5th Bao	546	11	3	The household head	石	连娃	Male	Snake	1917	12
Fujin Xiang 5th Bao				Daughter	罗	娃	Female	Tiger	1941	10
Fujin Xiang 5th Bao				The second daughter	石	女娃	Female	Monkey	1944	6
Fujin Xiang 5th Bao	547	11	4	The household head	石	含永	Male	Horse	1906	12
Fujin Xiang 5th Bao				The household head's wife	朱	赵秀	Female	Horse	1906	1
Fujin Xiang 5th Bao				Son	石	未生	Male	Pig	1947	3
Fujin Xiang 5th Bao				Younger brother	石	含义	Male	Dog	1909	7
Fujin Xiang 5th Bao				Younger brother's wife	谈	炉儿	Female	Monkey	1913	6
Fujin Xiang 5th Bao				Nephew	石	成生儿	Male	Snake	1941	10
Fujin Xiang 5th Bao	548	11	5	The household head	石	会明	Male	Monkey	1907	4

Date	The beginning time of inhabitation in this county	Educational level	Profession sector	Occupation	Marital status	Physical disability	Religion
1	Inhabited for generations	Illiteracy	Agriculture	Land-holding peasant		No	No religion
3	Inhabited for generations	Sishu 2 years	Agriculture	Land-holding peasant	Widowed	No	No religion
9	Inhabited for generations	Illiteracy	Agriculture	Land-holding peasant	Widowed	No	No religion
3	Inhabited for generations	Illiteracy	Agriculture	Land-holding peasant	Married	No	No religion
9	Inhabited for generations	Illiteracy	Agriculture	Land-holding peasant	Married	No	No religion
12	Inhabited for generations						
5	Inhabited for generations						
6	Inhabited for generations	Illiteracy	Agriculture	Land-holding peasant	Married	No	No religion
10	Inhabited for generations	Illiteracy	Agriculture	Land-holding peasant	Married	No	No religion
25	Inhabited for generations						
6	Inhabited for generations	Illiteracy	Agriculture	Land-holding peasant	Married	No	No religion
16	Inhabited for generations	Illiteracy	Agriculture	Land-holding peasant	Married	No	No religion
8	Inhabited for generations						
7	Inhabited for generations	Illiteracy	Agriculture	Land-holding peasant	Married	No	No religion
4	Inhabited for generations	Illiteracy	Agriculture	Land-holding peasant	Married	No	No religion
6	Inhabited for generations						
14	Inhabited for generations	Illiteracy	Agriculture	Land-holding peasant	Married	No	No religion

Appendix: Overview of Microdata for Wudu County (*cont.*)

	The no. of household	Jia	Household	Relationship with the household head	Surname	First name	Sex	The Chinese zodiac	Birthdate Year	Birthdate Month
Fujin Xiang 5th Bao				The household head's wife	崔	肖儿	Female	Monkey	1920	8
Fujin Xiang 5th Bao				Father	石	含撲	Male	Snake	1867	8
Fujin Xiang 5th Bao				Mother	成	氏	Female	Ox	1876	6
Fujin Xiang 5th Bao				Daughter	张	女	Female	Pig	1947	3
Fujin Xiang 5th Bao				Younger brother	石	会昌	Male	Pig	1910	4
Fujin Xiang 5th Bao				Younger brother's wife	刘	耀娃	Female	Pig	1910	8
Fujin Xiang 5th Bao				Niece	石	怌女子	Female	Horse	1942	8
Fujin Xiang 5th Bao				The second niece	石	才女子	Female	Pig	1947	4
Fujin Xiang 5th Bao				The third younger brother	石	会禄	Male	Rabbit	1915	11
Fujin Xiang 5th Bao				Younger brother's wife	徐	左女	Female	Rat	1912	7
Fujin Xiang 5th Bao				Uncle	石	含生	Male	Goat	1894	3
Fujin Xiang 5th Bao				Uncle's son	石	会转	Male	Rabbit	1915	6
Fujin Xiang 5th Bao				Uncle's son's wife	谈	早娃	Female	Horse	1918	5
Fujin Xiang 5th Bao	549	11	6	The household head	石	会举	Male	Snake	1904	11
Fujin Xiang 5th Bao				The household head's wife	何	廷	Female	Rat	1912	8
Fujin Xiang 5th Bao				Mother	谈	氏	Female	Snake	1888	9
Fujin Xiang 5th Bao				Son	石	均仓	Male	Rooster	1933	11

Date	The beginning time of inhabitation in this county	Educational level	Profession sector	Occupation	Marital status	Physical disability	Religion
4	Inhabited for generations	Illiteracy	Agriculture	Land-holding peasant	Married	No	No religion
1	Inhabited for generations						
1	Inhabited for generations						
27	Inhabited for generations						
1	Inhabited for generations	Illiteracy	Agriculture	Land-holding peasant	Married	No	No religion
26	Inhabited for generations	Illiteracy	Agriculture	Land-holding peasant	Married	No	No religion
12	Inhabited for generations						
23	Inhabited for generations						
7	Inhabited for generations	Illiteracy	Agriculture	Land-holding peasant	Married	No	No religion
1	Inhabited for generations	Illiteracy	Agriculture	Land-holding peasant	Married	No	No religion
1	Inhabited for generations	Illiteracy	Agriculture	Land-holding peasant	Widowed	No	No religion
4	Inhabited for generations	Illiteracy	Agriculture	Land-holding peasant	Married	No	No religion
27	Inhabited for generations	Illiteracy	Agriculture	Land-holding peasant	Married	No	No religion
16	Inhabited for generations	Illiteracy	Agriculture	Land-holding peasant	Married	No	No religion
27	Inhabited for generations	Illiteracy	Agriculture	Land-holding peasant	Married	No	No religion
1	Inhabited for generations		Agriculture			No	No religion
1	Inhabited for generations	Bao school 4 years					

Appendix: Overview of Microdata for Wudu County (*cont.*)

	The no. of household	Jia	Household	Relationship with the household head	Surname	First name	Sex	The Chinese zodiac	Birthdate	
									Year	Month
Fujin Xiang 5th Bao				Uncle's wife	陈	氏	Female	Rat	1943	11
Fujin Xiang 5th Bao				Uncle's son	石	会岫	Male	Tiger	1901	11
Fujin Xiang 5th Bao				Uncle's son's wife	李	双	Female	Rabbit	1915	2
Fujin Xiang 5th Bao				Nephew	石	乡成子	Male	Rooster	1933	3
Fujin Xiang 5th Bao				The second nephew	石	乡香	Female	Pig	1935	6
Fujin Xiang 5th Bao				Niece	石	明香	Female	Tiger	1938	12
Fujin Xiang 5th Bao				The second niece	石	水娃	Female	Dragon	1940	2
Fujin Xiang 5th Bao				Niece	石	艾香	Female	Dog	1946	1
Fujin Xiang 5th Bao	550	11	7	The household head	石	会荣	Male	Ox	1888	12
Fujin Xiang 5th Bao				The household head's wife	李	石姐	Female	Ox	1888	12
Fujin Xiang 5th Bao				Son	石	学儿	Male	Monkey	1907	7
Fujin Xiang 5th Bao				The second son	石	富学儿	Male	Goat	1913	12
Fujin Xiang 5th Bao				Daughter	石	么女子	Female	Snake	1941	2
Fujin Xiang 5th Bao	551	11	8	The household head	石	会花	Male	Snake	1892	10
Fujin Xiang 5th Bao				The household head's wife	李	金环	Female		1892	8
Fujin Xiang 5th Bao				Younger brother's wife	张	董哇	Female	Horse	1905	6
Fujin Xiang 5th Bao				Nephew	石	成学子	Male	Tiger	1926	2
Fujin Xiang 5th Bao				Nephew's wife	郭	贤哇	Female	Rabbit	1927	6

Date	The beginning time of inhabitation in this county	Educational level	Profession sector	Occupation	Marital status	Physical disability	Religion
1	Inhabited for generations				Widowed	No	No religion
1	Inhabited for generations	Illiteracy	Agricul-ture	Land-holding peasant	Married	No	No religion
2	Inhabited for generations	Illiteracy	Agricul-ture	Land-holding peasant	Married	No	No religion
7	Inhabited for generations	Bao school 4 years					
10	Inhabited for generations						
19	Inhabited for generations	Illiteracy					
16	Inhabited for generations						
12	Inhabited for generations						
22	Inhabited for generations	Illiteracy	Agricul-ture	Land-holding peasant	Married	No	No religion
26	Inhabited for generations	Illiteracy	Agricul-ture	Land-holding peasant	Married	No	No religion
6	Inhabited for generations	Illiteracy	Agricul-ture	Land-holding peasant	Widowed	No	No religion
18	Inhabited for generations	Illiteracy	Agricul-ture	Land-holding peasant	Unmarried	No	No religion
5	Inhabited for generations	Illiteracy					
21	Inhabited for generations	Illiteracy	Agricul-ture	Land-holding peasant	Married	No	No religion
23	Inhabited for generations	Illiteracy	Agricul-ture	Land-holding peasant	Married	No	No religion
23	Inhabited for generations	Illiteracy	Agricul-ture		Widowed	No	No religion
24	Inhabited for generations	Illiteracy	Agricul-ture	Land-holding peasant	Married	No	No religion
1	Inhabited for generations	Illiteracy	Agricul-ture	Land-holding peasant	Married	No	No religion

Appendix: Overview of Microdata for Wudu County (*cont.*)

	The no. of household	Jia	Household	Relationship with the household head	Surname	First name	Sex	The Chinese zodiac	Birthdate	
									Year	Month
Fujin Xiang 5th Bao				The second nephew	石	蛮哇	Male	Goat	1931	11
Fujin Xiang 5th Bao	552	11	9	The household head	石	会万	Male	Dragon	1891	8
Fujin Xiang 5th Bao				The household head's wife	王	惠哇	Female	Rabbit	1890	5
Fujin Xiang 5th Bao				Mother	陈	氏	Female	Dog	1873	3
Fujin Xiang 5th Bao				Son	石	金玉儿	Male	Tiger	1920	2
Fujin Xiang 5th Bao				Son's wife	石	扣儿	Female	Tiger	1914	3
Fujin Xiang 5th Bao				Grandson	石	知礼	Male	Ox	1937	10
Fujin Xiang 5th Bao				The second grandson	石	水哇	Female	Monkey	1944	2
Fujin Xiang 5th Bao				Granddaughter	石	知义	Female	Snake	1941	12
Fujin Xiang 5th Bao				The second son's wife	谈	环哇	Female	Dog	1922	10
Fujin Xiang 5th Bao				Uncle's son	石	转秀	Male	Rabbit	1902	11
Fujin Xiang 5th Bao				Younger brother's wife	石	桂女	Female	Tiger	1901	3
Fujin Xiang 5th Bao				Nephew	石	裉生儿	Male	Tiger	1938	7
Fujin Xiang 5th Bao				The second nephew	石	柱学儿	Male	Horse	1942	7
Fujin Xiang 5th Bao				Niece	石	崴庭儿	Female	Goat	1931	10
Fujin Xiang 5th Bao	553	11	10	The household head	石	学发	Male	Pig	1886	1
Fujin Xiang 5th Bao				Son	石	生生	Male	Goat	1931	12
Fujin Xiang 5th Bao				Son's wife	赵	娃娃	Female	Goat	1932	1

Date	The beginning time of inhabitation in this county	Educational level	Profession sector	Occupation	Marital status	Physical disability	Religion
19	Inhabited for generations	Illiteracy	Agriculture	Land-holding peasant	Unmarried	No	No religion
15	Inhabited for generations	Illiteracy	Agriculture	Land-holding peasant	Married	No	No religion
12	Inhabited for generations	Illiteracy	Agriculture	Land-holding peasant	Married	No	No religion
8	Inhabited for generations	Illiteracy					
27	Inhabited for generations	Illiteracy	Agriculture	Land-holding peasant	Married	No	No religion
29	Inhabited for generations	Illiteracy	Agriculture	Land-holding peasant	Married	No	No religion
23	Inhabited for generations	Bao school 1 year					
25	Inhabited for generations						
20	Inhabited for generations	Illiteracy					
28	Inhabited for generations	Illiteracy	Agriculture	Land-holding peasant	Widowed	No	No religion
19	Inhabited for generations	Illiteracy	Agriculture	Land-holding peasant	Married	No	No religion
10	Inhabited for generations	Illiteracy	Agriculture	Land-holding peasant	Married	No	No religion
5	Inhabited for generations	Illiteracy	Agriculture				
5	Inhabited for generations	Illiteracy					
11	Inhabited for generations	Illiteracy	Agriculture		Unmarried	No	No religion
25	Inhabited for generations	Illiteracy	Agriculture	Land-holding peasant	Widowed	No	No religion
10	Inhabited for generations	Illiteracy	Agriculture	Land-holding peasant	Married	No	No religion
17	Inhabited for generations	Illiteracy	Agriculture	Land-holding peasant	Married	No	No religion

Appendix: Overview of Microdata for Wudu County (*cont.*)

	The no. of household	Jia	Household	Relationship with the household head	Surname	First name	Sex	The Chinese zodiac	Birthdate Year	Birthdate Month
Fujin Xiang 5th Bao				The second son	石	毛人	Male	Rat	1936	3
Fujin Xiang 5th Bao				Uncle's son	石	寿娃	Male	Dog	1885	11
Fujin Xiang 5th Bao				Older brother's wife	石	屎儿	Female	Rat	1887	3
Fujin Xiang 5th Bao				Nephew	石	会满	Male	Rabbit	1902	11
Fujin Xiang 5th Bao				Nephew's wife	杨	环儿	Female	Rat	1899	8
Fujin Xiang 5th Bao				Grandson	顺	宿儿	Male	Rooster	1921	4
Fujin Xiang 5th Bao				Nephew's son's wife	石	早女	Female	Snake	1917	5
Fujin Xiang 5th Bao				Nephew's daughter	石	菊香	Female	Snake	1941	1
Fujin Xiang 5th Bao				The second great Grand-daughter	石	應女	Female	Dog	1946	9
Fujin Xiang 5th Bao	554	11	11	The household head	石	含宝	Male	Rabbit	1890	9
Fujin Xiang 5th Bao				The household head's wife	赵	路顺	Female	Goat	1894	9
Fujin Xiang 5th Bao				Son	石	等生儿	Male	Goat	1931	5
Fujin Xiang 5th Bao				Nephew	石	转有	Male	Rabbit	1901	7
Fujin Xiang 5th Bao				Nephew's wife	罗	秀儿	Female	Snake	1904	10
Fujin Xiang 5th Bao				Nephew's son	石	均科	Male	Rooster	1933	10
Fujin Xiang 5th Bao				Grand-daughter	石	社哇	Female	Ox	1934	7
Fujin Xiang 5th Bao	555	11	12	The household head	石	均山	Male	Pig	1910	4

Date	The beginning time of inhabitation in this county	Educational level	Profession sector	Occupation	Marital status	Physical disability	Religion
31	Inhabited for generations	Illiteracy					
26	Inhabited for generations	Illiteracy	Agriculture	Land-holding peasant	Married	No	No religion
11	Inhabited for generations	Illiteracy	Agriculture	Land-holding peasant	Married	No	No religion
8	Inhabited for generations	Illiteracy	Agriculture	Land-holding peasant	Married	No	No religion
29	Inhabited for generations	Illiteracy	Agriculture	Land-holding peasant	Married	No	No religion
29	Inhabited for generations	Illiteracy	Agriculture	Land-holding peasant	Married	No	No religion
16	Inhabited for generations	Illiteracy	Agriculture	Land-holding peasant	Married	No	No religion
20	Inhabited for generations	Illiteracy	Agriculture				
10	Inhabited for generations						
3	Inhabited for generations	Illiteracy	Agriculture	Land-holding peasant	Married	No	No religion
3	Inhabited for generations	Illiteracy	Agriculture	Land-holding peasant	Married	No	No religion
31	Inhabited for generations						
24	Inhabited for generations	Illiteracy	Agriculture	Land-holding peasant	Married	No	No religion
11	Inhabited for generations	Illiteracy	Agriculture	Land-holding peasant	Married	No	No religion
4	Inhabited for generations	Illiteracy	Agriculture				
7	Inhabited for generations	Illiteracy					
27	Inhabited for generations	Illiteracy	Agriculture	Land-holding peasant	Married	No	No religion

Appendix: Overview of Microdata for Wudu County (*cont.*)

	The no. of household	Jia	Household	Relationship with the household head	Surname	First name	Sex	The Chinese zodiac	Birthdate Year	Month
Fujin Xiang 5th Bao				The household head's wife	赵	翠儿	Female	Pig	1910	3
Fujin Xiang 5th Bao				Father	石	转清	Male	Dragon	1879	10
Fujin Xiang 5th Bao				Mother	郭	氏	Female	Tiger	1877	2
Fujin Xiang 5th Bao				Son	石	成代	Male	Tiger	1932	5
Fujin Xiang 5th Bao				Daughter	石	马女子	Female	Rabbit	1939	4
Fujin Xiang 5th Bao				Older brother	石	耀山	Male	Goat	1894	4
Fujin Xiang 5th Bao				Older brother's wife	蒋	文秀	Female	Dog	1909	3
Fujin Xiang 5th Bao				Uncle's wife	郭	氏	Female	Dragon	1867	6
Fujin Xiang 5th Bao	556	11	13	The household head	石	鸿儿	Male	Horse	1918	3
Fujin Xiang 5th Bao				The household head's wife	黄	柱女	Female	Horse	1918	4
Fujin Xiang 5th Bao				Mother	石	贱女	Female	Goat	1894	2
Fujin Xiang 5th Bao				Son	石	训练	Male	Monkey	1943	3
Fujin Xiang 5th Bao				The second son	石	成仁	Male	Dog	1946	10
Fujin Xiang 5th Bao				Daughter	石	乖娃	Female	Snake	1936	10
Fujin Xiang 5th Bao				Older brother	石	昇生儿	Male	Rabbit	1915	10
Fujin Xiang 5th Bao				Older brother's wife	根	宣儿	Female	Dragon	1918	6
Fujin Xiang 5th Bao				Niece	崴	明儿	Male	Dog	1946	9

Date	The beginning time of inhabitation in this county	Educational level	Profession sector	Occupation	Marital status	Physical disability	Religion
8	Inhabited for generations	Illiteracy	Agriculture	Land-holding peasant	Married	No	No religion
10	Inhabited for generations	Illiteracy	Agriculture	Land-holding peasant	Married	No	No religion
1	Inhabited for generations	Illiteracy	Agriculture	Land-holding peasant	Married	No	No religion
20	Inhabited for generations	Bao school 2 years					
3	Inhabited for generations	Illiteracy					
2	Inhabited for generations	Illiteracy	Agriculture	Land-holding peasant	Married	No	No religion
1	Inhabited for generations	Illiteracy	Agriculture	Land-holding peasant	Married	No	No religion
1	Inhabited for generations	Illiteracy					
7	Inhabited for generations	Illiteracy	Agriculture	Land-holding peasant	Married	No	No religion
6	Inhabited for generations	Illiteracy	Agriculture	Land-holding peasant	Married	No	No religion
2	Inhabited for generations	Illiteracy	Agriculture	Land-holding peasant	Widowed	No	No religion
10	Inhabited for generations						
19	Inhabited for generations						
5	Inhabited for generations	Illiteracy					
4	Inhabited for generations	Illiteracy	Agriculture	Land-holding peasant	Married	No	No religion
4	Inhabited for generations	Illiteracy	Agriculture	Land-holding peasant	Married	No	No religion
19	Inhabited for generations						

Appendix: Overview of Microdata for Wudu County (*cont.*)

	The no. of household	Jia	Household	Relationship with the household head	Surname	First name	Sex	The Chinese zodiac	Birthdate Year	Month
Fujin Xiang 5th Bao				Younger sister	石	有女子	Female	Dog	1934	12
Fujin Xiang 5th Bao				Younger brother	石	转生儿	Male	Pig	1923	2
Fujin Xiang 5th Bao	557	11	14	The household head	石	转林	Male	Rabbit	1902	3
Fujin Xiang 5th Bao				The household head's wife	浦	环儿	Female	Ox	1901	4
Fujin Xiang 5th Bao				Son	石	像儿	Male	Horse	1930	10
Fujin Xiang 5th Bao				The second son	陈	昇儿	Male	Rooster	1933	5
Fujin Xiang 5th Bao				The third son	石	双昇儿	Male	Snake	1941	5
Fujin Xiang 5th Bao				Daughter	石	慈娃	Female	Ox	1937	6
Fujin Xiang 5th Bao				The second daughter	周	女子	Female	Goat	1943	9
Fujin Xiang 5th Bao				Younger brother	石	特旺	Male	Ox	1925	11
Fujin Xiang 5th Bao				Younger brother's wife	杨	抢娃	Female	Dog	1942	2
Fujin Xiang 5th Bao				Niece	石	路女	Female	Goat	1943	12
Fujin Xiang 5th Bao	558	11	15	The household head	石	学秀	Male	Monkey	1884	1
Fujin Xiang 5th Bao				The household head's wife	欧	福香	Female	Tiger	1887	11
Fujin Xiang 5th Bao				Son	石	元儿	Male	Dog	1909	12
Fujin Xiang 5th Bao				Son's wife	赵	丑女	Female	Goat	1906	5
Fujin Xiang 5th Bao				Grandson	石	顺宿子	Male	Ox	1937	7
Fujin Xiang 5th Bao				The second grandson	石	鸡儿子	Male	Rooster	1945	12

Date	The beginning time of inhabitation in this county	Educational level	Profession sector	Occupation	Marital status	Physical disability	Religion	
15	Inhabited for generations							
24	Inhabited for generations							Serving in the army
10	Inhabited for generations	Illiteracy	Agriculture	Land-holding peasant	Married	No	No religion	
18	Inhabited for generations	Illiteracy	Agriculture	Land-holding peasant	Married	No	No religion	
2	Inhabited for generations	Illiteracy	Agriculture	Land-holding peasant	Unmarried	No	No religion	
26	Inhabited for generations	Bao school 3 years						
6	Inhabited for generations	Illiteracy						
21	Inhabited for generations	Illiteracy						
19	Inhabited for generations							
6	Inhabited for generations	Illiteracy	Agriculture	Land-holding peasant	Married	No	No religion	
16	Inhabited for generations	Illiteracy	Agriculture	Land-holding peasant	Married	No	No religion	
7	Inhabited for generations							
24	Inhabited for generations	Illiteracy	Agriculture	Land-holding peasant	Married	No	No religion	
14	Inhabited for generations	Illiteracy	Agriculture	Land-holding peasant	Married	No	No religion	
12	Inhabited for generations	Illiteracy	Agriculture	Land-holding peasant	Married	No	No religion	
24	Inhabited for generations	Illiteracy	Agriculture	Land-holding peasant	Married	No	No religion	
22	Inhabited for generations	Illiteracy						
6	Inhabited for generations							

Appendix: Overview of Microdata for Wudu County (*cont.*)

	The no. of household	Jia	Household	Relationship with the household head	Surname	First name	Sex	The Chinese zodiac	Birthdate Year	Month
Fujin Xiang 5th Bao				Granddaughter	石	路娃	Female	Monkey	1932	12
Fujin Xiang 5th Bao				The second granddaughter	石	苟女	Female	Horse	1942	6
Fujin Xiang 5th Bao	559	12	1	The household head	石	学敞	Male	Rooster	1896	2
Fujin Xiang 5th Bao				Father	石	文金	Male	Ox	1946	10
Fujin Xiang 5th Bao				Mother	黄	氏	Female	Ox	1876	12
Fujin Xiang 5th Bao				Younger brother	石	学三	Male	Goat	1906	6
Fujin Xiang 5th Bao				Younger brother's wife	成	义娃	Female	Rabbit	1915	3
Fujin Xiang 5th Bao				Nephew	苟	怪娃	Male	Monkey	1932	12
Fujin Xiang 5th Bao				Niece	石	清英	Female	Horse	1942	8
Fujin Xiang 5th Bao				Uncle's son's wife	李	罗氏	Female	Dog	1897	2
Fujin Xiang 5th Bao				Nephew	朱	拜子	Male	Rat	1924	11
Fujin Xiang 5th Bao				Nephew's wife	李	反娃	Female	Rat	1924	2
Fujin Xiang 5th Bao				Niece	银	廷儿	Female	Ox	1938	1
Fujin Xiang 5th Bao	560	12	2	The household head	石	学生	Male	Goat	1894	12
Fujin Xiang 5th Bao				The household head's wife	双	女子	Female	Rooster	1910	10
Fujin Xiang 5th Bao				Son	石	焦义	Male	Rooster	1934	1
Fujin Xiang 5th Bao				The second son	石	召才儿	Male	Ox	1938	1
Fujin Xiang 5th Bao				The third son	石	三信	Male	Dog	1946	7

Date	The beginning time of inhabitation in this county	Educational level	Profession sector	Occupation	Marital status	Physical disability	Religion
2	Inhabited for generations	Illiteracy	Agriculture	Land-holding peasant	Unmarried	No	No religion
22	Inhabited for generations	Illiteracy					
11	Inhabited for generations	Sishu 2 years	Agriculture	Land-holding peasant	Married	No	No religion
1	Inhabited for generations						
1	Inhabited for generations						
11	Inhabited for generations	Illiteracy	Agriculture	Land-holding peasant	Married	No	No religion
16	Inhabited for generations	Illiteracy	Agriculture	Land-holding peasant	Married	No	No religion
24	Inhabited for generations	Illiteracy	Agriculture	Land-holding peasant	Unmarried	No	No religion
29	Inhabited for generations						
8	Inhabited for generations	Illiteracy	Agriculture	Land-holding peasant	Widowed	No	No religion
28	Inhabited for generations	Illiteracy	Agriculture	Land-holding peasant	Married	No	No religion
12	Inhabited for generations	Illiteracy	Agriculture	Land-holding peasant	Married	No	No religion
2	Inhabited for generations	Illiteracy					
5	Inhabited for generations	Illiteracy	Agriculture	Land-holding peasant	Married	No	No religion
21	Inhabited for generations	Illiteracy	Agriculture	Land-holding peasant	Married	No	No religion
19	Inhabited for generations	Bao school 2 years					
3	Inhabited for generations	Bao school 1 year					
1	Inhabited for generations						

Appendix: Overview of Microdata for Wudu County (*cont.*)

	The no. of household	Jia	Household	Relationship with the household head	Surname	First name	Sex	The Chinese zodiac	Birthdate	
									Year	Month
Fujin Xiang 5th Bao				Daughter	石	旦娃	Female	Snake	1880	1
Fujin Xiang 5th Bao				Nephew	石	转文	Male	Ox	1887	3
Fujin Xiang 5th Bao				Nephew's wife	石	房娃	Female	Dragon	1891	5
Fujin Xiang 5th Bao				Nephew's son	石	均成	Male	Pig	1910	4
Fujin Xiang 5th Bao				Nephew's wife	谈	香儿	Female	Snake	1917	1
Fujin Xiang 5th Bao				Nephew's grandson	马	俊	Male	Monkey	1944	6
Fujin Xiang 5th Bao				Great Grand-daughter	柳	娃	Female	Rabbit	1939	11
Fujin Xiang 5th Bao	561	12	3	The household head	赵	玉清	Male	Dog	1909	6
Fujin Xiang 5th Bao				The household head's wife	石	长女子	Female	Rabbit	1902	2
Fujin Xiang 5th Bao				Mother	月	儿子	Female	Pig	1886	2
Fujin Xiang 5th Bao				Son	赵	江娃	Male	Rat	1924	3
Fujin Xiang 5th Bao				The second son	赵	韩信儿	Male	Pig	1931	3
Fujin Xiang 5th Bao				Uncle	赵	文银	Male	Goat	1894	3
Fujin Xiang 5th Bao				Uncle's son	赵	王境	Male	Rat	1925	1
Fujin Xiang 5th Bao				Younger brother's wife	廷	娃	Female	Rat	1924	9
Fujin Xiang 5th Bao				Nephew	赵	贵生	Male	Dog	1946	6
Fujin Xiang 5th Bao	562	12	4	The household head	赵	玉海	Male	Tiger	1901	11

Date	The beginning time of inhabitation in this county	Educational level	Profession sector	Occupation	Marital status	Physical disability	Religion
26	Inhabited for generations	Illiteracy					
19	Inhabited for generations	Illiteracy	Agriculture	Land-holding peasant	Married	No	No religion
22	Inhabited for generations	Illiteracy	Agriculture	Land-holding peasant	Married	No	No religion
29	Inhabited for generations	Illiteracy	Agriculture	Land-holding peasant	Married	No	No religion
23	Inhabited for generations	Illiteracy	Agriculture	Land-holding peasant	Married	No	No religion
2	Inhabited for generations						
27	Inhabited for generations	Illiteracy					
21	Inhabited for generations	Illiteracy	Agriculture	Land-holding peasant	Married	No	No religion
12	Inhabited for generations	Illiteracy	Agriculture	Land-holding peasant	Married	No	No religion
10	Inhabited for generations	Illiteracy	Agriculture	Land-holding peasant	Widowed	No	No religion
18	Inhabited for generations	Illiteracy	Agriculture	Land-holding peasant	Unmarried	No	No religion
8	Inhabited for generations	Illiteracy	Agriculture	Land-holding peasant	Unmarried	No	No religion
4	Inhabited for generations	Illiteracy	Agriculture	Land-holding peasant	Widowed	No	No religion
15	Inhabited for generations	Illiteracy	Agriculture	Land-holding peasant	Married	No	No religion
13	Inhabited for generations	Illiteracy	Agriculture	Land-holding peasant	Married	No	No religion
28	Inhabited for generations						
9	Inhabited for generations	Illiteracy	Agriculture	Land-holding peasant	Married	No	No religion

Appendix: Overview of Microdata for Wudu County (*cont.*)

	The no. of household	Jia	Household	Relationship with the household head	Surname	First name	Sex	The Chinese zodiac	Birthdate Year	Birthdate Month
Fujin Xiang 5th Bao				The household head's wife	石	林娃	Female	Horse	1905	6
Fujin Xiang 5th Bao				Son	赵	院娃	Male	Goat	1943	2
Fujin Xiang 5th Bao				Daughter	赵	连娃	Female	Rooster	1934	1
Fujin Xiang 5th Bao				The second daughter	赵	毕连	Female	Rabbit	1939	8
Fujin Xiang 5th Bao	563	12	5	The household head	赵	玉兴	Male	Horse	1918	5
Fujin Xiang 5th Bao				The household head's wife	柱	林娃	Female	Horse	1918	5
Fujin Xiang 5th Bao				Mother	有	节子	Female	Goat	1894	3
Fujin Xiang 5th Bao				Daughter	赵	银女	Female	Monkey	1945	1
Fujin Xiang 5th Bao				Grandfather's brother	赵	金喜	Male	Rat	1899	4
Fujin Xiang 5th Bao				Grandmother	赵	娃	Female	Goat	1905	1
Fujin Xiang 5th Bao				Farther's sister	列	娃	Female	Goat	1932	1
Fujin Xiang 5th Bao				Father's second sister	周	女	Female	Rabbit	1939	7
Fujin Xiang 5th Bao				Father's third sister	赵	仓儿	Female	Monkey	1944	7
Fujin Xiang 5th Bao				Father's fourth sister	四	女子	Female	Dog	1946	3
Fujin Xiang 5th Bao	564	12	6	The household head	赵	文章	Male	Rooster	1906	3
Fujin Xiang 5th Bao				The household head's wife	石	早娃	Female	Monkey	1894	9
Fujin Xiang 5th Bao				Son	杨	学娃	Male	Dog	1922	9

Date	The beginning time of inhabitation in this county	Educational level	Profession sector	Occupation	Marital status	Physical disability	Religion
29	Inhabited for generations	Illiteracy	Agriculture	Land-holding peasant	Married	No	No religion
21	Inhabited for generations						
30	Inhabited for generations	Illiteracy	Agriculture	Land-holding peasant			
16	Inhabited for generations	Illiteracy					
9	Inhabited for generations	Illiteracy	Agriculture	Land-holding peasant	Married	No	No religion
2	Inhabited for generations	Illiteracy	Agriculture	Land-holding peasant	Married	No	No religion
26	Inhabited for generations	Illiteracy	Agriculture	Land-holding peasant	Married	No	No religion
12	Inhabited for generations						
11	Inhabited for generations	Illiteracy	Agriculture	Land-holding peasant	Married	No	No religion
11	Inhabited for generations	Illiteracy	Agriculture	Land-holding peasant	Married	No	No religion
16	Inhabited for generations	Illiteracy	Agriculture	Land-holding peasant	Unmarried	No	No religion
1	Inhabited for generations	Illiteracy					
25	Inhabited for generations						
1	Inhabited for generations						
8	Inhabited for generations	Illiteracy	Agriculture	Land-holding peasant	Married	No	No religion
7	Inhabited for generations	Illiteracy	Agriculture	Land-holding peasant	Married	No	No religion
16	Inhabited for generations	Illiteracy	Agriculture	Land-holding peasant	Married	No	No religion

Appendix: Overview of Microdata for Wudu County (*cont.*)

	The no. of household	Jia	Household	Relationship with the household head	Surname	First name	Sex	The Chinese zodiac	Birthdate	
									Year	Month
Fujin Xiang 5th Bao				Son's wife	石	炉娃	Female	Dragon	1928	2
Fujin Xiang 5th Bao				Grandson	赵	清女	Female	Dog	1946	10
Fujin Xiang 5th Bao				The second son	赵	远娃	Male	Ox	1926	1
Fujin Xiang 5th Bao				Daughter	赵	崴柱娃	Female	Goat	1931	9
Fujin Xiang 5th Bao				Younger brother	赵	文回	Male	Ox	1900	10
Fujin Xiang 5th Bao				The second younger brother	赵	刘生儿	Male	Ox	1913	10
Fujin Xiang 5th Bao	565	12	7	The household head	石	变儿	Male	Rat	1924	6
Fujin Xiang 5th Bao				The household head's wife	李	拐娃	Female	Dragon	1928	10
Fujin Xiang 5th Bao	566	12	8	The household head	赵	金有	Male	Rooster	1908	10
Fujin Xiang 5th Bao				The household head's wife	石	禄儿	Female	Goat	1906	10
Fujin Xiang 5th Bao				Father	赵	登禄	Male	Dragon	1879	10
Fujin Xiang 5th Bao				Son	赵	文举	Male	Rooster	1933	10
Fujin Xiang 5th Bao				The second son	赵	岁社儿	Male	Ox	1937	11
Fujin Xiang 5th Bao				Daughter	赵	社女	Female	Goat	1943	9
Fujin Xiang 5th Bao				Nephew's wife	王	喜娃	Female	Rooster	1896	3
Fujin Xiang 5th Bao				Nephew's son	赵	王海儿	Male	Rooster	1921	4
Fujin Xiang 5th Bao				Nephew's son's wife	王	秀娃	Female	Tiger	1926	2
Fujin Xiang 5th Bao	567	12	9	The household head	赵	文田	Male	Dragon	1916	2

Date	The beginning time of inhabitation in this county	Educational level	Profession sector	Occupation	Marital status	Physical disability	Religion
28	Inhabited for generations	Illiteracy	Agriculture	Land-holding peasant	Married	No	No religion
16	Inhabited for generations						
14	Inhabited for generations	Illiteracy	Agriculture	Land-holding peasant	Unmarried	No	No religion
1	Inhabited for generations	Illiteracy	Agriculture	Land-holding peasant	Unmarried	No	No religion
20	Inhabited for generations	Illiteracy	Agriculture	Land-holding peasant	Widowed	No	No religion
20	Inhabited for generations	Illiteracy	Agriculture	Land-holding peasant	Unmarried	No	No religion
9	Inhabited for generations	Illiteracy	Agriculture	Land-holding peasant	Married	No	No religion
26	Inhabited for generations	Illiteracy	Agriculture	Land-holding peasant	Married	No	No religion
18	Inhabited for generations	Illiteracy	Agriculture	Land-holding peasant	Married	No	No religion
9	Inhabited for generations	Illiteracy	Agriculture	Land-holding peasant	Married	No	No religion
12	Inhabited for generations	Illiteracy					
8	Inhabited for generations	Bao school 3 years					
12	Inhabited for generations	Illiteracy					
29	Inhabited for generations						
28	Inhabited for generations	Illiteracy	Agriculture	Land-holding peasant	Widowed	No	No religion
10	Inhabited for generations	Illiteracy	Agriculture	Land-holding peasant	Married	No	No religion
15	Inhabited for generations	Illiteracy	Agriculture	Land-holding peasant	Married	No	No religion
5	Inhabited for generations	Illiteracy	Agriculture	Land-holding peasant	Married	No	No religion

Appendix: Overview of Microdata for Wudu County (*cont.*)

	The no. of household	Jia	Household	Relationship with the household head	Surname	First name	Sex	The Chinese zodiac	Birthdate Year	Month
Fujin Xiang 5th Bao				The household head's wife	炉	娃	Female	Pig	1910	3
Fujin Xiang 5th Bao				Daughter	赵	菜娃	Female	Rat	1936	11
Fujin Xiang 5th Bao	568	12	10	The household head	赵	文炳	Male	Rooster	1896	3
Fujin Xiang 5th Bao				The household head's wife	石	清娃	Female	Goat	1905	1
Fujin Xiang 5th Bao				Son	赵	董娃	Male	Monkey	1938	6
Fujin Xiang 5th Bao				Son's wife	石	坡女子	Female	Goat	1931	6
Fujin Xiang 5th Bao				The second son	赵	根娃	Male	Pig	1935	3
Fujin Xiang 5th Bao				The third son	赵	迟根儿	Male	Rabbit	1940	1
Fujin Xiang 5th Bao				The fourth son	赵	四娃	Male	Monkey	1944	2
Fujin Xiang 5th Bao				Daughter	赵	根女	Female	Snake	1941	5
Fujin Xiang 5th Bao				The second younger brother	赵	生娃	Male	Horse	1905	12
Fujin Xiang 5th Bao				The third younger brother	赵	双贵儿	Male	Horse	1918	5
Fujin Xiang 5th Bao				Uncle's son	赵	杨金儿	Male	Dragon	1903	11
Fujin Xiang 5th Bao	569	12	11	The household head	石	文三	Male	Goat	1906	7
Fujin Xiang 5th Bao				The household head's wife	张	等娃	Female	Dragon	1916	2
Fujin Xiang 5th Bao				Father	石	三磴眼	Male	Goat	1870	11
Fujin Xiang 5th Bao				Mother	鼠	姐子	Female	Rat	1894	12

Date	The beginning time of inhabitation in this county	Educational level	Profession sector	Occupation	Marital status	Physical disability	Religion
22	Inhabited for generations	Illiteracy	Agriculture	Land-holding peasant	Married	No	No religion
24	Inhabited for generations	Illiteracy					
9	Inhabited for generations	Illiteracy	Agriculture	Land-holding peasant	Married	No	No religion
26	Inhabited for generations	Illiteracy	Agriculture	Land-holding peasant	Married	No	No religion
1	Inhabited for generations	Illiteracy	Agriculture	Land-holding peasant			
20	Inhabited for generations	Illiteracy	Agriculture	Land-holding peasant			
18	Inhabited for generations	Bao school 2 years					
20	Inhabited for generations	Illiteracy					
1	Inhabited for generations						
6	Inhabited for generations	Illiteracy					
30	Inhabited for generations	Illiteracy	Agriculture	Land-holding peasant	Unmarried	No	No religion
21	Inhabited for generations	Illiteracy	Agriculture	Land-holding peasant	Unmarried	No	No religion
24	Inhabited for generations	Illiteracy	Agriculture		Widowed	No	No religion
17	Inhabited for generations	Illiteracy	Agriculture	Land-holding peasant	Married	No	No religion
1	Inhabited for generations	Illiteracy	Agriculture	Land-holding peasant	Married	No	No religion
1	Inhabited for generations						
5	Inhabited for generations						

Appendix: Overview of Microdata for Wudu County (*cont.*)

	The no. of household	Jia	Household	Relationship with the household head	Surname	First name	Sex	The Chinese zodiac	Birthdate	
									Year	Month
Fujin Xiang 5th Bao				Son	石	鸡唤儿	Male	Monkey	1944	11
Fujin Xiang 5th Bao				Younger brother	石	文进	Male	Tiger	1914	6
Fujin Xiang 5th Bao				Younger brother's wife	安	反娃	Female	Dragon	1916	10
Fujin Xiang 5th Bao				Nephew	石	才娃	Male	Rabbit	1939	8
Fujin Xiang 5th Bao				The second nephew	石	贾唤儿	Male	Goat	1943	5
Fujin Xiang 5th Bao				The third nephew	狗	善儿	Male	Dog	1946	6
Fujin Xiang 5th Bao	570	12	12	The household head	石	文奇	Male	Rooster	1884	11
Fujin Xiang 5th Bao				The household head's wife	石	田娃	Female	Dog	1885	1
Fujin Xiang 5th Bao				Son	石	长生儿	Male	Monkey	1915	12
Fujin Xiang 5th Bao				Son's wife	石	贵相儿	Female	Goat	1906	1
Fujin Xiang 5th Bao				The second son	石	董娃	Male	Rat	1924	12
Fujin Xiang 5th Bao				The second son's wife	赵	登哇	Female	Rat	1924	2
Fujin Xiang 5th Bao				Grandson	石	谢娃	Male	Rooster	1933	1
Fujin Xiang 5th Bao	571	12	13	The household head	石	含旺	Male	Horse	1918	7
Fujin Xiang 5th Bao				The household head's wife	腾	哇	Female	Horse	1918	12
Fujin Xiang 5th Bao				Son	石	小儿	Male	Snake	1941	10
Fujin Xiang 5th Bao				Daughter	石	苏女子	Female	Monkey	1943	3
Fujin Xiang 5th Bao	572	12	14	The household head	石	房儿	Female	Goat	1894	6

Date	The beginning time of inhabitation in this county	Educational level	Profession sector	Occupation	Marital status	Physical disability	Religion
27	Inhabited for generations						
2	Inhabited for generations	Illiteracy	Agriculture	Land-holding peasant	Married	No	No religion
21	Inhabited for generations	Illiteracy	Agriculture	Land-holding peasant	Married	No	No religion
23	Inhabited for generations	Illiteracy					
6	Inhabited for generations						
1	Inhabited for generations						
6	Inhabited for generations	Sishu 4 years	Agriculture	Land-holding peasant	Married	No	No religion
31	Inhabited for generations	Illiteracy	Agriculture	Land-holding peasant	Married	No	No religion
30	Inhabited for generations	Sishu 2 years	Agriculture	Land-holding peasant	Married	No	No religion
1	Inhabited for generations	Illiteracy	Agriculture	Land-holding peasant	Married	No	No religion
27	Inhabited for generations	Illiteracy	Agriculture	Land-holding peasant	Married	No	No religion
17	Inhabited for generations	Illiteracy	Agriculture	Land-holding peasant	Married	No	No religion
6	Inhabited for generations	Illiteracy	Agriculture		Unmarried		
20	Inhabited for generations	Illiteracy	Agriculture	Land-holding peasant	Married	No	No religion
7	Inhabited for generations	Illiteracy	Agriculture	Land-holding peasant	Married	No	No religion
20	Inhabited for generations	Illiteracy					
13	Inhabited for generations						
18	Inhabited for generations	Illiteracy	Agriculture	Land-holding peasant	Widowed	No	No religion

Appendix: Overview of Microdata for Wudu County (*cont.*)

	The no. of household	Jia	Household	Relationship with the household head	Surname	First name	Sex	The Chinese zodiac	Birthdate	
									Year	Month
Fujin Xiang 5th Bao				Son	石	仓儿	Male	Rat	1932	12
Fujin Xiang 5th Bao				Grandson	石	乘儿	Male	Dragon	1940	11
Fujin Xiang 5th Bao	573	12	15	The household head	石	文仓	Male	Monkey	1895	6
Fujin Xiang 5th Bao				The household head's wife	石	银娃	Female	Rat	1912	1
Fujin Xiang 5th Bao				Son	石	哇	Male	Monkey	1932	5
Fujin Xiang 5th Bao				Nephew	石	等成	Male	Tiger	1938	2
Fujin Xiang 5th Bao	574	13	1	The household head	赵	文进	Male	Dog	1885	2
Fujin Xiang 5th Bao				The household head's wife	李	吾儿	Female	Horse	1881	1
Fujin Xiang 5th Bao				Son	石	杨娃	Male	Dragon	1916	10
Fujin Xiang 5th Bao				Son's wife	石	腊气	Female	Snake	1917	10
Fujin Xiang 5th Bao				Grandson	赵	东娃	Male	Dog	1942	1
Fujin Xiang 5th Bao	575	13	2	The household head	石	文平	Male	Goat	1894	5
Fujin Xiang 5th Bao				The household head's wife	赵	祳秀	Female	Dog	1897	7
Fujin Xiang 5th Bao				Uncle's son	石	文昌	Male	Pig	1910	3
Fujin Xiang 5th Bao				Uncle's son's wife	安	著娃	Female	Rat	1912	7
Fujin Xiang 5th Bao				Nephew	石	时儿	Male	Rat	1945	6
Fujin Xiang 5th Bao				Niece	石	女儿	Female	Snake	1941	2
Fujin Xiang 5th Bao	576	13	3	The household head	石	文朱	Male	Goat	1894	5

Date	The beginning time of inhabitation in this county	Educational level	Profession sector	Occupation	Marital status	Physical disability	Religion
5	Inhabited for generations	Illiteracy	Agriculture				
5	Inhabited for generations	Illiteracy					
11	Inhabited for generations	Illiteracy	Agriculture	Land-holding peasant	Married	No	No religion
6	Inhabited for generations	Illiteracy	Agriculture	Land-holding peasant	Married	No	No religion
6	Inhabited for generations	Illiteracy	Agriculture	Land-holding peasant	Unmarried	No	No religion
9	Inhabited for generations	Illiteracy					
26	Inhabited for generations	Illiteracy	Agriculture	Land-holding peasant	Married	No	No religion
28	Inhabited for generations	Illiteracy	Agriculture	Land-holding peasant	Married	No	No religion
27	Inhabited for generations	Illiteracy	Agriculture	Land-holding peasant	Married	No	No religion
7	Inhabited for generations	Illiteracy	Agriculture	Land-holding peasant	Married	No	No religion
11	Inhabited for generations						
2	Inhabited for generations	Illiteracy	Agriculture	Land-holding peasant	Married	No	No religion
1	Inhabited for generations	Illiteracy	Agriculture	Land-holding peasant	Married	No	No religion
20	Inhabited for generations	Illiteracy	Agriculture	Land-holding peasant	Married	No	No religion
1	Inhabited for generations	Illiteracy	Agriculture	Land-holding peasant	Married	No	No religion
21	Inhabited for generations						
9	Inhabited for generations						
6	Inhabited for generations	Illiteracy					

Appendix: Overview of Microdata for Wudu County (*cont.*)

	The no. of house-hold	Jia	House-hold	Relation-ship with the house-hold head	Surname	First name	Sex	The Chinese zodiac	Birthdate	
									Year	Month
Fujin Xiang 5th Bao				Mother	赵	缠缠	Female	Dragon	1867	4
Fujin Xiang 5th Bao				Younger brother	石	文朝	Male	Rat	1899	12
Fujin Xiang 5th Bao				The second younger brother	石	文举	Male	Dog	1909	11
Fujin Xiang 5th Bao				The second younger brother's wife	陈	崴娃	Female	Rooster	1923	5
Fujin Xiang 5th Bao				Nephew	石	树生	Male	Rabbit	1935	10
Fujin Xiang 5th Bao				The house-hold head's wife's mother	陈	成女	Female	Snake	1892	10
Fujin Xiang 5th Bao	577	13	4	The house-hold head	石	含奇	Male	Rooster	1910	10
Fujin Xiang 5th Bao				The house-hold head's wife	崔	扣儿	Female	Pig	1910	8
Fujin Xiang 5th Bao				Father	石	文唤	Male		1877	12
Fujin Xiang 5th Bao				Son	石	徐姓儿	Male	Rabbit	1939	11
Fujin Xiang 5th Bao				Daughter	石	徐女	Female	Monkey	1944	6
Fujin Xiang 5th Bao				The second daughter	石	女徐	Female	Dog	1946	5
Fujin Xiang 5th Bao				Uncle's son's wife	石	著娃	Female	Rooster	1908	10
Fujin Xiang 5th Bao				Nephew	石	张存	Male	Tiger	1938	3
Fujin Xiang 5th Bao				Niece	石	柏儿	Female	Dragon	1936	5
Fujin Xiang 5th Bao				The second niece	石	女儿	Female	Goat	1943	5

Date	The beginning time of inhabitation in this county	Educational level	Profession sector	Occupation	Marital status	Physical disability	Religion
1	Inhabited for generations	Illiteracy					
22	Inhabited for generations	Illiteracy	Agriculture	Land-holding peasant	Widowed	No	No religion
1	Inhabited for generations	Illiteracy	Agriculture	Land-holding peasant	Married	No	No religion
30	Inhabited for generations	Illiteracy	Agriculture	Land-holding peasant	Married	No	No religion
1	Inhabited for generations	Illiteracy	Agriculture				
8	Inhabited for generations						
8	Inhabited for generations	Illiteracy	Agriculture	Land-holding peasant	Married	No	No religion
25	Inhabited for generations	Illiteracy	Agriculture	Land-holding peasant	Married	No	No religion
15	Inhabited for generations						
12	Inhabited for generations	Illiteracy					
3	Inhabited for generations						
3	Inhabited for generations						
16	Inhabited for generations	Illiteracy	Agriculture	Land-holding peasant	Widowed	No	No religion
13	Inhabited for generations	Illiteracy					
25	Inhabited for generations	Illiteracy					
30	Inhabited for generations						

Appendix: Overview of Microdata for Wudu County (*cont.*)

	The no. of household	Jia	Household	Relationship with the household head	Surname	First name	Sex	The Chinese zodiac	Birthdate Year	Month
Fujin Xiang 5th Bao	578	13	5	The household head	石	含科	Male	Pig	1910	2
Fujin Xiang 5th Bao				The household head's wife	黄	楼儿	Female	Pig	1910	3
Fujin Xiang 5th Bao				Son	石	杨时儿	Male	Horse	1942	10
Fujin Xiang 5th Bao				Daughter	石	早儿	Female	Horse	1923	2
Fujin Xiang 5th Bao				The second daughter	石	毕娃	Female	Tiger	1922	5
Fujin Xiang 5th Bao				The third daughter	石	妖女	Female	Snake	1926	2
Fujin Xiang 5th Bao	579	13	6	The household head	石	含富	Male	Pig	1887	5
Fujin Xiang 5th Bao				The household head's wife	赵	高姐子	Female	Ox	1888	4
Fujin Xiang 5th Bao				Son	石	富金儿	Male	Snake	1917	9
Fujin Xiang 5th Bao				Son's wife	石	富女	Female	Snake	1917	10
Fujin Xiang 5th Bao				Grandson	石	张张	Male	Pig	1947	9
Fujin Xiang 5th Bao				Granddaughter	石	乘女	Female	Rooster	1945	5
Fujin Xiang 5th Bao				The second son	石	正才儿	Male	Horse	1930	9
Fujin Xiang 5th Bao				Younger brother	石	朝银儿	Male	Tiger	1889	7
Fujin Xiang 5th Bao	580	13	7	The household head	石	会永	Male	Rabbit	1915	8
Fujin Xiang 5th Bao				The household head's wife	张	等娃	Female	Goat	1919	5
Fujin Xiang 5th Bao				Son	石	埋娃	Male	Horse	1931	1
Fujin Xiang 5th Bao				The second son	郭	有娃	Male	Monkey	1944	3

Date	The beginning time of inhabitation in this county	Educational level	Profession sector	Occupation	Marital status	Physical disability	Religion
4	Inhabited for generations	Illiteracy	Agriculture	Land-holding peasant	Married	No	No religion
4	Inhabited for generations	Illiteracy	Agriculture	Land-holding peasant	Married	No	No religion
15	Inhabited for generations						
4	Inhabited for generations	Illiteracy	Agriculture	Land-holding peasant	Unmarried	No	No religion
6	Inhabited for generations	Illiteracy	Agriculture	Land-holding peasant	Unmarried	No	No religion
8	Inhabited for generations	Illiteracy					
5	Inhabited for generations	Illiteracy	Agriculture	Land-holding peasant	Married	No	No religion
13	Inhabited for generations	Illiteracy	Agriculture	Land-holding peasant	Married	No	No religion
2	Inhabited for generations	Illiteracy	Agriculture	Land-holding peasant	Married	No	No religion
3	Inhabited for generations	Illiteracy	Agriculture	Land-holding peasant	Married	No	No religion
2	Inhabited for generations						
29	Inhabited for generations						
20	Inhabited for generations	Illiteracy	Agriculture				
22	Inhabited for generations	Illiteracy	Agriculture	Land-holding peasant	Widowed	No	No religion
23	Inhabited for generations	Illiteracy	Agriculture	Land-holding peasant	Married	No	No religion
1	Inhabited for generations	Illiteracy	Agriculture	Land-holding peasant	Married	No	No religion
31	Inhabited for generations	Illiteracy	Agriculture	Land-holding peasant	Unmarried	No	No religion
12	Inhabited for generations						

Appendix: Overview of Microdata for Wudu County (*cont.*)

	The no. of household	Jia	Household	Relationship with the household head	Surname	First name	Sex	The Chinese zodiac	Birthdate	
									Year	Month
Fujin Xiang 5th Bao				The third son	石	国健	Male	Pig	1947	5
Fujin Xiang 5th Bao	581	13	8	The household head	石	含生	Male	Dragon	1916	7
Fujin Xiang 5th Bao				The household head's wife	郭	董扣	Female	Rooster	1921	6
Fujin Xiang 5th Bao				Mother	陈	琴娃	Female	Rabbit	1944	4
Fujin Xiang 5th Bao				Son	石	仓儿	Male	Tiger	1938	1
Fujin Xiang 5th Bao				Older brother	石	含魁	Male	Monkey	1907	8
Fujin Xiang 5th Bao				Older brother's wife	安	庆娃	Female	Monkey	1907	7
Fujin Xiang 5th Bao				Nephew	石	高才儿	Male	Rat	1936	11
Fujin Xiang 5th Bao	582	13	9	The household head	石	含转	Male	Goat	1919	1
Fujin Xiang 5th Bao				The household head's wife	石	升女	Female	Goat	1919	4
Fujin Xiang 5th Bao				Father	石	文三	Male	Tiger	1933	11
Fujin Xiang 5th Bao				Daughter	石	永清	Female	Horse	1942	12
Fujin Xiang 5th Bao				The second daughter	石	成清	Female	Dog	1946	8
Fujin Xiang 5th Bao				Younger brother	石	称信儿	Male	Ox	1925	5
Fujin Xiang 5th Bao				The second younger brother	石	谢生儿	Male	Rabbit	1927	8
Fujin Xiang 5th Bao				The second younger brother's wife	石	进团儿	Female	Rabbit	1927	6

Date	The beginning time of inhabitation in this county	Educational level	Profession sector	Occupation	Marital status	Physical disability	Religion
19	Inhabited for generations						
5	Inhabited for generations	Illiteracy	Agriculture	Land-holding peasant	Married	No	No religion
21	Inhabited for generations	Illiteracy	Agriculture	Land-holding peasant	Married	No	No religion
8	Inhabited for generations						
24	Inhabited for generations						
15	Inhabited for generations	Illiteracy	Agriculture	Land-holding peasant	Married	No	No religion
7	Inhabited for generations	Illiteracy	Agriculture	Land-holding peasant	Married	No	No religion
28	Inhabited for generations	Illiteracy					
15	Inhabited for generations	Illiteracy	Agriculture	Land-holding peasant	Married	No	No religion
3	Inhabited for generations	Illiteracy	Agriculture	Land-holding peasant	Married	No	No religion
8	Inhabited for generations	Illiteracy	Agriculture	Land-holding peasant	Widowed	No	No religion
10	Inhabited for generations	Illiteracy					
14	Inhabited for generations						
31	Inhabited for generations	Illiteracy	Agriculture	Land-holding peasant	Unmarried	No	No religion
15	Inhabited for generations	Illiteracy	Agriculture	Land-holding peasant	Married	No	No religion
26	Inhabited for generations	Illiteracy	Agriculture	Land-holding peasant	Married	No	No religion

Appendix: Overview of Microdata for Wudu County (*cont.*)

	The no. of household	Jia	Household	Relationship with the household head	Surname	First name	Sex	The Chinese zodiac	Birthdate	
									Year	Month
Fujin Xiang 5th Bao				Older brother	石	含焕	Male	Pig	1910	6
Fujin Xiang 5th Bao				Older brother's wife	杨	扣儿	Female	Pig	1910	1
Fujin Xiang 5th Bao				Nephew	石	满娃	Male	Rooster	1933	9
Fujin Xiang 5th Bao				The second nephew	石	平娃	Female	Tiger	1938	8
Fujin Xiang 5th Bao				Uncle's son's wife	楼	儿	Female	Rabbit	1902	3
Fujin Xiang 5th Bao				Grand-daughter	石	女子	Female	Dragon	1928	10
Fujin Xiang 5th Bao				Uncle's son's great grandson	石	天时	Male	Dog	1946	12
Fujin Xiang 5th Bao				The second great Grand-daughter	石	兹娃	Female	Rabbit	1939	2
Fujin Xiang 5th Bao	583	13	10	The household head	石	含成	Male	Rooster	1896	2
Fujin Xiang 5th Bao				The household head's wife	李	大听娃	Female	Rat	1899	4
Fujin Xiang 5th Bao				Father	石	文德	Male	Rabbit	1866	7
Fujin Xiang 5th Bao				Mother	黄	氏	Female	Rabbit	1866	5
Fujin Xiang 5th Bao				Son	石	大满哇	Male	Pig	1910	4
Fujin Xiang 5th Bao				Son's wife	石	雪娃	Female	Tiger	1926	5
Fujin Xiang 5th Bao				Grand-daughter	石	早女	Female	Horse	1942	11
Fujin Xiang 5th Bao				The second son	石	双才	Male	Dragon	1928	2
Fujin Xiang 5th Bao				The second son's wife	蒙	生娃	Female	Rat	1924	3

Date	The beginning time of inhabitation in this county	Educational level	Profession sector	Occupation	Marital status	Physical disability	Religion
28	Inhabited for generations	Illiteracy	Agriculture	Land-holding peasant	Married	No	No religion
7	Inhabited for generations	Illiteracy	Agriculture	Land-holding peasant	Married	No	No religion
1	Inhabited for generations	Illiteracy	Agriculture	Land-holding peasant	Unmarried	No	No religion
21	Inhabited for generations	Illiteracy					
3	Inhabited for generations	Illiteracy	Agriculture	Land-holding peasant	Widowed	No	No religion
18	Inhabited for generations	Illiteracy	Agriculture	Land-holding peasant	Unmarried	No	No religion
9	Inhabited for generations						
21	Inhabited for generations	Illiteracy					
11	Inhabited for generations	Illiteracy	Agriculture	Land-holding peasant	Married	No	No religion
17	Inhabited for generations	Illiteracy	Agriculture	Land-holding peasant	Married	No	No religion
7	Inhabited for generations						
6	Inhabited for generations						
3	Inhabited for generations	Illiteracy	Agriculture	Land-holding peasant	Married	No	No religion
5	Inhabited for generations	Illiteracy	Agriculture	Land-holding peasant	Married	No	No religion
14	Inhabited for generations						
22	Inhabited for generations	Illiteracy	Agriculture	Land-holding peasant	Married	No	No religion
9	Inhabited for generations	Illiteracy	Agriculture	Land-holding peasant	Married	No	No religion

Appendix: Overview of Microdata for Wudu County (*cont.*)

									Birthdate	
	The no. of household	**Jia**	**Household**	**Relationship with the household head**	**Surname**	**First name**	**Sex**	**The Chinese zodiac**	**Year**	**Month**
Fujin Xiang 5th Bao				Grandson	杨	生	Male	Rooster	1944	11
Fujin Xiang 5th Bao				The second grandson	石	崴很	Male	Dog	1946	11
Fujin Xiang 5th Bao				The third grandson	石	金生	Male	Pig	1947	11
Fujin Xiang 5th Bao				Granddaughter	石	金平	Female	Rabbit	1939	11
Fujin Xiang 5th Bao				Younger brother	石	含清	Male	Goat	1917	3
Fujin Xiang 5th Bao				Younger brother's wife	李	还娃	Female	Ox	1913	6
Fujin Xiang 5th Bao				Nephew	石	学儿	Male	Rooster	1929	10
Fujin Xiang 5th Bao				The second nephew	石	董儿	Male	Rabbit	1939	12
Fujin Xiang 5th Bao				Niece	石	翠娃	Female	Snake	1941	5
Fujin Xiang 5th Bao	584	13	11	The household head	石	含泰	Male	Rooster	1895	12
Fujin Xiang 5th Bao				The household head's wife	李	房儿	Female	Rooster	1896	12
Fujin Xiang 5th Bao				Son	石	会章	Male	Ox	1925	
Fujin Xiang 5th Bao				Son's wife	赵	四乡	Female	Dog	1924	4
Fujin Xiang 5th Bao				Granddaughter	石	苏音	Female	Monkey	1944	2
Fujin Xiang 5th Bao				Daughter	石	吾儿	Female	Monkey	1921	1
Fujin Xiang 5th Bao	585	13	12	The household head	石	含德	Male	Pig	1885	1
Fujin Xiang 5th Bao				Son	石	会一	Male	Rabbit	1927	9
Fujin Xiang 5th Bao				Son's wife	石	赵娃	Female		1927	9

Date	The beginning time of inhabitation in this county	Educational level	Profession sector	Occupation	Marital status	Physical disability	Religion	
21	Inhabited for generations							
9	Inhabited for generations							
21	Inhabited for generations							
16	Inhabited for generations	Illiteracy						
3	Inhabited for generations	Sishu 3 years	Agriculture	Land-holding peasant	Married	No	No religion	
18	Inhabited for generations	Illiteracy	Agriculture	Land-holding peasant	Married	No	No religion	
13	Inhabited for generations	Illiteracy	Agriculture	Land-holding peasant	Unmarried	No	No religion	Serving in the army
24	Inhabited for generations	Illiteracy						
30	Inhabited for generations	Illiteracy						
21	Inhabited for generations	Illiteracy	Agriculture	Land-holding peasant	Married	No	No religion	
26	Inhabited for generations	Illiteracy	Agriculture	Land-holding peasant	Married	No	No religion	
2	Inhabited for generations	Illiteracy	Agriculture	Land-holding peasant	Married	No	No religion	
27	Inhabited for generations	Illiteracy	Agriculture	Land-holding peasant	Married	No	No religion	
2	Inhabited for generations							
19	Inhabited for generations					Dumb		
13	Inhabited for generations	Illiteracy	Agriculture	Land-holding peasant	Widowed	No	No religion	
24	Inhabited for generations	Illiteracy	Agriculture	Land-holding peasant	Married	No	No religion	
24	Inhabited for generations	Illiteracy	Agriculture	Land-holding peasant	Married	No	No religion	

Appendix: Overview of Microdata for Wudu County (*cont.*)

	The no. of household	Jia	Household	Relationship with the household head	Surname	First name	Sex	The Chinese zodiac	Birthdate	
									Year	Month
Fujin Xiang 5th Bao				Grandson	石	毕英	Female	Dog	1947	1
Fujin Xiang 5th Bao	586	13	13	The household head	李	珍科	Male	Pig	1910	7
Fujin Xiang 5th Bao				The household head's wife	石	清娃	Female	Rabbit	1915	6
Fujin Xiang 5th Bao				Son	李	银五儿	Male	Tiger	1938	5
Fujin Xiang 5th Bao				The second son	狮	子保	Male	Goat	1942	7
Fujin Xiang 5th Bao				The third son	李	高生	Male	Rooster	1945	11
Fujin Xiang 5th Bao				Daughter	李	银女儿	Female	Dragon	1940	3
Fujin Xiang 5th Bao	587	13	14	The household head	石	含宗	Male	Dragon	1903	2
Fujin Xiang 5th Bao				The household head's wife	崔	喜女子	Female	Rabbit	1902	2
Fujin Xiang 5th Bao				Mother	陈	氏	Female	Dragon	1879	10
Fujin Xiang 5th Bao				Son	石	女子	Male	Snake	1929	5
Fujin Xiang 5th Bao				Son's wife	岁	清哇	Female	Tiger	1926	12
Fujin Xiang 5th Bao				The second son	石	王怌儿	Male	Monkey	1932	7
Fujin Xiang 5th Bao				The third son	石	举才儿	Male	Rooster	1935	10
Fujin Xiang 5th Bao				Daughter	石	清娃	Female	Pig	1945	10
Fujin Xiang 5th Bao				Older brother's wife	赵	生娃	Female	Ox	1900	6
Fujin Xiang 5th Bao				Nephew	石	盖哇子	Male	Horse	1942	11
Fujin Xiang 5th Bao				Niece	石	女子	Female	Monkey	1932	6

Date	The beginning time of inhabitation in this county	Educational level	Profession sector	Occupation	Marital status	Physical disability	Religion
15	Inhabited for generations						
27	Inhabited for generations	Illiteracy	Agriculture	Land-holding peasant	Married	No	No religion
15	Inhabited for generations	Illiteracy	Agriculture	Land-holding peasant	Married	No	No religion
29	Inhabited for generations	Illiteracy					
7	Inhabited for generations						
17	Inhabited for generations						
16	Inhabited for generations	Illiteracy					
21	Inhabited for generations	Sishu 3 years	Agriculture	Land-holding peasant	Married	No	No religion
4	Inhabited for generations	Illiteracy	Agriculture	Land-holding peasant	Married	No	No religion
14	Inhabited for generations	Illiteracy					
30	Inhabited for generations	Illiteracy	Agriculture	Land-holding peasant	Married	No	No religion
14	Inhabited for generations	Illiteracy	Agriculture	Land-holding peasant	Married	No	No religion
11	Inhabited for generations	Illiteracy					
13	Inhabited for generations	Bao school 2 years					
13	Inhabited for generations						
21	Inhabited for generations	Illiteracy	Agriculture	Land-holding peasant	Widowed	No	No religion
21	Inhabited for generations						
21	Inhabited for generations	Illiteracy	Agriculture	Land-holding peasant	Unmarried		

Appendix: Overview of Microdata for Wudu County (*cont.*)

	The no. of household	Jia	Household	Relationship with the household head	Surname	First name	Sex	The Chinese zodiac	Birthdate Year	Month
Fujin Xiang 5th Bao	588	14	1	The household head	石	玉富	Male	Monkey	1907	6
Fujin Xiang 5th Bao				The household head's wife	李	王来儿	Female	Horse	1918	11
Fujin Xiang 5th Bao				Daughter	石	翠娃	Female	Pig	1935	7
Fujin Xiang 5th Bao				The second daughter	石	翠相	Female	Ox	1938	1
Fujin Xiang 5th Bao				The third daughter	石	银相儿	Female	Snake	1941	5
Fujin Xiang 5th Bao				The fourth daughter	石	四女子	Female	Goat	1943	4
Fujin Xiang 5th Bao	589	14	2	The household head	石	含周	Male	Goat	1906	11
Fujin Xiang 5th Bao				The household head's wife	李	登儿	Female	Rabbit	1915	7
Fujin Xiang 5th Bao				Daughter	石	居林	Female	Goat	1943	8
Fujin Xiang 5th Bao				The second daughter	石	清英	Female	Pig	1947	5
Fujin Xiang 5th Bao	590	14	3	The household head	石	玉韩	Male	Rabbit	1902	2
Fujin Xiang 5th Bao				Mother	赵	小贵	Female	Tiger	1877	10
Fujin Xiang 5th Bao				Younger brother	石	玉生	Male	Horse	1905	9
Fujin Xiang 5th Bao				Younger brother's wife	石	生生	Female	Goat	1906	10
Fujin Xiang 5th Bao				Niece	石	钱娃	Female	Rooster	1933	12
Fujin Xiang 5th Bao				The second daughter	石	送女	Female	Rabbit	1939	5
Fujin Xiang 5th Bao				The third daughter	石	清女	Female	Rooster	1943	6

Date	The beginning time of inhabitation in this county	Educational level	Profession sector	Occupation	Marital status	Physical disability	Religion
25	Inhabited for generations	Illiteracy	Agriculture	Land-holding peasant	Married	No	No religion
5	Inhabited for generations	Illiteracy	Agriculture	Land-holding peasant	Married	No	No religion
2	Inhabited for generations	Illiteracy	Agriculture	Land-holding peasant			
3	Inhabited for generations	Illiteracy					
3	Inhabited for generations	Illiteracy					
17	Inhabited for generations						
9	Inhabited for generations	Illiteracy	Agriculture	Land-holding peasant	Married	No	No religion
30	Inhabited for generations	Illiteracy	Agriculture	Land-holding peasant	Married	No	No religion
30	Inhabited for generations						
21	Inhabited for generations						
20	Inhabited for generations	Sishu 3 years	Agriculture	Land-holding peasant	Widowed	No	No religion
19	Inhabited for generations						
25	Inhabited for generations	Illiteracy	Agriculture	Land-holding peasant	Married	No	No religion
2	Inhabited for generations	Illiteracy	Agriculture	Land-holding peasant	Married	No	No religion
30	Inhabited for generations	Illiteracy	Agriculture	Land-holding peasant			
7	Inhabited for generations	Illiteracy					
16	Inhabited for generations						

Appendix: Overview of Microdata for Wudu County (*cont.*)

	The no. of household	Jia	Household	Relationship with the household head	Surname	First name	Sex	The Chinese zodiac	Birthdate Year	Month
Fujin Xiang 5th Bao				The third younger brother	石	岁进	Male	Monkey	1919	10
Fujin Xiang 5th Bao				The third younger brother's wife	狗	焦娃	Female	Monkey	1920	7
Fujin Xiang 5th Bao	591	14	4	The household head	李	石银	Male	Tiger	1888	2
Fujin Xiang 5th Bao				Older brother	李	富荣	Male	Goat	1894	5
Fujin Xiang 5th Bao				Nephew	李	三县	Male	Monkey	1937	12
Fujin Xiang 5th Bao				Nephew	李	三生娃	Male	Ox	1925	5
Fujin Xiang 5th Bao	592	14	5	The household head	李	文清	Male	Tiger	1915	8
Fujin Xiang 5th Bao				The household head's wife	石	三女子	Female	Rooster	1921	4
Fujin Xiang 5th Bao				Son	董	仓儿	Male	Dog	1946	3
Fujin Xiang 5th Bao				Daughter	李	东柱	Female	Rabbit		
Fujin Xiang 5th Bao				The second daughter	李	崴东儿	Female	Horse		
Fujin Xiang 5th Bao				Older brother's wife	李	龙儿	Female	Dog	1937	2
Fujin Xiang 5th Bao				Nephew	李	珍朝	Male	Rabbit	1915	4
Fujin Xiang 5th Bao				Nephew's wife	郭	哇	Female	Goat	1931	5
Fujin Xiang 5th Bao	593	14	6	The household head	李	文尚	Male	Horse	1905	2
Fujin Xiang 5th Bao				The household head's wife	谈	连哇	Female	Tiger	1914	6
Fujin Xiang 5th Bao				Son	陈	儿	Male	Ox	1937	7

Date	The beginning time of inhabitation in this county	Educational level	Profession sector	Occupation	Marital status	Physical disability	Religion
1	Inhabited for generations	Illiteracy	Agriculture	Land-holding peasant	Married	No	No religion
1	Inhabited for generations	Illiteracy	Agriculture	Land-holding peasant	Married	No	No religion
2	Inhabited for generations	Illiteracy	Agriculture	Land-holding peasant	Widowed	No	No religion
23	Inhabited for generations	Illiteracy	Agriculture	Land-holding peasant	Divorced	No	No religion
2	Inhabited for generations	Illiteracy	Agriculture	Land-holding peasant	Divorced	No	No religion
8	Inhabited for generations	Illiteracy	Agriculture	Land-holding peasant	Divorced	No	No religion
14	Inhabited for generations	Illiteracy	Agriculture	Land-holding peasant	Married	No	No religion
24	Inhabited for generations	Illiteracy	Agriculture	Land-holding peasant	Married	No	No religion
11	Inhabited for generations						
	Inhabited for generations	Illiteracy					
	Inhabited for generations	Illiteracy					
5	Inhabited for generations	Illiteracy	Agriculture	Land-holding peasant	Widowed	No	No religion
5	Inhabited for generations	Illiteracy	Agriculture	Land-holding peasant			
17	Inhabited for generations	Illiteracy	Agriculture	Land-holding peasant	Married	No	No religion
16	Inhabited for generations	Illiteracy	Agriculture	Land-holding peasant	Married	No	No religion
16	Inhabited for generations	Illiteracy	Agriculture	Land-holding peasant	Married	No	No religion
2	Inhabited for generations	Bao school 2 years					

Appendix: Overview of Microdata for Wudu County (*cont.*)

	The no. of household	Jia	Household	Relationship with the household head	Surname	First name	Sex	The Chinese zodiac	Birthdate	
									Year	Month
Fujin Xiang 5th Bao				The second son	李	陈陈	Male	Monkey	1944	5
Fujin Xiang 5th Bao				Daughter	李	早廷	Female		1932	7
Fujin Xiang 5th Bao				The second daughter	李	崴早	Female	Rabbit	1939	12
Fujin Xiang 5th Bao				The third daughter	李	兹女	Female	Horse	1942	4
Fujin Xiang 5th Bao				The fourth daughter	李	东英	Female	Dog	1946	11
Fujin Xiang 5th Bao				Uncle's wife	李	女子	Female	Monkey	1927	9
Fujin Xiang 5th Bao				Uncle's son	李	鸡生哇	Male	Ox	1925	3
Fujin Xiang 5th Bao	594	14	7	The household head	李	珍宽	Male	Dog	1922	12
Fujin Xiang 5th Bao				The household head's wife	赵	么哇	Female	Monkey	1920	6
Fujin Xiang 5th Bao				Grandfather	李	富举	Male	Monkey	1871	3
Fujin Xiang 5th Bao				Son	李	均生儿	Male	Snake	1941	6
Fujin Xiang 5th Bao				The second son	李	均英	Male	Rooster	1945	1
Fujin Xiang 5th Bao				Uncle	李	文彦	Male	Ox	1888	5
Fujin Xiang 5th Bao				Uncle's wife	谈	毕儿	Female	Rooster	1901	5
Fujin Xiang 5th Bao				Uncle's son	李	刘兹	Male	Rat	1936	4
Fujin Xiang 5th Bao				Uncle's wife	赵	贵女	Female	Monkey	1883	5
Fujin Xiang 5th Bao				Uncle's son	李	珍才	Male	Goat	1919	4
Fujin Xiang 5th Bao	595	14	8	The household head	李	珍田	Male	Goat	1906	5

Date	The beginning time of inhabitation in this county	Educational level	Profession sector	Occupation	Marital status	Physical disability	Religion
18	Inhabited for generations						
8	Inhabited for generations	Illiteracy	Agriculture	Land-holding peasant	Unmarried	No	No religion
1	Inhabited for generations	Illiteracy					
5	Inhabited for generations	Illiteracy					
13	Inhabited for generations						
21	Inhabited for generations	Illiteracy	Agriculture	Land-holding peasant	Widowed	No	No religion
1	Inhabited for generations	Illiteracy	Agriculture	Land-holding peasant	Unmarried	No	No religion
14	Inhabited for generations	Illiteracy	Agriculture	Land-holding peasant	Married	No	No religion
10	Inhabited for generations	Illiteracy	Agriculture	Land-holding peasant	Married	No	No religion
10	Inhabited for generations	Illiteracy	Agriculture				
23	Inhabited for generations						
21	Inhabited for generations						
21	Inhabited for generations	Illiteracy	Agriculture	Land-holding peasant	Married	No	No religion
2	Inhabited for generations	Illiteracy	Agriculture	Land-holding peasant	Married	No	No religion
11	Inhabited for generations	Bao school 1 year					
3	Inhabited for generations	Illiteracy	Agriculture	Land-holding peasant	Widowed	No	No religion
1	Inhabited for generations	Illiteracy	Agriculture	Land-holding peasant			
31	Inhabited for generations	Illiteracy	Agriculture	Land-holding peasant	Married	No	No religion

Appendix: Overview of Microdata for Wudu County (*cont.*)

	The no. of household	Jia	Household	Relationship with the household head	Surname	First name	Sex	The Chinese zodiac	Birthdate	
									Year	Month
Fujin Xiang 5th Bao				The household head's wife	石	杳哇	Female	Dragon	1903	4
Fujin Xiang 5th Bao				Son	李	明生	Male	Horse	1942	2
Fujin Xiang 5th Bao				Daughter	李	明哇	Female	Rat	1937	1
Fujin Xiang 5th Bao				Uncle	李	石三	Male	Rat	1887	8
Fujin Xiang 5th Bao				Uncle's son's wife	李	崴哇	Female	Rooster	1908	12
Fujin Xiang 5th Bao				Nephew	李	乖哇	Male	Ox	1937	10
Fujin Xiang 5th Bao				Niece	李	女女子	Female	Monkey	1933	10
Fujin Xiang 5th Bao	596	14	9	The household head	李	文熊	Male	Goat	1906	9
Fujin Xiang 5th Bao				The household head's wife	石	耀娃	Female	Horse	1918	11
Fujin Xiang 5th Bao				Son	李	前娃	Male	Rabbit	1939	8
Fujin Xiang 5th Bao				Older brother	李	文禄	Male	Tiger	1901	9
Fujin Xiang 5th Bao				Uncle's son's wife	石	桃娃子	Female	Dog	1885	2
Fujin Xiang 5th Bao				Nephew	李	珍清	Male	Monkey	1920	5
Fujin Xiang 5th Bao				Nephew's wife	赵	怌娃	Female	Goat	1919	9
Fujin Xiang 5th Bao				Nephew's son	杨	生	Male	Tiger		
Fujin Xiang 5th Bao				Granddaughter	李	落女	Female	Snake	1941	10
Fujin Xiang 5th Bao				The second granddaughter	李	落英	Female	Dog	1946	9
Fujin Xiang 5th Bao	597	14	10	The household head	李	珍永	Male	Ox	1913	10

Date	The beginning time of inhabitation in this county	Educational level	Profession sector	Occupation	Marital status	Physical disability	Religion
26	Inhabited for generations	Illiteracy	Agriculture	Land-holding peasant	Married	No	No religion
28	Inhabited for generations	Illiteracy	Agriculture	Land-holding peasant	Married	No	No religion
17	Inhabited for generations						
1	Inhabited for generations	Illiteracy	Agriculture	Land-holding peasant	Married	No	No religion
3	Inhabited for generations	Illiteracy	Agriculture	Land-holding peasant	Divorced	No	No religion
11	Inhabited for generations	Illiteracy					
12			Agriculture				
28	Inhabited for generations	Illiteracy	Agriculture	Land-holding peasant	Married	No	No religion
14	Inhabited for generations	Illiteracy	Agriculture	Land-holding peasant	Married	No	No religion
29	Inhabited for generations	Illiteracy					
28	Inhabited for generations	Illiteracy	Agriculture	Land-holding peasant	Widowed	No	No religion
10	Inhabited for generations	Illiteracy	Agriculture	Land-holding peasant	Widowed	No	No religion
17	Inhabited for generations	Illiteracy	Agriculture	Land-holding peasant	Married	No	No religion
12	Inhabited for generations	Illiteracy	Agriculture	Land-holding peasant	Married	No	No religion
	Inhabited for generations	Illiteracy					
18	Inhabited for generations	Illiteracy					
13	Inhabited for generations						
25	Inhabited for generations	Illiteracy	Agriculture	Land-holding peasant	Married	No	No religion

Appendix: Overview of Microdata for Wudu County (*cont.*)

	The no. of household	Jia	Household	Relationship with the household head	Surname	First name	Sex	The Chinese zodiac	Birthdate	
									Year	Month
Fujin Xiang 5th Bao				The household head's wife	石	进娃	Female	Tiger	1915	1
Fujin Xiang 5th Bao				Son	马	狗儿	Male	Horse	1942	6
Fujin Xiang 5th Bao				The second son	李	秋儿	Male	Monkey	1945	2
Fujin Xiang 5th Bao				Daughter	李	田田	Female	Dog	1947	1
Fujin Xiang 5th Bao				Older brother	李	珍泰	Male	Ox	1880	7
Fujin Xiang 5th Bao				Older brother's wife	石	银换子	Female	Dog	1897	2
Fujin Xiang 5th Bao				Nephew	李	保元	Male	Dog	1920	3
Fujin Xiang 5th Bao				Nephew's wife	石	还娃	Female	Monkey	1920	3
Fujin Xiang 5th Bao				Nephew's daughter	李	兵娃	Female	Horse	1942	4
Fujin Xiang 5th Bao				The second nephew	李	顺家	Male	Goat	1931	5
Fujin Xiang 5th Bao				The third nephew	谢	送	Male	Tiger	1938	12
Fujin Xiang 5th Bao				The fourth nephew	李	九九	Male	Horse	1942	2
Fujin Xiang 5th Bao				Niece	李	顺意	Female	Rooster	1933	5
Fujin Xiang 5th Bao	598	14	11	The household head	李	珍三	Male	Rabbit	1902	1
Fujin Xiang 5th Bao				The household head's wife	李	玉娃	Female	Horse	1905	9
Fujin Xiang 5th Bao				Son	李	高家	Male	Pig	1935	7
Fujin Xiang 5th Bao				Daughter	李	润娃	Female	Rooster	1933	5
Fujin Xiang 5th Bao				The second daughter	李	菊花	Female	Ox	1937	12

Date	The beginning time of inhabitation in this county	Educational level	Profession sector	Occupation	Marital status	Physical disability	Religion	
16	Inhabited for generations	Illiteracy	Agriculture	Land-holding peasant	Married	No	No religion	
20	Inhabited for generations	Bao school 1 year						
8	Inhabited for generations							
7	Inhabited for generations	Illiteracy						
29	Inhabited for generations	Illiteracy	Agriculture	Land-holding peasant	Married	No	No religion	
29	Inhabited for generations	Illiteracy	Agriculture	Land-holding peasant	Married	No	No religion	
3	Inhabited for generations	Illiteracy	Agriculture	Land-holding peasant	Married	No	No religion	Serving in the army
28	Inhabited for generations	Illiteracy	Agriculture	Land-holding peasant	Married	No	No religion	
21	Inhabited for generations	Illiteracy						
25	Inhabited for generations	Illiteracy	Agriculture	Land-holding peasant	Married	No	No religion	
24	Inhabited for generations	Illiteracy	Agriculture	Land-holding peasant	Married	No	No religion	
27	Inhabited for generations	Illiteracy						
18	Inhabited for generations	Illiteracy	Agriculture	Land-holding peasant	Unmarried	No	No religion	
6	Inhabited for generations	Illiteracy	Agriculture	Land-holding peasant	Married	No	No religion	
22	Inhabited for generations	Illiteracy	Agriculture	Land-holding peasant	Married	No	No religion	
7	Inhabited for generations	Bao school 1 year						
26	Inhabited for generations	Illiteracy	Agriculture	Land-holding peasant				
11	Inhabited for generations	Illiteracy						

Appendix: Overview of Microdata for Wudu County (*cont.*)

	The no. of household	Jia	Household	Relationship with the household head	Surname	First name	Sex	The Chinese zodiac	Birthdate	
									Year	Month
Fujin Xiang 5th Bao				The third daughter	李	双女	Female	Rabbit	1939	12
Fujin Xiang 5th Bao				The fourth daughter	李	五女子	Female	Horse	1942	11
Fujin Xiang 5th Bao				Uncle's son	李	珍有	Male	Ox	1888	11
Fujin Xiang 5th Bao	599	14	12	The household head	李	文邦	Male	Pig	1910	12
Fujin Xiang 5th Bao				The household head's wife	石	送娃子	Female	Rabbit	1915	12
Fujin Xiang 5th Bao				Mother	石	滕女子	Female	Horse	1881	1
Fujin Xiang 5th Bao				Son	李	仓保儿	Male	Horse	1942	9
Fujin Xiang 5th Bao				Daughter	李	香香	Female	Ox	1937	8
Fujin Xiang 5th Bao				The second daughter	李	双女	Female	Dragon	1940	3
Fujin Xiang 5th Bao				The third daughter	李	香清	Female	Rooster	1945	7
Fujin Xiang 5th Bao	600	15	1	The household head	李	珍荣	Male	Snake	1904	12
Fujin Xiang 5th Bao				Son	李	高家	Male	Tiger	1926	7
Fujin Xiang 5th Bao				Son's wife	黄	安明	Female	Rabbit	1927	6
Fujin Xiang 5th Bao				The second daughter	李	石娃	Female	Rooster	1932	12
Fujin Xiang 5th Bao				Older brother's wife	谈	么女子	Female	Goat	1903	4
Fujin Xiang 5th Bao				Nephew	李	秀林	Male	Tiger	1933	5
Fujin Xiang 5th Bao	601	15	2	The household head	李	彦林	Male	Goat	1906	11
Fujin Xiang 5th Bao				Mother	李	祁代娃	Female	Rabbit	1890	7
Fujin Xiang 5th Bao				Younger brother	李	彦章	Male	Monkey	1915	12

Date	The beginning time of inhabitation in this county	Educational level	Profession sector	Occupation	Marital status	Physical disability	Religion
11	Inhabited for generations	Illiteracy					
14	Inhabited for generations						
30	Inhabited for generations	Illiteracy	Agriculture	Land-holding peasant	Widowed	No	No religion
3	Inhabited for generations	Illiteracy	Agriculture	Land-holding peasant	Married	No	No religion
3	Inhabited for generations	Illiteracy	Agriculture	Land-holding peasant	Married	No	No religion
5	Inhabited for generations						
28	Inhabited for generations						
1	Inhabited for generations	Illiteracy					
30	Inhabited for generations	Illiteracy					
28	Inhabited for generations						
30	Inhabited for generations	Illiteracy	Agriculture	Land-holding peasant	Widowed	No	No religion
23	Inhabited for generations	Illiteracy	Agriculture	Land-holding peasant	Married	No	No religion
7	Inhabited for generations	Illiteracy	Agriculture	Land-holding peasant	Married	No	No religion
9	Inhabited for generations	Illiteracy	Agriculture	Land-holding peasant	Unmarried	No	No religion
5	Inhabited for generations	Illiteracy	Agriculture	Land-holding peasant	Divorced	No	No religion
4	Inhabited for generations	Illiteracy	Agriculture	Land-holding peasant	Widowed	No	No religion
27	Inhabited for generations	Illiteracy	Agriculture	Land-holding peasant	Widowed	No	No religion
11	Inhabited for generations	Illiteracy	Agriculture	Land-holding peasant	Widowed	No	No religion
23	Inhabited for generations	Illiteracy	Agriculture	Land-holding peasant	Married	No	No religion

Appendix: Overview of Microdata for Wudu County (*cont.*)

	The no. of house-hold	Jia	House-hold	Relation-ship with the house-hold head	Surname	First name	Sex	The Chinese zodiac	Birthdate	
									Year	Month
Fujin Xiang 5th Bao				Younger brother's wife	谈	桂娃	Female	Pig	1924	1
Fujin Xiang 5th Bao				Niece	李	鱼英	Female	Horse	1942	3
Fujin Xiang 5th Bao	602	15	3	The house-hold head	李	文永	Male	Goat	1894	8
Fujin Xiang 5th Bao				The house-hold head's wife	石	褪儿	Female	Goat	1894	11
Fujin Xiang 5th Bao				Son	李	崔儿	Male	Goat	1919	8
Fujin Xiang 5th Bao				Son's wife	石	车廷儿	Female	Dog	1922	8
Fujin Xiang 5th Bao				The second son	李	拜加	Male	Pig	1935	2
Fujin Xiang 5th Bao				Grand-mother	石	全娃	Female	Rabbit	1877	1
Fujin Xiang 5th Bao				Uncle	李	陈送儿	Male	Tiger	1914	4
Fujin Xiang 5th Bao				Uncle's wife	李	女娃	Female	Ox	1913	10
Fujin Xiang 5th Bao				Uncle's son	王	银娃	Male	Ox	1938	1
Fujin Xiang 5th Bao				The second younger brother	杨	树娃	Male	Dog	1946	11
Fujin Xiang 5th Bao				Uncle's daughter	李	董听儿	Female	Goat	1932	1
Fujin Xiang 5th Bao	603	15	4	The house-hold head	李	珍成	Male	Rat	1912	2
Fujin Xiang 5th Bao				The house-hold head's wife	石	团儿	Female	Rat	1912	4
Fujin Xiang 5th Bao				Father	李	文海	Male	Horse	1893	4
Fujin Xiang 5th Bao				Son	李	保三	Male	Rooster	1933	12

Date	The beginning time of inhabitation in this county	Educational level	Profession sector	Occupation	Marital status	Physical disability	Religion
22	Inhabited for generations	Illiteracy	Agriculture	Land-holding peasant	Married	No	No religion
25	Inhabited for generations	Illiteracy					
25	Inhabited for generations	Illiteracy	Agriculture	Land-holding peasant	Married	No	No religion
10	Inhabited for generations	Illiteracy	Agriculture	Land-holding peasant	Married	No	No religion
1	Inhabited for generations	Illiteracy	Agriculture	Land-holding peasant	Married	No	No religion
8	Inhabited for generations	Illiteracy	Agriculture	Land-holding peasant	Married	No	No religion
6	Inhabited for generations	Bao school 1 year					
15	Inhabited for generations						
1	Inhabited for generations	Illiteracy	Agriculture	Land-holding peasant	Married	No	No religion
5	Inhabited for generations	Illiteracy	Agriculture	Land-holding peasant	Married	No	No religion
13	Inhabited for generations	Bao school 1 year					
7	Inhabited for generations						
30	Inhabited for generations		Agriculture		Unmarried		
19	Inhabited for generations	Sishu 3 years	Agriculture	Land-holding peasant	Married	No	No religion
8	Inhabited for generations	Illiteracy	Agriculture	Land-holding peasant	Married	No	No religion
12	Inhabited for generations	Illiteracy	Agriculture	Land-holding peasant	Widowed	No	No religion
22	Inhabited for generations	Bao school 2 years					

Appendix: Overview of Microdata for Wudu County (*cont.*)

									Birthdate	
	The no. of household	Jia	Household	Relationship with the household head	Surname	First name	Sex	The Chinese zodiac	Year	Month
Fujin Xiang 5th Bao				The second son	李	观音代	Male	Dragon	1940	12
Fujin Xiang 5th Bao				The third son	李	三望儿	Male	Monkey	1944	7
Fujin Xiang 5th Bao				Uncle	李	石有	Male	Rooster	1896	5
Fujin Xiang 5th Bao				Uncle's wife	石	灯娃	Female	Pig	1910	2
Fujin Xiang 5th Bao				Younger brother	李	彦圳	Male	Rooster	1933	9
Fujin Xiang 5th Bao				Younger sister	李	改平儿	Female	Rat	1937	1
Fujin Xiang 5th Bao				The second daughter	李	双平儿	Female	Monkey	1944	1
Fujin Xiang 5th Bao	604	15	5	The household head	李	珍富	Male	Tiger	1889	9
Fujin Xiang 5th Bao				The household head's wife	赵	喜儿	Female	Dragon	1905	9
Fujin Xiang 5th Bao				Son	李	崴能儿	Male	Goat	1919	4
Fujin Xiang 5th Bao				Son's wife	崴	王儿	Female	Monkey	1920	4
Fujin Xiang 5th Bao				Grandson	李	长子	Male	Rooster	1945	4
Fujin Xiang 5th Bao	605	15	6	The household head	李	珍旺	Male	Goat	1906	11
Fujin Xiang 5th Bao				The household head's wife	石	柱娃	Female	Snake	1917	9
Fujin Xiang 5th Bao				Mother	黎	花	Female	Rat	1874	4
Fujin Xiang 5th Bao				Younger brother	李	尹两	Male	Snake	1917	3
Fujin Xiang 5th Bao				Younger brother's wife	石	进娃	Female	Snake	1917	12
Fujin Xiang 5th Bao				Nephew	石	讨口子	Male	Dragon	1940	10

Date	The beginning time of inhabitation in this county	Educational level	Profession sector	Occupation	Marital status	Physical disability	Religion
4	Inhabited for generations	Illiteracy					
26	Inhabited for generations						
6	Inhabited for generations	Illiteracy	Agriculture	Land-holding peasant	Married	No	No religion
1	Inhabited for generations	Illiteracy	Agriculture	Land-holding peasant	Married	No	No religion
25	Inhabited for generations	Bao school 2 years					
1	Inhabited for generations	Illiteracy					
26	Inhabited for generations						
17	Inhabited for generations	Illiteracy	Agriculture	Land-holding peasant	Married	No	No religion
17	Inhabited for generations	Illiteracy	Agriculture	Land-holding peasant	Married	No	No religion
5	Inhabited for generations	Illiteracy	Agriculture	Land-holding peasant	Married	No	No religion
24	Inhabited for generations	Illiteracy	Agriculture	Land-holding peasant	Married	No	No religion
24	Inhabited for generations						
6	Inhabited for generations	Illiteracy	Agriculture	Land-holding peasant	Married	No	No religion
26	Inhabited for generations	Illiteracy	Agriculture	Land-holding peasant	Married	No	No religion
15	Inhabited for generations	Illiteracy	Agriculture	Land-holding peasant	Widowed	No	No religion
2	Inhabited for generations	Illiteracy	Agriculture	Land-holding peasant	Married	No	No religion
24	Inhabited for generations	Illiteracy	Agriculture	Land-holding peasant	Married	No	No religion
2	Inhabited for generations	Bao school 1 year					

Appendix: Overview of Microdata for Wudu County (*cont.*)

	The no. of household	Jia	Household	Relationship with the household head	Surname	First name	Sex	The Chinese zodiac	Birthdate	
									Year	Month
Fujin Xiang 5th Bao				The second nephew	李	海生	Male	Monkey	1944	3
Fujin Xiang 5th Bao				Niece	李	女子	Female	Dog	1946	11
Fujin Xiang 5th Bao				Uncle's wife	石	杨银花	Female	Rabbit	1871	7
Fujin Xiang 5th Bao	606	15	7	The household head	李	珍和	Male	Ox	1900	7
Fujin Xiang 5th Bao				The household head's wife	赵	旦儿	Female	Goat	1913	5
Fujin Xiang 5th Bao				Son	李	高家	Male	Ox	1938	7
Fujin Xiang 5th Bao				The second son	李	高明	Male	Pig	1947	2
Fujin Xiang 5th Bao				Daughter	李	毛人	Female	Horse	1942	2
Fujin Xiang 5th Bao				Uncle's wife	石	小董儿	Female	Tiger	1889	2
Fujin Xiang 5th Bao				Uncle's son	李	珍学	Male	Horse	1918	8
Fujin Xiang 5th Bao	607	15	8	The household head	李	珍满	Male	Monkey	1895	7
Fujin Xiang 5th Bao				The household head's wife	郭	儿	Female	Rabbit	1902	5
Fujin Xiang 5th Bao				Son	李	许生	Male	Ox	1925	7
Fujin Xiang 5th Bao				Son's wife	赵	雪娃	Female	Rooster	1924	1
Fujin Xiang 5th Bao				Grandson	李	丑优儿	Male	Rooster	1945	10
Fujin Xiang 5th Bao				The second grandson	李	乡生	Male	Pig	1947	5
Fujin Xiang 5th Bao				The second son	李	进仓儿	Male	Horse	1931	1
Fujin Xiang 5th Bao				The third son	李	进陆儿	Male	Goat	1931	12

Date	The beginning time of inhabitation in this county	Educational level	Profession sector	Occupation	Marital status	Physical disability	Religion	
27	Inhabited for generations							
8	Inhabited for generations							
25	Inhabited for generations	Illiteracy	Agriculture	Land-holding peasant	Widowed	No	No religion	
16	Inhabited for generations	Illiteracy	Agriculture	Land-holding peasant	Married	No	No religion	
28	Inhabited for generations	Illiteracy	Agriculture	Land-holding peasant	Married	No	No religion	
8	Inhabited for generations	Bao school 1 year						
25	Inhabited for generations							
22	Inhabited for generations	Illiteracy						
23	Inhabited for generations	Illiteracy	Agriculture	Land-holding peasant	Widowed	No	No religion	
13	Inhabited for generations	Illiteracy	Agriculture	Land-holding peasant	Divorced	No	No religion	
15	Inhabited for generations	Illiteracy	Agriculture	Land-holding peasant	Married	No	No religion	
29	Inhabited for generations	Illiteracy	Agriculture	Land-holding peasant	Married	No	No religion	
27	Inhabited for generations	Illiteracy	Agriculture	Land-holding peasant	Married	No	No religion	Joined the army
16	Inhabited for generations	Illiteracy	Agriculture	Land-holding peasant	Married	No	No religion	
30	Inhabited for generations							
29	Inhabited for generations							
1	Inhabited for generations	Illiteracy	Agriculture	Land-holding peasant	Unmarried	No	No religion	
30	Inhabited for generations	Illiteracy	Agriculture	Land-holding peasant	Unmarried	No	No religion	

Appendix: Overview of Microdata for Wudu County (*cont.*)

	The no. of household	Jia	Household	Relationship with the household head	Surname	First name	Sex	The Chinese zodiac	Birthdate	
									Year	Month
Fujin Xiang 5th Bao				The fourth son	李	天赦	Male	Dog	1946	11
Fujin Xiang 5th Bao	608	15	9	The household head	李	文喜	Male	Goat	1905	1
Fujin Xiang 5th Bao				The household head's wife	崴	醒	Female	Rooster	1908	11
Fujin Xiang 5th Bao				Mother	李	惠娃	Female	Tiger	1885	11
Fujin Xiang 5th Bao				Son	李	禄成	Male	Tiger	1938	8
Fujin Xiang 5th Bao				The second son	李	修成	Male	Rooster	1945	5
Fujin Xiang 5th Bao				Younger brother	谢	生	Male	Rabbit	1915	3
Fujin Xiang 5th Bao				Younger brother's wife	石	进女子	Female	Monkey	1920	9
Fujin Xiang 5th Bao				Niece	李	兹娃	Female	Dragon	1940	4
Fujin Xiang 5th Bao				The third younger brother	李	文品	Male	Rabbit	1927	2
Fujin Xiang 5th Bao				Younger brother's wife	赵	三女	Female	Horse	1930	12
Fujin Xiang 5th Bao	609	15	10	The household head	李	文忠	Male	Goat	1906	5
Fujin Xiang 5th Bao				The household head's wife	赵	柱哇	Female	Rabbit	1902	12
Fujin Xiang 5th Bao				Father	李	富玉	Male	Pig	1874	3
Fujin Xiang 5th Bao				Son	李	保哇	Male	Pig	1935	7
Fujin Xiang 5th Bao				Older brother	李	文现	Male	Rat	1887	3
Fujin Xiang 5th Bao				Nephew	李	陈保	Male	Snake	1929	12

Date	The beginning time of inhabitation in this county	Educational level	Profession sector	Occupation	Marital status	Physical disability	Religion
18	Inhabited for generations						
5	Inhabited for generations	Illiteracy	Agriculture	Land-holding peasant	Married	No	No religion
21	Inhabited for generations	Illiteracy	Agriculture	Land-holding peasant	Married	No	No religion
21	Inhabited for generations	Illiteracy	Agriculture	Land-holding peasant	Widowed	No	No religion
23	Inhabited for generations	Illiteracy					
17	Inhabited for generations						
16	Inhabited for generations	Illiteracy	Agriculture	Land-holding peasant	Married	No	No religion
12	Inhabited for generations	Illiteracy	Agriculture	Land-holding peasant	Married	No	No religion
29	Inhabited for generations	Illiteracy					
26	Inhabited for generations	Illiteracy	Agriculture	Land-holding peasant	Married	No	No religion
1	Inhabited for generations	Illiteracy	Agriculture	Land-holding peasant	Married	No	No religion
4	Inhabited for generations	Illiteracy	Agriculture	Land-holding peasant	Married	No	No religion
13	Inhabited for generations	Illiteracy	Agriculture	Land-holding peasant	Married	No	No religion
1	Inhabited for generations						
7	Inhabited for generations	Bao school 2 years					
12	Inhabited for generations	Illiteracy	Agriculture	Land-holding peasant	Married	No	No religion
14	Inhabited for generations	Illiteracy	Agriculture	Land-holding peasant	Married	No	No religion

Appendix: Overview of Microdata for Wudu County (*cont.*)

	The no. of household	Jia	Household	Relationship with the household head	Surname	First name	Sex	The Chinese zodiac	Birthdate Year	Month
Fujin Xiang 5th Bao				Nephew's wife	石	席廷儿	Female	Horse	1930	12
Fujin Xiang 5th Bao				Older brother's wife	石	有姐子	Female	Horse	1881	3
Fujin Xiang 5th Bao				Nephew	李	称娃子	Male	Dog	1909	3
Fujin Xiang 5th Bao				Grandson	李	谢娃	Male	Pig	1935	9
Fujin Xiang 5th Bao	610	15	11	The household head	李	文柱	Male	Goat	1906	12
Fujin Xiang 5th Bao				The household head's wife	石	著娃	Female	Monkey	1907	3
Fujin Xiang 5th Bao				Son	李	彦全	Male	Pig	1935	5
Fujin Xiang 5th Bao				The second son	李	仓娃	Male	Rabbit	1939	3
Fujin Xiang 5th Bao				Daughter	李	前清儿	Female	Rooster	1945	10
Fujin Xiang 5th Bao				Older brother	李	文富	Male	Ox	1900	8
Fujin Xiang 5th Bao				Nephew	刘	学儿	Male	Monkey	1932	9
Fujin Xiang 5th Bao				The second nephew	李	称称	Male	Tiger	1938	3
Fujin Xiang 5th Bao				Niece	李	篩篩	Female	Horse	1942	10
Fujin Xiang 5th Bao	611	15	12	The household head	李	文正	Male	Dragon	1903	8
Fujin Xiang 5th Bao				Son	李	郭娃	Male	Ox	1925	4
Fujin Xiang 5th Bao				Son's wife	石	张廷	Female	Horse	1930	3
Fujin Xiang 5th Bao				Older brother's wife	石	岁桃娃	Female	Monkey	1883	3
Fujin Xiang 5th Bao				Niece	李	董女子	Female	Monkey	1920	3

Date	The beginning time of inhabitation in this county	Educational level	Profession sector	Occupation	Marital status	Physical disability	Religion
14	Inhabited for generations	Illiteracy	Agriculture	Land-holding peasant	Married	No	No religion
1	Inhabited for generations	Illiteracy	Agriculture	Land-holding peasant	Widowed	No	No religion
21	Inhabited for generations	Illiteracy	Agriculture	Land-holding peasant	Divorced	No	No religion
28	Inhabited for generations	Illiteracy	Agriculture	Land-holding peasant	Unmarried	No	No religion
22	Inhabited for generations	Illiteracy	Agriculture	Land-holding peasant	Married	No	No religion
6	Inhabited for generations	Illiteracy	Agriculture	Land-holding peasant	Married	No	No religion
2	Inhabited for generations	Bao school 3 years					
3	Inhabited for generations	Bao school 1 year					
13	Inhabited for generations						
27	Inhabited for generations	Illiteracy	Agriculture	Land-holding peasant	Widowed	No	No religion
5	Inhabited for generations	Illiteracy	Agriculture	Land-holding peasant	Unmarried	No	No religion
4	Inhabited for generations	Bao school 1 year					
17	Inhabited for generations						
21	Inhabited for generations	Illiteracy	Agriculture	Land-holding peasant	Widowed	No	No religion
8	Inhabited for generations	Illiteracy	Agriculture	Land-holding peasant	Married	No	No religion
10	Inhabited for generations	Illiteracy	Agriculture	Land-holding peasant	Married	No	No religion
8	Inhabited for generations	Illiteracy	Agriculture	Land-holding peasant	Widowed	No	No religion
11	Inhabited for generations	Illiteracy	Agriculture	Land-holding peasant	Widowed	No	No religion

Appendix: Overview of Microdata for Wudu County (*cont.*)

	The no. of household	Jia	Household	Relationship with the household head	Surname	First name	Sex	The Chinese zodiac	Birthdate	
									Year	Month
Fujin Xiang 5th Bao				Nephew's daughter	李	兵女子	Female	Goat	1943	5
Fujin Xiang 8th Bao	612	6	5	The household head	张	仲荣	Male	Ox	1900	6
Fujin Xiang 8th Bao				Son	张	高娃	Male	Snake	1941	7
Fujin Xiang 8th Bao	613	6	6	The household head	张	有洪	Male		1870	9
Fujin Xiang 8th Bao				Son	张	毛虫	Male	Rabbit	1915	6
Fujin Xiang 8th Bao				Younger brother's wife	諓	送节子	Female	Rabbit	1889	3
Fujin Xiang 8th Bao				Nephew	张	促谓	Male	Rooster	1921	9
Fujin Xiang 8th Bao				Nephew's son	张	才哇	Male	Rooster	1921	9
Fujin Xiang 8th Bao				Nephew's son's wife	杜	通哇	Female	Dragon	1928	5
Fujin Xiang 8th Bao	614	6	7	The household head	张	福才	Male	Rat	1912	8
Fujin Xiang 8th Bao				The household head's wife	赵	转女子	Female	Ox	1914	1
Fujin Xiang 8th Bao				Son	张	三民哇	Male	Rabbit	1939	3
Fujin Xiang 8th Bao				The second son	张	进民哇	Male	Dog	1946	7
Fujin Xiang 8th Bao				Uncle's wife	张	袓川儿	Female	Dragon	1902	1
Fujin Xiang 8th Bao				Uncle's son	张	来哇	Male	Ox	1925	11
Fujin Xiang 8th Bao				Uncle's son's wife	罗	连哇	Female	Pig	1924	10
Fujin Xiang 8th Bao				Uncle's second son	张	还人	Male	Rabbit	1940	1
Fujin Xiang 8th Bao	615	6	8	The household head	张	仲吕	Male	Monkey	1895	4

Date	The beginning time of inhabitation in this county	Educational level	Profession sector	Occupation	Marital status	Physical disability	Religion
11	Inhabited for generations						
4	Inhabited for generations	Illiteracy	Agriculture	Land-holding peasant	Widowed	No	No religion
9	Inhabited for generations	Illiteracy					
8	Inhabited for generations	Illiteracy	Unemployed	Land-holding peasant	Widowed		
10	Inhabited for generations	Illiteracy	Agriculture	Land-holding peasant	Unmarried	No	No religion
3	Inhabited for generations	Illiteracy	Agriculture	Land-holding peasant	Widowed	No	No religion
14	Inhabited for generations	Illiteracy	Agriculture	Land-holding peasant	Unmarried	Left eye blind	No religion
27	Inhabited for generations	Illiteracy	Agriculture	Land-holding peasant	Married	No	No religion
31	Inhabited for generations	Illiteracy	Agriculture	Land-holding peasant	Married	No	No religion
26	Inhabited for generations	Illiteracy	Agriculture	Land-holding peasant	Married	No	No religion
13	Inhabited for generations	Illiteracy	Agriculture	Land-holding peasant	Married	No	No religion
1	Inhabited for generations	Illiteracy					
11	Inhabited for generations						
3	Inhabited for generations	Illiteracy	Agriculture	Land-holding peasant	Widowed	No	No religion
24	Inhabited for generations	Illiteracy	Agriculture	Land-holding peasant	Married	No	No religion
18	Inhabited for generations	Illiteracy	Agriculture	Land-holding peasant	Married	No	No religion
18	Inhabited for generations	Illiteracy					
3	Inhabited for generations	Illiteracy	Agriculture	Land-holding peasant	Married	No	No religion

Appendix: Overview of Microdata for Wudu County (*cont.*)

	The no. of household	Jia	Household	Relationship with the household head	Surname	First name	Sex	The Chinese zodiac	Birthdate Year	Month
Fujin Xiang 8th Bao				The household head's wife	罗	永哇	Female	Snake	1892	7
Fujin Xiang 8th Bao				Son	罗	锁哇	Male	Horse	1918	9
Fujin Xiang 8th Bao				Son's wife	张	第哇	Female	Horse	1918	10
Fujin Xiang 8th Bao				The first Grand-daughter	张	露信	Female	Snake	1941	3
Fujin Xiang 8th Bao				The second granddaughter	张	民信	Female	Monkey	1944	3
Fujin Xiang 8th Bao				The third Grand-daughter	张	月哇	Female	Pig	1947	2
Fujin Xiang 8th Bao				Uncle	张	有堂	Male		1943	7
Fujin Xiang 8th Bao				Uncle's son	张	仲彦	Male	Rooster	1908	5
Fujin Xiang 8th Bao	616	6	9	The household head	张	仲田	Male		1880	10
Fujin Xiang 8th Bao				Son	张	文应	Female	Horse	1918	8
Fujin Xiang 8th Bao				The second son	张	二应	Male	Dog	1922	8
Fujin Xiang 8th Bao				The second son's wife	赵	睦哇	Female	Ox	1925	9
Fujin Xiang 8th Bao				The third son	张	董永	Male	Rabbit	1940	1
Fujin Xiang 8th Bao				Nephew	张	象哇	Male	Rat	1947	7
Fujin Xiang 8th Bao				The second nephew	张	曾应	Male	Tiger	1926	3
Fujin Xiang 8th Bao				The second nephew's wife	罗	照哇	Female	Tiger	1926	9
Fujin Xiang 8th Bao				Nephew's son	罗	啟福	Male	Rabbit	1915	11

Date	The beginning time of inhabitation in this county	Educational level	Profession sector	Occupation	Marital status	Physical disability	Religion
3	Inhabited for generations	Illiteracy	Agriculture	Land-holding peasant	Married	No	No religion
3	Inhabited for generations	Illiteracy	Agriculture	Land-holding peasant	Married	No	No religion
10	Inhabited for generations	Illiteracy	Agriculture	Land-holding peasant	Married	No	No religion
2	Inhabited for generations						
8	Inhabited for generations						
4	Inhabited for generations						
5	Inhabited for generations	Illiteracy	Unemployed		Widowed		
5	Inhabited for generations	Illiteracy	Agriculture	Land-holding peasant	Widowed	No	No religion
6	Inhabited for generations	Illiteracy	Unemployed				
18	Inhabited for generations	Illiteracy	Unemployed	Land-holding peasant	Unmarried	No	No religion
4	Inhabited for generations	Illiteracy	Agriculture	Land-holding peasant	Married	No	No religion
2	Inhabited for generations	Illiteracy	Agriculture	Land-holding peasant	Married	No	No religion
21	Inhabited for generations	Illiteracy					
2	Inhabited for generations	Illiteracy	Agriculture	Land-holding peasant			
20	Inhabited for generations	Illiteracy	Agriculture	Land-holding peasant	Married	No	No religion
17	Inhabited for generations	Illiteracy	Agriculture	Land-holding peasant	Married	No	No religion
29	Inhabited for generations	Illiteracy	Agriculture	Land-holding peasant	Unmarried	No	No religion

Appendix: Overview of Microdata for Wudu County (*cont.*)

	The no. of household	Jia	Household	Relationship with the household head	Surname	First name	Sex	The Chinese zodiac	Birthdate	
									Year	Month
Fujin Xiang 8th Bao	617	6	10	The household head	张	董儿	Male	Rabbit	1915	10
Fujin Xiang 8th Bao				The household head's wife	罗	毛旦	Female	Monkey	1920	9
Fujin Xiang 8th Bao				Mother	张	氏	Female		1876	8
Fujin Xiang 8th Bao				Daughter	张	月女	Female	Dog	1946	3
Fujin Xiang 8th Bao				Niece	张	虽拜	Female	Tiger	1899	10
Fujin Xiang 8th Bao	618	6	11	The household head	张	宽金儿	Male	Tiger	1926	5
Fujin Xiang 8th Bao				Older brother	张	福赵	Male	Horse	1918	5
Fujin Xiang 8th Bao				Older brother's wife	张	颖儿	Female	Horse	1918	7
Fujin Xiang 8th Bao				Nephew	张	转儿	Male	Rooster	1945	6
Fujin Xiang 8th Bao				Younger brother	张	董哇	Male	Tiger	1938	2
Fujin Xiang 8th Bao				Uncle's son	张	登哇	Male	Ox	1913	9
Fujin Xiang 8th Bao	619	6	12	The household head	张	信哇	Male	Monkey	1907	6
Fujin Xiang 8th Bao				The household head's wife	韩	注儿	Female	Rabbit	1902	4
Fujin Xiang 8th Bao				Father	张	仲武	Male		1876	2
Fujin Xiang 8th Bao				Mother	马	氏	Female		1877	4
Fujin Xiang 8th Bao				Son	张	庆哇	Male	Dragon	1928	3
Fujin Xiang 8th Bao				Son's wife	马	响哇	Female	Horse	1930	6
Fujin Xiang 8th Bao				Daughter	张	拜女	Female	Monkey	1932	6

Date	The beginning time of inhabitation in this county	Educational level	Profession sector	Occupation	Marital status	Physical disability	Religion	
12	Inhabited for generations	Illiteracy	Agriculture	Land-holding peasant	Married	No	No religion	
3	Inhabited for generations	Illiteracy	Agriculture	Land-holding peasant	Married	No	No religion	
17	Inhabited for generations		Unemployed		Widowed			
14	Inhabited for generations							
16	Inhabited for generations	Illiteracy	Agriculture	Land-holding peasant		No	No religion	
23	Inhabited for generations	Illiteracy	Agriculture	Land-holding peasant	Divorced	No	No religion	
2	Inhabited for generations	Illiteracy	Agriculture	Land-holding peasant	Married	No	No religion	Serving in the army
1	Inhabited for generations	Illiteracy	Agriculture	Land-holding peasant	Married	No	No religion	
20	Inhabited for generations							
10	Inhabited for generations	Illiteracy						
2	Inhabited for generations	Illiteracy	Agriculture	Land-holding peasant	Unmarried	No	No religion	
2	Inhabited for generations	Illiteracy	Agriculture	Land-holding peasant	Married	No	No religion	
9	Inhabited for generations	Illiteracy	Agriculture	Land-holding peasant	Married	No	No religion	
8	Inhabited for generations		Unemployed		Married			
19	Inhabited for generations		Unemployed		Married			
8	Inhabited for generations	Illiteracy	Agriculture	Land-holding peasant	Married	No	No religion	
1	Inhabited for generations	Illiteracy	Agriculture	Land-holding peasant	Married	No	No religion	
5	Inhabited for generations	Illiteracy	Agriculture	Land-holding peasant	Unmarried	No	No religion	

Appendix: Overview of Microdata for Wudu County (*cont.*)

	The no. of household	Jia	Household	Relationship with the household head	Surname	First name	Sex	The Chinese zodiac	Birthdate Year	Month
Fujin Xiang 8th Bao				The second daughter	张	赵哇	Female	Ox	1937	3
Fujin Xiang 8th Bao				The third daughter	张	田哇	Female	Rabbit	1939	4
Fujin Xiang 8th Bao	620	6	13	The household head	张	福转	Male	Monkey	1895	8
Fujin Xiang 8th Bao				The household head's wife	誸	晒儿	Female	Goat	1894	9
Fujin Xiang 8th Bao				Younger brother's wife	赵	糞哇	Female	Horse	1905	4
Fujin Xiang 8th Bao				The second younger brother	张	平儿	Male	Monkey	1920	8
Fujin Xiang 8th Bao				The second younger brother's wife	赵	洪女	Female	Monkey	1917	3
Fujin Xiang 8th Bao				Nephew	张	察哇	Male	Ox	1937	7
Fujin Xiang 8th Bao				The second nephew	张	杏儿	Male	Pig	1947	2
Fujin Xiang 8th Bao				Niece	张	贱女子	Female	Snake	1941	5
Fujin Xiang 8th Bao	621	7	1	The household head	赵	登满	Male	Goat	1894	5
Fujin Xiang 8th Bao				Son	赵	大瓜	Male	Pig	1924	1
Fujin Xiang 8th Bao				Son's wife	李	省哇	Female	Pig	1923	5
Fujin Xiang 8th Bao				Grandson	王	怌儿	Male	Horse	1942	3
Fujin Xiang 8th Bao				The second grandson	王	细怌	Male	Monkey	1944	7
Fujin Xiang 8th Bao				Granddaughter	贵	女	Female	Dog	1946	5
Fujin Xiang 8th Bao				The second son	赵	进才哇	Male	Dragon	1928	11

Date	The beginning time of inhabitation in this county	Educational level	Profession sector	Occupation	Marital status	Physical disability	Religion
22	Inhabited for generations						
24	Inhabited for generations						
9	Inhabited for generations	Illiteracy	Agriculture	Land-holding peasant	Married	No	No religion
8	Inhabited for generations	Illiteracy	Agriculture	Land-holding peasant	Married	No	No religion
22	Inhabited for generations	Illiteracy	Agriculture	Land-holding peasant	Widowed	No	No religion
28	Inhabited for generations	Illiteracy	Agriculture	Land-holding peasant	Married	No	No religion
1	Inhabited for generations	Illiteracy	Agriculture	Land-holding peasant	Married	No	No religion
19	Inhabited for generations						
28	Inhabited for generations						
5	Inhabited for generations						
10	Inhabited for generations	Illiteracy	Agriculture	Land-holding peasant	Widowed	No	No religion
28	Inhabited for generations	Illiteracy	Agriculture	Land-holding peasant	Married	No	No religion
2	Inhabited for generations	Illiteracy	Agriculture	Land-holding peasant	Married	No	No religion
31	Inhabited for generations						
21	Inhabited for generations						
12	Inhabited for generations						
21	Inhabited for generations	Illiteracy	Agriculture	Land-holding peasant	Unmarried	No	No religion

Appendix: Overview of Microdata for Wudu County (*cont.*)

	The no. of household	Jia	Household	Relationship with the household head	Surname	First name	Sex	The Chinese zodiac	Birthdate Year	Month
Fujin Xiang 8th Bao				The third son	赵	四儿	Male	Dragon	1940	8
Fujin Xiang 8th Bao				Younger brother	赵	登福	Male	Rooster	1896	4
Fujin Xiang 8th Bao				Nephew	赵	三哇	Male	Rat	1936	5
Fujin Xiang 8th Bao				Older brother's wife	张	九女	Female	Dog	1897	3
Fujin Xiang 8th Bao				Nephew	张	仁哇	Male	Rooster	1933	5
Fujin Xiang 8th Bao	622	7	2	The household head	赵	登琴	Male	Dog	1898	1
Fujin Xiang 8th Bao				The household head's wife	张	扣崴	Female	Rat	1899	3
Fujin Xiang 8th Bao				Son	赵	宗拜哇	Male	Rat	1936	4
Fujin Xiang 8th Bao				Son's wife	杨	等哇	Female	Pig	1934	5
Fujin Xiang 8th Bao				The second son	赵	宗女子	Male	Ox	1937	2
Fujin Xiang 8th Bao				The second daughter	赵	宗辈儿	Female	Rooster	1945	2
Fujin Xiang 8th Bao				Uncle	赵	金应	Male		1873	3
Fujin Xiang 8th Bao	623	7	3	The household head	赵	有太	Male	Goat	1896	10
Fujin Xiang 8th Bao				The household head's wife	张	孟哇	Female	Horse	1905	12
Fujin Xiang 8th Bao				Son	张	五斤	Male	Ox	1937	6
Fujin Xiang 8th Bao				Daughter	赵	检检	Female	Dog	1946	8
Fujin Xiang 8th Bao				Uncle	进	才儿	Male	Pig	1886	4
Fujin Xiang 8th Bao				Uncle's son	露	拜哇	Male	Goat	1919	11

Date	The beginning time of inhabitation in this county	Educational level	Profession sector	Occupation	Marital status	Physical disability	Religion
10	Inhabited for generations	Bao school 4 years					
7	Inhabited for generations	Illiteracy	Agriculture	Land-holding peasant	Widowed	No	No religion
21	Inhabited for generations	Illiteracy					
30	Inhabited for generations	Illiteracy	Agriculture	Land-holding peasant	Widowed	No	No religion
3	Inhabited for generations	Illiteracy	Agriculture	Land-holding peasant	Unmarried	No	No religion
3	Inhabited for generations	Illiteracy	Agriculture	Land-holding peasant	Married	No	No religion
15	Inhabited for generations	Illiteracy	Agriculture	Land-holding peasant	Married	No	No religion
5	Inhabited for generations	Bao school 2 years					
23	Inhabited for generations						
15	Inhabited for generations						
24	Inhabited for generations						
9	Inhabited for generations		Unemployed		Widowed		
20	Inhabited for generations	Illiteracy	Agriculture	Land-holding peasant	Married	No	No religion
13	Inhabited for generations	Illiteracy	Agriculture	Land-holding peasant	Married	No	No religion
19	Inhabited for generations						
20	Inhabited for generations						
29	Inhabited for generations	Illiteracy	Agriculture	Land-holding peasant	Widowed	No	No religion
2	Inhabited for generations	Illiteracy	Agriculture	Land-holding peasant	Married	No	No religion

Appendix: Overview of Microdata for Wudu County (*cont.*)

	The no. of household	Jia	Household	Relationship with the household head	Surname	First name	Sex	The Chinese zodiac	Birthdate	
									Year	Month
Fujin Xiang 8th Bao				Younger brother's wife	杨	款女	Female	Monkey	1920	8
Fujin Xiang 8th Bao	624	7	4	The household head	赵	登顺	Male	Rabbit	1902	4
Fujin Xiang 8th Bao				The household head's wife	丁	明哇	Female	Dragon	1891	9
Fujin Xiang 8th Bao				Son	赵	有印	Male	Ox	1913	6
Fujin Xiang 8th Bao				Son's wife	杨	海哇	Female	Monkey	1920	8
Fujin Xiang 8th Bao				Grandson	赵	三桃	Male	Goat	1943	5
Fujin Xiang 8th Bao				Granddaughter	赵	金牛	Female	Ox	1937	4
Fujin Xiang 8th Bao				The second granddaughter	赵	金桃	Female	Tiger	1938	5
Fujin Xiang 8th Bao				Uncle's son	赵	登文	Male	Snake	1892	4
Fujin Xiang 8th Bao				Older brother's wife	罗	岁子	Female	Rat	1891	6
Fujin Xiang 8th Bao	625	7	5	The household head	赵	登林	Male	Pig	1886	10
Fujin Xiang 8th Bao				The household head's wife	杨	秀儿	Female	Rat	1887	10
Fujin Xiang 8th Bao				Son	赵	有建	Male	Rooster	1908	4
Fujin Xiang 8th Bao				Son's wife	丁	毛旦	Female	Tiger	1914	10
Fujin Xiang 8th Bao				Grandson	赵	轮哇	Male	Rooster	1933	6
Fujin Xiang 8th Bao				Granddaughter	赵	香哇	Female	Ox	1937	3
Fujin Xiang 8th Bao	626	7	6	The household head	赵	金奎	Male	Dragon	1891	10

Date	The beginning time of inhabitation in this county	Educational level	Profession sector	Occupation	Marital status	Physical disability	Religion
20	Inhabited for generations	Illiteracy	Agriculture	Land-holding peasant	Married	No	No religion
9	Inhabited for generations	Illiteracy	Agriculture	Land-holding peasant	Married	No	No religion
1	Inhabited for generations	Illiteracy	Agriculture	Land-holding peasant	Married	No	No religion
22	Inhabited for generations	Illiteracy	Agriculture	Land-holding peasant	Married	No	No religion
14	Inhabited for generations	Illiteracy	Agriculture	Land-holding peasant	Married	No	No religion
3	Inhabited for generations						
12	Inhabited for generations						
10	Inhabited for generations						
3	Inhabited for generations	Illiteracy	Agriculture	Land-holding peasant	Married	No	No religion
25	Inhabited for generations	Illiteracy	Agriculture	Land-holding peasant	Married	No	No religion
18	Inhabited for generations	Illiteracy	Agriculture	Land-holding peasant	Married	No	No religion
16	Inhabited for generations	Illiteracy	Agriculture	Land-holding peasant	Married	No	No religion
29	Inhabited for generations	Illiteracy	Agriculture	Land-holding peasant	Married	No	No religion
30	Inhabited for generations	Illiteracy	Agriculture	Land-holding peasant	Married	No	No religion
1	Inhabited for generations	Illiteracy	Agriculture	Land-holding peasant	Unmarried	No	No religion
28	Inhabited for generations	Illiteracy					
9	Inhabited for generations	Illiteracy	Agriculture	Land-holding peasant	Married	No	No religion

Appendix: Overview of Microdata for Wudu County (*cont.*)

	The no. of household	Jia	Household	Relationship with the household head	Surname	First name	Sex	The Chinese zodiac	Birthdate	
									Year	Month
Fujin Xiang 8th Bao				The household head's wife	张	崴根	Female	Monkey	1895	8
Fujin Xiang 8th Bao				Son	赵	登旺	Male	Horse	1918	12
Fujin Xiang 8th Bao				Son's wife	杨	拜儿	Female	Pig	1923	10
Fujin Xiang 8th Bao				Grandson	赵	简哇	Male		1943	12
Fujin Xiang 8th Bao				Granddaughter		检女子	Female	Goat	1931	6
Fujin Xiang 8th Bao				Older brother	赵	金宝	Male	Rabbit	1890	6
Fujin Xiang 8th Bao				Older brother's wife	赵	住至	Female	Tiger	1889	7
Fujin Xiang 8th Bao				Nephew	赵	磁哇	Male	Goat	1931	11
Fujin Xiang 8th Bao	627	7	7	The household head	赵	登见	Male	Monkey	1920	12
Fujin Xiang 8th Bao				The household head's wife	赵	旦女	Female	Monkey	1920	11
Fujin Xiang 8th Bao				Father	赵	金德	Male		1882	10
Fujin Xiang 8th Bao				Nephew	赵	来哇	Male	Rooster	1934	1
Fujin Xiang 8th Bao				Uncle's wife	諓	岁连	Female	Ox	1900	3
Fujin Xiang 8th Bao	628	7	8	The household head	赵	登永	Male	Snake	1918	1
Fujin Xiang 8th Bao				The household head's wife	赵	肖儿	Female	Horse	1918	5
Fujin Xiang 8th Bao				Son	赵	高哇	Male	Rat	1937	1
Fujin Xiang 8th Bao				The second son		圈儿牛	Male	Rabbit	1940	1

Date	The beginning time of inhabitation in this county	Educational level	Profession sector	Occupation	Marital status	Physical disability	Religion
24	Inhabited for generations	Illiteracy	Agriculture	Land-holding peasant	Married	No	No religion
25	Inhabited for generations	Illiteracy	Agriculture	Land-holding peasant	Married	No	No religion
19	Inhabited for generations	Illiteracy	Agriculture	Land-holding peasant	Married	No	No religion
9	Inhabited for generations						
30	Inhabited for generations	Illiteracy	Agriculture	Land-holding peasant	Unmarried	No	No religion
27	Inhabited for generations	Illiteracy	Agriculture	Land-holding peasant	Married	No	No religion
24	Inhabited for generations	Illiteracy	Agriculture	Land-holding peasant	Married	No	No religion
28	Inhabited for generations	Illiteracy	Agriculture	Land-holding peasant	Unmarried	No	No religion
20	Inhabited for generations	Illiteracy	Agriculture	Land-holding peasant	Married	No	No religion
24	Inhabited for generations	Illiteracy	Agriculture	Land-holding peasant	Married	No	No religion
21	Inhabited for generations		Unemployed				
24	Inhabited for generations	Illiteracy	Agriculture	Land-holding peasant	Unmarried	No	No religion
27	Inhabited for generations	Illiteracy	Agriculture	Land-holding peasant	Widowed	No	No religion
27	Inhabited for generations	Illiteracy	Agriculture	Land-holding peasant	Married	No	No religion
27	Inhabited for generations	Illiteracy	Agriculture	Land-holding peasant	Married	No	No religion
31	Inhabited for generations	Bao school 2 years					
19	Inhabited for generations	Illiteracy					

Appendix: Overview of Microdata for Wudu County (*cont.*)

	The no. of house-hold	Jia	House-hold	Relation-ship with the house-hold head	Surname	First name	Sex	The Chinese zodiac	Birthdate	
									Year	Month
Fujin Xiang 8th Bao				Daughter		灵女子	Female	Horse	1942	5
Fujin Xiang 8th Bao				The second daughter		贱女子	Female	Rooster	1945	11
Fujin Xiang 8th Bao				Uncle		海生儿	Male	Ox	1913	4
Fujin Xiang 8th Bao				Younger sister	赵	菊哇	Female	Horse	1942	9
Fujin Xiang 8th Bao	629	7	9	The house-hold head	赵	登云	Male	Dog	1909	4
Fujin Xiang 8th Bao				The house-hold head's wife	张	岁女	Female	Horse	1905	5
Fujin Xiang 8th Bao				Son	赵	排哇	Male	Ox	1937	5
Fujin Xiang 8th Bao				The second son	赵	更旦	Male	Rabbit	1937	8
Fujin Xiang 8th Bao				The third son	赵	三旦	Male	Goat	1943	6
Fujin Xiang 8th Bao				Daughter		丑女子	Female	Dog	1946	7
Fujin Xiang 8th Bao				Older brother	赵	登海	Male	Rabbit	1902	6
Fujin Xiang 8th Bao				Nephew	赵	毛人	Male	Ox	1925	4
Fujin Xiang 8th Bao				Nephew's wife	韩	悬哇	Female	Tiger	1926	3
Fujin Xiang 8th Bao	630	7	10	The house-hold head	赵	万平	Male		1881	10
Fujin Xiang 8th Bao				The house-hold head's wife	杜	氏	Female		1875	10
Fujin Xiang 8th Bao				Son	赵	金仓	Male	Rabbit	1916	1
Fujin Xiang 8th Bao				Son's wife	杜	金容	Male	Horse	1918	9
Fujin Xiang 8th Bao				The second son	赵	金生	Male	Monkey	1920	8

Date	The beginning time of inhabitation in this county	Educational level	Profession sector	Occupation	Marital status	Physical disability	Religion
25	Inhabited for generations						
9	Inhabited for generations						
24	Inhabited for generations	Illiteracy	Agriculture	Land-holding peasant	Divorced	No	No religion
1	Inhabited for generations						
4	Inhabited for generations	Illiteracy	Agriculture	Land-holding peasant	Married	No	No religion
6	Inhabited for generations	Illiteracy	Agriculture	Land-holding peasant	Married	No	No religion
29	Inhabited for generations	Bao school 2 years					
13	Inhabited for generations	Illiteracy					
9	Inhabited for generations						
6	Inhabited for generations						
3	Inhabited for generations	Illiteracy	Agriculture	Land-holding peasant	Widowed	No	No religion
18	Inhabited for generations	Illiteracy	Agriculture	Land-holding peasant	Married	No	No religion
28	Inhabited for generations	Illiteracy	Agriculture	Land-holding peasant	Married	No	No religion
8	Inhabited for generations		Unemployed		Married		
5	Inhabited for generations		Unemployed		Married		
24	Inhabited for generations	Illiteracy	Agriculture	Land-holding peasant	Married	No	No religion
2	Inhabited for generations	Illiteracy	Agriculture	Land-holding peasant	Married	No	No religion
15	Inhabited for generations	Illiteracy	Agriculture	Land-holding peasant	Married	No	No religion

Appendix: Overview of Microdata for Wudu County (*cont.*)

	The no. of household	Jia	Household	Relationship with the household head	Surname	First name	Sex	The Chinese zodiac	Birthdate	
									Year	Month
Fujin Xiang 8th Bao				The second son's wife	罗	吴儿	Female	Horse	1930	4
Fujin Xiang 8th Bao				Nephew	赵	金宽	Male		1876	12
Fujin Xiang 8th Bao				Nephew's son	赵	登应	Male	Goat	1919	5
Fujin Xiang 8th Bao				Nephew's son's wife	赵	金儿	Female	Dog	1909	12
Fujin Xiang 8th Bao				The second great grandson	董	才儿	Male	Snake	1942	1
Fujin Xiang 8th Bao				The third Grand-daughter	董	谢	Female	Dog	1946	12
Fujin Xiang 8th Bao				Great Grand-daughter		三女	Female	Goat	1944	1
Fujin Xiang 8th Bao	631	8	1	The household head	杨	进朝	Male	Tiger	1889	8
Fujin Xiang 8th Bao				The household head's wife	曹	根秀	Female	Snake	1892	6
Fujin Xiang 8th Bao				Son		脆生哇	Male	Rat	1923	11
Fujin Xiang 8th Bao				The second son		三生哇	Male	Rabbit	1927	2
Fujin Xiang 8th Bao				The second son's wife	詨	岁招	Female	Ox	1925	3
Fujin Xiang 8th Bao	632	8	2	The household head	杨	万龙	Male	Goat	1906	8
Fujin Xiang 8th Bao				The household head's wife	马	崔哇	Female	Dog	1909	9
Fujin Xiang 8th Bao				Father	杨	进桂	Male		1881	8
Fujin Xiang 8th Bao				Mother	杨	氏	Female		1880	8
Fujin Xiang 8th Bao				Daughter	刘	女子	Female	Dog	1934	4

Date	The beginning time of inhabitation in this county	Educational level	Profession sector	Occupation	Marital status	Physical disability	Religion	
19	Inhabited for generations	Illiteracy	Agriculture	Land-holding peasant	Married	No	No religion	
2	Inhabited for generations		Unemployed					
4	Inhabited for generations	Illiteracy	Agriculture	Land-holding peasant	Married	No	No religion	
13	Inhabited for generations	Illiteracy	Agriculture	Land-holding peasant	Married	No	No religion	
26	Inhabited for generations							
12	Inhabited for generations							
20	Inhabited for generations							
28	Inhabited for generations	Illiteracy	Agriculture	Land-holding peasant	Married	No	No religion	
29	Inhabited for generations	Illiteracy	Agriculture	Land-holding peasant	Married	No	No religion	
29	Inhabited for generations	Illiteracy	Agriculture	Land-holding peasant	Unmarried	No	No religion	Serving in the army
16	Inhabited for generations	Illiteracy	Agriculture	Land-holding peasant	Married	No	No religion	
4	Inhabited for generations	Illiteracy	Agriculture	Land-holding peasant	Married	No	No religion	
2	Inhabited for generations	Illiteracy	Agriculture	Land-holding peasant	Married	No	No religion	
2	Inhabited for generations	Illiteracy	Agriculture	Land-holding peasant	Married	No	No religion	
13	Inhabited for generations		Unemployed		Married			
10	Inhabited for generations		Unemployed		Married			
6	Inhabited for generations	Illiteracy	Agriculture	Land-holding peasant	Unmarried	No	No religion	

Appendix: Overview of Microdata for Wudu County (*cont.*)

	The no. of household	Jia	Household	Relationship with the household head	Surname	First name	Sex	The Chinese zodiac	Birthdate	
									Year	Month
Fujin Xiang 8th Bao				The second daughter	杨	乖哇	Female	Dragon	1940	9
Fujin Xiang 8th Bao				The third daughter	杨	改哇	Female	Dog	1946	8
Fujin Xiang 8th Bao				Younger brother	杨	万有	Male	Dragon	1916	7
Fujin Xiang 8th Bao				Younger brother's wife	张	毛女	Female	Pig	1923	8
Fujin Xiang 8th Bao	633	8	3	The household head	杨	文拜	Male	Monkey	1907	11
Fujin Xiang 8th Bao				The household head's wife	石	润哇	Female	Dragon	1916	4
Fujin Xiang 8th Bao				Grandfather's brother	杨	万科	Male	Rabbit	1902	12
Fujin Xiang 8th Bao				Grandfather's brother's wife	李	传富	Female	Dragon	1903	4
Fujin Xiang 8th Bao				Uncle	杨	代哇	Male	Ox	1925	12
Fujin Xiang 8th Bao				Uncle's wife	罗	简哇	Female	Goat	1937	12
Fujin Xiang 8th Bao				Grandfather's brother's wife	赵	灯哇	Female	Dragon	1891	11
Fujin Xiang 8th Bao	634	8	4	The household head	杨	万李	Male	Horse	1893	10
Fujin Xiang 8th Bao				The household head's wife	李	必儿	Female	Ox	1913	10
Fujin Xiang 8th Bao				Son		大射哇	Male	Horse	1930	8
Fujin Xiang 8th Bao				Son's wife	张	茶哇	Female	Dragon	1928	2

Date	The beginning time of inhabitation in this county	Educational level	Profession sector	Occupation	Marital status	Physical disability	Religion
29	Inhabited for generations						
22	Inhabited for generations						
29	Inhabited for generations	Illiteracy	Agriculture	Land-holding peasant	Married	No	No religion
20	Inhabited for generations	Illiteracy	Agriculture	Land-holding peasant	Married	No	No religion
11	Inhabited for generations	Illiteracy	Agriculture	Land-holding peasant	Married	No	No religion
30	Inhabited for generations	Illiteracy	Agriculture	Land-holding peasant	Married	No	No religion
12	Inhabited for generations	Illiteracy	Agriculture	Land-holding peasant	Married	No	No religion
14	Inhabited for generations	Illiteracy	Agriculture	Land-holding peasant	Married	No	No religion
9	Inhabited for generations	Illiteracy	Agriculture	Land-holding peasant	Married	No	No religion
11	Inhabited for generations	Illiteracy	Agriculture	Land-holding peasant	Married	No	No religion
30	Inhabited for generations	Illiteracy	Agriculture	Land-holding peasant	Widowed	No	No religion
1	Inhabited for generations	Illiteracy	Agriculture	Land-holding peasant	Married	No	No religion
5	Inhabited for generations	Illiteracy	Agriculture	Land-holding peasant	Married	No	No religion
31	Inhabited for generations	Illiteracy	Agriculture	Land-holding peasant	Married	No	No religion
21	Inhabited for generations	Illiteracy	Agriculture	Land-holding peasant	Married	No	No religion

Appendix: Overview of Microdata for Wudu County (*cont.*)

	The no. of house-hold	Jia	House-hold	Relation-ship with the house-hold head	Surname	First name	Sex	The Chinese zodiac	Birthdate	
									Year	Month
Fujin Xiang 8th Bao				The second son		双射哇	Male	Rooster	1933	6
Fujin Xiang 8th Bao				The third son		三射哇	Male	Dragon	1940	9
Fujin Xiang 8th Bao				The fourth son	杨	腰哇	Male	Dog	1946	5
Fujin Xiang 8th Bao	635	8	5	The house-hold head	杨	文李	Male	Rat	1912	11
Fujin Xiang 8th Bao				The house-hold head's wife	杨	别哇	Female	Tiger	1914	8
Fujin Xiang 8th Bao				Mother	杨	氏	Female	Rooster	1882	10
Fujin Xiang 8th Bao				Uncle's wife	张	拜儿	Female	Dog	1900	11
Fujin Xiang 8th Bao				Uncle's son	杨	文有	Male	Tiger	1938	8
Fujin Xiang 8th Bao	636	8	6	The house-hold head	杨	生田	Male	Rabbit	1924	6
Fujin Xiang 8th Bao				The house-hold head's wife	李	陈子	Female	Tiger	1926	4
Fujin Xiang 8th Bao				Mother	杨	传儿	Female	Tiger	1891	3
Fujin Xiang 8th Bao				Older brother	杨	生全	Male	Rat	1912	3
Fujin Xiang 8th Bao				Younger brother	杨	毛儿	Male	Rooster	1933	10
Fujin Xiang 8th Bao	637	8	7	The house-hold head	杨	万明	Male	Rooster	1882	4
Fujin Xiang 8th Bao				The house-hold head's wife	誴	有儿	Female	Tiger	1889	6
Fujin Xiang 8th Bao				Son	杨	生世	Male	Monkey	1920	8
Fujin Xiang 8th Bao				Son's wife	誴	佐福	Female	Rooster	1908	6
Fujin Xiang 8th Bao				Grand-daughter	冬	且儿	Female	Horse	1942	3

Date	The beginning time of inhabitation in this county	Educational level	Profession sector	Occupation	Marital status	Physical disability	Religion
4	Inhabited for generations	Bao school 2 years			Unmarried	No	No religion
5	Inhabited for generations	Illiteracy					
6	Inhabited for generations						
19	Inhabited for generations	Illiteracy	Agriculture	Land-holding peasant	Unmarried	No	No religion
9	Inhabited for generations	Illiteracy	Agriculture	Land-holding peasant	Unmarried	No	No religion
1	Inhabited for generations		Unemployed				
30	Inhabited for generations	Illiteracy	Agriculture	Land-holding peasant	Widowed	No	No religion
5	Inhabited for generations	Illiteracy					
9	Inhabited for generations	Illiteracy	Agriculture	Land-holding peasant	Married	No	No religion
23	Inhabited for generations	Illiteracy	Agriculture	Land-holding peasant	Married	No	No religion
20	Inhabited for generations	Illiteracy	Agriculture	Land-holding peasant	Widowed	No	No religion
9	Inhabited for generations	Illiteracy	Agriculture	Land-holding peasant	Widowed	No	No religion
7	Inhabited for generations	Bao school 2 years					
13	Inhabited for generations	Illiteracy	Agriculture	Land-holding peasant	Married	No	No religion
17	Inhabited for generations	Illiteracy	Agriculture	Land-holding peasant	Married	No	No religion
27	Inhabited for generations	Illiteracy	Agriculture	Land-holding peasant	Married	No	No religion
28	Inhabited for generations	Illiteracy	Agriculture	Land-holding peasant	Married	No	No religion
1	Inhabited for generations						

Appendix: Overview of Microdata for Wudu County (*cont.*)

	The no. of household	Jia	Household	Relationship with the household head	Surname	First name	Sex	The Chinese zodiac	Birthdate	
									Year	Month
Fujin Xiang 8th Bao				The second granddaughter	冬	连儿	Female	Rooster	1945	11
Fujin Xiang 8th Bao				The second son	杨	生起	Male	Ox	1926	5
Fujin Xiang 8th Bao				The second son's wife	马	尕哇	Male	Tiger	1925	9
Fujin Xiang 8th Bao				Nephew	苟	让哇	Male	Ox	1925	1
Fujin Xiang 8th Bao				Nephew's wife	丁	谢娃	Female	Dog	1922	5
Fujin Xiang 8th Bao				Uncle's son's wife	丁	节哇	Male	Goat	1919	11
Fujin Xiang 8th Bao	638	8	8	The household head	杨	万海	Male	Horse	1893	6
Fujin Xiang 8th Bao				The household head's wife	韩	当儿	Female	Dragon	1891	2
Fujin Xiang 8th Bao				Son	王	女子	Male	Horse	1930	1
Fujin Xiang 8th Bao				The second son	杨	满哇	Male	Tiger	1938	10
Fujin Xiang 8th Bao	639	8	9	The household head	杨	万林	Male	Rooster	1882	5
Fujin Xiang 8th Bao				The household head's wife	罗	富儿	Female	Rooster	1887	8
Fujin Xiang 8th Bao				Son	杨	生才	Male	Rat	1912	8
Fujin Xiang 8th Bao				Son's wife	张	踉赛	Female	Goat	1919	10
Fujin Xiang 8th Bao	640	8	10	The household head	杨	生见	Male	Goat	1905	1
Fujin Xiang 8th Bao				The household head's wife	罗	苏家	Female	Rooster	1908	4
Fujin Xiang 8th Bao				Son	杨	王家	Male	Monkey	1932	3

Date	The beginning time of inhabitation in this county	Educational level	Profession sector	Occupation	Marital status	Physical disability	Religion	
14	Inhabited for generations							
21	Inhabited for generations	Illiteracy	Agriculture	Land-holding peasant	Married	No	No religion	Serving in the army
16	Inhabited for generations	Illiteracy	Agriculture	Land-holding peasant	Married	No	No religion	
29	Inhabited for generations	Illiteracy	Agriculture	Land-holding peasant	Married	No	No religion	
4	Inhabited for generations	Illiteracy	Agriculture	Land-holding peasant	Married	No	No religion	
30	Inhabited for generations	Illiteracy	Agriculture	Land-holding peasant	Widowed	No	No religion	
15	Inhabited for generations	Illiteracy	Agriculture	Land-holding peasant	Married	No	No religion	
20	Inhabited for generations	Illiteracy	Agriculture	Land-holding peasant	Married	No	No religion	
2	Inhabited for generations	Illiteracy	Agriculture	Land-holding peasant	Unmarried	No	No religion	
11	Inhabited for generations	Bao school 1 year						
2	Inhabited for generations		Unemployed		Married	No	No religion	
17	Inhabited for generations	Illiteracy	Agriculture	Land-holding peasant	Married	No	No religion	
17	Inhabited for generations	Illiteracy	Agriculture	Land-holding peasant	Married	No	No religion	
27	Inhabited for generations	Illiteracy	Agriculture	Land-holding peasant	Married	No	No religion	
2	Inhabited for generations	Illiteracy	Agriculture	Land-holding peasant	Married	No	No religion	
2	Inhabited for generations	Illiteracy	Agriculture	Land-holding peasant	Married	No	No religion	
14	Inhabited for generations	Sishu 2 years			Unmarried	No	No religion	

Appendix: Overview of Microdata for Wudu County (*cont.*)

	The no. of household	Jia	Household	Relationship with the household head	Surname	First name	Sex	The Chinese zodiac	Birthdate	
									Year	Month
Fujin Xiang 8th Bao				The second son	杨	王才儿	Male	Tiger	1938	6
Fujin Xiang 8th Bao				Daughter	罗	铁桂儿	Female	Monkey	1944	3
Fujin Xiang 8th Bao				The second daughter	杨	王才女	Female	Dog	1946	12
Fujin Xiang 8th Bao	641	8	11	The household head	杨	万世	Male	Pig	1910	12
Fujin Xiang 8th Bao				Son	杨	谢哇	Male	Pig	1935	4
Fujin Xiang 8th Bao				Daughter	杨	双女	Female	Goat	1943	5
Fujin Xiang 8th Bao	642	8	12	The household head	杨	万清	Male	Monkey	1895	5
Fujin Xiang 8th Bao				The household head's wife	赵	转节	Female	Ox	1888	7
Fujin Xiang 8th Bao				Son	杨	建连	Male	Snake	1917	12
Fujin Xiang 8th Bao				The second son		进儿子	Male	Ox	1925	4
Fujin Xiang 8th Bao				The second son's wife	赵	玉娃	Female	Rabbit	1924	6
Fujin Xiang 8th Bao				Grandson	杨	吴儿	Male	Pig	1947	4
Fujin Xiang 8th Bao				The third son		缠生哇	Male	Horse	1930	4
Fujin Xiang 8th Bao				Younger brother's wife	丁	转富	Female	Goat	1894	7
Fujin Xiang 8th Bao				Nephew	赵	哇	Male	Ox	1925	11
Fujin Xiang 8th Bao	643	8	13	The household head	杨	万云	Male	Monkey	1920	5
Fujin Xiang 8th Bao				Father	杨	进双	Male		1872	12
Fujin Xiang 8th Bao				Mother	杨	氏	Female		1882	9

Date	The beginning time of inhabitation in this county	Educational level	Profession sector	Occupation	Marital status	Physical disability	Religion	
7	Inhabited for generations	Bao school 1 year						
2	Inhabited for generations							
24	Inhabited for generations							
17	Inhabited for generations	Illiteracy	Agriculture	Land-holding peasant	Widowed	No	No religion	
18	Inhabited for generations	Bao school 1 year						
22	Inhabited for generations							
11	Inhabited for generations	Illiteracy	Agriculture	Land-holding peasant	Married	No	No religion	
2	Inhabited for generations	Illiteracy	Agriculture	Land-holding peasant	Married	No	No religion	
5	Inhabited for generations	Illiteracy	Agriculture	Land-holding peasant	Married	No	No religion	Serving in the army
22	Inhabited for generations	Illiteracy	Agriculture	Land-holding peasant	Married	Blind	No religion	
26	Inhabited for generations	Illiteracy	Agriculture	Land-holding peasant	Married	No	No religion	
20	Inhabited for generations							
30	Inhabited for generations	Bao school 2 years			Unmarried	No	No religion	Serving in the army
12	Inhabited for generations	Illiteracy	Agriculture	Land-holding peasant	Widowed	No	No religion	
6	Inhabited for generations	Illiteracy	Agriculture	Land-holding peasant	Unmarried	No	No religion	
1	Inhabited for generations	Sishu 2 years	Agriculture	Land-holding peasant	Widowed	No	No religion	
1	Inhabited for generations		Unemployed		Married			
9	Inhabited for generations		Unemployed		Married			

Appendix: Overview of Microdata for Wudu County (*cont.*)

	The no. of household	Jia	Household	Relationship with the household head	Surname	First name	Sex	The Chinese zodiac	Birthdate Year	Month
Fujin Xiang 8th Bao				Older brother	杨	万昇	Male	Dragon	1891	4
Fujin Xiang 8th Bao				Older brother's wife	赵	亲女	Female	Rat	1899	11
Fujin Xiang 8th Bao				Nephew		羊女子	Male	Pig	1947	4
Fujin Xiang 8th Bao	644	8	14	The household head	杨	生义	Male	Monkey	1906	1
Fujin Xiang 8th Bao				The household head's wife	諔	田女	Female	Monkey	1907	11
Fujin Xiang 8th Bao				Son	金	有儿	Male	Ox	1937	3
Fujin Xiang 8th Bao				The second son	金	仓儿	Male	Snake	1941	3
Fujin Xiang 8th Bao				Daughter	林	女子	Female	Monkey	1944	3
Fujin Xiang 8th Bao				Younger brother	王	海儿	Male	Rabbit	1915	8
Fujin Xiang 8th Bao	645	8	15	The household head	杨	生仓	Male	Snake	1917	2
Fujin Xiang 8th Bao				The household head's wife	諔	罗金	Female	Ox	1913	4
Fujin Xiang 8th Bao				Mother	杨	氏	Female		1878	3
Fujin Xiang 8th Bao				Uncle	杨	万平	Male	Dragon	1891	12
Fujin Xiang 8th Bao				Uncle's wife	赵	建哇	Female	Goat	1893	1
Fujin Xiang 8th Bao	646	8	16	The household head	杨	文举	Male	Dog	1896	2
Fujin Xiang 8th Bao				Father	杨	生富	Male		1875	10
Fujin Xiang 8th Bao				Mother	杜	氏	Female		1874	9
Fujin Xiang 8th Bao				Son	杨	有明	Male	Pig	1924	1

Date	The beginning time of inhabitation in this county	Educational level	Profession sector	Occupation	Marital status	Physical disability	Religion
3	Inhabited for generations	Illiteracy	Agriculture	Land-holding peasant	Married	No	No religion
10	Inhabited for generations	Illiteracy	Agriculture	Land-holding peasant	Married	No	No religion
28	Inhabited for generations						
16	Inhabited for generations	Illiteracy	Agriculture	Land-holding peasant	Married	No	No religion
8	Inhabited for generations	Illiteracy	Agriculture	Land-holding peasant	Married	No	No religion
4	Inhabited for generations	Bao school 1 year					
19	Inhabited for generations						
3	Inhabited for generations						
23	Inhabited for generations	Illiteracy	Agriculture	Land-holding peasant	Unmarried	No	No religion
13	Inhabited for generations	Illiteracy	Agriculture	Land-holding peasant	Married	No	No religion
24	Inhabited for generations	Illiteracy	Agriculture	Land-holding peasant	Married	No	No religion
19	Inhabited for generations		Unemployed				
27	Inhabited for generations	Illiteracy	Agriculture	Land-holding peasant	Married	No	No religion
26	Inhabited for generations	Illiteracy	Agriculture	Land-holding peasant	Married	No	No religion
1	Inhabited for generations	Illiteracy	Agriculture	Land-holding peasant	Widowed	No	No religion
11	Inhabited for generations		Unemployed				
8	Inhabited for generations						
8	Inhabited for generations	Illiteracy	Agriculture	Land-holding peasant	Married	No	No religion

Appendix: Overview of Microdata for Wudu County (*cont.*)

	The no. of household	Jia	Household	Relationship with the household head	Surname	First name	Sex	The Chinese zodiac	Birthdate Year	Month
Fujin Xiang 8th Bao				Son's wife	马	相娃	Female	Horse	1942	1
Fujin Xiang 8th Bao				Granddaughter		信女子	Female	Rooster	1945	5
Fujin Xiang 8th Bao	647	8	17	The household head	杨	羊拜	Male	Ox	1925	12
Fujin Xiang 8th Bao				The household head's wife	张	崴新	Female	Goat	1920	1
Fujin Xiang 8th Bao				Father	杨	文海	Male	Rat	1912	9
Fujin Xiang 8th Bao				Mother	杨	连娃	Female	Snake	1904	12
Fujin Xiang 8th Bao				Uncle	杨	都哇	Male	Rabbit	1915	8
Fujin Xiang 8th Bao				Younger sister	董	女子	Female	Tiger	1938	8
Fujin Xiang 8th Bao	648	8	18	The household head	杨	万登	Male	Rat	1887	12
Fujin Xiang 8th Bao				The household head's wife	张	崴建	Female	Dog	1896	5
Fujin Xiang 8th Bao				Son	杨	永哇	Male	Pig	1923	7
Fujin Xiang 8th Bao				Son's wife	杨	海	Female	Snake	1929	5
Fujin Xiang 8th Bao	649	9	1	The household head	杨	登云	Male	Horse	1919	1
Fujin Xiang 8th Bao				The household head's wife	赵	同儿	Female	Tiger	1914	9
Fujin Xiang 8th Bao				Father	杨	生气	Male		1881	12
Fujin Xiang 8th Bao				Daughter	杨	冬英	Female	Snake	1942	5
Fujin Xiang 8th Bao				The second daughter	崴	冬英	Female	Goat	1943	5
Fujin Xiang 8th Bao				Older brother	杨	登义	Male	Dog	1897	9

Date	The beginning time of inhabitation in this county	Educational level	Profession sector	Occupation	Marital status	Physical disability	Religion
30	Inhabited for generations	Illiteracy	Agriculture	Land-holding peasant	Married	No	No religion
25	Inhabited for generations						
15	Inhabited for generations	Illiteracy	Agriculture	Land-holding peasant	Married	No	No religion
10	Inhabited for generations	Illiteracy	Agriculture	Land-holding peasant	Married	No	No religion
28	Inhabited for generations	Illiteracy	Agriculture	Land-holding peasant	Married	No	No religion
30	Inhabited for generations	Illiteracy	Agriculture	Land-holding peasant	Married	No	No religion
3	Inhabited for generations	Illiteracy	Agriculture	Land-holding peasant	Unmarried	No	No religion
29	Inhabited for generations	Illiteracy					
31	Inhabited for generations	Illiteracy	Agriculture	Land-holding peasant	Married	No	No religion
7	Inhabited for generations	Illiteracy	Others		Married	No	No religion
1	Inhabited for generations	Illiteracy	Agriculture	Land-holding peasant	Married	No	No religion
22	Inhabited for generations	Illiteracy	Others		Married	No	No religion
21	Inhabited for generations	Illiteracy	Agriculture	Land-holding peasant	Married	No	No religion
17	Inhabited for generations	Illiteracy	Agriculture	Land-holding peasant	Married	No	No religion
9	Inhabited for generations		Unemployed		Widowed		
11	Inhabited for generations	Illiteracy					
24	Inhabited for generations						
30	Inhabited for generations	Illiteracy	Agriculture	Land-holding peasant	Married	No	No religion

Appendix: Overview of Microdata for Wudu County (*cont.*)

	The no. of household	Jia	Household	Relationship with the household head	Surname	First name	Sex	The Chinese zodiac	Birthdate	
									Year	Month
Fujin Xiang 8th Bao				Older brother's wife	罗	两儿	Female	Dog	1897	10
Fujin Xiang 8th Bao				Nephew		大祥儿	Male	Rooster	1933	3
Fujin Xiang 8th Bao				Nephew's wife	张	泒哇	Female	Horse	1930	8
Fujin Xiang 8th Bao				Nephew's son	杨	早怌	Male	Dog	1946	5
Fujin Xiang 8th Bao	650	9	2	The household head	杨	生双	Male	Tiger	1901	11
Fujin Xiang 8th Bao				The household head's wife	罗	双清儿	Female	Rooster	1908	5
Fujin Xiang 8th Bao				Mother	李	氏	Female		1874	7
Fujin Xiang 8th Bao				Son	杨	神哇	Male	Rabbit	1938	12
Fujin Xiang 8th Bao				Daughter	杨	杜女	Female	Dog	1946	11
Fujin Xiang 8th Bao	651	9	3	The household head	张	仲乡	Male	Dragon	1885	11
Fujin Xiang 8th Bao				The household head's wife	罗	麦哇	Female	Horse	1905	7
Fujin Xiang 8th Bao				Son	张	福峻	Male	Tiger	1926	5
Fujin Xiang 8th Bao				Son's wife	张	新哇	Female	Rabbit	1927	5
Fujin Xiang 8th Bao				The second son	张	锁	Male	Rabbit	1943	11
Fujin Xiang 8th Bao				Younger brother	张	文科	Male	Monkey	1920	10
Fujin Xiang 8th Bao				Younger brother's wife	杜	毛女	Female	Monkey	1919	7
Fujin Xiang 8th Bao	652	9	4	The household head	杨	生林	Male	Pig	1886	3

Date	The beginning time of inhabitation in this county	Educational level	Profession sector	Occupation	Marital status	Physical disability	Religion
1	Inhabited for generations	Illiteracy	Agriculture	Land-holding peasant	Married	No	No religion
13	Inhabited for generations	Illiteracy	Agriculture	Land-holding peasant	Married	No	No religion
26	Inhabited for generations	Illiteracy	Agriculture	Land-holding peasant	Married	No	No religion
5	Inhabited for generations						
24	Inhabited for generations	Illiteracy	Agriculture	Land-holding peasant	Married	No	No religion
18	Inhabited for generations	Illiteracy	Agriculture	Land-holding peasant	Married	No	No religion
8	Inhabited for generations	Illiteracy	Agriculture	Land-holding peasant	Widowed	No	No religion
28	Inhabited for generations	Illiteracy					
28	Inhabited for generations						
6	Inhabited for generations	Illiteracy	Agriculture	Land-holding peasant	Married	No	No religion
6	Inhabited for generations	Illiteracy	Agriculture	Land-holding peasant	Married	No	No religion
6	Inhabited for generations	Illiteracy	Agriculture	Land-holding peasant	Married	No	No religion
8	Inhabited for generations	Illiteracy	Agriculture	Land-holding peasant	Married	No	No religion
9	Inhabited for generations						
8	Inhabited for generations	Illiteracy	Agriculture	Land-holding peasant	Married	No	No religion
9	Inhabited for generations	Illiteracy	Agriculture	Land-holding peasant	Married	No	No religion
22	Inhabited for generations	Illiteracy	Agriculture	Land-holding peasant	Married	No	No religion

Appendix: Overview of Microdata for Wudu County (*cont.*)

	The no. of household	Jia	Household	Relationship with the household head	Surname	First name	Sex	The Chinese zodiac	Birthdate Year	Month
Fujin Xiang 8th Bao				The household head's wife	赵	贤哇	Female	Dog	1897	12
Fujin Xiang 8th Bao				Son	杨	登升	Male	Monkey	1920	8
Fujin Xiang 8th Bao				Son's wife	赵	冯哇	Female	Monkey	1920	5
Fujin Xiang 8th Bao				Granddaughter	杨	简哇	Female	Horse	1942	8
Fujin Xiang 8th Bao	653	9	5	The household head	杨	赵金儿	Male	Dog	1908	1
Fujin Xiang 8th Bao				The household head's wife	杨	崴儿	Female	Dragon	1916	4
Fujin Xiang 8th Bao				Father	杨	万九	Male	Pig	1885	1
Fujin Xiang 8th Bao				Mother	杨	傅儿	Female	Pig	1885	6
Fujin Xiang 8th Bao				Adopted daughter	罗	水音	Female	Goat	1931	5
Fujin Xiang 8th Bao	654	9	6	The household head	杨	生尚	Male	Monkey	1906	2
Fujin Xiang 8th Bao				The household head's wife	赵	双悬	Female	Rat	1918	6
Fujin Xiang 8th Bao				Son	杨	番儿	Male	Dog	1934	9
Fujin Xiang 8th Bao				The second son	杨	长哇	Male	Snake	1941	5
Fujin Xiang 8th Bao				Older brother	杨	生旺	Male	Goat	1906	5
Fujin Xiang 8th Bao				Older brother's wife	罗	白哇	Female	Dragon	1916	12
Fujin Xiang 8th Bao				Niece		绑女子	Female	Monkey	1944	5
Fujin Xiang 8th Bao				The second niece	杨	月女	Female	Dog	1946	6

Date	The beginning time of inhabitation in this county	Educational level	Profession sector	Occupation	Marital status	Physical disability	Religion
4	Inhabited for generations	Illiteracy	Agriculture	Land-holding peasant	Married	No	No religion
17	Inhabited for generations	Illiteracy	Agriculture	Land-holding peasant	Married	No	No religion
25	Inhabited for generations	Illiteracy	Agriculture	Land-holding peasant	Married	No	No religion
5	Inhabited for generations						
19	Inhabited for generations	Illiteracy	Agriculture	Land-holding peasant	Married	No	No religion
11	Inhabited for generations	Illiteracy	Agriculture	Land-holding peasant	Married	No	No religion
6	Inhabited for generations	Illiteracy	Agriculture	Land-holding peasant	Married	No	No religion
21	Inhabited for generations	Illiteracy	Agriculture	Land-holding peasant	Married	No	No religion
15	Inhabited for generations	Illiteracy	Agriculture	Land-holding peasant	Unmarried	No	No religion
8	Inhabited for generations	Illiteracy	Agriculture	Land-holding peasant	Married	No	No religion
10	Inhabited for generations	Illiteracy	Agriculture	Land-holding peasant	Married	No	No religion
15	Inhabited for generations						
10	Inhabited for generations						
25	Inhabited for generations	Illiteracy	Agriculture	Land-holding peasant	Married	No	No religion
29	Inhabited for generations	Illiteracy	Agriculture	Land-holding peasant	Married	No	No religion
13	Inhabited for generations						
4	Inhabited for generations						

Appendix: Overview of Microdata for Wudu County (*cont.*)

	The no. of household	Jia	Household	Relationship with the household head	Surname	First name	Sex	The Chinese zodiac	Birthdate	
									Year	Month
Fujin Xiang 8th Bao				Younger brother	杨	郭哇	Male	Horse	1918	8
Fujin Xiang 8th Bao	655	9	7	The household head	杨	生明	Male	Rat	1887	2
Fujin Xiang 8th Bao				The household head's wife	丁	润秀	Female	Tiger	1901	10
Fujin Xiang 8th Bao				Son	杨	大学哇	Male	Tiger	1926	8
Fujin Xiang 8th Bao				Son's wife	杨	菜娃	Female	Rat	1924	5
Fujin Xiang 8th Bao				Nephew's son	杨	仲留	Male	Rat	1912	6
Fujin Xiang 8th Bao				The second grandson	杨	仲仓	Male	Ox	1925	9
Fujin Xiang 8th Bao				The second grandson's wife	赵	早哇	Female	Horse	1930	8
Fujin Xiang 8th Bao	656	9	8	The household head	杨	文响	Male	Dog	1909	8
Fujin Xiang 8th Bao				The household head's wife	罗	斌哇	Female	Ox	1913	10
Fujin Xiang 8th Bao				Mother	杨	吉儿	Female	Tiger	1889	5
Fujin Xiang 8th Bao				Son	双	平儿	Male	Tiger	1938	6
Fujin Xiang 8th Bao				The second son	杨	毕儿	Male	Dog	1946	8
Fujin Xiang 8th Bao				Daughter	杨	寇哇	Female	Pig	1935	11
Fujin Xiang 8th Bao	657	9	9	The household head	杨	生成	Male	Dog	1885	6
Fujin Xiang 8th Bao				The household head's wife	赵	崴富	Female	Dog	1885	7
Fujin Xiang 8th Bao				Son	杨	文建	Male	Rooster	1908	2

Date	The beginning time of inhabitation in this county	Educational level	Profession sector	Occupation	Marital status	Physical disability	Religion	
23	Inhabited for generations	Illiteracy	Agriculture	Land-holding peasant	Unmarried	No	No religion	Serving in the army
27	Inhabited for generations	Illiteracy	Agriculture	Land-holding peasant	Married	No	No religion	
31	Inhabited for generations	Illiteracy	Agriculture	Land-holding peasant	Married	No	No religion	
10	Inhabited for generations	Illiteracy	Agriculture	Land-holding peasant	Married	No	No religion	
19	Inhabited for generations	Illiteracy	Agriculture	Land-holding peasant	Married	No	No religion	
16	Inhabited for generations	Illiteracy	Agriculture	Land-holding peasant	Married	No	No religion	
23	Inhabited for generations	Illiteracy	Agriculture	Land-holding peasant	Married	No	No religion	
22	Inhabited for generations	Illiteracy	Agriculture	Land-holding peasant	Married	No	No religion	
26	Inhabited for generations	Illiteracy	Agriculture	Land-holding peasant	Married	No	No religion	
18	Inhabited for generations	Illiteracy	Agriculture	Land-holding peasant	Married	No	No religion	
29	Inhabited for generations	Illiteracy	Agriculture	Land-holding peasant	Widowed	No	No religion	
25	Inhabited for generations	Illiteracy						
3	Inhabited for generations							
16	Inhabited for generations							
16	Inhabited for generations	Illiteracy	Agriculture	Land-holding peasant	Married	No	No religion	
11	Inhabited for generations	Illiteracy	Agriculture	Land-holding peasant	Married	No	No religion	
22	Inhabited for generations	Illiteracy	Agriculture	Land-holding peasant	Unmarried	No	No religion	

Appendix: Overview of Microdata for Wudu County (*cont.*)

	The no. of household	Jia	Household	Relationship with the household head	Surname	First name	Sex	The Chinese zodiac	Birthdate	
									Year	Month
Fujin Xiang 8th Bao				The second son	杨	文仲	Male	Horse	1930	4
Fujin Xiang 8th Bao				Son's wife	李	腰女	Female	Goat	1931	7
Fujin Xiang 8th Bao				Uncle's son	杨	生炳	Male	Snake	1892	8
Fujin Xiang 8th Bao				Younger brother's wife	赵	海儿	Female	Snake	1892	5
Fujin Xiang 8th Bao				Nephew	杨	文信	Male	Rabbit	1927	2
Fujin Xiang 8th Bao				Nephew's wife	罗	射娃	Female	Tiger	1933	5
Fujin Xiang 8th Bao	658	9	10	The household head	杨	文忠	Male	Monkey	1907	2
Fujin Xiang 8th Bao				The household head's wife	石	银娃	Female	Rabbit	1902	5
Fujin Xiang 8th Bao				Father	杨	生彦	Male		1869	1
Fujin Xiang 8th Bao				Mother	丁	氏	Female		1879	4
Fujin Xiang 8th Bao				Son	杨	仲容	Male	Goat	1919	11
Fujin Xiang 8th Bao				Son's wife	丁	扣娃	Female	Horse	1918	11
Fujin Xiang 8th Bao				The second son	杨	才哇	Male	Tiger	1938	4
Fujin Xiang 8th Bao				Daughter	杨	苏哇	Female	Snake	1941	2
Fujin Xiang 8th Bao				The second daughter	杨	安女	Female	Rooster	1945	10
Fujin Xiang 8th Bao	659	9	11	The household head	杨	落儿	Female	Dragon	1903	8
Fujin Xiang 8th Bao				The household head's husband	杨	喜哇	Male	Dragon	1903	7
Fujin Xiang 8th Bao				Son	杨	哑学子	Male	Rat	1924	11

Date	The beginning time of inhabitation in this county	Educational level	Profession sector	Occupation	Marital status	Physical disability	Religion	
9	Inhabited for generations	Illiteracy	Agriculture	Land-holding peasant	Married	No	No religion	
17	Inhabited for generations	Illiteracy	Agriculture	Land-holding peasant	Married	No	No religion	
19	Inhabited for generations	Illiteracy	Agriculture	Land-holding peasant	Married	No	No religion	
14	Inhabited for generations	Illiteracy	Agriculture	Land-holding peasant	Married	No	No religion	
15	Inhabited for generations	Illiteracy	Agriculture	Land-holding peasant	Married	No	No religion	
30	Inhabited for generations	Illiteracy	Agriculture	Land-holding peasant	Married	No	No religion	
9	Inhabited for generations	Illiteracy	Agriculture	Land-holding peasant	Married	No	No religion	
31	Inhabited for generations	Illiteracy	Agriculture	Land-holding peasant	Married	No	No religion	
5	Inhabited for generations		Unemployed		Married			
28	Inhabited for generations		Unemployed		Married			
17	Inhabited for generations	Illiteracy	Agriculture	Land-holding peasant	Married	No	No religion	
15	Inhabited for generations	Illiteracy	Agriculture	Land-holding peasant	Married	No	No religion	
4	Inhabited for generations							
2	Inhabited for generations							
8	Inhabited for generations							
31	Inhabited for generations	Illiteracy	Agriculture	Land-holding peasant	Married	No	No religion	
21	Inhabited for generations	Illiteracy	Agriculture	Land-holding peasant	Married	No	No religion	Serving in the army
4	Inhabited for generations	Illiteracy	Agriculture	Land-holding peasant	Unmarried	No	No religion	

Appendix: Overview of Microdata for Wudu County (*cont.*)

	The no. of household	Jia	Household	Relationship with the household head	Surname	First name	Sex	The Chinese zodiac	Birthdate	
									Year	Month
Fujin Xiang 8th Bao				The second son	杨	马哇	Male	Horse	1930	6
Fujin Xiang 8th Bao				Family member	誐	栢家	Male	Dragon	1902	3
Fujin Xiang 8th Bao	660	9	12	The household head	杨	万勇	Male	Horse	1892	2
Fujin Xiang 8th Bao				Son	杨	海哇	Male	Monkey	1920	8
Fujin Xiang 8th Bao				Son's wife	赵	君哇	Female	Horse	1918	12
Fujin Xiang 8th Bao				Grandson		大洋哇	Male	Dragon	1940	2
Fujin Xiang 8th Bao				The second grandson		崴羊哇	Male	Horse	1942	5
Fujin Xiang 8th Bao				The third grandson	杨	栢哇	Male	Dog	1946	6
Fujin Xiang 8th Bao	661	9	13	The household head	杨	万动	Male	Ox	1913	7
Fujin Xiang 8th Bao				The household head's wife	杨	佐女	Female	Tiger	1914	11
Fujin Xiang 8th Bao				Daughter		萧哇	Female	Monkey	1944	5
Fujin Xiang 8th Bao				The second daughter		崴萧女	Female	Pig	1947	5
Fujin Xiang 8th Bao	662	9	14	The household head	杨	万宽	Male	Tiger	1926	9
Fujin Xiang 8th Bao				The household head's wife	张	甦娃	Female	Ox	1925	4
Fujin Xiang 8th Bao				Mother	杜	深儿	Female	Snake	1892	4
Fujin Xiang 8th Bao				Son	张	怌	Male	Rat	1932	7
Fujin Xiang 8th Bao				Older brother	杨	催哇	Male	Goat	1894	9
Fujin Xiang 8th Bao				Older brother's wife	罗	刘哇	Female	Goat	1896	11

Date	The beginning time of inhabitation in this county	Educational level	Profession sector	Occupation	Marital status	Physical disability	Religion
6	Inhabited for generations	Illiteracy	Agriculture	Land-holding peasant	Unmarried	No	No religion
19	Inhabited for generations	Illiteracy	Agriculture	Land-holding peasant	Widowed	No	No religion
2	Inhabited for generations	Illiteracy	Agriculture	Land-holding peasant	Widowed	No	No religion
5	Inhabited for generations	Illiteracy	Agriculture	Land-holding peasant	Widowed	No	No religion
2	Inhabited for generations	Illiteracy	Agriculture	Land-holding peasant	Married	No	No religion
11	Inhabited for generations	Illiteracy					
19	Inhabited for generations						
10	Inhabited for generations						
9	Inhabited for generations	Illiteracy	Agriculture	Land-holding peasant	Married	No	No religion
24	Inhabited for generations	Illiteracy	Agriculture	Land-holding peasant	Married	No	No religion
5	Inhabited for generations						
25	Inhabited for generations						
9	Inhabited for generations	Illiteracy	Agriculture	Land-holding peasant	Married	No	No religion
6	Inhabited for generations	Illiteracy	Agriculture	Land-holding peasant	Married	No	No religion
6	Inhabited for generations	Illiteracy	Agriculture	Land-holding peasant	Widowed	No	No religion
1	Inhabited for generations	Illiteracy	Agriculture	Land-holding peasant	Unmarried	No	No religion
24	Inhabited for generations	Illiteracy	Agriculture	Land-holding peasant	Married	No	No religion
27	Inhabited for generations	Illiteracy	Agriculture	Land-holding peasant	Married	No	No religion

Appendix: Overview of Microdata for Wudu County (*cont.*)

	The no. of household	Jia	Household	Relationship with the household head	Surname	First name	Sex	The Chinese zodiac	Birthdate	
									Year	Month
Fujin Xiang 8th Bao				Nephew	杨	生有	Male	Rat	1912	12
Fujin Xiang 8th Bao				Nephew's wife	罗	响哇	Female	Rat	1924	12
Fujin Xiang 8th Bao				Nephew's son	苏	联	Male	Horse	1943	2
Fujin Xiang 8th Bao				Niece		腾哇	Female	Rooster	1914	6
Fujin Xiang 8th Bao				Uncle's son	万	禄	Male	Monkey	1907	5
Fujin Xiang 8th Bao				Older brother's wife	罗	响哇	Female	Rat	1912	7
Fujin Xiang 8th Bao				Uncle's second son		崴諎哇	Male	Rat	1924	4
Fujin Xiang 8th Bao	663	9	15	The household head	誜	龙保	Male	Rooster	1896	3
Fujin Xiang 8th Bao				The household head's wife	杨	德哇	Female	Goat	1894	12
Fujin Xiang 8th Bao				Son	石	代哇	Male	Dog	1922	11
Fujin Xiang 8th Bao				Son's wife	誜	二哇	Female	Dog	1922	6
Fujin Xiang 8th Bao	664	9	16	The household head	杨	生银	Male	Horse	1893	8
Fujin Xiang 8th Bao				Son	王	拜子	Male	Pig	1923	11
Fujin Xiang 8th Bao				Son's wife	誜	糴哇	Female	Rat	1924	8
Fujin Xiang 8th Bao				Younger brother	罗	毛人	Male	Monkey	1915	3
Fujin Xiang 8th Bao				Younger brother's wife		四女子	Female	Rooster	1908	2
Fujin Xiang 8th Bao				Uncle's wife	马	氏	Female		1875	12
Fujin Xiang 8th Bao				Nephew	杨	谢哇	Male	Tiger	1926	8

Date	The beginning time of inhabitation in this county	Educational level	Profession sector	Occupation	Marital status	Physical disability	Religion
14	Inhabited for generations	Illiteracy	Agriculture	Land-holding peasant	Married	No	No religion
16	Inhabited for generations	Illiteracy	Agriculture	Land-holding peasant	Married	No	No religion
3	Inhabited for generations						
10	Inhabited for generations	Illiteracy	Agriculture	Land-holding peasant	Unmarried	No	No religion
7	Inhabited for generations	Illiteracy	Agriculture	Land-holding peasant	Married	No	No religion
4	Inhabited for generations	Illiteracy	Agriculture	Land-holding peasant	Married	No	No religion
18	Inhabited for generations	Illiteracy	Agriculture	Land-holding peasant	Unmarried	No	No religion
20	Inhabited for generations	Illiteracy	Agriculture	Land-holding peasant	Married	No	No religion
15	Inhabited for generations	Illiteracy	Agriculture	Land-holding peasant	Married	No	No religion
23	Inhabited for generations	Illiteracy	Agriculture	Land-holding peasant	Married	No	No religion
14	Inhabited for generations	Illiteracy	Agriculture	Land-holding peasant	Married	No	No religion
18	Inhabited for generations	Illiteracy	Agriculture	Land-holding peasant	Married	No	No religion
9	Inhabited for generations	Illiteracy	Agriculture	Land-holding peasant	Married	No	No religion
7	Inhabited for generations	Illiteracy	Agriculture	Land-holding peasant	Married	No	No religion
20	Inhabited for generations	Illiteracy	Agriculture	Land-holding peasant	Married	No	No religion
22	Inhabited for generations	Illiteracy	Agriculture	Land-holding peasant	Married	No	No religion
2	Inhabited for generations		Unemployed		Widowed		
2	Inhabited for generations	Illiteracy	Agriculture	Land-holding peasant	Unmarried	No	No religion

Appendix: Overview of Microdata for Wudu County (*cont.*)

	The no. of household	Jia	Household	Relationship with the household head	Surname	First name	Sex	The Chinese zodiac	Birthdate	
									Year	Month
Fujin Xiang 8th Bao				The second nephew	杨	抑哇	Female	Rabbit	1937	8
Fujin Xiang 8th Bao				Niece	杨	贱哇	Female	Dragon	1940	3
Fujin Xiang 8th Bao				The second niece	进	文子	Female	Goat	1944	1
Fujin Xiang 8th Bao	665	9	17	The household head	杨	生刚	Male	Rooster	1899	3
Fujin Xiang 8th Bao				The household head's wife	罗	毛女	Female	Goat	1919	2
Fujin Xiang 8th Bao				Son	杨	宝谢	Male	Ox	1937	12
Fujin Xiang 8th Bao				Daughter	杨	篇鞭	Female	Snake	1942	1
Fujin Xiang 8th Bao				The second daughter	杨	迷米	Female	Goat	1943	6
Fujin Xiang 8th Bao	666	10	1	The household head	李	兴章	Male	Horse	1893	2
Fujin Xiang 8th Bao				The household head's wife	謏	刘儿	Female	Snake	1892	7
Fujin Xiang 8th Bao				Son	李	颖哇	Male	Ox	1925	1
Fujin Xiang 8th Bao				Daughter	李	一哇	Female	Ox	1937	6
Fujin Xiang 8th Bao	667	10	2	The household head	李	文库	Male		1882	2
Fujin Xiang 8th Bao				The household head's wife	韩	氏	Female		1898	3
Fujin Xiang 8th Bao				Son	金	焕子	Male	Ox	1913	10
Fujin Xiang 8th Bao				Son's wife	謏	焕儿	Female	Ox	1913	12
Fujin Xiang 8th Bao				Granddaughter	李	光哇	Female	Tiger	1938	3
Fujin Xiang 8th Bao				The second granddaughter	李	居哇	Female	Dragon	1940	4

Date	The beginning time of inhabitation in this county	Educational level	Profession sector	Occupation	Marital status	Physical disability	Religion
8	Inhabited for generations	Illiteracy	Agriculture	Land-holding peasant	Unmarried	No	No religion
10	Inhabited for generations						
17	Inhabited for generations						
24	Inhabited for generations	Illiteracy	Agriculture	Land-holding peasant	Married	No	No religion
12	Inhabited for generations	Illiteracy	Agriculture	Land-holding peasant	Married	No	No religion
6	Inhabited for generations	Illiteracy					
1	Inhabited for generations						
4	Inhabited for generations						
8	Inhabited for generations	Illiteracy	Agriculture	Land-holding peasant	Married	No	No religion
14	Inhabited for generations	Illiteracy	Agriculture	Land-holding peasant	Married	No	No religion
29	Inhabited for generations	Illiteracy	Agriculture	Land-holding peasant	Unmarried	No	No religion
23	Inhabited for generations	Illiteracy					
1	Inhabited for generations	Illiteracy					
10	Inhabited for generations	Illiteracy					
13	Inhabited for generations	Illiteracy	Agriculture	Land-holding peasant	Married	No	No religion
2	Inhabited for generations	Illiteracy	Agriculture	Land-holding peasant	Married	No	No religion
27	Inhabited for generations						
21	Inhabited for generations						

Appendix: Overview of Microdata for Wudu County (*cont.*)

	The no. of household	Jia	Household	Relationship with the household head	Surname	First name	Sex	The Chinese zodiac	Birthdate Year	Month
Fujin Xiang 8th Bao				The third grandson	李	迷哇	Female	Monkey	1944	3
Fujin Xiang 8th Bao	668	10	3	The household head	李	荣朝	Male	Horse	1918	9
Fujin Xiang 8th Bao				The household head's wife	誏	和儿	Female	Horse	1918	6
Fujin Xiang 8th Bao				Grandmother	赵	氏	Female		1882	9
Fujin Xiang 8th Bao				Younger sister	李	宜哇	Female	Tiger	1926	12
Fujin Xiang 8th Bao				Uncle	张	哇	Male	Horse	1918	11
Fujin Xiang 8th Bao				Grandfather's brother's wife	韩	氏	Female		1882	6
Fujin Xiang 8th Bao				The second uncle	李	禄哇	Male	Tiger	1926	3
Fujin Xiang 8th Bao	669	10	4	The household head	杜	万建	Male	Dog	1922	1
Fujin Xiang 8th Bao				The household head's wife	李	住哇	Female	Rabbit	1927	8
Fujin Xiang 8th Bao				Mother	张	女子	Female	Rooster	1896	6
Fujin Xiang 8th Bao				Son	郭	信	Male	Pig	1947	5
Fujin Xiang 8th Bao				Older brother	杜	万代	Male	Horse	1918	12
Fujin Xiang 8th Bao				Older brother's wife	杜	香哇	Female	Rabbit	1927	3
Fujin Xiang 8th Bao				Niece	杜	笼哇	Female	Dog	1946	12
Fujin Xiang 8th Bao	670	10	5	The household head	杜	万发	Male	Rabbit	1902	1
Fujin Xiang 8th Bao				The household head's wife	石	节哇	Female	Goat	1906	3

Date	The beginning time of inhabitation in this county	Educational level	Profession sector	Occupation	Marital status	Physical disability	Religion
5	Inhabited for generations						
10	Inhabited for generations	Illiteracy	Agriculture	Land-holding peasant	Married	No	No religion
10	Inhabited for generations	Illiteracy	Agriculture	Land-holding peasant	Married	No	No religion
1	Inhabited for generations		Unemployed		Widowed		
11	Inhabited for generations	Illiteracy	Agriculture	Land-holding peasant	Unmarried	No	No religion
11	Inhabited for generations	Illiteracy	Agriculture	Land-holding peasant	Unmarried	No	No religion
4	Inhabited for generations	Illiteracy	Agriculture	Land-holding peasant	Widowed	No	No religion
18	Inhabited for generations	Illiteracy	Agriculture	Land-holding peasant	Unmarried	No	No religion
27	Inhabited for generations	Illiteracy	Agriculture	Land-holding peasant	Married	No	No religion
16	Inhabited for generations	Illiteracy	Agriculture	Land-holding peasant	Married	No	No religion
30	Inhabited for generations	Illiteracy	Agriculture	Land-holding peasant	Married	No	No religion
17	Inhabited for generations						
10	Inhabited for generations	Illiteracy	Agriculture	Land-holding peasant	Married	No	No religion
23	Inhabited for generations	Illiteracy	Agriculture	Land-holding peasant	Married	No	No religion
4	Inhabited for generations						
26	Inhabited for generations	Illiteracy	Agriculture	Land-holding peasant	Married	No	No religion
31	Inhabited for generations	Illiteracy	Agriculture	Land-holding peasant	Married	No	No religion

Appendix: Overview of Microdata for Wudu County (*cont.*)

	The no. of household	Jia	Household	Relationship with the household head	Surname	First name	Sex	The Chinese zodiac	Birthdate	
									Year	Month
Fujin Xiang 8th Bao				Son	杜	宽哇	Male	Dragon	1941	1
Fujin Xiang 8th Bao				The second son		崴宽哇	Male	Goat	1944	1
Fujin Xiang 8th Bao				Younger brother	杜	万双	Male	Ox	1914	1
Fujin Xiang 8th Bao				Son's wife	李	旦儿	Female	Horse	1892	1
Fujin Xiang 8th Bao				Nephew		四有哇	Male	Ox	1925	10
Fujin Xiang 8th Bao				Nephew's wife	杜	买哇	Female	Dragon	1928	8
Fujin Xiang 8th Bao	671	10	6	The household head	杜	文宽	Male	Horse	1893	4
Fujin Xiang 8th Bao				The household head's wife	马	转儿	Female	Tiger	1888	1
Fujin Xiang 8th Bao				Son	杜	万谢	Male	Horse	1918	12
Fujin Xiang 8th Bao				Son's wife	杜	身哇	Female	Goat	1919	7
Fujin Xiang 8th Bao				Nephew's wife	誮	代翠	Female	Snake	1892	4
Fujin Xiang 8th Bao	672	10	7	The household head	杜	文应	Male	Ox	1888	5
Fujin Xiang 8th Bao				The household head's wife	罗	将戎子	Female		1888	2
Fujin Xiang 8th Bao				Son		夘儿子	Male	Ox	1925	4
Fujin Xiang 8th Bao				Son's wife	张	端哇	Female	Horse	1930	5
Fujin Xiang 8th Bao	673	10	8	The household head	杜	万举	Male	Rabbit	1902	12
Fujin Xiang 8th Bao				The household head's wife	马	谢儿	Female	Rooster	1921	10
Fujin Xiang 8th Bao				Daughter	杜	胡娃	Female	Dragon	1940	6
Fujin Xiang 8th Bao				The second daughter		崴胡哇	Female	Goat	1943	6

Date	The beginning time of inhabitation in this county	Educational level	Profession sector	Occupation	Marital status	Physical disability	Religion
17	Inhabited for generations						
7	Inhabited for generations						
4	Inhabited for generations	Illiteracy	Agriculture	Land-holding peasant	Married	No	No religion
7	Inhabited for generations	Illiteracy	Agriculture	Land-holding peasant	Married	No	No religion
22	Inhabited for generations	Illiteracy	Agriculture	Land-holding peasant	Married	No	No religion
25	Inhabited for generations	Illiteracy	Agriculture	Land-holding peasant	Married	No	No religion
25	Inhabited for generations	Illiteracy	Agriculture	Land-holding peasant	Married	No	No religion
20	Inhabited for generations	Illiteracy	Agriculture	Land-holding peasant	Married	No	No religion
8	Inhabited for generations	Illiteracy	Agriculture	Land-holding peasant	Married	No	No religion
31	Inhabited for generations	Illiteracy	Agriculture	Land-holding peasant	Married	No	No religion
25	Inhabited for generations	Illiteracy	Agriculture	Land-holding peasant	Married	No	No religion
2	Inhabited for generations	Illiteracy	Agriculture	Land-holding peasant	Married	No	No religion
3	Inhabited for generations	Illiteracy			Married	No	
14	Inhabited for generations	Illiteracy			Married	No	No religion
29	Inhabited for generations	Illiteracy	Agriculture	Land-holding peasant	Married	No	No religion
12	Inhabited for generations	Illiteracy	Agriculture	Land-holding peasant	Married	No	No religion
20	Inhabited for generations	Illiteracy	Agriculture	Land-holding peasant		No	No religion
22	Inhabited for generations						
22	Inhabited for generations						

Appendix: Overview of Microdata for Wudu County (*cont.*)

	The no. of household	Jia	Household	Relationship with the household head	Surname	First name	Sex	The Chinese zodiac	Birthdate Year	Month
Fujin Xiang 8th Bao				Older brother	杜	万焕	Male	Rooster	1908	8
Fujin Xiang 8th Bao	674	10	9	The household head	杜	万黄	Male	Pig	1898	12
Fujin Xiang 8th Bao				The household head's wife	李	早儿	Female	Tiger	1901	10
Fujin Xiang 8th Bao				Daughter	杜	丑哇	Female	Monkey	1932	10
Fujin Xiang 8th Bao				The second daughter	杜	保仁	Female	Rabbit	1940	4
Fujin Xiang 8th Bao				The third daughter	杜	官娃	Female	Horse	1942	3
Fujin Xiang 8th Bao	675	10	10	The household head	杜	文怌	Male	Rabbit	1890	7
Fujin Xiang 8th Bao				The household head's wife	赵	润哇	Female	Rabbit	1890	4
Fujin Xiang 8th Bao				Son	杜	万益	Male	Horse	1930	11
Fujin Xiang 8th Bao				Younger brother	杜	文富	Male	Snake	1892	6
Fujin Xiang 8th Bao				Younger brother's wife	李	无戎子	Female	Tiger	1889	11
Fujin Xiang 8th Bao				Older brother's wife	赵	买哇子	Female		1880	10
Fujin Xiang 8th Bao				Nephew	杜	张哇	Male	Tiger	1880	11
Fujin Xiang 8th Bao	676	10	11	The household head	杜	万林	Male	Monkey	1907	3
Fujin Xiang 8th Bao				The household head's wife	赵	成哇	Female	Horse		
Fujin Xiang 8th Bao				Son	张	团	Male	Rat	1940	2
Fujin Xiang 8th Bao				The second son	杜	团哇	Male	Goat	1943	3
Fujin Xiang 8th Bao				Daughter	杜	省哇	Female	Monkey	1932	3

Date	The beginning time of inhabitation in this county	Educational level	Profession sector	Occupation	Marital status	Physical disability	Religion
29	Inhabited for generations	Illiteracy	Agriculture	Land-holding peasant	Unmarried	No	No religion
12	Inhabited for generations	Illiteracy	Agriculture	Land-holding peasant	Married	No	No religion
21	Inhabited for generations	Illiteracy	Agriculture	Land-holding peasant	Married	No	No religion
28	Inhabited for generations	Illiteracy	Agriculture	Land-holding peasant	Unmarried	No	No religion
10	Inhabited for generations						
28	Inhabited for generations						
15	Inhabited for generations	Illiteracy	Agriculture	Land-holding peasant	Married	No	No religion
2	Inhabited for generations	Illiteracy	Agriculture	Land-holding peasant	Married	No	No religion
30	Inhabited for generations	Bao school 4 years	Agriculture	Land-holding peasant	Married	No	No religion
5	Inhabited for generations	Illiteracy	Agriculture	Land-holding peasant	Married	No	No religion
30	Inhabited for generations	Illiteracy	Agriculture	Land-holding peasant	Married	No	No religion
1	Inhabited for generations		Unemployed				
9	Inhabited for generations	Illiteracy	Agriculture	Land-holding peasant	Widowed		No religion
6	Inhabited for generations	Illiteracy	Agriculture	Land-holding peasant	Widowed		No religion
	Inhabited for generations	Illiteracy	Agriculture	Land-holding peasant	Married		No religion
4	Inhabited for generations						
12	Inhabited for generations						
24	Inhabited for generations	Illiteracy	Agriculture	Land-holding peasant	Unmarried	No	No religion

Appendix: Overview of Microdata for Wudu County (*cont.*)

	The no. of household	Jia	Household	Relationship with the household head	Surname	First name	Sex	The Chinese zodiac	Birthdate	
									Year	Month
Fujin Xiang 8th Bao				Younger brother	刘	哇	Male	Ox	1913	7
Fujin Xiang 8th Bao				Younger brother's wife	赵	崴佳儿	Female	Dragon	1916	5
Fujin Xiang 8th Bao	677	10	12	The household head	杜	万科	Male	Pig	1924	1
Fujin Xiang 8th Bao				Grandfather	杜	生官	Male		1875	4
Fujin Xiang 8th Bao				Father	杜	文李	Male	Tiger	1901	6
Fujin Xiang 8th Bao				Mother	杜	毛旦	Female	Rabbit	1902	2
Fujin Xiang 8th Bao				Grandmother	杨	氏	Female		1873	6
Fujin Xiang 8th Bao				Younger brother	杜	韩哇	Male	Rabbit	1939	3
Fujin Xiang 8th Bao				Younger sister	杜	萧哇	Female	Pig	1935	5
Fujin Xiang 8th Bao	678	10	13	The household head	杜	万宽	Male	Goat	1906	10
Fujin Xiang 8th Bao				The household head's wife	李	两哇	Female	Snake	1908	10
Fujin Xiang 8th Bao				Mother	赵	氏	Female		1882	5
Fujin Xiang 8th Bao				Daughter	杜	戊哇	Female	Dog	1934	7
Fujin Xiang 8th Bao				The second daughter	杜	巧哇	Female	Ox	1938	1
Fujin Xiang 8th Bao				The third daughter	杜	悦哇	Female	Goat	1943	9
Fujin Xiang 8th Bao				Uncle	洪	生武	Male	Ox	1925	8
Fujin Xiang 8th Bao				The second uncle	讨	建儿	Male	Goat	1919	5
Fujin Xiang 8th Bao	679	10	14	The household head	杜	万颖	Male	Pig	1923	11

Date	The beginning time of inhabitation in this county	Educational level	Profession sector	Occupation	Marital status	Physical disability	Religion
13	Inhabited for generations	Illiteracy	Agriculture	Land-holding peasant	Married	No	No religion
9	Inhabited for generations	Illiteracy	Agriculture	Land-holding peasant	Married	No	No religion
20	Inhabited for generations	Illiteracy	Agriculture	Land-holding peasant	Married	No	No religion
3	Inhabited for generations		Unemployed		Married		
23	Inhabited for generations	Illiteracy	Agriculture	Land-holding peasant	Married	No	No religion
18	Inhabited for generations	Illiteracy	Agriculture	Land-holding peasant	Married	No	No religion
5	Inhabited for generations		Unemployed		Married		
11	Inhabited for generations						
2	Inhabited for generations						
28	Inhabited for generations	Illiteracy	Agriculture	Land-holding peasant	Married	No	No religion
7	Inhabited for generations	Illiteracy	Agriculture	Land-holding peasant	Married	No	No religion
26	Inhabited for generations		Unemployed		Widowed		
1	Inhabited for generations						
6	Inhabited for generations						
12	Inhabited for generations						
28	Inhabited for generations	Illiteracy	Agriculture	Land-holding peasant	Married	No	No religion
20	Inhabited for generations	Illiteracy	Agriculture	Land-holding peasant	Married	No	No religion
14	Inhabited for generations	Illiteracy	Agriculture	Land-holding peasant	Married	No	No religion

Appendix: Overview of Microdata for Wudu County (*cont.*)

	The no. of household	Jia	Household	Relationship with the household head	Surname	First name	Sex	The Chinese zodiac	Birthdate Year	Month
Fujin Xiang 8th Bao				The household head's wife	赵	成哇	Female	Pig	1923	9
Fujin Xiang 8th Bao				Mother	李	崴连	Female	Pig	1923	2
Fujin Xiang 8th Bao				Older brother's wife	杜	松哇	Female	Rat	1912	10
Fujin Xiang 8th Bao				Nephew	杜	登汪	Male	Pig	1935	4
Fujin Xiang 8th Bao				The second nephew	諛	人	Male	Goat	1943	12
Fujin Xiang 8th Bao				Niece		美哇	Female	Ox	1934	12
Fujin Xiang 8th Bao				The second niece		二女儿	Female	Dragon	1940	5
Fujin Xiang 8th Bao	680	10	15	The household head	杜	文义	Male	Goat	1919	4
Fujin Xiang 8th Bao				The household head's wife	諛	聒哇	Female	Goat	1919	8
Fujin Xiang 8th Bao				Older brother	杜	文奎	Male	Ox	1913	10
Fujin Xiang 8th Bao				Older brother's wife	马	丑女子	Female	Ox	1913	5
Fujin Xiang 8th Bao				Nephew's son	杜	胡排哇	Male	Rooster	1945	11
Fujin Xiang 8th Bao	681	10	16	The household head	杜	文科	Male	Rabbit	1915	3
Fujin Xiang 8th Bao				The household head's wife	諛	能女子	Female	Monkey	1920	10
Fujin Xiang 8th Bao				Mother	赵	氏	Female		1940	11
Fujin Xiang 8th Bao				Son		双玄儿	Male	Rooster	1945	11
Fujin Xiang 8th Bao				The second son	杜	尚哇	Male	Pig	1947	2

Date	The beginning time of inhabitation in this county	Educational level	Profession sector	Occupation	Marital status	Physical disability	Religion
27	Inhabited for generations	Illiteracy	Agriculture	Land-holding peasant	Married	No	No religion
9	Inhabited for generations	Illiteracy	Agriculture	Land-holding peasant	Widowed	No	No religion
21	Inhabited for generations	Illiteracy	Agriculture	Land-holding peasant	Widowed	No	No religion
13	Inhabited for generations	Illiteracy	Agriculture	Land-holding peasant	Unmarried	No	No religion
4	Inhabited for generations						
3	Inhabited for generations						
18	Inhabited for generations						
12	Inhabited for generations	Illiteracy	Agriculture	Land-holding peasant	Married	No	No religion
18	Inhabited for generations	Illiteracy	Agriculture	Land-holding peasant	Married	No	No religion
12	Inhabited for generations	Illiteracy	Agriculture	Land-holding peasant	Married	No	No religion
10	Inhabited for generations	Illiteracy	Agriculture	Land-holding peasant	Married	No	No religion
11	Inhabited for generations						
17	Inhabited for generations	Illiteracy	Agriculture	Land-holding peasant	Married	No	No religion
20	Inhabited for generations	Illiteracy	Agriculture	Land-holding peasant	Married	No	No religion
6	Inhabited for generations						
6	Inhabited for generations						
29	Inhabited for generations						

(*cont.*)

	The no. of household	Jia	Household	Relationship with the household head	Surname	First name	Sex	The Chinese zodiac	Birthdate	
									Year	Month
Fujin Xiang 8th Bao				Daughter	杜	怌哇	Female	Rabbit	1939	4
Fujin Xiang 8th Bao				Nephew	杜	万虎	Male	Rat	1924	4
Fujin Xiang 8th Bao				Nephew's wife	赵	贤哇	Female	Pig	1923	5
Fujin Xiang 8th Bao				The second nephew	杜	万保	Male	Dragon	1916	2
Fujin Xiang 8th Bao				The third nephew	杜	万登	Male	Ox	1925	7
Fujin Xiang 8th Bao				Uncle's son's wife	赵	采儿	Female	Pig	1887	6
Fujin Xiang 8th Bao	682	10	17	The household head	杜	文明	Male	Snake	1904	5
Fujin Xiang 8th Bao				The household head's wife	杜	拜女子	Female	Snake	1904	11
Fujin Xiang 8th Bao				Son		鹹菜哇	Male	Rabbit	1939	3
Fujin Xiang 8th Bao				The second son		崴兹儿	Male	Monkey	1944	4
Fujin Xiang 8th Bao				The third son		三响儿	Male	Pig	1947	2
Fujin Xiang 8th Bao				Daughter	杜	能哇	Female	Dog	1934	5
Fujin Xiang 8th Bao				The second daughter		崴响儿	Female	Rooster	1945	8
Fujin Xiang 8th Bao	683	10	18	The household head	杜	万财	Male	Rabbit	1902	15
Fujin Xiang 8th Bao				The household head's wife	杜	双罗儿	Female	Horse	1904	1
Fujin Xiang 8th Bao				Daughter	杜	薄哇	Female	Tiger	1938	7
Fujin Xiang 8th Bao				Uncle's wife	誜	能哇	Female	Tiger	1889	7
Fujin Xiang 8th Bao				Uncle's son	杜	万山	Male	Ox	1925	7
Fujin Xiang 8th Bao	684	11	1	The household head	罗	义明	Male	Ox	1913	4

Date	The beginning time of inhabitation in this county	Educational level	Profession sector	Occupation	Marital status	Physical disability	Religion
23	Inhabited for generations						
20	Inhabited for generations	Illiteracy	Agriculture	Land-holding peasant	Married	No	No religion
26	Inhabited for generations	Illiteracy	Agriculture	Land-holding peasant	Married	No	No religion
21	Inhabited for generations	Illiteracy	Agriculture	Land-holding peasant	Widowed	No	No religion
23	Inhabited for generations	Illiteracy	Agriculture	Land-holding peasant	Unmarried	No	No religion
2	Inhabited for generations	Illiteracy	Agriculture	Land-holding peasant	Widowed	No	No religion
12	Inhabited for generations	Illiteracy	Agriculture	Land-holding peasant	Married	No	No religion
21	Inhabited for generations	Illiteracy	Agriculture	Land-holding peasant	Married	No	No religion
1	Inhabited for generations	Illiteracy					
26	Inhabited for generations						
4	Inhabited for generations						
7	Inhabited for generations						
16	Inhabited for generations						
12	Inhabited for generations	Illiteracy	Agriculture	Land-holding peasant	Married	No	No religion
29	Inhabited for generations	Illiteracy	Agriculture	Land-holding peasant	Married	No	No religion
3	Inhabited for generations						
1	Inhabited for generations	Illiteracy	Agriculture	Land-holding peasant	Widowed	No	No religion
8	Inhabited for generations	Illiteracy	Agriculture	Land-holding peasant	Unmarried	No	No religion
25	Inhabited for generations	Sishu 2 years	Agriculture	Land-holding peasant	Married	No	No religion

(*cont.*)

	The no. of house-hold	Jia	House-hold	Relation-ship with the house-hold head	Surname	First name	Sex	The Chinese zodiac	Birthdate	
									Year	Month
Fujin Xiang 8th Bao				The house-hold head's wife	杜	毛旦	Female	Snake	1917	9
Fujin Xiang 8th Bao				Father	韩	登罗	Male	Dog	1885	5
Fujin Xiang 8th Bao				Mother	石	氏	Female		1879	9
Fujin Xiang 8th Bao				Daughter	罗	桥女	Female	Goat	1943	6
Fujin Xiang 8th Bao				The second daughter		崴徐哇	Female	Dog	1946	6
Fujin Xiang 8th Bao	685	11	2	The house-hold head	罗	牛家	Male	Rabbit	1927	9
Fujin Xiang 8th Bao				Mother	杨	转儿	Female	Goat	1916	8
Fujin Xiang 8th Bao				The second son		大平哇	Male	Rabbit	1939	10
Fujin Xiang 8th Bao				Daughter	罗	满哇	Female	Goat	1943	9
Fujin Xiang 8th Bao				The second daughter	罗	朱家	Female	Rooster	1945	11
Fujin Xiang 8th Bao				The third daughter		崴平哇	Female	Pig	1946	9
Fujin Xiang 8th Bao	686	11	3	The house-hold head	罗	长哇	Male	Snake	1917	3
Fujin Xiang 8th Bao				The house-hold head's wife	赵	新哇	Female	Pig	1923	8
Fujin Xiang 8th Bao				Mother	赵	马苏	Female	Goat	1882	2
Fujin Xiang 8th Bao				Son	罗	三信	Male	Goat	1943	11
Fujin Xiang 8th Bao				Younger brother	罗	省哇	Male	Pig	1923	5
Fujin Xiang 8th Bao				Younger brother's wife	赵	新哇	Female	Dog	1922	6
Fujin Xiang 8th Bao				Nephew	罗	吴信	Male	Pig	1947	2

Date	The beginning time of inhabitation in this county	Educational level	Profession sector	Occupation	Marital status	Physical disability	Religion
14	Inhabited for generations	Illiteracy	Agriculture	Land-holding peasant	Married	No	No religion
10	Inhabited for generations	Illiteracy	Agriculture	Land-holding peasant	Married	No	No religion
4	Inhabited for generations		Unemployed		Married		
24	Inhabited for generations						
17	Inhabited for generations						
15	Inhabited for generations	Illiteracy	Agriculture	Land-holding peasant	Divorced	No	No religion
20	Inhabited for generations	Illiteracy	Others		Widowed	No	No religion
30	Inhabited for generations	Illiteracy					
23	Inhabited for generations						
17	Inhabited for generations						
4	Inhabited for generations						
1	Inhabited for generations	Illiteracy	Agriculture	Land-holding peasant	Married	No	No religion
23	Inhabited for generations	Illiteracy	Agriculture	Land-holding peasant	Married	No	No religion
10	Inhabited for generations		Unemployed		Widowed		
9	Inhabited for generations						
3	Inhabited for generations	Illiteracy	Agriculture	Land-holding peasant	Married	No	No religion
13	Inhabited for generations	Illiteracy	Agriculture	Land-holding peasant	Married	No	No religion
21	Inhabited for generations						

(*cont.*)

	The no. of household	Jia	Household	Relationship with the household head	Surname	First name	Sex	The Chinese zodiac	Birthdate	
									Year	Month
Fujin Xiang 8th Bao				The second younger brother	罗	才哇	Male	Rabbit	1927	3
Fujin Xiang 8th Bao				The second younger brother's wife	张	顺哇	Female	Ox	1927	7
Fujin Xiang 8th Bao				Nephew	罗	杨成	Male	Dog	1946	12
Fujin Xiang 8th Bao	687	11	4	The household head	罗	兴富	Male	Ox	1900	4
Fujin Xiang 8th Bao				The household head's wife	赵	孝儿	Female	Rabbit	1902	3
Fujin Xiang 8th Bao				Son	张	崇哇	Male	Goat	1931	2
Fujin Xiang 8th Bao				Son's wife	赵	采哇	Female	Horse		
Fujin Xiang 8th Bao				The second son	正	崇哇	Male	Rat	1936	3
Fujin Xiang 8th Bao				The third son	罗	倍仁	Male	Snake	1947	8
Fujin Xiang 8th Bao				The fourth son	罗	干人	Male	Snake	1941	8
Fujin Xiang 8th Bao	688	11	5	The household head	罗	义韩	Male	Rooster	1896	2
Fujin Xiang 8th Bao				The household head's wife	罗	三傅	Female	Monkey	1895	12
Fujin Xiang 8th Bao				Son	罗	杨生	Male	Dragon	1940	11
Fujin Xiang 8th Bao	689	11	6	The household head	罗	义生	Male	Rat	1899	12
Fujin Xiang 8th Bao				The household head's wife	誒	转建	Female	Rabbit	1902	4
Fujin Xiang 8th Bao				Son		大时哇	Male	Goat	1943	4
Fujin Xiang 8th Bao				The second son		崴时哇	Male	Dog	1946	3

Date	The beginning time of inhabitation in this county	Educational level	Profession sector	Occupation	Marital status	Physical disability	Religion
17	Inhabited for generations	Illiteracy	Agriculture	Land-holding peasant	Married	No	No religion
31	Inhabited for generations	Illiteracy	Agriculture	Land-holding peasant	Married	No	No religion
5	Inhabited for generations						
7	Inhabited for generations	Illiteracy	Agriculture	Land-holding peasant	Married	No	No religion
20	Inhabited for generations	Illiteracy	Agriculture	Land-holding peasant	Married	No	No religion
2	Inhabited for generations	Bao school 1 year			Married	No	No religion
	Inhabited for generations	Illiteracy	Agriculture	Land-holding peasant	Married	No	No religion
12	Inhabited for generations						
2	Inhabited for generations						
2	Inhabited for generations						
14	Inhabited for generations	Illiteracy	Agriculture	Land-holding peasant	Married	No	No religion
7	Inhabited for generations	Illiteracy	Agriculture	Land-holding peasant	Married	No	No religion
25	Inhabited for generations	Illiteracy					
21	Inhabited for generations	Illiteracy	Agriculture	Land-holding peasant	Married	No	No religion
17	Inhabited for generations	Illiteracy	Agriculture	Land-holding peasant	Married	No	No religion
27	Inhabited for generations						
31	Inhabited for generations						

(*cont.*)

	The no. of house-hold	Jia	House-hold	Relation-ship with the house-hold head	Surname	First name	Sex	The Chinese zodiac	Birthdate Year	Month
Fujin Xiang 8th Bao				Younger brother	罗	义仲	Male	Pig	1910	2
Fujin Xiang 8th Bao				Younger brother's wife	諓	还女子	Female	Goat	1919	7
Fujin Xiang 8th Bao				Nephew		射桥儿	Male	Pig	1947	1
Fujin Xiang 8th Bao	690	11	7	The house-hold head	罗	登吕	Male	Tiger	1901	2
Fujin Xiang 8th Bao				The house-hold head's wife	丁	根哇	Female	Rat	1899	5
Fujin Xiang 8th Bao				Son	石	女子	Male	Goat	1931	10
Fujin Xiang 8th Bao				Son's wife	丁	黑哇	Female	Tiger	1926	11
Fujin Xiang 8th Bao				Grand-daughter		早建儿	Female	Pig	1947	2
Fujin Xiang 8th Bao				Daughter		等女子	Female	Ox	1937	5
Fujin Xiang 8th Bao				Uncle	罗	成库	Male		1874	4
Fujin Xiang 8th Bao	691	11	87	The house-hold head	罗	登财	Male	Dragon	1927	2
Fujin Xiang 8th Bao				The house-hold head's wife	李	绸儿	Female	Pig	1923	8
Fujin Xiang 8th Bao				Mother	李	氏	Female		1880	2
Fujin Xiang 8th Bao				Son		建並有	Male	Pig	1947	5
Fujin Xiang 8th Bao	692	11	9	The house-hold head	罗	登富	Male	Dragon	1903	4
Fujin Xiang 8th Bao				Son		新凤儿	Male	Dragon	1940	5
Fujin Xiang 8th Bao	693	11	10	The house-hold head	罗	义建	Male	Rabbit	1915	7
Fujin Xiang 8th Bao				The house-hold head's wife	张	倫哇	Female	Dragon	1916	3

Date	The beginning time of inhabitation in this county	Educational level	Profession sector	Occupation	Marital status	Physical disability	Religion
5	Inhabited for generations	Illiteracy	Agriculture	Land-holding peasant	Married	No	No religion
10	Inhabited for generations	Illiteracy	Agriculture	Land-holding peasant	Married	No	No religion
23	Inhabited for generations						
26	Inhabited for generations	Illiteracy	Agriculture	Land-holding peasant	Married	No	No religion
17	Inhabited for generations	Illiteracy	Agriculture	Land-holding peasant	Married	No	No religion
6	Inhabited for generations	Illiteracy	Agriculture	Land-holding peasant	Married	No	No religion
18	Inhabited for generations	Illiteracy	Agriculture	Land-holding peasant	Married	No	No religion
10	Inhabited for generations						
31	Inhabited for generations						
6	Inhabited for generations						
6	Inhabited for generations	Illiteracy	Agriculture	Land-holding peasant	Married	No	No religion
24	Inhabited for generations	Illiteracy	Agriculture	Land-holding peasant	Married	No	No religion
1	Inhabited for generations						
4	Inhabited for generations						
7	Inhabited for generations	Illiteracy	Agriculture	Land-holding peasant	Married	No	No religion
11	Inhabited for generations	Illiteracy					
29	Inhabited for generations	Illiteracy	Agriculture	Land-holding peasant	Married	No	No religion
15	Inhabited for generations	Illiteracy	Agriculture	Land-holding peasant	Married	No	No religion

(*cont.*)

	The no. of household	Jia	Household	Relationship with the household head	Surname	First name	Sex	The Chinese zodiac	Birthdate Year	Month
Fujin Xiang 8th Bao				Son	郭	都哇	Male	Rabbit	1939	4
Fujin Xiang 8th Bao				Daughter		未春哇	Female	Dog	1946	5
Fujin Xiang 8th Bao				The second son		进宝哇	Male	Dog	1922	8
Fujin Xiang 8th Bao				The second son's wife	马	兹娃	Female	Dragon	1928	8
Fujin Xiang 8th Bao				Grandson	罗	春哇	Male	Monkey	1944	2
Fujin Xiang 8th Bao	694	11	11	The household head	罗	义德	Male	Dog	1909	3
Fujin Xiang 8th Bao				The household head's wife	赵	冯秀	Female	Goat	1906	7
Fujin Xiang 8th Bao				Mother	赵	建哇	Female	Snake	1892	6
Fujin Xiang 8th Bao				Son		注生儿	Male	Dog	1946	10
Fujin Xiang 8th Bao				Daughter	李	女子	Female	Ox	1937	10
Fujin Xiang 8th Bao				The second daughter		印女子	Female	Snake	1945	11
Fujin Xiang 8th Bao				Younger brother	罗	义尚	Male	Snake	1917	5
Fujin Xiang 8th Bao				Younger brother's wife	李	来哇	Female	Pig	1923	4
Fujin Xiang 8th Bao				Nephew		天时儿	Male	Dog	1946	11
Fujin Xiang 8th Bao	695	11	12	The household head	罗	柏家	Female	Tiger	1926	2
Fujin Xiang 8th Bao				Mother	马	菜儿	Female	Horse	1893	7
Fujin Xiang 8th Bao				Daughter	罗	建儿	Female	Dog	1946	2
Fujin Xiang 8th Bao	696	11	13	The household head	罗	登平	Male	Pig	1886	8
Fujin Xiang 8th Bao				The household head's wife	张	拜拜	Female	Dog	1937	3

Date	The beginning time of inhabitation in this county	Educational level	Profession sector	Occupation	Marital status	Physical disability	Religion
15	Inhabited for generations	Bao school 2 years					
31	Inhabited for generations						
30	Inhabited for generations	Illiteracy	Agricul-ture	Land-holding peasant	Married	No	No religion
13	Inhabited for generations	Illiteracy	Agricul-ture	Land-holding peasant	Married	No	No religion
2	Inhabited for generations						
26	Inhabited for generations	Illiteracy	Agricul-ture	Land-holding peasant	Married	No	No religion
22	Inhabited for generations	Illiteracy	Agricul-ture	Land-holding peasant	Married	No	No religion
1	Inhabited for generations	Illiteracy	Agricul-ture	Land-holding peasant	Married	No	No religion
1	Inhabited for generations						
18	Inhabited for generations						
17	Inhabited for generations						
10	Inhabited for generations	Illiteracy	Agricul-ture	Land-holding peasant	Married	No	No religion
18	Inhabited for generations	Illiteracy	Agricul-ture	Land-holding peasant	Married	No	No religion
3	Inhabited for generations						
28	Inhabited for generations	Illiteracy		Others	Widowed	No	No religion
3	Inhabited for generations	Illiteracy		Others	Widowed	No	No religion
3	Inhabited for generations						
6	Inhabited for generations	Illiteracy	Agricul-ture	Land-holding peasant	Married	No	No religion
16	Inhabited for generations	Illiteracy	Agricul-ture	Land-holding peasant	Married	No	No religion

(*cont.*)

	The no. of household	Jia	Household	Relationship with the household head	Surname	First name	Sex	The Chinese zodiac	Birthdate	
									Year	Month
Fujin Xiang 8th Bao				Son		转运哇	Male	Goat	1919	5
Fujin Xiang 8th Bao				Son's wife	马	二哇	Female	Monkey	1920	8
Fujin Xiang 8th Bao				Grandson	刘	信哇	Male	Dog	1946	9
Fujin Xiang 8th Bao				The second son	罗	义林	Male	Dragon	1928	6
Fujin Xiang 8th Bao				The second son's wife	赵	二娃	Female	Rat	1912	3
Fujin Xiang 8th Bao				Grandson	罗	具娃	Male	Ox	1937	4
Fujin Xiang 8th Bao				Granddaughter	朱	建儿	Female	Dog	1946	6
Fujin Xiang 8th Bao	697	12	1	The household head	李	建义	Male	Ox	1913	11
Fujin Xiang 8th Bao				The household head's wife	尚	生哇	Female	Rabbit	1915	8
Fujin Xiang 8th Bao				Older brother's wife	王	怌儿	Female	Tiger	1914	8
Fujin Xiang 8th Bao				Nephew	李	学儿	Male	Monkey	1932	8
Fujin Xiang 8th Bao				Older brother	李	建林	Male	Rat	1899	4
Fujin Xiang 8th Bao				Older brother's wife	赵	崔儿	Female	Rabbit	1902	6
Fujin Xiang 8th Bao				Nephew	李	仁哇	Male	Rat	1936	2
Fujin Xiang 8th Bao				Niece		银女子	Female	Tiger	1938	5
Fujin Xiang 8th Bao	698	12	2	The household head	李	文均	Male	Rabbit	1915	5
Fujin Xiang 8th Bao				Mother	赵	氏	Female		1877	4
Fujin Xiang 8th Bao				Older brother	文	章	Male	Rooster	1908	2

Date	The beginning time of inhabitation in this county	Educational level	Profession sector	Occupation	Marital status	Physical disability	Religion
7	Inhabited for generations	Illiteracy	Agriculture	Land-holding peasant	Married	No	No religion
28	Inhabited for generations	Illiteracy	Agriculture	Land-holding peasant	Married	No	No religion
24	Inhabited for generations						
22	Inhabited for generations	Illiteracy	Agriculture	Land-holding peasant	Married	No	No religion
26	Inhabited for generations	Illiteracy	Agriculture	Land-holding peasant	Married	No	No religion
28	Inhabited for generations	Bao school 2 years					
26	Inhabited for generations						
10	Inhabited for generations	Illiteracy	Agriculture	Land-holding peasant	Married	No	No religion
24	Inhabited for generations	Illiteracy	Agriculture	Land-holding peasant	Married	No	No religion
1	Inhabited for generations	Illiteracy	Agriculture	Land-holding peasant	Widowed	No	No religion
18	Inhabited for generations	Illiteracy	Agriculture	Land-holding peasant	Unmarried	No	No religion
8	Inhabited for generations	Illiteracy	Agriculture	Land-holding peasant	Married	No	No religion
1	Inhabited for generations	Illiteracy	Agriculture	Land-holding peasant	Married	No	No religion
28	Inhabited for generations	Illiteracy					
8	Inhabited for generations	Illiteracy					
4	Inhabited for generations	Illiteracy		Land-holding peasant	Divorced	No	No religion
17	Inhabited for generations		Unemployed		Widowed		
2	Inhabited for generations	Illiteracy	Agriculture	Land-holding peasant	Widowed	No	No religion

(*cont.*)

	The no. of household	Jia	Household	Relationship with the household head	Surname	First name	Sex	The Chinese zodiac	Birthdate	
									Year	Month
Fujin Xiang 8th Bao	699	12	3	The household head	赵	登王	Male	Rabbit	1915	8
Fujin Xiang 8th Bao				The household head's wife	赵	招招	Female	Rooster	1908	4
Fujin Xiang 8th Bao				Father	赵	金尚	Male	Rat	1887	6
Fujin Xiang 8th Bao				Son	赵	都哇	Male	Snake	1929	10
Fujin Xiang 8th Bao				The second son	赵	宗哇	Male	Rat	1936	6
Fujin Xiang 8th Bao				The third son	赵	学儿	Male	Tiger	1938	8
Fujin Xiang 8th Bao				The fourth son	赵	谢哇	Male	Dog	1946	6
Fujin Xiang 8th Bao	700	12	4	The household head	赵	登宣	Male	Dragon	1916	11
Fujin Xiang 8th Bao				The household head's wife	石	节哇	Female	Snake	1917	6
Fujin Xiang 8th Bao				Son	赵	才儿	Male	Dog	1946	11
Fujin Xiang 8th Bao				Daughter	赵	良娃	Female	Goat	1943	9
Fujin Xiang 8th Bao				Father	赵	金喜	Male		1882	5
Fujin Xiang 8th Bao				Older brother	赵	登祥	Male	Dog	1910	9
Fujin Xiang 8th Bao	701	12	5	The household head	赵	金福	Male	Dog	1897	7
Fujin Xiang 8th Bao				The household head's wife	石	杜娃	Female	Rabbit	1902	6
Fujin Xiang 8th Bao				Son	赵	登科	Male	Horse	1918	7
Fujin Xiang 8th Bao				Son's wife	王	生娃	Female	Monkey	1920	6
Fujin Xiang 8th Bao				The second son		细崴拜	Male	Dragon	1919	9

Date	The beginning time of inhabitation in this county	Educational level	Profession sector	Occupation	Marital status	Physical disability	Religion
26	Inhabited for generations	Illiteracy	Agriculture	Land-holding peasant	Married	No	No religion
30	Inhabited for generations	Illiteracy	Agriculture	Land-holding peasant	Married	No	No religion
13	Inhabited for generations	Illiteracy	Agriculture	Land-holding peasant	Widowed	No	No religion
10	Inhabited for generations	Illiteracy	Agriculture	Land-holding peasant	Unmarried	No	No religion
3	Inhabited for generations						
10	Inhabited for generations						
6	Inhabited for generations						
5	Inhabited for generations	Illiteracy	Agriculture	Land-holding peasant	Married	No	No religion
9	Inhabited for generations	Illiteracy	Others		Married	No	No religion
3	Inhabited for generations						
19	Inhabited for generations						
8	Inhabited for generations						
17	Inhabited for generations	Illiteracy	Agriculture	Land-holding peasant	Widowed	No	No religion
1	Inhabited for generations	Sishu 2 years	Agriculture	Land-holding peasant	Married	No	No religion
25	Inhabited for generations	Illiteracy	Agriculture	Land-holding peasant	Married	No	No religion
10	Inhabited for generations	Illiteracy	Agriculture	Land-holding peasant	Married	No	No religion
3	Inhabited for generations	Illiteracy	Agriculture	Land-holding peasant	Married	No	No religion
7	Inhabited for generations	Illiteracy	Agriculture	Land-holding peasant	Married	No	No religion

(*cont.*)

	The no. of household	Jia	Household	Relationship with the household head	Surname	First name	Sex	The Chinese zodiac	Birthdate Year	Month
Fujin Xiang 8th Bao				The second son's wife	李	等哇	Female	Dog	1909	5
Fujin Xiang 8th Bao				Grandson	赵	望来	Male	Goat	1931	9
Fujin Xiang 8th Bao				Daughter		二姐娃	Female	Goat	1932	1
Fujin Xiang 8th Bao				Older brother	赵	金发	Male	Goat	1894	9
Fujin Xiang 8th Bao	702	12	6	The household head	赵	登陆	Male	Dragon	1928	8
Fujin Xiang 8th Bao				The household head's wife	张	省哇子	Female	Snake	1929	8
Fujin Xiang 8th Bao				Father	赵	金双	Male	Dragon	1891	6
Fujin Xiang 8th Bao				Younger sister	赵	登盖	Female	Horse	1942	9
Fujin Xiang 8th Bao				Uncle's wife	蒋	氏	Female		1882	3
Fujin Xiang 8th Bao				Older brother	赵	卢哇	Male	Horse	1905	5
Fujin Xiang 8th Bao				Older brother's wife	张	连儿	Female	Horse	1892	2
Fujin Xiang 8th Bao				Nephew		进才哇	Male	Goat	1919	11
Fujin Xiang 8th Bao				The second nephew		根生儿	Male	Pig	1923	5
Fujin Xiang 8th Bao	703	12	7	The household head	赵	登正	Male	Ox	1888	8
Fujin Xiang 8th Bao				The household head's wife	张	刘儿子	Female	Ox	1888	10
Fujin Xiang 8th Bao				Son		六代哇	Male	Rooster	1908	3
Fujin Xiang 8th Bao				Son's wife	张	生哇	Female	Ox	1913	4
Fujin Xiang 8th Bao				Grandson	赵	门儿	Male	Horse	1942	3

Date	The beginning time of inhabitation in this county	Educational level	Profession sector	Occupation	Marital status	Physical disability	Religion
24	Inhabited for generations	Illiteracy	Agriculture	Land-holding peasant	Married	No	No religion
3	Inhabited for generations	Illiteracy	Agriculture	Land-holding peasant	Unmarried	No	No religion
6	Inhabited for generations	Illiteracy	Agriculture	Land-holding peasant	Unmarried	No	No religion
23	Inhabited for generations	Illiteracy	Agriculture	Land-holding peasant	Widowed	No	No religion
15	Inhabited for generations	Illiteracy	Agriculture	Land-holding peasant	Widowed	No	No religion
8	Inhabited for generations	Illiteracy	Others		Married	No	No religion
12	Inhabited for generations	Illiteracy	Agriculture	Land-holding peasant	Widowed	No	No religion
16	Inhabited for generations						
22	Inhabited for generations		Unemployed		Widowed		
2	Inhabited for generations	Illiteracy	Agriculture	Land-holding peasant	Married	No	No religion
7	Inhabited for generations	Illiteracy	Agriculture	Land-holding peasant	Married	No	No religion
22	Inhabited for generations	Illiteracy	Agriculture	Land-holding peasant	Married	No	No religion
6	Inhabited for generations	Illiteracy	Agriculture	Land-holding peasant	Unmarried	No	No religion
31	Inhabited for generations	Illiteracy	Agriculture	Land-holding peasant	Married	No	No religion
8	Inhabited for generations	Illiteracy	Agriculture	Land-holding peasant	Married	No	No religion
4	Inhabited for generations	Illiteracy	Agriculture	Land-holding peasant	Married	No	No religion
17	Inhabited for generations	Illiteracy	Agriculture	Land-holding peasant	Married	No	No religion
21	Inhabited for generations						

(*cont.*)

	The no. of household	Jia	Household	Relationship with the household head	Surname	First name	Sex	The Chinese zodiac	Birthdate Year	Month
Fujin Xiang 8th Bao				Granddaughter		冬女子	Female	Rooster	1945	12
Fujin Xiang 8th Bao				The second son		拜银儿	Male	Ox	1913	3
Fujin Xiang 8th Bao	704	12	8	The household head	赵	有东	Male	Dog	1909	11
Fujin Xiang 8th Bao				The household head's wife	张	贤儿	Female	Dragon	1916	7
Fujin Xiang 8th Bao				Son	赵	石代	Male	Ox	1937	8
Fujin Xiang 8th Bao				Daughter	赵	傍女	Female	Goat	1943	10
Fujin Xiang 8th Bao				Younger brother	赵	有明	Male	Snake	1917	11
Fujin Xiang 8th Bao				The second son	赵	坡儿	Male	Pig	1947	6
Fujin Xiang 8th Bao	705	12	9	The household head	赵	登杜	Male	Tiger	1901	9
Fujin Xiang 8th Bao				The household head's wife	余	润相子	Female	Rabbit	1902	11
Fujin Xiang 8th Bao				Mother	石	氏	Female		1874	10
Fujin Xiang 8th Bao				Son	赵	有宗	Male	Monkey	1932	6
Fujin Xiang 8th Bao	706	12	10	The household head	赵	登云	Male	Dragon	1903	7
Fujin Xiang 8th Bao				The household head's wife	马	难哇	Female	Horse	1905	7
Fujin Xiang 8th Bao				Son	赵	石平	Male	Dragon	1940	4
Fujin Xiang 8th Bao				Uncle's son	赵	登生	Male	Dragon	1903	5
Fujin Xiang 8th Bao				Uncle's son	赵	登海	Male	Horse	1904	7

Date	The beginning time of inhabitation in this county	Educational level	Profession sector	Occupation	Marital status	Physical disability	Religion
5	Inhabited for generations						
15	Inhabited for generations	Illiteracy	Agriculture	Land-holding peasant	Unmarried	No	No religion
12	Inhabited for generations	Illiteracy	Agriculture	Land-holding peasant	Married	No	No religion
17	Inhabited for generations	Illiteracy	Agriculture	Land-holding peasant	Married	No	No religion
15	Inhabited for generations	Illiteracy					
10	Inhabited for generations						
7	Inhabited for generations	Illiteracy	Agriculture	Land-holding peasant	Unmarried	No	No religion
2	Inhabited for generations						
7	Inhabited for generations	Illiteracy	Agriculture	Land-holding peasant	Married	No	No religion
14	Inhabited for generations	Illiteracy	Agriculture	Land-holding peasant	Married	No	No religion
20	Inhabited for generations		Unemployed		Widowed		
15	Inhabited for generations	Illiteracy	Agriculture	Land-holding peasant	Unmarried	No	No religion
25	Inhabited for generations	Illiteracy	Agriculture	Land-holding peasant	Married	No	No religion
28	Inhabited for generations	Illiteracy	Agriculture	Land-holding peasant	Married	No	No religion
14	Inhabited for generations	Illiteracy					
21	Inhabited for generations	Illiteracy	Agriculture	Land-holding peasant	Married	No	No religion
24	Inhabited for generations	Illiteracy	Agriculture	Land-holding peasant	Married	No	No religion

(*cont.*)

	The no. of house-hold	Jia	House-hold	Relation-ship with the house-hold head	Surname	First name	Sex	The Chinese zodiac	Birthdate	
									Year	Month
Fujin Xiang 8th Bao				Younger brother's wife	丁	佳哇	Female	Horse	1905	4
Fujin Xiang 8th Bao				Nephew	赵	根虎儿	Male	Dog	1946	9
Fujin Xiang 8th Bao				Uncle's second son	赵	登法	Male	Rabbit	1915	4
Fujin Xiang 8th Bao	707	12	11	The house-hold head	傅	举荣	Male	Monkey	1895	8
Fujin Xiang 8th Bao				The house-hold head's wife	张	翠银	Female	Horse	1893	2
Fujin Xiang 8th Bao				Son	傅	金海	Male	Ox	1913	10
Fujin Xiang 8th Bao				Son's wife	赵	能儿	Female	Goat	1919	12
Fujin Xiang 8th Bao				The second son	傅	代生哇	Male	Ox	1925	10
Fujin Xiang 8th Bao				The second son's wife	赵	田筛哇	Female	Ox	1925	11
Fujin Xiang 8th Bao				Grandson	傅	端平儿	Male	Rooster	1945	4
Fujin Xiang 8th Bao	708	12	12	The house-hold head	张	大生子	Male	Ox	1900	7
Fujin Xiang 8th Bao				Son	张	桃娃	Male	Rooster	1933	1
Fujin Xiang 8th Bao				Uncle	张	登怀	Male	Rat	1899	8
Fujin Xiang 8th Bao				Uncle's wife	杨	拜子	Female	Ox	1900	3
Fujin Xiang 8th Bao				Uncle's son	张	高娃	Male	Rooster	1933	8
Fujin Xiang 8th Bao	709	12	13	The house-hold head	李	文永	Male	Tiger	1901	2
Fujin Xiang 8th Bao				The house-hold head's wife	赵	路儿	Female	Tiger	1901	6
Fujin Xiang 8th Bao				Son	李	德娃	Male	Tiger	1926	3

Date	The beginning time of inhabitation in this county	Educational level	Profession sector	Occupation	Marital status	Physical disability	Religion
6	Inhabited for generations	Illiteracy	Agriculture	Land-holding peasant	Married	No	No religion
29	Inhabited for generations						
18	Inhabited for generations	Illiteracy	Agriculture	Land-holding peasant	Unmarried	No	No religion
25	Inhabited for generations	Illiteracy	Agriculture	Land-holding peasant	Married	No	No religion
11	Inhabited for generations	Illiteracy	Agriculture	Land-holding peasant	Married	No	No religion
4	Inhabited for generations	Illiteracy	Agriculture	Land-holding peasant	Married	No	No religion
18	Inhabited for generations	Illiteracy	Agriculture	Land-holding peasant	Married	No	No religion
26	Inhabited for generations	Illiteracy	Agriculture	Land-holding peasant	Married	No	No religion
18	Inhabited for generations	Illiteracy	Agriculture	Land-holding peasant	Married	No	No religion
16	Inhabited for generations						
18	Inhabited for generations	Illiteracy	Agriculture	Land-holding peasant	Widowed	No	No religion
21	Inhabited for generations	Illiteracy	Agriculture	Land-holding peasant	Unmarried	No	No religion
28	Inhabited for generations	Illiteracy	Agriculture	Land-holding peasant	Married	No	No religion
31	Inhabited for generations	Illiteracy	Agriculture	Land-holding peasant	Married	No	No religion
2	Inhabited for generations	Illiteracy	Agriculture	Land-holding peasant	Unmarried	No	No religion
20	Inhabited for generations	Illiteracy	Agriculture	Land-holding peasant	Married	No	No religion
4	Inhabited for generations	Illiteracy	Agriculture	Land-holding peasant	Married	No	No religion
4	Inhabited for generations	Illiteracy	Agriculture	Land-holding peasant	Married	No	No religion

(*cont.*)

	The no. of household	Jia	Household	Relationship with the household head	Surname	First name	Sex	The Chinese zodiac	Birthdate	
									Year	Month
Fujin Xiang 8th Bao				Son's wife	罗	进林哇	Female	Rat	1924	12
Fujin Xiang 8th Bao				The second son	李	成敞	Male	Monkey	1922	10
Fujin Xiang 8th Bao				The third son	李	成钦儿	Male	Pig	1935	2
Fujin Xiang 8th Bao	710	12	14	The household head	丁	敞荣	Male	Horse	1917	12
Fujin Xiang 8th Bao				The household head's wife	杨	大早哇	Female	Snake	1904	5
Fujin Xiang 8th Bao				Father	丁	发上	Male		1871	8
Fujin Xiang 8th Bao				Mother	赵	氏	Female		1883	4
Fujin Xiang 8th Bao				Son	丁	仅生哇	Male	Ox	1937	10
Fujin Xiang 8th Bao				Daughter	丁	种谢	Female	Rabbit	1939	8
Fujin Xiang 8th Bao				The second daughter	丁	贱女子	Female	Goat	1944	1
Fujin Xiang 8th Bao				Older brother's wife	石	山秀	Female	Pig	1890	9
Fujin Xiang 8th Bao				Nephew	丁	秀山	Male	Rat	1912	10
Fujin Xiang 8th Bao				Nephew's wife	李	雪罗	Female	Snake	1904	6
Fujin Xiang 8th Bao				Grandson	丁	王学	Male	Monkey	1944	11
Fujin Xiang 8th Bao				The second son	丁	石全	Male	Dog	1946	11
Fujin Xiang 8th Bao				Niece	丁	捡娃	Female	Dog	1946	4
Fujin Xiang 8th Bao	711	12	15	The household head	丁	四来娃	Male	Rooster	1908	6
Fujin Xiang 8th Bao				The household head's wife	丁	跑娃	Female	Snake	1917	5

Date	The beginning time of inhabitation in this county	Educational level	Profession sector	Occupation	Marital status	Physical disability	Religion
17	Inhabited for generations	Illiteracy	Agriculture	Land-holding peasant	Married	No	No religion
31	Inhabited for generations	Illiteracy	Agriculture	Land-holding peasant	Married	No	No religion
4	Inhabited for generations	Illiteracy					
17	Inhabited for generations	Illiteracy	Agriculture	Land-holding peasant	Married	No	No religion
14	Inhabited for generations	Illiteracy	Agriculture	Land-holding peasant	Married	No	No religion
14	Inhabited for generations		Unemployed		Married		
8	Inhabited for generations		Unemployed		Married		
14	Inhabited for generations					Right arm and right leg lost	
5	Inhabited for generations	Illiteracy					
9	Inhabited for generations						
6	Inhabited for generations	Illiteracy	Agriculture	Land-holding peasant	Widowed	No	No religion
29	Inhabited for generations	Illiteracy	Agriculture	Land-holding peasant	Married	No	No religion
14	Inhabited for generations	Illiteracy	Agriculture	Land-holding peasant	Married	No	No religion
4	Inhabited for generations						
10	Inhabited for generations						
5	Inhabited for generations						
29	Inhabited for generations	Illiteracy	Agriculture	Land-holding peasant	Married	No	No religion
27	Inhabited for generations	Illiteracy	Agriculture	Land-holding peasant	Married	No	No religion

(*cont.*)

	The no. of household	Jia	Household	Relationship with the household head	Surname	First name	Sex	The Chinese zodiac	Birthdate	
									Year	Month
Fujin Xiang 8th Bao				Grandfather	丁	发奎	Male		1876	6
Fujin Xiang 8th Bao				Son	丁	王有	Male	Goat	1943	4
Fujin Xiang 8th Bao				Younger brother	丁	宝仓	Male	Dog	1922	5
Fujin Xiang 8th Bao				Younger brother's wife	韩	生娃	Female	Rooster	1921	3
Fujin Xiang 8th Bao				Nephew	丁	刘学	Male	Rooster	1943	8
Fujin Xiang 8th Bao				Uncle's wife	石	四秀娃	Female		1862	9
Fujin Xiang 8th Bao				Uncle	丁	起洪	Male	Tiger	1901	6
Fujin Xiang 8th Bao				Uncle's wife	赵	进娃	Female	Dragon	1903	6
Fujin Xiang 8th Bao				Uncle's son	丁	兹娃	Male	Ox	1937	5
Fujin Xiang 8th Bao				Uncle's daughter	丁	忠女	Female	Snake	1941	1
Fujin Xiang 8th Bao	712	12	16	The household head	丁	啟运	Male		1882	12
Fujin Xiang 8th Bao				The household head's wife	丁	傅耀	Female	Ox	1888	6
Fujin Xiang 8th Bao				Son	丁	彩山	Male	Goat	1906	8
Fujin Xiang 8th Bao				Son's wife	宋	毛女	Female	Goat	1906	7
Fujin Xiang 8th Bao				Granddaughter	丁	番包子	Female	Rooster	1945	2
Fujin Xiang 8th Bao				The second son	丁	银山	Male	Tiger	1915	1
Fujin Xiang 8th Bao				The second son's wife	杨	赛哇	Female	Pig	1923	1
Fujin Xiang 8th Bao				The third son	丁	五代哇	Male	Ox	1925	1
Fujin Xiang 8th Bao				The fourth son	丁	狗哇	Male	Ox	1933	1

Date	The beginning time of inhabitation in this county	Educational level	Profession sector	Occupation	Marital status	Physical disability	Religion	
1	Inhabited for generations		Unemployed		Widowed			
29	Inhabited for generations							
14	Inhabited for generations	Illiteracy	Agriculture	Land-holding peasant	Married	No	No religion	
20	Inhabited for generations	Illiteracy	Agriculture	Land-holding peasant	Married	No	No religion	
22	Inhabited for generations							
16	Inhabited for generations		Unemployed		Widowed			
4	Inhabited for generations	Illiteracy	Agriculture	Land-holding peasant	Married	No	No religion	
24	Inhabited for generations	Illiteracy	Agriculture	Land-holding peasant	Married	No	No religion	
19	Inhabited for generations	Illiteracy						
20	Inhabited for generations							
6	Inhabited for generations		Unemployed		Married	No	No religion	
28	Inhabited for generations	Illiteracy	Agriculture	Land-holding peasant	Married	No	No religion	
3	Inhabited for generations	Illiteracy	Agriculture	Land-holding peasant	Married	No	No religion	
22	Inhabited for generations	Illiteracy	Agriculture	Land-holding peasant	Married	No	No religion	
8	Inhabited for generations							
14	Inhabited for generations	Illiteracy	Agriculture	Land-holding peasant	Married	No	No religion	
3	Inhabited for generations	Illiteracy	Agriculture	Land-holding peasant	Married	No	No religion	
18	Inhabited for generations	Illiteracy	Agriculture	Land-holding peasant	Married	No	No religion	Serving in the army
11	Inhabited for generations	Illiteracy	Agriculture	Land-holding peasant	Unmarried	No	No religion	

(*cont.*)

	The no. of household	Jia	Household	Relationship with the household head	Surname	First name	Sex	The Chinese zodiac	Birthdate	
									Year	Month
Fujin Xiang 8th Bao	713	12	17	The household head	丁	保山	Male	Rabbit	1902	11
Fujin Xiang 8th Bao				The household head's wife	李	拜娃	Female	Snake	1904	5
Fujin Xiang 8th Bao				Mother	李	氏	Female		1872	2
Fujin Xiang 8th Bao				Son	丁	庆哇	Male	Monkey	1932	12
Fujin Xiang 8th Bao				The second son	丁	才儿	Male	Dragon	1940	9
Fujin Xiang 8th Bao				Daughter	丁	进女	Female	Dog	1934	10
Fujin Xiang 8th Bao				The second daughter	丁	东清	Female	Snake	1942	1
Fujin Xiang 8th Bao	714	12	18	The household head	丁	赵登元	Male	Dragon	1916	2
Fujin Xiang 8th Bao				The household head's wife	张	连儿	Female	Ox	1913	6
Fujin Xiang 8th Bao				Mother	丁	氏	Female		1876	9
Fujin Xiang 8th Bao				Son	丁	高林	Male	Dog	1946	7
Fujin Xiang 8th Bao				Daughter	丁	椒椒	Female	Ox	1937	7
Fujin Xiang 8th Bao	715	12	19	The household head	丁	全山	Male	Horse	1918	11
Fujin Xiang 8th Bao				The household head's wife	宋	祥哇	Female	Horse	1918	9
Fujin Xiang 8th Bao				Son	丁	甫英	Male	Rooster	1945	7
Fujin Xiang 8th Bao				The second son	丁	苏才	Male	Snake	1941	1
Fujin Xiang 8th Bao				Daughter	丁	苏女	Female	Goat	1943	8
Fujin Xiang 8th Bao				Grandfather	丁	发宗	Male	Snake	1880	8

Date	The beginning time of inhabitation in this county	Educational level	Profession sector	Occupation	Marital status	Physical disability	Religion
3	Inhabited for generations	Illiteracy	Agriculture	Land-holding peasant	Married	No	No religion
25	Inhabited for generations	Illiteracy	Agriculture	Land-holding peasant	Married	No	No religion
1	Inhabited for generations		Unemployed		Widowed		
23	Inhabited for generations	Illiteracy	Agriculture	Land-holding peasant	Unmarried	No	No religion
15	Inhabited for generations	Illiteracy					
29	Inhabited for generations	Illiteracy					
12	Inhabited for generations						
12	Inhabited for generations	Illiteracy	Agriculture	Land-holding peasant	Married	No	No religion
18	Inhabited for generations	Illiteracy	Agriculture	Land-holding peasant	Married	No	No religion
28	Inhabited for generations		Unemployed		Widowed		
24	Inhabited for generations						
19	Inhabited for generations	Illiteracy					
5	Inhabited for generations	Illiteracy	Agriculture	Land-holding peasant	Married	No	No religion
21	Inhabited for generations	Illiteracy	Others		Married	No	No religion
13	Inhabited for generations						
22	Inhabited for generations						
27	Inhabited for generations						
7	Inhabited for generations						

(*cont.*)

	The no. of household	Jia	Household	Relationship with the household head	Surname	First name	Sex	The Chinese zodiac	Birthdate	
									Year	Month
Fujin Xiang 8th Bao	716	13	1	The household head	张	仲前	Male	Pig	1886	4
Fujin Xiang 8th Bao				The household head's wife	王	交儿	Female	Dog	1885	11
Fujin Xiang 8th Bao				Son	张	啟元	Male	Rooster	1908	8
Fujin Xiang 8th Bao				Son's wife	石	月儿	Female	Horse	1918	8
Fujin Xiang 8th Bao				Grandson	张	连儿	Male	Horse	1942	7
Fujin Xiang 8th Bao				Granddaughter	张	怀儿	Female	Rabbit	1939	5
Fujin Xiang 8th Bao				Nephew	张	进哇	Male	Monkey	1920	11
Fujin Xiang 8th Bao				Nephew's wife	张	宝哇	Female	Monkey	1920	7
Fujin Xiang 8th Bao				Nephew's son	张	柏女子	Female	Dog	1946	10
Fujin Xiang 8th Bao	717	13	2	The household head	张	啟明	Male	Snake	1904	6
Fujin Xiang 8th Bao				The household head's wife	尚	焕儿	Female	Snake	1904	8
Fujin Xiang 8th Bao				Son	张	良哇	Male	Snake	1941	9
Fujin Xiang 8th Bao				Younger brother	张	早来哇	Male	Goat	1906	4
Fujin Xiang 8th Bao				Younger brother's wife	赵	杨秀	Female	Pig	1910	11
Fujin Xiang 8th Bao				Nephew	张	逞娃	Male	Dog	1934	11
Fujin Xiang 8th Bao				Niece	张	薄娃	Female	Monkey	1944	7
Fujin Xiang 8th Bao	718	13	3	The household head	张	仲成	Male	Horse	1905	5
Fujin Xiang 8th Bao				The household head's wife	李	三秀	Female	Horse	1905	3

Date	The beginning time of inhabitation in this county	Educational level	Profession sector	Occupation	Marital status	Physical disability	Religion
24	Inhabited for generations	Illiteracy	Agriculture	Land-holding peasant	Married	No	No religion
17	Inhabited for generations	Illiteracy	Agriculture	Land-holding peasant	Married	No	No religion
12	Inhabited for generations	Illiteracy	Agriculture	Land-holding peasant	Married	No	No religion
12	Inhabited for generations	Illiteracy	Agriculture	Land-holding peasant	Married	No	No religion
3	Inhabited for generations						
25	Inhabited for generations	Illiteracy					
12	Inhabited for generations	Illiteracy	Agriculture	Land-holding peasant	Married	No	No religion
22	Inhabited for generations	Illiteracy	Agriculture	Land-holding peasant	Married	No	No religion
5	Inhabited for generations						
14	Inhabited for generations	Illiteracy	Agriculture	Land-holding peasant	Married	No	No religion
18	Inhabited for generations	Illiteracy	Agriculture	Land-holding peasant	Married	No	No religion
2	Inhabited for generations	Illiteracy					
6	Inhabited for generations	Illiteracy	Agriculture	Land-holding peasant	Married	No	No religion
2	Inhabited for generations	Illiteracy	Agriculture	Land-holding peasant	Married	No	No religion
11	Inhabited for generations						
5	Inhabited for generations						
5	Inhabited for generations	Illiteracy	Agriculture	Land-holding peasant	Married	No	No religion
12	Inhabited for generations	Illiteracy	Agriculture	Land-holding peasant	Married	No	No religion

(*cont.*)

	The no. of house-hold	Jia	House-hold	Relation-ship with the house-hold head	Surname	First name	Sex	The Chinese zodiac	Birthdate Year	Month
Fujin Xiang 8th Bao				Mother	李	氏	Female		1882	9
Fujin Xiang 8th Bao				Son	张	陈生娃	Male	Rat	1936	2
Fujin Xiang 8th Bao				The second son	张	住学儿	Male	Pig	1947	1
Fujin Xiang 8th Bao				Daughter	张	陈女	Female	Dragon	1940	4
Fujin Xiang 8th Bao				Uncle's son	张	仲雷	Male	Snake	1904	12
Fujin Xiang 8th Bao				Older brother's wife	李	长哇	Female	Monkey	1907	10
Fujin Xiang 8th Bao				Niece	张	傅女子	Female	Dragon	1940	5
Fujin Xiang 8th Bao	719	13	4	The house-hold head	张	仲尚	Male	Rooster	1884	4
Fujin Xiang 8th Bao				Son	张	啟生	Male	Horse	1905	7
Fujin Xiang 8th Bao				The second son	张	啟平	Male	Monkey	1920	12
Fujin Xiang 8th Bao				Nephew's wife	罗	落儿	Female	Dragon	1891	5
Fujin Xiang 8th Bao				Nephew's son	张	仓娃	Male	Rooster	1917	10
Fujin Xiang 8th Bao	720	13	5	The house-hold head	张	仲赵	Male	Tiger	1901	3
Fujin Xiang 8th Bao				The house-hold head's wife	龙	润英	Female	Goat	1894	8
Fujin Xiang 8th Bao				Son	张	田娃	Male	Ox	1925	4
Fujin Xiang 8th Bao				Son's wife	王	妖女	Female	Rat	1924	3
Fujin Xiang 8th Bao				Grandson	张	树生儿	Male	Dog	1946	1
Fujin Xiang 8th Bao				Grand-daughter	张	送女子	Female	Horse	1942	1

Date	The beginning time of inhabitation in this county	Educational level	Profession sector	Occupation	Marital status	Physical disability	Religion
20	Inhabited for generations		Unemployed		Widowed		
7	Inhabited for generations	Illiteracy					
26	Inhabited for generations						
15	Inhabited for generations	Illiteracy					
24	Inhabited for generations	Illiteracy	Agriculture	Land-holding peasant	Married	No	No religion
12	Inhabited for generations	Illiteracy	Agriculture	Land-holding peasant	Married	No	No religion
22	Inhabited for generations	Illiteracy					
21	Inhabited for generations	Illiteracy	Agriculture	Land-holding peasant	Widowed	No	No religion
3	Inhabited for generations	Illiteracy	Agriculture	Land-holding peasant	Widowed	No	No religion
29	Inhabited for generations	Illiteracy	Agriculture	Land-holding peasant	Widowed	No	No religion
19	Inhabited for generations	Illiteracy	Agriculture	Land-holding peasant	Widowed	No	No religion
20	Inhabited for generations	Illiteracy	Agriculture	Land-holding peasant	Married	No	No religion
20	Inhabited for generations	Illiteracy	Agriculture	Land-holding peasant	Married	No	No religion
31	Inhabited for generations	Illiteracy	Agriculture	Land-holding peasant	Married	No	No religion
8	Inhabited for generations	Illiteracy	Agriculture	Land-holding peasant	Married	No	No religion
24	Inhabited for generations	Illiteracy	Agriculture	Land-holding peasant	Married	No	No religion
9	Inhabited for generations						
11	Inhabited for generations						

(*cont.*)

	The no. of household	Jia	Household	Relationship with the household head	Surname	First name	Sex	The Chinese zodiac	Birthdate	
									Year	Month
Fujin Xiang 8th Bao				The second son	张	狗生娃	Male	Rooster	1933	2
Fujin Xiang 8th Bao				The second son's wife	丁	朝哇	Female	Horse	1930	3
Fujin Xiang 8th Bao				The third son	张	腰哇	Male	Dog	1934	11
Fujin Xiang 8th Bao				Uncle's wife	赵	氏	Female		1877	7
Fujin Xiang 8th Bao				Uncle's son	张	仲全	Male	Tiger	1914	8
Fujin Xiang 8th Bao	721	13	6	The household head	张	啟法	Male	Pig	1898	9
Fujin Xiang 8th Bao				The household head's wife	卢	赠娃	Female	Pig	1898	3
Fujin Xiang 8th Bao				Father	张	仲隆	Male		1877	2
Fujin Xiang 8th Bao				Son	张	仅哇	Male	Rooster	1914	5
Fujin Xiang 8th Bao				Son's wife	石	耀哇	Female	Monkey	1920	6
Fujin Xiang 8th Bao				Granddaughter	张	徐儿	Female	Goat	1943	11
Fujin Xiang 8th Bao				Younger brother	张	啟仓	Male	Ox	1913	3
Fujin Xiang 8th Bao				Younger brother's wife	李	等哇	Female	Ox	1913	2
Fujin Xiang 8th Bao				Nephew	张	官官	Male	Rooster	1945	3
Fujin Xiang 8th Bao	722	13	7	The household head	张	啟才	Male	Pig	1886	11
Fujin Xiang 8th Bao				The household head's wife	黄	秀娃	Female	Snake	1892	3
Fujin Xiang 8th Bao				Son	张	来娃	Male	Ox	1913	11

Date	The beginning time of inhabitation in this county	Educational level	Profession sector	Occupation	Marital status	Physical disability	Religion
	Inhabited for generations	Illiteracy	Agriculture	Peasant	Married	Right leg lame	No religion
5	Inhabited for generations	Illiteracy	Agriculture	Land-holding peasant	Married	No	No religion
11	Inhabited for generations	Illiteracy	Agriculture	Land-holding peasant	Unmarried	No	No religion
2	Inhabited for generations		Unemployed		Widowed		
2	Inhabited for generations	Illiteracy	Agriculture	Land-holding peasant	Divorced	No	No religion
22	Inhabited for generations	Illiteracy	Agriculture	Land-holding peasant	Married	No	No religion
18	Inhabited for generations	Illiteracy	Agriculture	Land-holding peasant	Married	No	No religion
13	Inhabited for generations		Unemployed		Widowed		
25	Inhabited for generations	Illiteracy	Agriculture	Land-holding peasant	Married	No	No religion
18	Inhabited for generations	Illiteracy	Agriculture	Land-holding peasant	Married	No	No religion
16	Inhabited for generations						
16	Inhabited for generations	Illiteracy	Agriculture	Land-holding peasant	Married	No	No religion
12	Inhabited for generations	Illiteracy	Agriculture	Land-holding peasant	Married	No	No religion
26	Inhabited for generations						
21	Inhabited for generations	Illiteracy	Agriculture	Land-holding peasant	Married	No	No religion
25	Inhabited for generations	Illiteracy	Agriculture	Land-holding peasant	Married	No	No religion
9	Inhabited for generations	Illiteracy	Agriculture	Land-holding peasant	Married	No	No religion

(*cont.*)

	The no. of house-hold	Jia	House-hold	Relation-ship with the house-hold head	Surname	First name	Sex	The Chinese zodiac	Birthdate	
									Year	Month
Fujin Xiang 8th Bao				The second son	张	宽娃	Male	Dog	1922	2
Fujin Xiang 8th Bao				The second son's wife	赵	欢节	Female	Ox	1925	3
Fujin Xiang 8th Bao				Nephew	张	文侄	Male	Rooster	1897	11
Fujin Xiang 8th Bao				Nephew's wife	马	苏哇	Female	Rooster	1908	8
Fujin Xiang 8th Bao				Nephew's daughter	张	神拜	Female	Goat	1943	10
Fujin Xiang 8th Bao	723	13	8	The house-hold head	张	啟祥	Male	Monkey	1907	11
Fujin Xiang 8th Bao				Mother	赵	氏	Female	Pig	1886	12
Fujin Xiang 8th Bao				Younger brother	张	啟全	Male	Horse	1918	8
Fujin Xiang 8th Bao				Younger brother's wife	丁	代秀	Female	Ox	1925	12
Fujin Xiang 8th Bao				Nephew	张	产生娃	Male	Dog	1946	8
Fujin Xiang 8th Bao				The second younger brother	张	柒哇	Male	Goat	1931	4
Fujin Xiang 8th Bao				Younger sister	张	狗娃	Female	Rat	1936	7
Fujin Xiang 8th Bao				Uncle's son	张	啟朝	Male	Snake	1904	8
Fujin Xiang 8th Bao				Older brother's wife	黄	毛还	Female	Dragon	1903	10
Fujin Xiang 8th Bao				Uncle's wife	赵	氏	Female		1882	9
Fujin Xiang 8th Bao				Uncle's son	张	啟顺	Male	Pig	1908	10
Fujin Xiang 8th Bao	724	13	9	The house-hold head	张	啟荣	Male	Rooster	1907	10

Date	The beginning time of inhabitation in this county	Educational level	Profession sector	Occupation	Marital status	Physical disability	Religion
2	Inhabited for generations	Illiteracy	Agriculture	Land-holding peasant	Married	No	No religion
8	Inhabited for generations	Illiteracy	Agriculture	Land-holding peasant	Married	No	No religion
16	Inhabited for generations	Illiteracy	Agriculture	Land-holding peasant	Married	No	No religion
23	Inhabited for generations	Illiteracy	Agriculture	Land-holding peasant	Married	No	No religion
10	Inhabited for generations						
22	Inhabited for generations	Illiteracy	Agriculture	Land-holding peasant	Widowed	No	No religion
25	Inhabited for generations		Unemployed		Widowed	No	
12	Inhabited for generations	Illiteracy	Agriculture	Land-holding peasant	Married	No	No religion
16	Inhabited for generations	Illiteracy	Agriculture	Land-holding peasant	Married	No	No religion
2	Inhabited for generations						
16	Inhabited for generations	Illiteracy	Agriculture	Land-holding peasant	Unmarried	No	No religion
9	Inhabited for generations	Illiteracy					
25	Inhabited for generations	Illiteracy	Agriculture	Land-holding peasant	Married	No	No religion
14	Inhabited for generations	Illiteracy	Agriculture	Land-holding peasant	Married	No	No religion
5	Inhabited for generations		Unemployed		Widowed		
4	Inhabited for generations	Illiteracy	Agriculture	Land-holding peasant	Married	No	No religion
1	Inhabited for generations	Illiteracy	Agriculture	Land-holding peasant	Married	No	No religion

(*cont.*)

	The no. of household	Jia	Household	Relationship with the household head	Surname	First name	Sex	The Chinese zodiac	Birthdate Year	Month
Fujin Xiang 8th Bao				The household head's wife	黄	迁迁	Female	Goat	1906	10
Fujin Xiang 8th Bao				Mother	张	氏	Female		1882	2
Fujin Xiang 8th Bao				Younger brother	张	敔汉	Male	Rat	1912	11
Fujin Xiang 8th Bao				Younger brother's wife	韩	王节	Female	Pig	1910	5
Fujin Xiang 8th Bao				Son	张	田进儿	Male	Ox	1885	1
Fujin Xiang 8th Bao				Niece	张	黑女子	Female	Rabbit	1944	11
Fujin Xiang 8th Bao	725	13	10	The household head	张	敔万	Male	Horse	1905	3
Fujin Xiang 8th Bao				The household head's wife	赵	崔哇	Female	Horse	1905	3
Fujin Xiang 8th Bao				Son	杨	谢儿	Male	Rooster	1933	8
Fujin Xiang 8th Bao				Younger brother	张	敔科	Male	Goat	1919	12
Fujin Xiang 8th Bao				Younger brother's wife	李	田儿	Female	Rabbit	1915	5
Fujin Xiang 8th Bao				Nephew	张	俊哇	Male	Dog	1934	11
Fujin Xiang 8th Bao				Niece	张	进女子	Female	Rooster	1945	5
Ganquan Xiang 7th Bao	726	1	1	The household head	董	维汉	Male	Dog	1896	1
Ganquan Xiang 7th Bao				Son	董	禄哇	Male	Rooster	1921	7

Date	The beginning time of inhabitation in this county	Educational level	Profession sector	Occupation	Marital status	Physical disability	Religion
29	Inhabited for generations	Illiteracy	Agriculture	Land-holding peasant	Married	No	No religion
3	Inhabited for generations		Unemployed		Widowed		
6	Inhabited for generations	Illiteracy	Agriculture	Land-holding peasant	Married	No	No religion
10	Inhabited for generations	Illiteracy	Agriculture	Land-holding peasant	Married	No	No religion
4	Inhabited for generations	Illiteracy	Agriculture	Land-holding peasant	Married	No	No religion
5	Inhabited for generations						
7	Inhabited for generations	Illiteracy	Agriculture	Land-holding peasant	Married	No	No religion
16	Inhabited for generations	Illiteracy	Agriculture	Land-holding peasant	Married	No	No religion
30	Inhabited for generations	Illiteracy	Agriculture	Land-holding peasant	Unmarried	No	No religion
20	Inhabited for generations	Illiteracy	Agriculture	Land-holding peasant	Married	No	No religion
17	Inhabited for generations	Illiteracy	Agriculture	Land-holding peasant	Married	No	No religion
10	Inhabited for generations	Illiteracy	Agriculture	Land-holding peasant	Unmarried	Mouth crooked	No religion
12	Inhabited for generations						
18	Inhabited for generations	Illiteracy	Agriculture	Land-holding peasant	Widowed		Buddhism
14	Inhabited for generations	Illiteracy	Agriculture	Land-holding peasant	Married		No religion

(*cont.*)

	The no. of house-hold	Jia	House-hold	Relation-ship with the house-hold head	Surname	First name	Sex	The Chinese zodiac	Birthdate	
									Year	Month
Ganquan Xiang 7th Bao				Son's wife	杜	毛哇	Female	Rabbit	1927	5
Ganquan Xiang 7th Bao				Daughter	董	川哇	Female	Dog	1934	8
Ganquan Xiang 7th Bao				The second daughter	董	检哇	Female	Ox	1937	3
Ganquan Xiang 7th Bao	727	1	2	The house-hold head	张	文学	Male	Horse	1893	3
Ganquan Xiang 7th Bao				The house-hold head's wife	崔	氏	Female	Snake	1892	6
Ganquan Xiang 7th Bao	728	1	3	The house-hold head	董	万芳	Male	Ox	1900	9
Ganquan Xiang 7th Bao				The house-hold head's wife	梁	转儿	Female	Monkey	1895	12
Ganquan Xiang 7th Bao				Son	董	建福	Male	Monkey	1920	4
Ganquan Xiang 7th Bao				Son's wife	梁	兹儿	Female	Tiger	1926	11
Ganquan Xiang 7th Bao				The second son	董	建财	Male	Tiger	1926	8
Ganquan Xiang 7th Bao				The second son's wife	王	庆哇	Female	Tiger	1926	4
Ganquan Xiang 7th Bao				The third son	董	建科	Male	Pig	1935	6
Ganquan Xiang 7th Bao				The fourth son	董	建禧	Male	Tiger	1938	12

Date	The beginning time of inhabitation in this county	Educational level	Profession sector	Occupation	Marital status	Physical disability	Religion
1	Inhabited for generations	Illiteracy	Agriculture	Land-holding peasant	Married		No religion
11	Inhabited for generations	Illiteracy	Agriculture	Land-holding peasant	Unmarried		No religion
5	Inhabited for generations	Illiteracy					
5	Inhabited for generations	Illiteracy	Agriculture	Land-holding peasant	Married		Buddhism
2	Inhabited for generations	Illiteracy	Agriculture	Land-holding peasant	Married		Buddhism
7	Inhabited for generations	Illiteracy	Agriculture	Land-holding peasant	Married		No religion
24	Inhabited for generations	Illiteracy	Agriculture	Land-holding peasant	Married	Eye disease	No religion
8	Inhabited for generations	Illiteracy	Agriculture	Land-holding peasant	Married	Leg injury	No religion
17	Inhabited for generations	Illiteracy	Agriculture	Land-holding peasant	Married		No religion
26	Inhabited for generations	Illiteracy	Agriculture	Land-holding peasant	Married		No religion
3	Inhabited for generations	Illiteracy	Agriculture	Land-holding peasant	Married		No religion
3	Inhabited for generations	Primary school 2 years					
14	Inhabited for generations	Illiteracy					

(*cont.*)

	The no. of household	Jia	Household	Relationship with the household head	Surname	First name	Sex	The Chinese zodiac	Birthdate	
									Year	Month
Ganquan Xiang 7th Bao				The fifth son	董	贵义	Male	Rooster	1945	2
Ganquan Xiang 7th Bao				Younger brother	董	万金	Male	Horse	1905	12
Ganquan Xiang 7th Bao				Younger brother's wife	石	永哇	Female	Goat	1906	4
Ganquan Xiang 7th Bao				Nephew	董	苏伍	Male	Rabbit	1939	6
Ganquan Xiang 7th Bao				Niece	董	仓仓	Female	Rooster	1945	3
Ganquan Xiang 7th Bao	729	1	4	The household head	董	维耀	Male	Snake	1917	4
Ganquan Xiang 7th Bao				The household head's wife	张	讲哇	Female	Rooster	1921	3
Ganquan Xiang 7th Bao				Mother	韩	氏	Female	Rooster	1872	10
Ganquan Xiang 7th Bao				Son	董	问信	Male	Rooster	1945	2
Ganquan Xiang 7th Bao	730	1	5	The household head	董	维川	Male	Snake	1892	1
Ganquan Xiang 7th Bao				The household head's wife	王	氏	Female	Snake	1893	8
Ganquan Xiang 7th Bao				Son	董	万勇	Male	Horse	1918	6
Ganquan Xiang 7th Bao				Son's wife	张	七哇子	Female	Rabbit	1915	8

Date	The beginning time of inhabitation in this county	Educational level	Profession sector	Occupation	Marital status	Physical disability	Religion
21	Inhabited for generations						
10	Inhabited for generations	Illiteracy	Agriculture	Land-holding peasant	Married	Blind	No religion
25	Inhabited for generations	Illiteracy	Agriculture	Land-holding peasant	Married		No religion
20	Inhabited for generations	Illiteracy					
18							
30	Inhabited for generations	Illiteracy	Agriculture	Land-holding peasant	Married		No religion
10	Inhabited for generations	Illiteracy	Agriculture	Land-holding peasant	Married		No religion
2	Inhabited for generations	Illiteracy	Unemployed		Widowed		No religion
24	Inhabited for generations						
3	Inhabited for generations	Illiteracy	Agriculture	Land-holding peasant	Married		No religion
4	Inhabited for generations	Illiteracy	Agriculture	Land-holding peasant	Married		No religion
12	Inhabited for generations	Illiteracy	Agriculture	Land-holding peasant	Married		No religion
11	Inhabited for generations	Illiteracy	Agriculture	Land-holding peasant	Married		No religion

(*cont.*)

	The no. of household	Jia	Household	Relationship with the household head	Surname	First name	Sex	The Chinese zodiac	Birthdate Year	Month
Ganquan Xiang 7th Bao				Granddaughter	董	苏女	Female	Dragon	1941	1
Ganquan Xiang 7th Bao				The second granddaughter	董	跟女	Female	Dog	1946	4
Ganquan Xiang 7th Bao				The second son	董	万良	Male	Rat	1925	1
Ganquan Xiang 7th Bao				The second son's wife	雍	三女	Female	Tiger	1926	5
Ganquan Xiang 7th Bao	731	1	6	The household head	董	奠川	Male	Horse	1905	2
Ganquan Xiang 7th Bao				The household head's wife	胡	秀儿	Female	Dragon	1903	12
Ganquan Xiang 7th Bao				Son	董	万进	Male	Dragon	1927	10
Ganquan Xiang 7th Bao				Son's wife	胡	秀哇	Female	Ox	1924	9
Ganquan Xiang 7th Bao				The second son	董	万富	Male	Rooster	1933	7
Ganquan Xiang 7th Bao				Daughter	董	辉哇	Female	Ox	1937	4
Ganquan Xiang 7th Bao	732	1	7	The household head	董	慧川	Male	Rabbit	1902	8
Ganquan Xiang 7th Bao				The household head's wife	王	春哇	Female	Rat	1912	3
Ganquan Xiang 7th Bao				Son	董	万礼	Male	Ox	1925	12

Date	The beginning time of inhabitation in this county	Educational level	Profession sector	Occupation	Marital status	Physical disability	Religion
2	Inhabited for generations						
3							
5	Inhabited for generations	Illiteracy	Unemployed		Married		No religion
16	Inhabited for generations	Illiteracy	Agriculture	Land-holding peasant	Married		No religion
4	Inhabited for generations	Illiteracy	Agriculture	Land-holding peasant	Married		No religion
22	Inhabited for generations	Illiteracy	Agriculture	Land-holding peasant	Married		No religion
2	Inhabited for generations	Illiteracy	Agriculture	Land-holding peasant	Married		No religion
15	Inhabited for generations	Illiteracy	Agriculture	Land-holding peasant	Married		No religion
9	Inhabited for generations	Primary school 4 years	Agriculture	Land-holding peasant			
1	Inhabited for generations	Illiteracy					
11	Inhabited for generations	Illiteracy	Agriculture	Land-holding peasant	Married		No religion
3	Inhabited for generations	Illiteracy	Agriculture	Land-holding peasant	Married		No religion
24	Inhabited for generations	Illiteracy	Agriculture	Land-holding peasant	Married		No religion

(*cont.*)

	The no. of household	Jia	Household	Relationship with the household head	Surname	First name	Sex	The Chinese zodiac	Birthdate	
									Year	Month
Ganquan Xiang 7th Bao				Son's wife	苟	革哇	Female	Horse	1930	5
Ganquan Xiang 7th Bao	733	1	8	The household head	董	维永	Male	Rat	1912	12
Ganquan Xiang 7th Bao				The household head's wife	柏	胖哇	Female	Rabbit	1902	10
Ganquan Xiang 7th Bao				Father	董	学义	Male	Dragon	1879	9
Ganquan Xiang 7th Bao				Mother	雍	氏	Female	Dragon	1879	9
Ganquan Xiang 7th Bao				Son	董	张姓子	Male	Dragon	1940	8
Ganquan Xiang 7th Bao	734	1	9	The household head	董	学敏	Male	Monkey	1895	5
Ganquan Xiang 7th Bao				The household head's wife	张	改哇	Female	Ox	1900	5
Ganquan Xiang 7th Bao				Son	董	玄玄子	Male	Pig	1923	6
Ganquan Xiang 7th Bao				The second son	董	地哇	Male	Rabbit	1928	1
Ganquan Xiang 7th Bao				Daughter	董	凤凤	Female	Goat	1931	4
Ganquan Xiang 7th Bao				The second daughter	董	腰女	Female	Dog	1934	4
Ganquan Xiang 7th Bao	735	1	10	The household head	董	学泰	Male	Goat	1882	3
Ganquan Xiang 7th Bao				The household head's wife	苟	氏	Female	Rabbit	1878	10

Date	The beginning time of inhabitation in this county	Educational level	Profession sector	Occupation	Marital status	Physical disability	Religion
13	Inhabited for generations	Illiteracy	Agriculture	Land-holding peasant	Married		No religion
11	Inhabited for generations	Illiteracy	Agriculture	Land-holding peasant	Married		No religion
28	Inhabited for generations	Illiteracy	Agriculture	Land-holding peasant	Married		No religion
28	Inhabited for generations	Illiteracy	Unemployed		Married		No religion
2	Inhabited for generations	Illiteracy	Unemployed		Married		No religion
3	Inhabited for generations	Illiteracy					
30	Inhabited for generations	Illiteracy	Agriculture	Land-holding peasant	Married		No religion
6	Inhabited for generations	Illiteracy	Agriculture	Land-holding peasant	Married		No religion
3	Inhabited for generations	Illiteracy	Agriculture	Land-holding peasant	Unmarried		No religion
14	Inhabited for generations	Illiteracy	Agriculture	Land-holding peasant	Unmarried		No religion
30	Inhabited for generations	Illiteracy	Agriculture	Land-holding peasant	Unmarried		No religion
29	Inhabited for generations	Illiteracy	Agriculture	Land-holding peasant	Unmarried		No religion
30	Inhabited for generations	Illiteracy	Agriculture	Land-holding peasant	Married		Buddhism
11	Inhabited for generations	Illiteracy	Agriculture	Land-holding peasant	Married		No religion

(*cont.*)

	The no. of house-hold	Jia	House-hold	Relation-ship with the house-hold head	Surname	First name	Sex	The Chinese zodiac	Birthdate	
									Year	Month
Ganquan Xiang 7th Bao				Son	董	维顺	Male	Rabbit	1915	9
Ganquan Xiang 7th Bao				Grandson	董	信信	Male	Pig	1935	2
Ganquan Xiang 7th Bao				The second grandson	大	斤	Male	Dragon	1940	3
Ganquan Xiang 7th Bao				The third grandson	董	拜儿子	Male	Goat	1943	5
Ganquan Xiang 7th Bao	736	1	11	The house-hold head	董	学仓	Male	Horse	1893	10
Ganquan Xiang 7th Bao				The house-hold head's wife	田	缠哇	Female	Snake	1892	5
Ganquan Xiang 7th Bao				Son	董	维禧	Male	Ox	1925	3
Ganquan Xiang 7th Bao				Son's wife	苟	崇性	Female	Rabbit	1927	8
Ganquan Xiang 7th Bao	737	1	12	The house-hold head	董	维库	Male	Monkey	1895	2
Ganquan Xiang 7th Bao				The house-hold head's wife	崔	东哇	Female	Monkey	1895	10
Ganquan Xiang 7th Bao				Son	董	万明	Male	Pig	1923	5
Ganquan Xiang 7th Bao				Son's wife	张	扣儿	Female	Horse	1918	4
Ganquan Xiang 7th Bao				Grandson	董	五斤	Male	Rooster	1945	8

Date	The beginning time of inhabitation in this county	Educational level	Profession sector	Occupation	Marital status	Physical disability	Religion
25	Inhabited for generations	Illiteracy	Agriculture	Land-holding peasant	Widowed		No religion
29	Inhabited for generations	Illiteracy	Agriculture	Land-holding peasant			
27	Inhabited for generations	Illiteracy					
7							
28	Inhabited for generations	Illiteracy	Agriculture	Land-holding peasant	Married	Eye disease	No religion
30	Inhabited for generations	Illiteracy	Agriculture	Land-holding peasant	Married		No religion
23	Inhabited for generations	Illiteracy	Agriculture	Land-holding peasant	Married		No religion
25	Inhabited for generations	Illiteracy	Agriculture	Land-holding peasant	Married		No religion
30	Inhabited for generations	Illiteracy	Agriculture	Land-holding peasant	Married		No religion
15	Inhabited for generations	Illiteracy	Agriculture	Land-holding peasant	Married		No religion
29	Inhabited for generations	Illiteracy	Unemployed		Married		No religion
1	Inhabited for generations	Illiteracy	Agriculture	Land-holding peasant	Married		No religion
28							

(*cont.*)

									Birthdate	
	The no. of household	**Jia**	**Household**	**Relationship with the household head**	**Surname**	**First name**	**Sex**	**The Chinese zodiac**	**Year**	**Month**
Ganquan Xiang 7th Bao				The second granddaughter	董	欵哇	Female	Ox	1937	10
Ganquan Xiang 7th Bao				The third Granddaughter	董	根款	Female	Rabbit	1939	4
Ganquan Xiang 7th Bao				The second son	董	张才	Male	Monkey	1932	3
Ganquan Xiang 7th Bao				Nephew	董	禄哇	Male	Goat	1932	1
Ganquan Xiang 7th Bao	738	1	13	The household head	董	万德	Male	Snake	1917	3
Ganquan Xiang 7th Bao				The household head's wife	樊	崴女	Female	Rabbit	1915	7
Ganquan Xiang 7th Bao				Father	董	维仓	Male	Dog	1885	8
Ganquan Xiang 7th Bao				Mother	焦	氏	Female	Pig	1886	2
Ganquan Xiang 7th Bao				Younger brother	董	万章	Male	Monkey	1920	11
Ganquan Xiang 7th Bao				Younger brother's wife	胡	锡锡	Female	Snake	1917	8
Ganquan Xiang 7th Bao				Niece	董	杜哇子	Female	Snake	1942	1
Ganquan Xiang 7th Bao				The second niece	董	廷廷	Female	Goat	1943	3
Ganquan Xiang 7th Bao				The second younger brother	董	万昌	Male	Goat	1927	7

Date	The beginning time of inhabitation in this county	Educational level	Profession sector	Occupation	Marital status	Physical disability	Religion
21	Inhabited for generations	Illiteracy					
1	Inhabited for generations	Illiteracy					
16	Inhabited for generations	Illiteracy	Agriculture	Land-holding peasant	Unmarried		No religion
10	Inhabited for generations	Illiteracy	Agriculture	Land-holding peasant	Unmarried		No religion
24	Inhabited for generations	Illiteracy	Agriculture	Land-holding peasant	Married		No religion
24	Inhabited for generations	Illiteracy	Agriculture	Land-holding peasant	Married		No religion
3	Inhabited for generations	Illiteracy	Agriculture	Land-holding peasant	Married		No religion
14	Inhabited for generations	Illiteracy	Agriculture	Land-holding peasant	Married		Buddhism
27	Inhabited for generations	Illiteracy	Agriculture	Land-holding peasant	Married		No religion
29	Inhabited for generations	Illiteracy	Agriculture	Land-holding peasant	Married		No religion
2	Inhabited for generations						
30							
8	Inhabited for generations	Junior middle school 1 year	Agriculture	Land-holding peasant	Unmarried		No religion

(*cont.*)

	The no. of household	Jia	Household	Relationship with the household head	Surname	First name	Sex	The Chinese zodiac	Birthdate	
									Year	Month
Ganquan Xiang 7th Bao	739	1	14	The household head	董	学才	Male	Ox	1913	2
Ganquan Xiang 7th Bao				The household head's wife	梁	开哇	Female	Dragon	1916	12
Ganquan Xiang 7th Bao				Mother	张	条明	Female	Tiger	1890	3
Ganquan Xiang 7th Bao				Daughter	董	菜女	Female	Horse	1943	1
Ganquan Xiang 7th Bao				The second daughter	董	宗家	Female	Dog	1946	12
Ganquan Xiang 7th Bao	740	1	15	The household head	董	维堂	Male	Dog	1909	12
Ganquan Xiang 7th Bao				The household head's wife	后	氏	Female	Rat	1899	2
Ganquan Xiang 7th Bao				Son	董	后姓	Male	Rat	1936	11
Ganquan Xiang 7th Bao	741	2	1	The household head	董	万全	Male	Dragon	1909	5
Ganquan Xiang 7th Bao				The household head's wife	张	著哇	Female	Snake	1904	8
Ganquan Xiang 7th Bao				Father	董	志红	Male	Dragon	1865	4
Ganquan Xiang 7th Bao				Son	董	建清	Male	Rat	1936	10
Ganquan Xiang 7th Bao				Daughter	董	德女	Female	Dog	1947	1

Date	The beginning time of inhabitation in this county	Educational level	Profession sector	Occupation	Marital status	Physical disability	Religion
12	Inhabited for generations	Illiteracy	Agriculture	Land-holding peasant	Married		No religion
28	Inhabited for generations	Illiteracy	Agriculture	Land-holding peasant	Married		No religion
30	Inhabited for generations	Illiteracy	Agriculture	Land-holding peasant	Widowed		No religion
2							
21							
29	Inhabited for generations	Illiteracy	Agriculture	Land-holding peasant	Married		No religion
28	Inhabited for generations	Illiteracy	Agriculture	Land-holding peasant	Married		No religion
25	Inhabited for generations	Primary school 2 years					
2	Inhabited for generations	Illiteracy	Agriculture	Land-holding peasant	Married		No religion
5	Inhabited for generations	Illiteracy	Agriculture	Land-holding peasant	Married		No religion
10	Inhabited for generations	Illiteracy	Unemployed		Widowed		No religion
24	Inhabited for generations	Primary school 1 year					
23							

(*cont.*)

	The no. of household	Jia	Household	Relationship with the household head	Surname	First name	Sex	The Chinese zodiac	Birthdate	
									Year	Month
Ganquan Xiang 7th Bao				Younger brother	董	万禄	Male	Snake	1918	1
Ganquan Xiang 7th Bao				Younger brother's wife	董	童女	Female	Ox	1925	3
Ganquan Xiang 7th Bao				Niece	董	菊花	Female	Dog	1946	6
Ganquan Xiang 7th Bao	742	2	2	The household head	董	学会	Male	Monkey	1900	11
Ganquan Xiang 7th Bao				The household head's wife	李	冯儿	Female	Monkey	1900	11
Ganquan Xiang 7th Bao				Son	董	维俊	Male	Monkey	1932	4
Ganquan Xiang 7th Bao				Son's wife	胡	仁仁	Female	Horse	1930	2
Ganquan Xiang 7th Bao				The second son	董	维均	Male	Pig	1935	12
Ganquan Xiang 7th Bao				Daughter	董	浪哇子	Female	Goat	1943	4
Ganquan Xiang 7th Bao				The second daughter	董	崴女子	Female	Rooster	1945	8
Ganquan Xiang 7th Bao				Younger brother's wife	后	水哇	Female	Goat	1919	4
Ganquan Xiang 7th Bao				Nephew	董	二顺子	Male	Pig	1935	10
Ganquan Xiang 7th Bao				Niece	董	华女子	Female	Rooster	1945	3

Date	The beginning time of inhabitation in this county	Educational level	Profession sector	Occupation	Marital status	Physical disability	Religion
2	Inhabited for generations	Illiteracy	Agriculture	Land-holding peasant	Married		No religion
28	Inhabited for generations	Illiteracy	Agriculture	Land-holding peasant	Married		No religion
9							
28	Inhabited for generations	Illiteracy	Agriculture	Land-holding peasant	Married		No religion
30	Inhabited for generations	Illiteracy	Agriculture	Land-holding peasant	Married		No religion
3	Inhabited for generations	Illiteracy	Agriculture	Land-holding peasant	Married		No religion
30	Inhabited for generations	Illiteracy	Agriculture	Land-holding peasant	Married		No religion
8	Inhabited for generations	Primary school 4 years	Agriculture	Land-holding peasant			
14							
1							
19	Inhabited for generations	Illiteracy	Agriculture	Land-holding peasant	Widowed		No religion
27	Inhabited for generations	Illiteracy	Agriculture	Land-holding peasant			
14							

(*cont.*)

	The no. of household	Jia	Household	Relationship with the household head	Surname	First name	Sex	The Chinese zodiac	Birthdate	
									Year	Month
Ganquan Xiang 7th Bao				Older brother	董	虽玉子	Male	Pig	1886	8
Ganquan Xiang 7th Bao	743	2	3	The household head	董	维田	Male	Rabbit	1902	4
Ganquan Xiang 7th Bao				The household head's wife	孟	其哇	Female	Ox	1913	4
Ganquan Xiang 7th Bao				Mother	王	氏	Female	Rabbit	1877	12
Ganquan Xiang 7th Bao				Son	董	胡家	Male	Rat	1936	4
Ganquan Xiang 7th Bao				Daughter	董	仁女	Female	Dog	1946	3
Ganquan Xiang 7th Bao				The second daughter	董	绑女	Female	Dragon	1940	8
Ganquan Xiang 7th Bao				Older brother	董	彦川	Male	Pig	1898	3
Ganquan Xiang 7th Bao				Older brother's wife	后	庚子	Female	Goat	1906	10
Ganquan Xiang 7th Bao				Nephew	董	仁义子	Male	Ox	1937	10
Ganquan Xiang 7th Bao				Younger brother	董	维刚	Male	Snake	1917	4
Ganquan Xiang 7th Bao				Younger brother's wife	张	欵哇	Female	Pig	1923	7
Ganquan Xiang 7th Bao	744	2	4	The household head	赵	长义	Male	Monkey	1907	6

Date	The beginning time of inhabitation in this county	Educational level	Profession sector	Occupation	Marital status	Physical disability	Religion
27	Inhabited for generations	Illiteracy	Agriculture	Land-holding peasant	Widowed		No religion
1	Inhabited for generations	Illiteracy	Agriculture	Land-holding peasant	Married		No religion
12	Inhabited for generations	Illiteracy	Agriculture	Land-holding peasant	Married		No religion
29	Inhabited for generations	Illiteracy	Agriculture	Land-holding peasant	Widowed		Buddhism
17	Inhabited for generations	Illiteracy					
5							
8	Inhabited for generations	Primary school 3 years					
30	Inhabited for generations	Illiteracy	Agriculture	Land-holding peasant	Married		No religion
8	Inhabited for generations	Illiteracy	Agriculture	Land-holding peasant	Married		No religion
16	Inhabited for generations	Illiteracy					
26	Inhabited for generations	Illiteracy	Agriculture	Land-holding peasant	Married		No religion
18	Inhabited for generations	Illiteracy	Agriculture	Land-holding peasant	Married		No religion
4	Inhabited for generations	Illiteracy	Agriculture	Land-holding peasant	Married		No religion

(*cont.*)

	The no. of household	Jia	Household	Relationship with the household head	Surname	First name	Sex	The Chinese zodiac	Birthdate	
									Year	Month
Ganquan Xiang 7th Bao				The household head's wife	董	会哇	Female	Monkey	1907	7
Ganquan Xiang 7th Bao				Son	赵	桃款子	Male	Rabbit	1939	5
Ganquan Xiang 7th Bao				Daughter	赵	苟女子	Female	Dragon	1940	4
Ganquan Xiang 7th Bao				Cohabitant	尹	玉林哇	Male	Snake	1941	2
Ganquan Xiang 7th Bao	745	2	5	The household head	董	学治	Male	Tiger	1879	6
Ganquan Xiang 7th Bao				The household head's wife	赵	转儿	Female	Monkey	1895	6
Ganquan Xiang 7th Bao				Son	董	杨姓	Male	Rooster	1933	6
Ganquan Xiang 7th Bao				Son's wife	王	园女	Female	Snake	1929	5
Ganquan Xiang 7th Bao				The second son	董	苏信	Male	Ox	1937	12
Ganquan Xiang 7th Bao				Younger brother	董	学禄	Male	Horse	1905	3
Ganquan Xiang 7th Bao				Younger brother's wife	张	草哇	Female	Rat	1912	10
Ganquan Xiang 7th Bao				Niece	董	能能	Female	Monkey	1944	3
Ganquan Xiang 7th Bao	746	2	6	The household head	董	维廷	Male	Pig	1923	10

Date	The beginning time of inhabitation in this county	Educational level	Profession sector	Occupation	Marital status	Physical disability	Religion
6	Inhabited for generations	Illiteracy	Agriculture	Land-holding peasant	Married		No religion
24	Inhabited for generations	Primary school 2 years					
12	Inhabited for generations	Illiteracy					
30	Inhabited for generations	Illiteracy					
27	Inhabited for generations	Illiteracy	Agriculture	Land-holding peasant	Married		No religion
8	Inhabited for generations	Illiteracy	Agriculture	Land-holding peasant	Married		No religion
22	Inhabited for generations	Illiteracy	Agriculture	Land-holding peasant	Married		No religion
1	Inhabited for generations	Illiteracy	Agriculture	Land-holding peasant	Married		No religion
26	Inhabited for generations	Primary school 2 years					
25	Inhabited for generations	Illiteracy	Agriculture	Land-holding peasant	Married		No religion
3	Inhabited for generations	Illiteracy	Agriculture	Land-holding peasant	Married		No religion
12							
2	Inhabited for generations	Illiteracy	Agriculture	Land-holding peasant	Married		No religion

(*cont.*)

	The no. of household	Jia	Household	Relationship with the household head	Surname	First name	Sex	The Chinese zodiac	Birthdate	
									Year	Month
Ganquan Xiang 7th Bao				The household head's wife	魏	冯哇	Female	Ox	1913	8
Ganquan Xiang 7th Bao				Mother	苟	氏	Female	Rat	1887	12
Ganquan Xiang 7th Bao				Son	董	榜儿	Male	Horse	1942	1
Ganquan Xiang 7th Bao				The second son	董	田信	Male	Rooster	1945	3
Ganquan Xiang 7th Bao				The third son	董	张勤	Male	Pig	1947	3
Ganquan Xiang 7th Bao				Younger brother	董	维德	Male	Dragon	1928	7
Ganquan Xiang 7th Bao				Younger brother's wife	贾	抢哇	Female	Horse	1930	10
Ganquan Xiang 7th Bao	747	2	7	The household head	董	维庆	Male	Rabbit	1902	12
Ganquan Xiang 7th Bao				The household head's wife	吴	冬香	Female	Dog	1897	2
Ganquan Xiang 7th Bao				Son	董	顺义	Male	Horse	1893	3
Ganquan Xiang 7th Bao	748	2	8	The household head	董	维昌	Male	Goat	1906	12
Ganquan Xiang 7th Bao				The household head's wife	李	勤儿	Female	Horse	1905	3
Ganquan Xiang 7th Bao				Father	董	学生	Male	Rabbit	1866	4

Date	The beginning time of inhabitation in this county	Educational level	Profession sector	Occupation	Marital status	Physical disability	Religion
13	Inhabited for generations	Illiteracy	Agriculture	Land-holding peasant	Married		No religion
22	Inhabited for generations	Illiteracy	Unemployed		Widowed		No religion
16							
20							
13							
26	Inhabited for generations	Illiteracy	Agriculture	Land-holding peasant	Married		No religion
6	Inhabited for generations	Illiteracy	Agriculture	Land-holding peasant	Married		No religion
26	Inhabited for generations	Illiteracy	Agriculture	Land-holding peasant	Married		No religion
7	Inhabited for generations	Illiteracy	Agriculture	Land-holding peasant	Married		No religion
15	Inhabited for generations	Illiteracy	Agriculture	Land-holding peasant	Unmarried		No religion
19	Inhabited for generations	Illiteracy	Agriculture	Land-holding peasant	Married		No religion
12	Inhabited for generations	Illiteracy	Agriculture	Land-holding peasant	Married		No religion
5	Inhabited for generations	Illiteracy	Unemployed		Married		No religion

(*cont.*)

	The no. of household	Jia	Household	Relationship with the household head	Surname	First name	Sex	The Chinese zodiac	Birthdate	
									Year	Month
Ganquan Xiang 7th Bao				Mother	章	氏	Female	Rabbit	1876	5
Ganquan Xiang 7th Bao				Son	董	黑哇子	Male	Ox	1925	11
Ganquan Xiang 7th Bao				Son's wife	曹	女儿	Female	Rabbit	1927	12
Ganquan Xiang 7th Bao				The second son	董	招才	Male	Dog	1946	12
Ganquan Xiang 7th Bao				Daughter	董	胡哇	Female	Rat	1937	1
Ganquan Xiang 7th Bao				The second daughter	董	崴胡哇	Female	Snake	1941	4
Ganquan Xiang 7th Bao	749	2	9	The household head	董	维海	Male	Pig	1910	5
Ganquan Xiang 7th Bao				The household head's wife	曹	勤勤	Female	Pig	1923	3
Ganquan Xiang 7th Bao				Son	董	宿宿	Male	Pig	1923	5
Ganquan Xiang 7th Bao				Son's wife	张	冬庆	Female	Horse	1930	2
Ganquan Xiang 7th Bao	750	2	10	The household head	董	维清	Male	Rabbit	1902	6
Ganquan Xiang 7th Bao				The household head's wife	胡	冯哇	Female	Rat	1899	9
Ganquan Xiang 7th Bao				Son	董	敢敢	Male	Ox	1925	7

Date	The beginning time of inhabitation in this county	Educational level	Profession sector	Occupation	Marital status	Physical disability	Religion
5	Inhabited for generations	Illiteracy	Unemployed		Married		No religion
27	Inhabited for generations	Illiteracy	Agriculture	Land-holding peasant	Married		No religion
25	Inhabited for generations	Illiteracy	Agriculture	Land-holding peasant	Married		No religion
18							
2	Inhabited for generations	Illiteracy					
14							
3	Inhabited for generations	Illiteracy	Agriculture	Land-holding peasant	Married		No religion
25	Inhabited for generations	Illiteracy	Agriculture	Land-holding peasant	Married		No religion
29	Inhabited for generations	Illiteracy	Agriculture	Land-holding peasant	Married		No religion
8	Inhabited for generations	Illiteracy	Agriculture	Land-holding peasant	Married		No religion
5	Inhabited for generations	Illiteracy	Agriculture	Land-holding peasant	Married		No religion
26	Inhabited for generations	Illiteracy	Agriculture	Land-holding peasant	Married		No religion
27	Inhabited for generations	Illiteracy	Agriculture	Land-holding peasant	Married		No religion

(*cont.*)

	The no. of household	Jia	Household	Relationship with the household head	Surname	First name	Sex	The Chinese zodiac	Birthdate	
									Year	Month
Ganquan Xiang 7th Bao				Son's wife	曹	国儿	Female	Rabbit	1927	4
Ganquan Xiang 7th Bao				Grandson	董	节用	Male	Dog	1947	2
Ganquan Xiang 7th Bao	751	2	11	The household head	董	维忠	Male	Horse	1904	1
Ganquan Xiang 7th Bao				Mother	朱	氏	Female	Dragon	1879	2
Ganquan Xiang 7th Bao				Younger brother	董	维秀	Male	Pig	1910	8
Ganquan Xiang 7th Bao				Younger brother's wife	李	扣儿	Female	Tiger	1914	11
Ganquan Xiang 7th Bao				The second younger brother	董	维生	Male	Monkey	1913	3
Ganquan Xiang 7th Bao				Younger brother's wife	苏	香哇	Female	Monkey	1920	3
Ganquan Xiang 7th Bao				Nephew	董	万喜	Male	Rat	1936	10
Ganquan Xiang 7th Bao				Niece	董	夺哇	Female	Rabbit	1939	5
Ganquan Xiang 7th Bao				The second niece	董	崴夺哇	Female	Horse	1941	12
Ganquan Xiang 7th Bao				The third niece	董	三女	Female	Rooster	1945	10
Ganquan Xiang 7th Bao	752	2	12	The household head	董	维建	Male	Pig	1923	7

Date	The beginning time of inhabitation in this county	Educational level	Profession sector	Occupation	Marital status	Physical disability	Religion
21	Inhabited for generations	Illiteracy	Agriculture	Land-holding peasant	Married		No religion
1							
19	Inhabited for generations	Illiteracy	Agriculture	Land-holding peasant	Unmarried		No religion
15	Inhabited for generations	Illiteracy	Unemployed		Widowed		No religion
27	Inhabited for generations	Illiteracy	Agriculture	Land-holding peasant	Married		No religion
7	Inhabited for generations	Illiteracy	Agriculture	Land-holding peasant	Married		No religion
25	Inhabited for generations	Illiteracy	Agriculture	Land-holding peasant	Married		No religion
20	Inhabited for generations	Illiteracy	Agriculture	Land-holding peasant	Married		No religion
4	Inhabited for generations	Illiteracy					
3	Inhabited for generations	Illiteracy					
3							
25							
11	Inhabited for generations	Illiteracy	Agriculture	Land-holding peasant	Divorced		No religion

(*cont.*)

	The no. of household	Jia	Household	Relationship with the household head	Surname	First name	Sex	The Chinese zodiac	Birthdate	
									Year	Month
Ganquan Xiang 7th Bao				Father	董	学银	Male	Horse	1881	12
Ganquan Xiang 7th Bao				Mother	胡	氏	Female	Horse	1881	10
Ganquan Xiang 7th Bao	753	2	13	The household head	董	园哇	Male	Rabbit	1927	9
Ganquan Xiang 7th Bao				Niece	董	能女	Female	Monkey	1944	8
Ganquan Xiang 7th Bao	754	2	14	The household head	董	生岐	Male	Goat	1920	2
Ganquan Xiang 7th Bao				The household head's wife	梁	占哇	Female	Dragon	1917	1
Ganquan Xiang 7th Bao				Mother	雍	焦哇	Female	Rabbit	1890	11
Ganquan Xiang 7th Bao				Younger brother	董	生林	Male	Dog	1934	8
Ganquan Xiang 7th Bao				Son	董	谢谢	Male	Monkey	1944	12
Ganquan Xiang 7th Bao				Daughter	董	兵女	Female	Dragon	1940	3
Ganquan Xiang 7th Bao				The second daughter	董	纳女	Female	Dog	1946	9
Ganquan Xiang 7th Bao	755	2	15	The household head	焦	万刘	Male	Tiger	1901	9
Ganquan Xiang 7th Bao				The household head's wife	赵	梅还	Female	Dragon	1903	4

Date	The beginning time of inhabitation in this county	Educational level	Profession sector	Occupation	Marital status	Physical disability	Religion
30	Inhabited for generations	Illiteracy	Agriculture	Land-holding peasant	Married		No religion
20	Inhabited for generations	Illiteracy	Agriculture	Land-holding peasant	Married		No religion
29	Inhabited for generations	Illiteracy	Agriculture	Land-holding peasant	Married		No religion
21	Inhabited for generations						
17	Inhabited for generations	Illiteracy	Agriculture	Land-holding peasant	Married		No religion
23	Inhabited for generations	Illiteracy	Agriculture	Land-holding peasant	Married		No religion
22	Inhabited for generations	Illiteracy	Agriculture	Land-holding peasant	Widowed		No religion
31	Inhabited for generations	Primary school 2 years	Agriculture	Land-holding peasant			
18							
16	Inhabited for generations	Illiteracy					
24							
17	Inhabited for generations	Illiteracy	Agriculture	Land-holding peasant	Married		No religion
11	Inhabited for generations	Illiteracy	Agriculture	Land-holding peasant	Married		No religion

(*cont.*)

	The no. of household	Jia	Household	Relationship with the household head	Surname	First name	Sex	The Chinese zodiac	Birthdate	
									Year	Month
Ganquan Xiang 7th Bao				Son	焦	中信子	Male	Rooster	1933	8
Ganquan Xiang 7th Bao				The second son	焦	三哇子	Male	Dragon	1940	2
Ganquan Xiang 7th Bao				Daughter	焦	董羌子	Female	Horse	1942	4
Ganquan Xiang 7th Bao				The second daughter	董	三羌子	Female	Dog	1946	5
Ganquan Xiang 7th Bao	756	3	1	The household head	焦	万祥	Male	Ox	1900	6
Ganquan Xiang 7th Bao				The household head's wife	董	腰女	Female	Rabbit	1902	9
Ganquan Xiang 7th Bao				Father	焦	仲福	Male	Dragon	1876	1
Ganquan Xiang 7th Bao				Son	焦	问家	Male	Monkey	1944	12
Ganquan Xiang 7th Bao				The second son	焦	四儿	Male	Dog	1946	9
Ganquan Xiang 7th Bao				Daughter	焦	崔女	Female	Pig	1923	8
Ganquan Xiang 7th Bao				The second daughter	焦	乘香	Female	Snake	1941	12
Ganquan Xiang 7th Bao	757	3	2	The household head	柏	学化	Male	Ox	1900	7
Ganquan Xiang 7th Bao				The household head's wife	盘	披哇	Female	Tiger	1915	1

Date	The beginning time of inhabitation in this county	Educational level	Profession sector	Occupation	Marital status	Physical disability	Religion
21	Inhabited for generations	Illiteracy	Agriculture	Land-holding peasant			
19	Inhabited for generations	Illiteracy					
2							
3							
21	Inhabited for generations	Illiteracy	Agriculture	Land-holding peasant	Married		No religion
26	Inhabited for generations	Illiteracy	Agriculture	Land-holding peasant	Married		No religion
2	Inhabited for generations	Illiteracy	Agriculture	Land-holding peasant	Widowed		No religion
23							
18							
17	Inhabited for generations	Illiteracy	Agriculture	Land-holding peasant	Unmarried	Dumb	No religion
21	Inhabited for generations	Illiteracy					
30	Inhabited for generations	Illiteracy	Agriculture	Land-holding peasant	Married		No religion
26	Inhabited for generations	Illiteracy	Agriculture	Land-holding peasant	Married		No religion

(*cont.*)

	The no. of household	Jia	Household	Relationship with the household head	Surname	First name	Sex	The Chinese zodiac	Birthdate	
									Year	Month
Ganquan Xiang 7th Bao				Mother	胡	氏	Female	Pig	1886	2
Ganquan Xiang 7th Bao				Son	柏	问代	Male	Dog	1934	10
Ganquan Xiang 7th Bao				The second son	柏	根问	Male	Ox	1937	6
Ganquan Xiang 7th Bao				Daughter	柏	勤勤	Female	Rabbit	1939	4
Ganquan Xiang 7th Bao				The second daughter	柏	三勤	Female	Rooster	1945	11
Ganquan Xiang 7th Bao	758	3	3	The household head	柏	学义	Male	Horse	1904	1
Ganquan Xiang 7th Bao				The household head's wife	董	香香	Female	Pig	1909	1
Ganquan Xiang 7th Bao				Son	柏	榜榜	Male	Dog	1946	7
Ganquan Xiang 7th Bao				Daughter	柏	毛人	Female	Pig	1935	3
Ganquan Xiang 7th Bao				The second daughter	柏	款款	Female	Dragon	1940	11
Ganquan Xiang 7th Bao				The third daughter	柏	根欵	Female	Horse	1942	6
Ganquan Xiang 7th Bao	759	3	4	The household head	柏	文阴	Male	Monkey	1883	7
Ganquan Xiang 7th Bao				The household head's wife	尹	氏	Female	Dog	1885	4

Date	The beginning time of inhabitation in this county	Educational level	Profession sector	Occupation	Marital status	Physical disability	Religion
12	Inhabited for generations	Illiteracy	Agriculture	Land-holding peasant	Widowed		No religion
28	Inhabited for generations	Illiteracy	Agriculture	Land-holding peasant			
30	Inhabited for generations	Illiteracy					
22	Inhabited for generations	Illiteracy					
2							
6	Inhabited for generations	Illiteracy	Agriculture	Land-holding peasant	Married		No religion
5	Inhabited for generations	Illiteracy	Agriculture	Land-holding peasant	Married		No religion
4							
21	Inhabited for generations	Illiteracy	Agriculture	Land-holding peasant			
4	Inhabited for generations	Illiteracy					
28							
23	Inhabited for generations	Illiteracy	Agriculture	Land-holding peasant	Married		No religion
22	Inhabited for generations	Illiteracy	Agriculture	Land-holding peasant	Married		No religion

(*cont.*)

	The no. of house-hold	Jia	House-hold	Relation-ship with the house-hold head	Surname	First name	Sex	The Chinese zodiac	Birthdate	
									Year	Month
Ganquan Xiang 7th Bao				Son	柏	学升	Male	Monkey	1920	10
Ganquan Xiang 7th Bao				Son's wife	王	刚女	Female	Rooster	1922	1
Ganquan Xiang 7th Bao				Grand-daughter	柏	能能	Female	Horse	1942	9
Ganquan Xiang 7th Bao				The second granddaugh-ter	柏	朵朵	Female	Dog	1946	6
Ganquan Xiang 7th Bao	760	3	5	The house-hold head	柏	学稳	Male	Tiger	1901	10
Ganquan Xiang 7th Bao				The house-hold head's wife	崔	党儿	Female	Goat	1906	12
Ganquan Xiang 7th Bao				Father	柏	文宋	Male	Ox	1874	1
Ganquan Xiang 7th Bao				Son	柏	仁义子	Male	Rat	1936	10
Ganquan Xiang 7th Bao				The second son	柏	登全	Male	Tiger	1939	2
Ganquan Xiang 7th Bao				The third son	柏	刘信	Male	Dog	1946	8
Ganquan Xiang 7th Bao				Daughter	柏	根勤	Female	Snake	1941	7
Ganquan Xiang 7th Bao				The second daughter	柏	根秉	Female	Goat	1943	9
Ganquan Xiang 7th Bao	761	3	6	The house-hold head	柏	登云	Male	Goat	1919	4

Date	The beginning time of inhabitation in this county	Educational level	Profession sector	Occupation	Marital status	Physical disability	Religion
28	Inhabited for generations	Illiteracy	Agriculture	Land-holding peasant	Married		No religion
21	Inhabited for generations	Illiteracy	Agriculture	Land-holding peasant	Married		No religion
26							
3							
22	Inhabited for generations	Illiteracy	Agriculture	Land-holding peasant	Married		No religion
27	Inhabited for generations	Illiteracy	Agriculture	Land-holding peasant	Married		No religion
25	Inhabited for generations	Sishu 2 years	Unemployed		Widowed		No religion
10	Inhabited for generations	Illiteracy					
12	Inhabited for generations	Primary school 3 years					
2							
18	Inhabited for generations	Illiteracy					
3							
2	Inhabited for generations	Sishu 2 years	Agriculture	Land-holding peasant	Divorced		No religion

(*cont.*)

	The no. of household	Jia	Household	Relationship with the household head	Surname	First name	Sex	The Chinese zodiac	Birthdate	
									Year	Month
Ganquan Xiang 7th Bao				Father	柏	学禄	Male	Horse	1893	11
Ganquan Xiang 7th Bao				Mother	董	腰哇	Female	Rat	1899	3
Ganquan Xiang 7th Bao				Younger brother	柏	高應	Male	Tiger	1926	5
Ganquan Xiang 7th Bao				Younger brother's wife	坐	乖儿	Female	Goat	1931	10
Ganquan Xiang 7th Bao				Younger sister	柏	根乖儿	Female	Dog	1935	1
Ganquan Xiang 7th Bao				The second younger brother	柏	三乖儿	Female	Tiger	1938	7
Ganquan Xiang 7th Bao				Uncle	柏	学生	Male	Pig	1874	1
Ganquan Xiang 7th Bao	762	3	7	The household head	杨	园哇	Male	Rabbit	1927	7
Ganquan Xiang 7th Bao				The household head's wife	梁	菜家	Female	Ox	1925	5
Ganquan Xiang 7th Bao				Father	杨	生奇	Male	Goat	1894	2
Ganquan Xiang 7th Bao				Mother	梁	虽建	Female	Pig	1898	3
Ganquan Xiang 7th Bao				Younger sister	杨	孔哇子	Female	Ox	1925	4
Ganquan Xiang 7th Bao				The second younger sister	杨	董信哇	Female	Pig	1935	10

Date	The beginning time of inhabitation in this county	Educational level	Profession sector	Occupation	Marital status	Physical disability	Religion
13	Inhabited for generations	Illiteracy	Agriculture	Land-holding peasant	Married		No religion
16	Inhabited for generations	Illiteracy	Agriculture	Land-holding peasant	Married		No religion
11	Inhabited for generations	Illiteracy	Agriculture	Land-holding peasant	Married		No religion
23	Inhabited for generations	Illiteracy	Agriculture	Land-holding peasant	Married		No religion
29	Inhabited for generations	Illiteracy	Agriculture	Land-holding peasant			
8	Inhabited for generations	Illiteracy					
8	Inhabited for generations	Illiteracy	Agriculture	Land-holding peasant	Widowed		No religion
13	Inhabited for generations	Illiteracy	Agriculture	Land-holding peasant	Married		No religion
6	Inhabited for generations	Illiteracy	Agriculture	Land-holding peasant	Married		No religion
10	Inhabited for generations	Illiteracy	Agriculture	Land-holding peasant	Married		No religion
27	Inhabited for generations	Illiteracy	Agriculture	Land-holding peasant	Married		No religion
11	Inhabited for generations	Illiteracy	Agriculture	Land-holding peasant	Unmarried		No religion
6	Inhabited for generations	Illiteracy	Agriculture	Land-holding peasant			

(*cont.*)

	The no. of household	Jia	Household	Relationship with the household head	Surname	First name	Sex	The Chinese zodiac	Birthdate Year	Month
Ganquan Xiang 7th Bao	763	3	8	The household head	杨	创哇	Male	Rat	1912	3
Ganquan Xiang 7th Bao				Father	杨	生俊	Male	Dog	1885	4
Ganquan Xiang 7th Bao				Son	杨	禀禀子	Male	Horse	1942	5
Ganquan Xiang 7th Bao				Daughter	杨	能能	Female	Pig	1935	6
Ganquan Xiang 7th Bao	764	3	9	The household head	王	玄哇子	Male	Ox	1925	7
Ganquan Xiang 7th Bao				The household head's wife	张	赛女	Female	Dragon	1928	8
Ganquan Xiang 7th Bao				Mother	董	强哇子	Female	Snake	1892	9
Ganquan Xiang 7th Bao				Uncle	王	生忠	Male	Goat	1894	9
Ganquan Xiang 7th Bao				Cohabitant	王	马顺子	Male	Horse	1930	1
Ganquan Xiang 7th Bao	765	3	10	The household head	石	彦堂	Male	Rabbit	1915	7
Ganquan Xiang 7th Bao				The household head's wife	张	开哇	Female	Rabbit	1915	10
Ganquan Xiang 7th Bao				Son	石	锡甲	Male	Goat	1944	1
Ganquan Xiang 7th Bao				Daughter	石	旦女	Female	Snake	1941	2

Date	The beginning time of inhabitation in this county	Educational level	Profession sector	Occupation	Marital status	Physical disability	Religion
19	Inhabited for generations	Illiteracy	Agriculture	Land-holding peasant	Divorced		No religion
6	Inhabited for generations	Illiteracy	Agriculture	Land-holding peasant	Widowed		No religion
16	Inhabited for generations						
4	Inhabited for generations	Illiteracy	Agriculture	Land-holding peasant			
2	Inhabited for generations	Illiteracy	Agriculture	Land-holding peasant	Married		No religion
26	Inhabited for generations	Illiteracy	Agriculture	Land-holding peasant	Married		No religion
26	Inhabited for generations	Illiteracy	Agriculture	Land-holding peasant	Widowed		No religion
25	Inhabited for generations	Illiteracy	Agriculture	Land-holding peasant	Widowed		No religion
5	Inhabited for generations	Illiteracy	Agriculture	Land-holding peasant	Unmarried		No religion
21	Inhabited for generations	Sishu 3 years	Agriculture	Land-holding peasant	Married		No religion
10	Inhabited for generations	Illiteracy	Agriculture	Land-holding peasant	Married		No religion
11							
29	Inhabited for generations	Illiteracy					

(*cont.*)

	The no. of house-hold	Jia	House-hold	Relation-ship with the house-hold head	Surname	First name	Sex	The Chinese zodiac	Birthdate	
									Year	Month
Ganquan Xiang 7th Bao				The second daughter	石	焦女	Female	Dog	1946	11
Ganquan Xiang 7th Bao				Uncle	石	爱生	Male	Rabbit	1876	6
Ganquan Xiang 7th Bao				The second uncle	石	海生	Male	Horse	1881	5
Ganquan Xiang 7th Bao	766	3	11	The house-hold head	石	彦泰	Male	Dog	1909	8
Ganquan Xiang 7th Bao				The house-hold head's wife	石	香哇	Female	Ox	1913	6
Ganquan Xiang 7th Bao				Son	石	双保	Male	Monkey	1944	2
Ganquan Xiang 7th Bao				Daughter	石	中玉	Female	Pig	1935	4
Ganquan Xiang 7th Bao				The second daughter	石	佐佐	Female	Dog	1946	3
Ganquan Xiang 7th Bao	767	3	12	The house-hold head	王	万忠	Male	Pig	1910	12
Ganquan Xiang 7th Bao				The house-hold head's wife	苟	海哇	Female	Dragon	1916	11
Ganquan Xiang 7th Bao				Son	王	丑哇	Male	Snake	1941	6
Ganquan Xiang 7th Bao				Younger brother	王	雨雨子	Male	Monkey	1932	3
Ganquan Xiang 7th Bao				Younger brother's wife	董	厥哇	Female	Snake	1929	2

Date	The beginning time of inhabitation in this county	Educational level	Profession sector	Occupation	Marital status	Physical disability	Religion
1							
1	Inhabited for generations	Illiteracy	Unemployed		Widowed		No religion
1	Inhabited for generations	Illiteracy	Unemployed		Widowed		No religion
30	Inhabited for generations	Illiteracy	Agriculture	Land-holding peasant	Married		No religion
6	Inhabited for generations	Illiteracy	Agriculture	Land-holding peasant	Married		No religion
23							
7	Inhabited for generations	Illiteracy	Agriculture	Land-holding peasant			
6							
29	Inhabited for generations	Illiteracy	Agriculture	Land-holding peasant	Married		No religion
27	Inhabited for generations	Illiteracy	Agriculture	Land-holding peasant	Married		No religion
20	Inhabited for generations	Illiteracy					
18	Inhabited for generations	Illiteracy	Agriculture	Land-holding peasant	Married		No religion
30	Inhabited for generations	Illiteracy	Agriculture	Land-holding peasant	Married		No religion

(*cont.*)

	The no. of household	Jia	Household	Relationship with the household head	Surname	First name	Sex	The Chinese zodiac	Birthdate	
									Year	Month
Ganquan Xiang 7th Bao	768	3	13	The household head	王	彦清	Male	Tiger	1933	5
Ganquan Xiang 7th Bao				The household head's wife	雍	惠儿	Female	Pig	1886	11
Ganquan Xiang 7th Bao				Son	王	万昌	Male	Pig	1910	5
Ganquan Xiang 7th Bao				Son's wife	董	香香	Female	Rooster	1921	8
Ganquan Xiang 7th Bao				The second son	王	万才	Male	Ox	1913	12
Ganquan Xiang 7th Bao				The second son's wife	张	勤儿	Female	Dragon	1916	12
Ganquan Xiang 7th Bao				Grandson	王	德贤	Male	Rabbit	1939	9
Ganquan Xiang 7th Bao	769	3	14	The household head	贾	路代	Male	Monkey	1907	2
Ganquan Xiang 7th Bao				The household head's wife	梁	利玉	Female	Rat	1912	4
Ganquan Xiang 7th Bao				Mother	梁	崴命	Female	Ox	1888	5
Ganquan Xiang 7th Bao				Son	贾	节用	Male	Rabbit	1927	10
Ganquan Xiang 7th Bao				Son's wife	苟	爱哇	Female	Dragon	1928	12
Ganquan Xiang 7th Bao				Daughter	贾	菊香	Female	Monkey	1932	1

Date	The beginning time of inhabitation in this county	Educational level	Profession sector	Occupation	Marital status	Physical disability	Religion
19	Inhabited for generations	Sishu 2 years	Agriculture	Land-holding peasant	Married		No religion
21	Inhabited for generations	Illiteracy	Agriculture	Land-holding peasant	Married		No religion
13	Inhabited for generations	Illiteracy	Agriculture	Land-holding peasant	Married		No religion
7	Inhabited for generations	Illiteracy	Agriculture	Land-holding peasant	Married		No religion
5	Inhabited for generations	Illiteracy	Agriculture	Land-holding peasant	Married		No religion
23	Inhabited for generations	Illiteracy	Agriculture	Land-holding peasant	Married		No religion
18	Inhabited for generations	Primary school 2 years					
20	Inhabited for generations	Illiteracy	Agriculture	Land-holding peasant	Married		No religion
2	Inhabited for generations	Illiteracy	Agriculture	Land-holding peasant	Married		No religion
14	Inhabited for generations	Illiteracy	Agriculture	Land-holding peasant	Widowed		No religion
4	Inhabited for generations	Illiteracy	Agriculture	Land-holding peasant	Married		No religion
14	Inhabited for generations	Illiteracy	Agriculture	Land-holding peasant	Married		No religion
11	Inhabited for generations	Illiteracy	Agriculture	Land-holding peasant	Unmarried		No religion

(*cont.*)

	The no. of household	Jia	Household	Relationship with the household head	Surname	First name	Sex	The Chinese zodiac	Birthdate	
									Year	Month
Ganquan Xiang 7th Bao				Younger brother	贾	双玄	Male	Rabbit	1915	4
Ganquan Xiang 7th Bao				Younger brother's wife	梁	陆哇	Female	Rabbit	1915	4
Ganquan Xiang 7th Bao				Nephew	贾	过哇	Male	Pig	1935	7
Ganquan Xiang 7th Bao				The second nephew	贾	郭炳	Male	Monkey	1946	1
Ganquan Xiang 7th Bao				Niece	贾	水女	Female	Snake	1941	1
Ganquan Xiang 7th Bao				The second niece	贾	佐女	Female	Horse	1943	12
Ganquan Xiang 7th Bao	770	3	15	The household head	王	赠子	Male	Ox	1913	4
Ganquan Xiang 7th Bao				The household head's wife	苟	虽女	Female	Rabbit	1927	4
Ganquan Xiang 7th Bao				Mother	刘	具哇	Female	Monkey	1883	2
Ganquan Xiang 7th Bao	771	3	16	The household head	贾	尚雲	Male	Pig	1910	11
Ganquan Xiang 7th Bao				The household head's wife	王	梁家	Female	Dragon	1916	9
Ganquan Xiang 7th Bao				Mother	张	忙姐	Female	Goat	1882	4
Ganquan Xiang 7th Bao				Son	贾	仁子	Male	Horse	1942	11

Date	The beginning time of inhabitation in this county	Educational level	Profession sector	Occupation	Marital status	Physical disability	Religion
25	Inhabited for generations	Illiteracy	Agriculture	Land-holding peasant	Married		No religion
25	Inhabited for generations	Illiteracy	Agriculture	Land-holding peasant	Married		No religion
2	Inhabited for generations	Illiteracy	Agriculture	Land-holding peasant			
8							
19	Inhabited for generations	Illiteracy					
11							
9	Inhabited for generations	Illiteracy	Agriculture	Land-holding peasant	Married		No religion
8	Inhabited for generations	Illiteracy	Agriculture	Land-holding peasant	Married		No religion
28	Inhabited for generations	Illiteracy	Agriculture	Land-holding peasant	Widowed		No religion
22	Inhabited for generations	Illiteracy	Agriculture	Land-holding peasant	Married		No religion
30	Inhabited for generations	Illiteracy	Agriculture	Land-holding peasant	Married		No religion
8	Inhabited for generations	Illiteracy	Agriculture	Land-holding peasant	Widowed		No religion
21							

(*cont.*)

	The no. of household	Jia	Household	Relationship with the household head	Surname	First name	Sex	The Chinese zodiac	Birthdate	
									Year	Month
Ganquan Xiang 7th Bao				Daughter	贾	菜哇子	Female	Snake	1942	2
Ganquan Xiang 7th Bao				The second daughter	贾	虽菜	Female	Pig	1947	2
Ganquan Xiang 7th Bao				Younger brother	贾	得贵	Male	Snake	1917	11
Ganquan Xiang 7th Bao				Nephew	贾	仓仓	Male	Pig	1934	11
Ganquan Xiang 7th Bao				Uncle's wife	董	崔还	Female	Snake	1892	6
Ganquan Xiang 7th Bao				The second younger brother	贾	雍雍	Male	Horse	1918	4
Ganquan Xiang 7th Bao				Younger brother's wife	雍	抢哇	Female	Dragon	1928	2
Ganquan Xiang 7th Bao	772	4	1	The household head	王	生明	Male	Tiger	1901	7
Ganquan Xiang 7th Bao				The household head's wife	杜	兵连	Female	Dragon	1916	4
Ganquan Xiang 7th Bao				Father	王	在堂	Male	Monkey	1881	2
Ganquan Xiang 7th Bao				Son	王	陈家	Male	Snake	1941	5
Ganquan Xiang 7th Bao				Daughter	王	佐家	Female	Rat	1936	11
Ganquan Xiang 7th Bao	773	4	2	The household head	王	万禄	Male	Dragon	1916	7

Date	The beginning time of inhabitation in this county	Educational level	Profession sector	Occupation	Marital status	Physical disability	Religion
21							
3							
16	Inhabited for generations	Illiteracy	Agriculture	Land-holding peasant	Unmarried		No religion
10							
21	Inhabited for generations	Illiteracy	Agriculture	Land-holding peasant	Widowed		No religion
11	Inhabited for generations	Illiteracy	Agriculture	Land-holding peasant	Married		No religion
24	Inhabited for generations	Illiteracy	Agriculture	Land-holding peasant	Married		No religion
29	Inhabited for generations	Illiteracy	Agriculture	Land-holding peasant	Married		No religion
10	Inhabited for generations	Illiteracy	Agriculture	Land-holding peasant	Married		No religion
16	Inhabited for generations	Illiteracy	Agriculture	Land-holding peasant	Widowed		No religion
20	Inhabited for generations	Illiteracy					
3	Inhabited for generations	Illiteracy	Agriculture	Land-holding peasant			
22	Inhabited for generations	Illiteracy	Agriculture	Land-holding peasant	Divorced		No religion

(*cont.*)

	The no. of household	Jia	Household	Relationship with the household head	Surname	First name	Sex	The Chinese zodiac	Birthdate	
									Year	Month
Ganquan Xiang 7th Bao				Mother	魏	转哇	Female	Monkey	1895	7
Ganquan Xiang 7th Bao				Daughter	王	均女	Female	Pig	1936	8
Ganquan Xiang 7th Bao				Younger brother	王	万龙	Male	Pig	1935	12
Ganquan Xiang 7th Bao	774	4	3	The household head	尹	凤德	Male	Pig	1898	4
Ganquan Xiang 7th Bao				The household head's wife	李	调儿	Female	Ox	1913	6
Ganquan Xiang 7th Bao				Mother	辛	氏	Female	Rabbit	1863	11
Ganquan Xiang 7th Bao				Son	尹	长兴	Male	Goat	1931	11
Ganquan Xiang 7th Bao				The second son	尹	长长	Male	Pig	1935	12
Ganquan Xiang 7th Bao				The third son	尹	灰哇	Male	Snake	1941	1
Ganquan Xiang 7th Bao				The fourth son	尹	信成	Male	Goat	1943	4
Ganquan Xiang 7th Bao				The fifth son	尹	五信	Male	Dog	1946	4
Ganquan Xiang 7th Bao				Daughter	尹	长哇	Female	Rooster	1933	8
Ganquan Xiang 7th Bao				The second daughter	尹	早哇	Female	Dragon	1940	6

Date	The beginning time of inhabitation in this county	Educational level	Profession sector	Occupation	Marital status	Physical disability	Religion
5	Inhabited for generations	Illiteracy	Agriculture	Land-holding peasant	Widowed		No religion
15	Inhabited for generations	Illiteracy					
6	Inhabited for generations	Illiteracy	Agriculture	Land-holding peasant			
1	Inhabited for generations	Illiteracy	Agriculture	Tenant farmer	Married		No religion
4	Inhabited for generations	Illiteracy	Agriculture	Tenant farmer	Married		No religion
18	Inhabited for generations	Illiteracy	Unemployed		Widowed		No religion
24	Inhabited for generations	Illiteracy	Agriculture	Tenant farmer	Married	Eye disease	No religion
24	Inhabited for generations	Primary school 2 years	Agriculture	Tenant farmer			
21	Inhabited for generations	Illiteracy					
20							
20							
27	Inhabited for generations	Illiteracy	Agriculture	Tenant farmer			
10	Inhabited for generations	Illiteracy					

(*cont.*)

	The no. of house-hold	Jia	House-hold	Relation-ship with the house-hold head	Surname	First name	Sex	The Chinese zodiac	Birthdate	
									Year	Month
Ganquan Xiang 7th Bao				Younger brother	尹	凤清	Male	Tiger	1901	8
Ganquan Xiang 7th Bao				Son's wife	李	川哇	Female	Goat	1931	3
Ganquan Xiang 7th Bao	775	4	4	The house-hold head	尹	奐生	Male	Monkey	1895	8
Ganquan Xiang 7th Bao				The house-hold head's wife	魏	莊儿	Female	Rabbit	1890	3
Ganquan Xiang 7th Bao				Son	尹	應哇子	Male	Rabbit	1915	11
Ganquan Xiang 7th Bao				Son's wife	魏	招哇	Female	Rabbit	1915	10
Ganquan Xiang 7th Bao				The second son	尹	王顺	Male	Horse	1918	4
Ganquan Xiang 7th Bao				The third son	尹	抢哇	Male	Dog	1922	3
Ganquan Xiang 7th Bao				The fourth son	尹	四哇	Male	Ox	1937	2
Ganquan Xiang 7th Bao				Son's wife	魏	雨雨	Female	Ox	1925	1
Ganquan Xiang 7th Bao				Grandson	尹	杨财	Male	Dog	1946	3
Ganquan Xiang 7th Bao				Nephew's wife	石	消哇	Female	Snake	1918	6
Ganquan Xiang 7th Bao				Grandson	尹	刘家	Male	Dragon	1940	9

Date	The beginning time of inhabitation in this county	Educational level	Profession sector	Occupation	Marital status	Physical disability	Religion
9	Inhabited for generations	Illiteracy	Agriculture	Tenant farmer	Unmarried		No religion
2	Inhabited for generations	Illiteracy	Agriculture	Tenant farmer	Married		No religion
2	Inhabited for generations	Illiteracy	Agriculture	Tenant farmer	Married		No religion
19	Inhabited for generations	Illiteracy	Agriculture	Tenant farmer	Married		No religion
25	Inhabited for generations	Illiteracy	Agriculture	Tenant farmer	Married		No religion
15	Inhabited for generations	Illiteracy	Agriculture	Tenant farmer	Married		No religion
26	Inhabited for generations	Illiteracy	Agriculture	Tenant farmer	Unmarried		No religion
17	Inhabited for generations	Illiteracy	Agriculture	Tenant farmer	Unmarried		No religion
28	Inhabited for generations	Illiteracy	Agriculture	Tenant farmer	Married		No religion
15	Inhabited for generations	Illiteracy	Agriculture	Tenant farmer	Married		No religion
12							
1	Inhabited for generations	Illiteracy	Agriculture	Tenant farmer	Widowed		No religion
30	Inhabited for generations	Illiteracy					

(*cont.*)

	The no. of house-hold	Jia	House-hold	Relation-ship with the house-hold head	Surname	First name	Sex	The Chinese zodiac	Birthdate	
									Year	Month
Ganquan Xiang 7th Bao				The second grandson	尹	菜叶	Female	Pig	1935	10
Ganquan Xiang 7th Bao				Uncle	尹	craving堂	Male	Rooster	1881	7
Ganquan Xiang 7th Bao				Younger brother	尹	许哇	Male	Snake	1917	5
Ganquan Xiang 7th Bao	776	4	5	The house-hold head	尹	三考儿	Male	Horse	1894	1
Ganquan Xiang 7th Bao				The house-hold head's wife	尹	侯调女	Female	Rat	1887	3
Ganquan Xiang 7th Bao				Son	尹	米哇子	Male	Tiger	1914	5
Ganquan Xiang 7th Bao				Son's wife	王	开哇子	Female	Horse	1918	5
Ganquan Xiang 7th Bao				Grand-daughter	尹	兹女	Female	Goat	1943	5
Ganquan Xiang 7th Bao				Older brother's wife	曹	金明子	Female	Rooster	1896	3
Ganquan Xiang 7th Bao				Nephew	尹	生清	Male	Ox	1937	10
Ganquan Xiang 7th Bao	777	4	6	The house-hold head	尹	生岐	Male	Monkey	1907	7
Ganquan Xiang 7th Bao				The house-hold head's wife	王	谷哇	Female	Horse	1905	4
Ganquan Xiang 7th Bao				Mother	董	麦唤	Female	Dog	1885	7

Date	The beginning time of inhabitation in this county	Educational level	Profession sector	Occupation	Marital status	Physical disability	Religion
11	Inhabited for generations	Illiteracy	Agriculture	Land-holding peasant			
1	Inhabited for generations	Illiteracy	Agriculture	Land-holding peasant	Widowed		No religion
27	Inhabited for generations	Illiteracy	Agriculture	Land-holding peasant	Unmarried		No religion
6	Inhabited for generations	Illiteracy	Agriculture	Land-holding peasant	Married	Eye disease	No religion
7	Inhabited for generations	Illiteracy	Agriculture	Land-holding peasant	Married		No religion
18	Inhabited for generations	Illiteracy	Agriculture	Land-holding peasant	Married		No religion
23	Inhabited for generations	Illiteracy	Agriculture	Land-holding peasant	Married		No religion
29							
1	Inhabited for generations	Illiteracy	Agriculture	Land-holding peasant	Widowed		No religion
5	Inhabited for generations	Primary school 1 year					
14	Inhabited for generations	Illiteracy	Agriculture	Land-holding peasant	Married		No religion
29	Inhabited for generations	Illiteracy	Agriculture	Land-holding peasant	Married		No religion
28	Inhabited for generations	Illiteracy	Agriculture	Land-holding peasant	Widowed		No religion

(*cont.*)

	The no. of household	Jia	Household	Relationship with the household head	Surname	First name	Sex	The Chinese zodiac	Birthdate	
									Year	**Month**
Ganquan Xiang 7th Bao				Son	尹	胡家	Male	Dog	1900	2
Ganquan Xiang 7th Bao				Son's wife	尹	秉哇	Female	Dragon	1928	11
Ganquan Xiang 7th Bao				The second son	尹	雍家	Male	Ox	1925	2
Ganquan Xiang 7th Bao				Daughter	尹	乖乖	Female	Rabbit	1940	12
Ganquan Xiang 7th Bao				The second daughter	尹	关关	Female	Snake	1942	1
Ganquan Xiang 7th Bao				The third daughter	尹	长长	Female	Rooster	1945	8
Ganquan Xiang 7th Bao				Younger brother	尹	款哇	Male	Rooster	1921	11
Ganquan Xiang 7th Bao				Younger brother's wife	苏	郝哇	Female	Rat	1924	5
Ganquan Xiang 7th Bao	778	4	7	The household head	尹	生义	Male	Tiger	1901	7
Ganquan Xiang 7th Bao				The household head's wife	董	秀哇	Female	Ox	1900	4
Ganquan Xiang 7th Bao				Son	杜	才子	Male	Monkey	1921	1
Ganquan Xiang 7th Bao				The second son	尹	刘才	Male	Goat	1931	8
Ganquan Xiang 7th Bao				Daughter	尹	保保	Female	Monkey	1944	9

Date	The beginning time of inhabitation in this county	Educational level	Profession sector	Occupation	Marital status	Physical disability	Religion
1	Inhabited for generations	Illiteracy	Agriculture	Land-holding peasant	Married		No religion
13	Inhabited for generations	Illiteracy	Agriculture	Land-holding peasant	Married		No religion
9	Inhabited for generations	Illiteracy	Agriculture	Land-holding peasant	Unmarried		No religion
21	Inhabited for generations	Illiteracy					
16							
29							
7	Inhabited for generations	Illiteracy	Agriculture	Land-holding peasant	Married		No religion
6	Inhabited for generations	Illiteracy	Agriculture	Land-holding peasant	Married		No religion
9	Inhabited for generations	Illiteracy	Agriculture	Land-holding peasant	Married		No religion
1	Inhabited for generations	Illiteracy	Agriculture	Land-holding peasant	Married		No religion
29	Inhabited for generations	Illiteracy	Agriculture	Land-holding peasant	Unmarried		No religion
14	Inhabited for generations	Illiteracy	Agriculture	Land-holding peasant			
11							

(*cont.*)

	The no. of household	Jia	Household	Relationship with the household head	Surname	First name	Sex	The Chinese zodiac	Birthdate	
									Year	Month
Ganquan Xiang 7th Bao	779	4	8	The household head	尹	生进	Male	Monkey	1895	7
Ganquan Xiang 7th Bao				The household head's wife	李	尕女	Female	Horse	1918	8
Ganquan Xiang 7th Bao				Son	尹	文才	Male	Monkey	1944	9
Ganquan Xiang 7th Bao	780	4	9	The household head	尹	纪未	Male	Rooster	1921	10
Ganquan Xiang 7th Bao				Father	尹	老熊	Male	Rooster	1896	5
Ganquan Xiang 7th Bao				Mother	孟	抢哇	Female	Dragon	1891	1
Ganquan Xiang 7th Bao				Older brother's wife	孟	佐佐	Female	Tiger	1926	10
Ganquan Xiang 7th Bao	781	4	10	The household head	尹	焕平	Male	Goat	1894	11
Ganquan Xiang 7th Bao				The household head's wife	雍	乔儿	Female	Snake	1904	10
Ganquan Xiang 7th Bao				Son	尹	玄玄	Male	Pig	1916	2
Ganquan Xiang 7th Bao				Daughter	尹	香哇	Female	Snake	1941	5
Ganquan Xiang 7th Bao				Nephew	尹	董绪	Male	Goat	1919	11
Ganquan Xiang 7th Bao				Nephew's wife	龙	丙子	Female	Dog	1922	7

Date	The beginning time of inhabitation in this county	Educational level	Profession sector	Occupation	Marital status	Physical disability	Religion
12	Inhabited for generations	Illiteracy	Agriculture	Land-holding peasant	Married		No religion
1	Inhabited for generations	Illiteracy	Agriculture	Land-holding peasant	Married		No religion
1							
23	Inhabited for generations	Illiteracy	Agriculture	Land-holding peasant	Widowed		No religion
6	Inhabited for generations	Illiteracy	Agriculture	Land-holding peasant	Married		No religion
24	Inhabited for generations	Illiteracy	Agriculture	Land-holding peasant	Married		No religion
5	Inhabited for generations	Illiteracy	Agriculture	Land-holding peasant	Widowed		No religion
4	Inhabited for generations	Illiteracy	Agriculture	Land-holding peasant	Married		No religion
23	Inhabited for generations	Illiteracy	Agriculture	Land-holding peasant	Married		No religion
10	Inhabited for generations	Illiteracy	Agriculture	Land-holding peasant	Unmarried		No religion
5	Inhabited for generations	Illiteracy					
21	Inhabited for generations	Illiteracy	Agriculture	Land-holding peasant	Married		No religion
15	Inhabited for generations	Illiteracy	Agriculture	Land-holding peasant	Married		No religion

(*cont.*)

	The no. of household	Jia	Household	Relationship with the household head	Surname	First name	Sex	The Chinese zodiac	Birthdate	
									Year	Month
Ganquan Xiang 7th Bao				The second nephew	尹	回园子	Male	Monkey	1932	10
Ganquan Xiang 7th Bao				Cohabitant	尹	焕清	Male	Tiger	1895	8
Ganquan Xiang 7th Bao	782	4	11	The household head	尹	生富	Male	Goat	1906	5
Ganquan Xiang 7th Bao				The household head's wife	魏	召女	Female	Ox	1909	6
Ganquan Xiang 7th Bao	783	4	12	The household head	尹	来绪子	Male	Rat	1912	11
Ganquan Xiang 7th Bao				The household head's wife	张	曾哇	Female	Rat	1912	1
Ganquan Xiang 7th Bao				Son	尹	卯家	Male	Rooster	1933	7
Ganquan Xiang 7th Bao				The second son	尹	二卯	Male	Ox	1937	7
Ganquan Xiang 7th Bao				The third son	尹	三卯	Male	Horse	1942	6
Ganquan Xiang 7th Bao				The fourth son	尹	四儿子	Male	Monkey	1944	4
Ganquan Xiang 7th Bao	784	4	13	The household head	尹	宗堂	Male	Horse	1893	2
Ganquan Xiang 7th Bao				The household head's wife	魏	冬哇	Female	Dragon	1902	1
Ganquan Xiang 7th Bao				Daughter	尹	等等	Female	Rooster	1933	9

Date	The beginning time of inhabitation in this county	Educational level	Profession sector	Occupation	Marital status	Physical disability	Religion
12	Inhabited for generations	Illiteracy	Agriculture	Land-holding peasant	Unmarried		No religion
6	Inhabited for generations	Illiteracy	Agriculture	Land-holding peasant	Divorced		No religion
25	Inhabited for generations	Illiteracy	Agriculture	Land-holding peasant	Married		No religion
1	Inhabited for generations	Illiteracy	Agriculture	Land-holding peasant	Married		No religion
1	Inhabited for generations	Illiteracy	Agriculture	Land-holding peasant	Married		No religion
7	Inhabited for generations	Illiteracy	Agriculture	Land-holding peasant	Married		No religion
26	Inhabited for generations	Illiteracy	Agriculture	Land-holding peasant			
20	Inhabited for generations	Illiteracy					
28							
30							
8	Inhabited for generations	Illiteracy	Agriculture	Land-holding peasant	Married		No religion
13	Inhabited for generations	Illiteracy	Agriculture	Land-holding peasant	Married		No religion
17	Inhabited for generations	Illiteracy	Agriculture	Land-holding peasant			

(*cont.*)

	The no. of household	Jia	Household	Relationship with the household head	Surname	First name	Sex	The Chinese zodiac	Birthdate	
									Year	Month
Ganquan Xiang 7th Bao				The second daughter	尹	二哇子	Female	Tiger	1938	5
Ganquan Xiang 7th Bao				The third daughter	赵	女子	Female	Rooster	1945	4
Ganquan Xiang 7th Bao				Older brother	尹	宗会	Male	Ox	1934	15
Ganquan Xiang 7th Bao				Nephew	尹	羊喜	Male	Rabbit	1927	3
Ganquan Xiang 7th Bao				Nephew's wife	梁	走哇	Female	Horse	1930	3
Ganquan Xiang 7th Bao	785	4	14	The household head	尹	均有	Male	Dragon	1891	12
Ganquan Xiang 7th Bao				The household head's wife	马	氏	Female	Horse	1881	8
Ganquan Xiang 7th Bao				Son	尹	冯生	Male	Horse	1904	1
Ganquan Xiang 7th Bao				The second son	尹	生哇子	Male	Rabbit	1915	10
Ganquan Xiang 7th Bao	786	4	15	The household head	尹	凤昌	Male	Rat	1899	11
Ganquan Xiang 7th Bao				The household head's wife	董	建哇	Female	Pig	1897	2
Ganquan Xiang 7th Bao				Son	尹	才才子	Male	Ox	1938	1
Ganquan Xiang 7th Bao				The second son	尹	陈陈哇子	Male	Dog	1946	7

Date	The beginning time of inhabitation in this county	Educational level	Profession sector	Occupation	Marital status	Physical disability	Religion
12	Inhabited for generations	Illiteracy					
20							
5	Inhabited for generations	Illiteracy	Agriculture	Land-holding peasant	Widowed		No religion
8	Inhabited for generations	Illiteracy	Agriculture	Land-holding peasant	Married		No religion
3	Inhabited for generations	Illiteracy	Agriculture	Land-holding peasant	Married		No religion
4	Inhabited for generations	Illiteracy	Agriculture	Land-holding peasant	Married		No religion
2	Inhabited for generations	Illiteracy	Agriculture	Land-holding peasant	Married		Buddhism
4	Inhabited for generations	Illiteracy	Agriculture	Land-holding peasant	Unmarried		No religion
10	Inhabited for generations	Illiteracy	Agriculture	Land-holding peasant	Unmarried		No religion
29	Inhabited for generations	Illiteracy	Agriculture	Land-holding peasant	Married		No religion
4	Inhabited for generations	Illiteracy	Agriculture	Land-holding peasant	Married		No religion
5	Inhabited for generations	Illiteracy					
27							

(*cont.*)

	The no. of household	Jia	Household	Relationship with the household head	Surname	First name	Sex	The Chinese zodiac	Birthdate	
									Year	Month
Ganquan Xiang 7th Bao				Daughter	尹	香女子	Female	Pig	1947	2
Ganquan Xiang 7th Bao				The second daughter	尹	才女	Female	Goat	1943	9
Ganquan Xiang 7th Bao	787	4	16	The household head	尹	凤收	Male	Dog	1885	8
Ganquan Xiang 7th Bao				The household head's wife	辛	苏哇	Female	Rat	1887	3
Ganquan Xiang 7th Bao				Son	尹	生全	Male	Rat	1912	4
Ganquan Xiang 7th Bao				Son's wife	梁	赵哇	Female	Monkey	1907	5
Ganquan Xiang 7th Bao				Granddaughter	尹	红女	Female	Horse	1942	4
Ganquan Xiang 7th Bao				The second granddaughter	尹	红菊	Female	Dog	1946	10
Ganquan Xiang 7th Bao	788	4	17	The household head	尹	凤祥	Male	Goat	1905	1
Ganquan Xiang 7th Bao				The household head's wife	王	约儿	Female	Goat	1906	12
Ganquan Xiang 7th Bao				Father	尹	宗库	Male	Tiger	1877	6
Ganquan Xiang 7th Bao				Mother	王	氏	Female	Horse	1881	6
Ganquan Xiang 7th Bao				Son	尹	进财儿	Male	Rooster	1933	3

Date	The beginning time of inhabitation in this county	Educational level	Profession sector	Occupation	Marital status	Physical disability	Religion
25							
5							
19	Inhabited for generations	Illiteracy	Agriculture	Land-holding peasant	Married		No religion
6	Inhabited for generations	Illiteracy	Agriculture	Land-holding peasant	Married		No religion
17	Inhabited for generations	Illiteracy	Agriculture	Land-holding peasant	Married		No religion
30	Inhabited for generations	Illiteracy	Agriculture	Land-holding peasant	Married		No religion
17							
24							
28	Inhabited for generations	Illiteracy	Agriculture	Land-holding peasant	Married		No religion
30	Inhabited for generations	Illiteracy	Agriculture	Land-holding peasant	Married		No religion
7	Inhabited for generations	Illiteracy	Unemployed		Married		Buddhism
29	Inhabited for generations	Illiteracy	Unemployed		Married		Buddhism
15	Inhabited for generations	Illiteracy	Agriculture	Land-holding peasant			

(*cont.*)

	The no. of household	Jia	Household	Relationship with the household head	Surname	First name	Sex	The Chinese zodiac	Birthdate	
									Year	Month
Ganquan Xiang 7th Bao				The second son	尹	张信	Male	Dragon	1940	3
Ganquan Xiang 7th Bao				The third son	尹	掌财	Male	Rooster	1945	5
Ganquan Xiang 7th Bao				Daughter	尹	冯女	Female	Pig	1935	5
Ganquan Xiang 7th Bao				Cohabitant	尹	马哇子	Female	Goat	1918	11
Ganquan Xiang 7th Bao	789	5	1	The household head	董	虽升	Male	Goat	1919	12
Ganquan Xiang 7th Bao				The household head's wife	刘	改信	Female	Dragon	1928	1
Ganquan Xiang 7th Bao				Father	董	仲成	Male	Snake	1880	9
Ganquan Xiang 7th Bao				Cohabitant	董	仲明	Male	Pig	1882	8
Ganquan Xiang 7th Bao	790	5	2	The household head	董	玉哇	Male	Tiger	1901	4
Ganquan Xiang 7th Bao				The household head's wife	雍	良女	Female	Horse	1905	9
Ganquan Xiang 7th Bao				Son	董	长才	Male	Monkey	1944	2
Ganquan Xiang 7th Bao				Daughter	董	朵哇	Female	Monkey	1933	1
Ganquan Xiang 7th Bao				The second daughter	董	浦哇	Female	Rat	1935	8

Date	The beginning time of inhabitation in this county	Educational level	Profession sector	Occupation	Marital status	Physical disability	Religion
31	Inhabited for generations	Illiteracy					
21							
22	Inhabited for generations	Illiteracy	Agriculture	Land-holding peasant			
7	Inhabited for generations	Illiteracy	Agriculture	Land-holding peasant	Unmarried		No religion
1	Inhabited for generations	Illiteracy	Agriculture	Land-holding peasant	Married		No religion
2	Inhabited for generations	Illiteracy	Agriculture	Land-holding peasant	Married		No religion
25	Inhabited for generations	Illiteracy	Unemployed		Widowed		No religion
5	Inhabited for generations	Illiteracy	Unemployed		Widowed		No religion
3	Inhabited for generations	Illiteracy	Agriculture	Land-holding peasant	Married		No religion
29	Inhabited for generations	Illiteracy	Agriculture	Land-holding peasant	Married		No religion
18							
25	Inhabited for generations	Illiteracy	Agriculture	Land-holding peasant			
23	Inhabited for generations	Illiteracy					

(*cont.*)

	The no. of house-hold	Jia	House-hold	Relation-ship with the house-hold head	Surname	First name	Sex	The Chinese zodiac	Birthdate	
									Year	Month
Ganquan Xiang 7th Bao				The third daughter	董	虽朵	Female	Rat	1936	8
Ganquan Xiang 7th Bao				The fourth daughter	董	交哇	Female	Snake	1942	1
Ganquan Xiang 7th Bao				The fifth daughter	董	巨香	Female	Dog	1946	2
Ganquan Xiang 7th Bao	791	5	3	The house-hold head	董	仲收	Male	Dog	1898	2
Ganquan Xiang 7th Bao				The house-hold head's wife	冯	玉儿	Female	Rooster	1896	5
Ganquan Xiang 7th Bao				Son	董	维财	Male	Monkey	1932	5
Ganquan Xiang 7th Bao				The second son	董	王平	Male	Rooster	1933	10
Ganquan Xiang 7th Bao				Son's wife	冯	顺哇	Female	Rooster	1933	9
Ganquan Xiang 7th Bao				Grandson	董	存长	Male	Pig	1947	3
Ganquan Xiang 7th Bao				The second granddaugh-ter	董	肖肖	Female	Monkey	1944	9
Ganquan Xiang 7th Bao				The third Grand-daughter	董	虽明	Female	Rat	1936	8
Ganquan Xiang 7th Bao	792	5	4	The house-hold head	侯	玉清	Male	Horse	1905	11
Ganquan Xiang 7th Bao				The house-hold head's wife	焦	刘泰	Female	Dog	1909	11

Date	The beginning time of inhabitation in this county	Educational level	Profession sector	Occupation	Marital status	Physical disability	Religion
23	Inhabited for generations	Illiteracy					
7							
27							
1	Inhabited for generations	Illiteracy	Agriculture	Land-holding peasant	Married		No religion
9	Inhabited for generations	Illiteracy	Agriculture	Land-holding peasant	Married		No religion
1	Inhabited for generations	Illiteracy	Agriculture	Land-holding peasant	Unmarried		No religion
20	Inhabited for generations	Illiteracy	Agriculture	Land-holding peasant	Married		No religion
2	Inhabited for generations	Illiteracy	Agriculture	Land-holding peasant	Married		No religion
1							
15							
22	Inhabited for generations	Illiteracy					
27	Inhabited for generations	Illiteracy	Agriculture	Land-holding peasant	Married		No religion
11	Inhabited for generations	Illiteracy	Agriculture	Land-holding peasant	Married		No religion

(*cont.*)

	The no. of house-hold	Jia	House-hold	Relation-ship with the house-hold head	Surname	First name	Sex	The Chinese zodiac	Birthdate	
									Year	Month
Ganquan Xiang 7th Bao				Son	侯	永龙	Male	Dog	1934	10
Ganquan Xiang 7th Bao				The second son	侯	顺才	Male	Dragon	1940	4
Ganquan Xiang 7th Bao				Daughter	侯	香女	Female	Ox	1937	6
Ganquan Xiang 7th Bao				The second daughter	侯	三香	Female	Horse	1943	1
Ganquan Xiang 7th Bao				The third daughter	侯	金香	Female	Rooster	1935	10
Ganquan Xiang 7th Bao				Younger brother	侯	占清	Male	Tiger	1914	5
Ganquan Xiang 7th Bao				Younger brother's wife	曹	林林	Female	Tiger	1914	4
Ganquan Xiang 7th Bao				Nephew	侯	永花	Male	Pig	1936	1
Ganquan Xiang 7th Bao				Niece	侯	张才	Female	Rooster	1945	4
Ganquan Xiang 7th Bao				The second niece	侯	合合	Female	Dragon	1940	8
Ganquan Xiang 7th Bao	793	5	5	The house-hold head	侯	纪清	Male	Rabbit	1902	5
Ganquan Xiang 7th Bao				The house-hold head's wife	刘	看登	Female	Goat	1906	2
Ganquan Xiang 7th Bao				Son	侯	跟信	Male	Dog	1934	12

Date	The beginning time of inhabitation in this county	Educational level	Profession sector	Occupation	Marital status	Physical disability	Religion
22	Inhabited for generations	Primary school 2 years	Agriculture	Land-holding peasant			
31	Inhabited for generations	Illiteracy					
24	Inhabited for generations	Illiteracy					
15							
21	Inhabited for generations	Illiteracy	Agriculture	Land-holding peasant			
8	Inhabited for generations	Illiteracy	Agriculture	Land-holding peasant	Married		No religion
7	Inhabited for generations	Illiteracy	Agriculture	Land-holding peasant	Married		No religion
17	Inhabited for generations	Primary school 1 year	Agriculture	Land-holding peasant			
16							
4	Inhabited for generations	Illiteracy					
1	Inhabited for generations	Illiteracy	Agriculture	Land-holding peasant	Married		No religion
15	Inhabited for generations	Illiteracy	Agriculture	Land-holding peasant	Married		No religion
27	Inhabited for generations	Illiteracy	Agriculture	Land-holding peasant			

(*cont.*)

	The no. of household	Jia	Household	Relationship with the household head	Surname	First name	Sex	The Chinese zodiac	Birthdate	
									Year	Month
Ganquan Xiang 7th Bao				Daughter	侯	意哇	Female	Goat	1931	12
Ganquan Xiang 7th Bao				The second daughter	侯	顺意	Female	Ox	1937	11
Ganquan Xiang 7th Bao				The third daughter	侯	乖香	Female	Rooster	1945	12
Ganquan Xiang 7th Bao	794	5	6	The household head	侯	四明	Male	Pig	1898	3
Ganquan Xiang 7th Bao				The household head's wife	盟	明哇	Female	Goat	1895	1
Ganquan Xiang 7th Bao				Son	侯	存存	Male	Rat	1924	10
Ganquan Xiang 7th Bao				Daughter	侯	长哇	Female	Rabbit	1927	2
Ganquan Xiang 7th Bao				The second daughter	侯	虽祥	Female	Monkey	1932	11
Ganquan Xiang 7th Bao				Nephew	侯	恩儿	Male	Rabbit	1914	4
Ganquan Xiang 7th Bao				Nephew's wife	王	海哇	Female	Tiger	1914	3
Ganquan Xiang 7th Bao				Granddaughter	侯	月义	Female	Ox	1937	4
Ganquan Xiang 7th Bao				The second granddaughter	侯	桃應	Female	Goat	1943	7
Ganquan Xiang 7th Bao	795	5	7	The household head	侯	永才	Male	Rabbit	1915	4

Date	The beginning time of inhabitation in this county	Educational level	Profession sector	Occupation	Marital status	Physical disability	Religion
6	Inhabited for generations	Illiteracy	Agriculture	Land-holding peasant	Unmarried		No religion
10	Inhabited for generations	Illiteracy					
2							
8	Inhabited for generations	Illiteracy	Agriculture	Land-holding peasant	Married		No religion
28	Inhabited for generations	Illiteracy	Agriculture	Land-holding peasant	Married		No religion
16	Inhabited for generations	Illiteracy	Agriculture	Land-holding peasant			
5	Inhabited for generations	Illiteracy	Agriculture	Land-holding peasant			
25	Inhabited for generations	Illiteracy	Agriculture	Land-holding peasant			
22	Inhabited for generations	Illiteracy	Agriculture	Land-holding peasant	Married		
24	Inhabited for generations	Illiteracy	Agriculture	Land-holding peasant	Married		
9	Inhabited for generations	Illiteracy					
4	Inhabited for generations						
24	Inhabited for generations	Illiteracy	Agriculture	Land-holding peasant	Married		No religion

(*cont.*)

	The no. of household	Jia	Household	Relationship with the household head	Surname	First name	Sex	The Chinese zodiac	Birthdate	
									Year	Month
Ganquan Xiang 7th Bao				The household head's wife	王	巧哇子	Female	Goat	1919	5
Ganquan Xiang 7th Bao				Father	侯	建清	Male	Rooster	1884	8
Ganquan Xiang 7th Bao				Mother	转	杨儿	Female	Rabbit	1890	11
Ganquan Xiang 7th Bao				Son	侯	里桂儿	Male	Horse	1930	3
Ganquan Xiang 7th Bao				Younger brother	侯	永长	Male	Dragon	1916	12
Ganquan Xiang 7th Bao				Younger brother's wife	董	吉祥	Female	Tiger	1914	2
Ganquan Xiang 7th Bao				Nephew	侯	调信哇	Male	Goat	1914	1
Ganquan Xiang 7th Bao				Niece	侯	乖乖	Female	Dragon	1940	9
Ganquan Xiang 7th Bao				The second younger brother	侯	永库	Male	Horse	1930	7
Ganquan Xiang 7th Bao				Younger brother's wife	陈	雲哇	Female	Tiger	1926	8
Ganquan Xiang 7th Bao	796	5	8	The household head	侯	彦进	Male	Horse	1905	11
Ganquan Xiang 7th Bao				The household head's wife	苟	秋儿	Female	Rabbit	1914	1
Ganquan Xiang 7th Bao				Son	侯	顺顺	Male	Pig	1923	11

Date	The beginning time of inhabitation in this county	Educational level	Profession sector	Occupation	Marital status	Physical disability	Religion
12	Inhabited for generations	Illiteracy	Agriculture	Land-holding peasant	Married		No religion
10	Inhabited for generations	Illiteracy	Agriculture	Land-holding peasant	Married		No religion
2	Inhabited for generations	Illiteracy	Agriculture	Land-holding peasant	Married		No religion
1	Inhabited for generations	Illiteracy	Agriculture	Land-holding peasant			
7	Inhabited for generations	Illiteracy	Agriculture	Land-holding peasant	Married		
1	Inhabited for generations	Illiteracy	Agriculture	Land-holding peasant	Married		
6	Inhabited for generations	Illiteracy	Agriculture	Land-holding peasant			
3	Inhabited for generations	Illiteracy					
3	Inhabited for generations	Illiteracy	Agriculture	Land-holding peasant	Married		
19	Inhabited for generations	Illiteracy	Agriculture	Land-holding peasant	Married		
26	Inhabited for generations	Illiteracy	Agriculture	Land-holding peasant	Married		No religion
30	Inhabited for generations	Illiteracy	Agriculture	Land-holding peasant	Married		No religion
19	Inhabited for generations	Illiteracy	Agriculture	Land-holding peasant	Married		No religion

(*cont.*)

	The no. of household	Jia	Household	Relationship with the household head	Surname	First name	Sex	The Chinese zodiac	Birthdate Year	Month
Ganquan Xiang 7th Bao				Son's wife	侯	乘哇	Female	Goat	1931	5
Ganquan Xiang 7th Bao				Daughter	侯	松松	Female	Rooster	1933	11
Ganquan Xiang 7th Bao				The second daughter	侯	香梅	Female	Monkey	1944	8
Ganquan Xiang 7th Bao				Older brother	侯	石生哇	Male	Goat	1886	12
Ganquan Xiang 7th Bao				Nephew	侯	庆哇	Male	Ox	1913	7
Ganquan Xiang 7th Bao				Nephew's wife	王	细哇子	Female	Goat	1919	5
Ganquan Xiang 7th Bao				Granddaughter	侯	田儿	Female	Rabbit	1939	9
Ganquan Xiang 7th Bao				The second granddaughter	侯	虽贤	Female	Horse	1942	10
Ganquan Xiang 7th Bao				The third Granddaughter	侯	坡哇	Female	Rooster	1945	6
Ganquan Xiang 7th Bao				Cohabitant	侯	石拜	Male	Rooster	1908	6
Ganquan Xiang 7th Bao				Cohabitant	曹	氏	Female	Horse	1905	4
Ganquan Xiang 7th Bao				Cohabitant	楼	哇	Female	Rooster	1933	2
Ganquan Xiang 7th Bao	797	5	9	The household head	侯	建云	Male	Monkey	1906	4

Date	The beginning time of inhabitation in this county	Educational level	Profession sector	Occupation	Marital status	Physical disability	Religion
28	Inhabited for generations	Illiteracy	Agriculture	Land-holding peasant	Married		No religion
29	Inhabited for generations	Illiteracy	Agriculture	Land-holding peasant	Married		No religion
13	Inhabited for generations						
10	Inhabited for generations	Illiteracy	Agriculture	Land-holding peasant	Widowed		
20	Inhabited for generations	Illiteracy	Agriculture	Land-holding peasant	Married		
23	Inhabited for generations	Illiteracy	Agriculture	Land-holding peasant	Married		
4	Inhabited for generations	Illiteracy					
25	Inhabited for generations						
4	Inhabited for generations						
26	Inhabited for generations	Illiteracy	Agriculture	Land-holding peasant	Married		
25	Inhabited for generations	Illiteracy	Agriculture	Land-holding peasant	Married		
26							
27	Inhabited for generations	Illiteracy	Agriculture	Land-holding peasant	Married		No religion

(*cont.*)

	The no. of household	Jia	Household	Relationship with the household head	Surname	First name	Sex	The Chinese zodiac	Birthdate	
									Year	Month
Ganquan Xiang 7th Bao				The household head's wife	田	桃儿	Female	Horse	1905	11
Ganquan Xiang 7th Bao				Son	侯	坡哇	Male	Rat	1924	4
Ganquan Xiang 7th Bao				The second son	侯	马哇	Male	Rooster	1945	11
Ganquan Xiang 7th Bao				Daughter	侯	称秤	Female	Monkey	1932	7
Ganquan Xiang 7th Bao				The second daughter	侯	四儿	Female	Snake	1942	4
Ganquan Xiang 7th Bao				Younger brother	侯	平哇	Male	Dog	1922	8
Ganquan Xiang 7th Bao				Younger brother's wife	陈	毛人	Female	Snake	1929	8
Ganquan Xiang 7th Bao	798	5	10	The household head	侯	永清	Male	Goat	1894	12
Ganquan Xiang 7th Bao				The household head's wife	董	叩还	Female	Pig	1898	5
Ganquan Xiang 7th Bao				Son	侯	廷哇	Male	Snake	1929	6
Ganquan Xiang 7th Bao				Son's wife	曹	桃哇	Female	Dragon	1928	5
Ganquan Xiang 7th Bao				The second son	侯	永均	Male	Pig	1935	11
Ganquan Xiang 7th Bao				The second son's wife	尹	绑哇	Female	Snake	1929	12

Date	The beginning time of inhabitation in this county	Educational level	Profession sector	Occupation	Marital status	Physical disability	Religion
30	Inhabited for generations	Illiteracy	Agriculture	Land-holding peasant	Married		No religion
16	Inhabited for generations	Illiteracy	Agriculture	Land-holding peasant			No religion
11	Inhabited for generations						
9	Inhabited for generations	Illiteracy	Agriculture	Land-holding peasant			
1	Inhabited for generations						
23	Inhabited for generations	Illiteracy	Agriculture	Land-holding peasant	Married		
1	Inhabited for generations	Illiteracy	Agriculture	Land-holding peasant	Married		
30	Inhabited for generations	Illiteracy	Agriculture	Land-holding peasant	Married		No religion
9	Inhabited for generations	Illiteracy	Agriculture	Land-holding peasant	Married		No religion
3	Inhabited for generations	Illiteracy	Agriculture	Land-holding peasant	Married		No religion
1	Inhabited for generations	Illiteracy	Agriculture	Land-holding peasant	Married		No religion
29	Inhabited for generations	Illiteracy	Agriculture	Land-holding peasant	Married		No religion
6	Inhabited for generations	Illiteracy	Agriculture	Land-holding peasant	Married		No religion

(*cont.*)

	The no. of household	Jia	Household	Relationship with the household head	Surname	First name	Sex	The Chinese zodiac	Birthdate Year	Month
Ganquan Xiang 7th Bao				The third son	侯	正才	Male	Monkey	1945	1
Ganquan Xiang 7th Bao				Daughter	侯	月信	Female	Pig	1935	5
Ganquan Xiang 7th Bao				The second daughter	侯	灰哇	Female	Snake	1941	12
Ganquan Xiang 7th Bao	799	5	11	The household head	侯	举哇	Male	Monkey	1907	12
Ganquan Xiang 7th Bao				The household head's wife	尹	反哇	Female	Pig	1910	10
Ganquan Xiang 7th Bao				Son	侯	马才	Male	Ox	1937	6
Ganquan Xiang 7th Bao				The second son	侯	二才	Male	Snake	1941	12
Ganquan Xiang 7th Bao				Daughter	侯	金金	Female	Dog	1946	7
Ganquan Xiang 7th Bao				Uncle	侯	彦来	Male	Dragon	1891	9
Ganquan Xiang 7th Bao				Uncle's wife	李	胡生	Female	Ox	1888	4
Ganquan Xiang 7th Bao				Younger brother	侯	田吴儿	Male	Goat	1919	2
Ganquan Xiang 7th Bao				Younger brother's wife	辛	党儿	Female	Horse	1919	2
Ganquan Xiang 7th Bao				Nephew	侯	天时儿	Male	Rooster	1946	1

Date	The beginning time of inhabitation in this county	Educational level	Profession sector	Occupation	Marital status	Physical disability	Religion
27	Inhabited for generations						
12	Inhabited for generations	Illiteracy					
20	Inhabited for generations	Illiteracy					
8	Inhabited for generations	Illiteracy	Agriculture	Land-holding peasant	Married		No religion
29	Inhabited for generations	Illiteracy	Agriculture	Land-holding peasant	Married		No religion
16	Inhabited for generations	Illiteracy					
18	Inhabited for generations	Illiteracy					
28	Inhabited for generations						
6	Inhabited for generations	Illiteracy	Agriculture	Land-holding peasant	Married		
7	Inhabited for generations	Illiteracy	Agriculture	Land-holding peasant	Married		
11	Inhabited for generations	Illiteracy	Agriculture	Land-holding peasant	Married		
16	Inhabited for generations	Illiteracy	Agriculture	Land-holding peasant	Married		
29	Inhabited for generations						

(*cont.*)

	The no. of household	Jia	Household	Relationship with the household head	Surname	First name	Sex	The Chinese zodiac	Birthdate	
									Year	Month
Ganquan Xiang 7th Bao				Niece	侯	補哇	Female	Pig	1935	10
Ganquan Xiang 7th Bao	800	5	12	The household head	侯	建仓	Male	Rabbit	1915	9
Ganquan Xiang 7th Bao				The household head's wife	尹	三顺	Female	Goat	1919	5
Ganquan Xiang 7th Bao				Father	侯	彦富	Male	Dragon	1879	1
Ganquan Xiang 7th Bao				Son	侯	混才	Male	Monkey	1944	11
Ganquan Xiang 7th Bao				Daughter	侯	灰哇	Female	Snake	1942	2
Ganquan Xiang 7th Bao	801	6	1	The household head	侯	永昌	Male	Ox	1926	1
Ganquan Xiang 7th Bao				The household head's wife	曹	應應	Female	Rat	1924	12
Ganquan Xiang 7th Bao				Father	侯	建喜	Male	Pig	1898	5
Ganquan Xiang 7th Bao				Mother	杜	买卖	Female	Horse	1904	1
Ganquan Xiang 7th Bao				Son	侯	西气	Male	Goat	1943	5
Ganquan Xiang 7th Bao	802	8	2	The household head	侯	绪娃	Male	Monkey	1907	11
Ganquan Xiang 7th Bao				The household head's wife	董	良良	Female	Rabbit	1915	4

Date	The beginning time of inhabitation in this county	Educational level	Profession sector	Occupation	Marital status	Physical disability	Religion
21	Inhabited for generations	Illiteracy					
21	Inhabited for generations	Illiteracy	Agricul-ture	Land-holding peasant	Married		No religion
19	Inhabited for generations	Illiteracy	Agricul-ture	Land-holding peasant	Married		No religion
25	Inhabited for generations	Illiteracy	Unem-ployed		Widowed		No religion
16							
9							
27	Inhabited for generations	Illiteracy	Agricul-ture	Land-holding peasant	Married		No religion
22	Inhabited for generations	Illiteracy	Agricul-ture	Land-holding peasant	Married		No religion
19	Inhabited for generations	Illiteracy	Agricul-ture	Land-holding peasant	Married		No religion
10	Inhabited for generations	Illiteracy	Agricul-ture	Land-holding peasant	Married		No religion
21	Inhabited for generations	Illiteracy	Agricul-ture	Land-holding peasant	Married		No religion
1	Inhabited for generations	Illiteracy	Agricul-ture	Land-holding peasant	Married		No religion
21	Inhabited for generations	Illiteracy	Agricul-ture	Land-holding peasant	Married		No religion

(*cont.*)

	The no. of house-hold	Jia	House-hold	Relation-ship with the house-hold head	Surname	First name	Sex	The Chinese zodiac	Birthdate	
									Year	Month
Ganquan Xiang 7th Bao				Daughter	侯	焦焦	Female	Snake	1941	12
Ganquan Xiang 7th Bao				The second daughter	侯	崴女	Female	Monkey	1914	3
Ganquan Xiang 7th Bao				The third daughter	侯	女儿	Female	Dog	1946	7
Ganquan Xiang 7th Bao				Older brother	侯	像娃	Male	Horse	1905	5
Ganquan Xiang 7th Bao				Older brother's wife	曹	林林	Female	Pig	1910	4
Ganquan Xiang 7th Bao				Nephew	侯	尚娃	Male	Pig	1935	10
Ganquan Xiang 7th Bao				The second niece	侯	左娃	Female	Monkey	1944	9
Ganquan Xiang 7th Bao	803	6	3	The house-hold head	侯	满清	Male	Monkey	1907	6
Ganquan Xiang 7th Bao				The house-hold head's wife	曹	陈花	Female	Monkey	1907	5
Ganquan Xiang 7th Bao				Son	侯	永陞	Male	Monkey	1932	9
Ganquan Xiang 7th Bao				The second son	侯	永德	Male	Dragon	1940	5
Ganquan Xiang 7th Bao				Daughter	侯	崴女	Female	Snake	1941	1
Ganquan Xiang 7th Bao				The second daughter	侯	四女	Female	Monkey	1944	9

Date	The beginning time of inhabitation in this county	Educational level	Profession sector	Occupation	Marital status	Physical disability	Religion
25	Inhabited for generations	Illiteracy					
10	Inhabited for generations						
23	Inhabited for generations						
4	Inhabited for generations	Illiteracy	Agriculture	Land-holding peasant	Married		No religion
10	Inhabited for generations	Illiteracy	Agriculture	Land-holding peasant	Married		No religion
10	Inhabited for generations	Illiteracy	Agriculture	Land-holding peasant	Unmarried		No religion
2	Inhabited for generations						
26	Inhabited for generations	Illiteracy	Agriculture	Land-holding peasant	Married		No religion
27	Inhabited for generations	Illiteracy	Agriculture	Land-holding peasant	Married		No religion
8	Inhabited for generations	Primary school 2 years	Agriculture	Land-holding peasant			
28	Inhabited for generations	Primary school 1 year	Agriculture	Land-holding peasant			
30	Inhabited for generations						
25	Inhabited for generations						

(*cont.*)

	The no. of household	Jia	Household	Relationship with the household head	Surname	First name	Sex	The Chinese zodiac	Birthdate	
									Year	Month
Ganquan Xiang 7th Bao				Cohabitant	侯	良台	Male	Rabbit	1927	4
Ganquan Xiang 7th Bao				Cohabitant	曹	长女	Female	Rabbit	1927	8
Ganquan Xiang 7th Bao				Cohabitant	刘	生	Male	Ox	1925	10
Ganquan Xiang 7th Bao	804	6	3	The household head	侯	郭林	Male	Dragon	1891	7
Ganquan Xiang 7th Bao				Son	侯	功良	Male	Rabbit	1927	11
Ganquan Xiang 7th Bao				Son's wife	曹	科娃	Female	Horse	1930	5
Ganquan Xiang 7th Bao				Younger brother	侯	刘西娃	Male	Horse	1893	5
Ganquan Xiang 7th Bao				Younger brother's wife	雍	赙娃	Female	Monkey	1920	5
Ganquan Xiang 7th Bao				Grandson	侯	才儿	Male	Horse	1942	1
Ganquan Xiang 7th Bao				The second son	侯	二力	Male	Goat	1943	8
Ganquan Xiang 7th Bao				Cohabitant	雍	平儿	Female	Dragon	1891	5
Ganquan Xiang 7th Bao				Cohabitant	侯	锁哇	Male	Pig	1935	9
Ganquan Xiang 7th Bao	805	6	5	The household head	侯	建平	Male	Rooster	1896	10

Date	The beginning time of inhabitation in this county	Educational level	Profession sector	Occupation	Marital status	Physical disability	Religion
16	Inhabited for generations	Illiteracy	Agriculture	Land-holding peasant	Married		No religion
9	Inhabited for generations	Illiteracy	Agriculture	Land-holding peasant	Married		No religion
26	Inhabited for generations	Illiteracy	Agriculture	Land-holding peasant			
8	Inhabited for generations	Illiteracy	Agriculture	Land-holding peasant	Widowed		No religion
8	Inhabited for generations	Illiteracy	Agriculture	Land-holding peasant	Married		No religion
4	Inhabited for generations	Illiteracy	Agriculture	Land-holding peasant	Married		No religion
19	Inhabited for generations	Illiteracy	Agriculture	Land-holding peasant	Married		No religion
4	Inhabited for generations	Illiteracy	Agriculture	Land-holding peasant	Married		No religion
6	Inhabited for generations						
1	Inhabited for generations						
19	Inhabited for generations	Illiteracy	Agriculture	Land-holding peasant	Widowed		No religion
1	Inhabited for generations	Illiteracy	Agriculture	Land-holding peasant			No religion
27	Inhabited for generations	Illiteracy	Agriculture	Land-holding peasant	Married		No religion

(*cont.*)

	The no. of house-hold	Jia	House-hold	Relation-ship with the house-hold head	Surname	First name	Sex	The Chinese zodiac	Birthdate	
									Year	Month
Ganquan Xiang 7th Bao				The house-hold head's wife	尹	琴儿	Female	Tiger	1901	4
Ganquan Xiang 7th Bao				Son	侯	七娃	Male	Rabbit	1927	3
Ganquan Xiang 7th Bao				Son's wife	尹	均娃	Female	Dragon	1928	9
Ganquan Xiang 7th Bao				The second son	侯	斑娃	Male	Rat	1936	6
Ganquan Xiang 7th Bao				Daughter	侯	怀哇	Female	Snake	1941	10
Ganquan Xiang 7th Bao	806	6	6	The house-hold head	冯	满全	Male	Ox	1913	8
Ganquan Xiang 7th Bao				The house-hold head's wife	補	召儿	Female	Tiger	1914	4
Ganquan Xiang 7th Bao				Mother	幸	惠儿	Female	Rat	1887	6
Ganquan Xiang 7th Bao				Son	冯	各达	Male	Goat	1943	11
Ganquan Xiang 7th Bao				The second son	冯	凤娃	Male	Dog	1946	11
Ganquan Xiang 7th Bao				Cohabitant	冯	生清	Male	Ox	1901	1
Ganquan Xiang 7th Bao				Cohabitant	崔	氏	Female	Dog	1875	7
Ganquan Xiang 7th Bao	807	6	7	The house-hold head	冯	生彦	Male	Tiger	1889	7

Date	The beginning time of inhabitation in this county	Educational level	Profession sector	Occupation	Marital status	Physical disability	Religion
25	Inhabited for generations	Illiteracy	Agriculture	Land-holding peasant	Married		No religion
22	Inhabited for generations	Illiteracy	Agriculture	Land-holding peasant	Married		No religion
11	Inhabited for generations	Illiteracy	Agriculture	Land-holding peasant	Married		No religion
18	Inhabited for generations	Primary school 2 years	Agriculture	Land-holding peasant	Married		No religion
11	Inhabited for generations						
24	Inhabited for generations	Illiteracy	Agriculture	Land-holding peasant	Married		No religion
13	Inhabited for generations	Illiteracy	Agriculture	Land-holding peasant	Married		No religion
4	Inhabited for generations	Illiteracy	Agriculture	Land-holding peasant	Widowed		No religion
4	Inhabited for generations						
18	Inhabited for generations						
21	Inhabited for generations	Illiteracy	Agriculture	Land-holding peasant	Widowed		No religion
1	Inhabited for generations	Illiteracy	Agriculture	Land-holding peasant	Widowed		No religion
31	Inhabited for generations	Illiteracy	Agriculture	Land-holding peasant	Married		No religion

(*cont.*)

	The no. of household	Jia	Household	Relationship with the household head	Surname	First name	Sex	The Chinese zodiac	Birthdate	
									Year	Month
Ganquan Xiang 7th Bao				The household head's wife	张	莹童	Female	Pig	1898	4
Ganquan Xiang 7th Bao				Son	冯	吉祥	Male	Monkey	1932	10
Ganquan Xiang 7th Bao				Son's wife	王	草草	Female	Horse	1930	9
Ganquan Xiang 7th Bao				The second son	冯	赛谢	Male	Rat	1936	6
Ganquan Xiang 7th Bao				Cohabitant	曹	邓娃	Female	Rat	1899	12
Ganquan Xiang 7th Bao				Cohabitant	冯	月香	Female	Rat	1936	2
Ganquan Xiang 7th Bao	808	6	8	The household head	冯	官官	Male	Rat	1924	11
Ganquan Xiang 7th Bao				Father	冯	生平	Male	Rat	1899	3
Ganquan Xiang 7th Bao				Mother	张	崔平	Female	Horse	1905	5
Ganquan Xiang 7th Bao				Younger brother	冯	挪娃	Male	Horse	1931	10
Ganquan Xiang 7th Bao	809	6	9	The household head	冯	建明	Male	Horse	1905	5
Ganquan Xiang 7th Bao				The household head's wife	冯	吓哇	Female	Tiger	1914	5
Ganquan Xiang 7th Bao				Father	冯	永平	Male	Snake	1879	7

Date	The beginning time of inhabitation in this county	Educational level	Profession sector	Occupation	Marital status	Physical disability	Religion
19	Inhabited for generations	Illiteracy	Agriculture	Land-holding peasant	Married		No religion
12	Inhabited for generations	Illiteracy	Agriculture	Land-holding peasant	Married		No religion
1	Inhabited for generations	Illiteracy	Agriculture	Land-holding peasant	Married		No religion
29	Inhabited for generations	Illiteracy	Agriculture				
18	Inhabited for generations	Illiteracy	Agriculture	Land-holding peasant			No religion
25	Inhabited for generations	Illiteracy			Widowed		No religion
4	Inhabited for generations	Illiteracy	Agriculture	Land-holding peasant	Unmarried		No religion
11	Inhabited for generations	Illiteracy	Agriculture	Land-holding peasant	Married		No religion
5	Inhabited for generations	Illiteracy	Agriculture	Land-holding peasant	Married		No religion
6	Inhabited for generations	Primary school 2 years			Unmarried		
29	Inhabited for generations	Illiteracy	Agriculture	Land-holding peasant	Married		No religion
6	Inhabited for generations	Illiteracy	Agriculture	Land-holding peasant	Married		No religion
15	Inhabited for generations						

(*cont.*)

	The no. of household	Jia	Household	Relationship with the household head	Surname	First name	Sex	The Chinese zodiac	Birthdate	
									Year	Month
Ganquan Xiang 7th Bao				Son	冯	安吉	Male	Rooster	1933	4
Ganquan Xiang 7th Bao				The second son	冯	杨家	Male	Rat	1936	12
Ganquan Xiang 7th Bao				The third son	冯	才才	Male	Dragon	1940	3
Ganquan Xiang 7th Bao				The fourth son	冯	成成	Male	Dog	1946	12
Ganquan Xiang 7th Bao				Daughter	冯	才女	Female	Goat	1943	4
Ganquan Xiang 7th Bao				Cohabitant	赵	氏	Female	Pig	1898	4
Ganquan Xiang 7th Bao	810	6	10	The household head	冯	生贵	Male	Rat	1899	5
Ganquan Xiang 7th Bao				The household head's wife	张	冯冯	Female	Pig	1898	6
Ganquan Xiang 7th Bao				Son	冯	礼拜	Male	Dragon	1928	4
Ganquan Xiang 7th Bao				Son's wife	魏	跟响子	Female	Pig	1923	3
Ganquan Xiang 7th Bao				Younger brother	冯	杨谢	Male	Horse	1905	10
Ganquan Xiang 7th Bao				The second younger brother	冯	三娃	Male	Rat	1912	4
Ganquan Xiang 7th Bao	811	6	11	The household head	曹	王孙子	Male	Snake	1917	8

Date	The beginning time of inhabitation in this county	Educational level	Profession sector	Occupation	Marital status	Physical disability	Religion
13	Inhabited for generations	Primary school 3 years	Agriculture	Land-holding peasant			No religion
11	Inhabited for generations	Illiteracy					
20	Inhabited for generations	Illiteracy					
24	Inhabited for generations						
14	Inhabited for generations						
14	Inhabited for generations	Illiteracy	Agriculture	Land-holding peasant	Widowed		No religion
6	Inhabited for generations	Illiteracy	Agriculture	Land-holding peasant	Married		No religion
11	Inhabited for generations	Illiteracy	Agriculture	Land-holding peasant	Married		No religion
22	Inhabited for generations	Illiteracy	Agriculture	Land-holding peasant	Married		No religion
19	Inhabited for generations	Illiteracy	Agriculture	Land-holding peasant	Married		No religion
29	Inhabited for generations	Illiteracy	Agriculture	Land-holding peasant			No religion
1	Inhabited for generations	Illiteracy	Agriculture	Land-holding peasant			No religion
18	Inhabited for generations	Illiteracy	Agriculture	Land-holding peasant	Married		No religion

(*cont.*)

	The no. of household	Jia	Household	Relationship with the household head	Surname	First name	Sex	The Chinese zodiac	Birthdate	
									Year	Month
Ganquan Xiang 7th Bao				The household head's wife	冯	氏	Female	Horse	1918	5
Ganquan Xiang 7th Bao				Son	曹	娃	Male	Snake	1941	6
Ganquan Xiang 7th Bao				The second son	曹	虽哇	Male	Goat	1943	8
Ganquan Xiang 7th Bao	812	6	12	The household head	冯	高高	Male	Horse	1905	5
Ganquan Xiang 7th Bao				The household head's wife	庞	利娃	Female	Dragon	1916	9
Ganquan Xiang 7th Bao				Son	冯	應平	Male	Tiger	1938	4
Ganquan Xiang 7th Bao				Younger brother	冯	功良	Male	Horse	1930	4
Ganquan Xiang 7th Bao	813	6	13	The household head	冯	呱子	Male	Dragon	1916	3
Ganquan Xiang 7th Bao				The household head's wife	陈	改娃	Female	Monkey	1920	5
Ganquan Xiang 7th Bao				Younger brother	冯	建功	Male	Rabbit	1927	3
Ganquan Xiang 7th Bao				Younger brother's wife	曹	过娃	Female	Rabbit	1927	9
Ganquan Xiang 7th Bao	814	7	1	The household head	苏	万奇	Male	Dragon	1916	6
Ganquan Xiang 7th Bao				The household head's wife	梁	翠儿	Female	Goat	1919	12

Date	The beginning time of inhabitation in this county	Educational level	Profession sector	Occupation	Marital status	Physical disability	Religion
15	Inhabited for generations	Illiteracy	Agriculture	Land-holding peasant	Married		No religion
25	Inhabited for generations	Illiteracy					
2	Inhabited for generations	Illiteracy					
1	Inhabited for generations	Illiteracy	Agriculture	Land-holding peasant	Married		No religion
20	Inhabited for generations	Illiteracy	Agriculture	Land-holding peasant	Married		No religion
5	Inhabited for generations	Primary school 1 year					
10	Inhabited for generations	Illiteracy	Agriculture	Land-holding peasant	Unmarried		No religion
22	Inhabited for generations	Illiteracy	Agriculture	Land-holding peasant	Married		No religion
28	Inhabited for generations	Illiteracy	Agriculture	Land-holding peasant	Married		No religion
10	Inhabited for generations	Illiteracy	Agriculture	Land-holding peasant	Married		No religion
10	Inhabited for generations	Illiteracy	Agriculture	Land-holding peasant	Married		No religion
11	Inhabited for generations	Illiteracy	Agriculture	Land-holding peasant	Married		No religion
13	Inhabited for generations	Illiteracy	Agriculture	Land-holding peasant	Married		No religion

(*cont.*)

	The no. of household	Jia	Household	Relationship with the household head	Surname	First name	Sex	The Chinese zodiac	Birthdate	
									Year	Month
Ganquan Xiang 7th Bao				Father	苏	奠川	Male	Ox	1888	5
Ganquan Xiang 7th Bao				Son	苏	宝宝	Male	Snake	1942	1
Ganquan Xiang 7th Bao				The second son	苏	二宝	Male	Monkey	1944	10
Ganquan Xiang 7th Bao	815	7	2	The household head	曹	金地	Male	Dog	1909	3
Ganquan Xiang 7th Bao				The household head's wife	孟	永哇	Female	Rooster	1921	4
Ganquan Xiang 7th Bao				Father	曹	登芳	Male	Rat	1887	10
Ganquan Xiang 7th Bao				Mother	王	能女子	Female	Rabbit	1890	5
Ganquan Xiang 7th Bao				Younger brother	曹	金喜	Male	Tiger	1924	10
Ganquan Xiang 7th Bao				Nephew	曹	成全	Male	Horse	1942	4
Ganquan Xiang 7th Bao	816	7	3	The household head	田	福得	Male	Goat	1920	2
Ganquan Xiang 7th Bao				The household head's wife	曹	改面子	Female	Rooster	1921	8
Ganquan Xiang 7th Bao				Son	田	毛哇	Male	Monkey	1946	8
Ganquan Xiang 7th Bao				Daughter	田	春哇	Female	Monkey	1944	2

Date	The beginning time of inhabitation in this county	Educational level	Profession sector	Occupation	Marital status	Physical disability	Religion
14	Inhabited for generations	Sishu 2 years	Agriculture	Land-holding peasant	Widowed		No religion
27	Inhabited for generations						
8	Inhabited for generations						
2	Inhabited for generations	Illiteracy	Agriculture	Land-holding peasant	Married		No religion
30	Inhabited for generations	Illiteracy	Agriculture	Land-holding peasant	Married		No religion
18	Inhabited for generations	Illiteracy	Agriculture	Land-holding peasant	Married		No religion
28	Inhabited for generations	Illiteracy	Agriculture	Land-holding peasant	Married		No religion
13	Inhabited for generations	Primary school 4 years	Agriculture	Land-holding peasant	Widowed		No religion
17							
5	Inhabited for generations	Illiteracy	Agriculture	Land-holding peasant	Married		No religion
4	Inhabited for generations	Illiteracy	Agriculture	Land-holding peasant	Married		No religion
11	Inhabited for generations						
8	Inhabited for generations						

(*cont.*)

									Birthdate	
	The no. of household	Jia	Household	Relationship with the household head	Surname	First name	Sex	The Chinese zodiac	Year	Month
Ganquan Xiang 7th Bao	817	7	4	The household head	马	登福	Male	Tiger	1875	2
Ganquan Xiang 7th Bao				Son's wife	赵	考子	Female	Ox	1900	3
Ganquan Xiang 7th Bao				Son	马	满仓	Male	Tiger	1926	11
Ganquan Xiang 7th Bao	818	7	5	The household head	曹	金明	Male	Ox	1913	3
Ganquan Xiang 7th Bao				The household head's wife	王	仲哇	Female	Goat	1919	2
Ganquan Xiang 7th Bao				Father	曹	登满	Male	Horse	1893	6
Ganquan Xiang 7th Bao				Son	曹	成人	Male	Rooster	1945	6
Ganquan Xiang 7th Bao				Daughter	曹	桃哇	Female	Tiger	1938	2
Ganquan Xiang 7th Bao				The second daughter	曹	易哇	Female	Snake	1941	6
Ganquan Xiang 7th Bao	819	7	6	The household head	曹	仲堂	Male	Snake	1880	1
Ganquan Xiang 7th Bao				Son	曹	抢哇	Male	Dragon	1916	2
Ganquan Xiang 7th Bao				Son's wife	胡	闹闹	Female	Dog	1932	9
Ganquan Xiang 7th Bao				Grandson	曹	川哇	Male	Goat	1931	4

Date	The beginning time of inhabitation in this county	Educational level	Profession sector	Occupation	Marital status	Physical disability	Religion
2	Inhabited for generations	Illiteracy	Agricul-ture	Land-holding peasant	Widowed		No religion
13	Inhabited for generations	Illiteracy	Agricul-ture	Land-holding peasant	Widowed		No religion
15	Inhabited for generations	Primary school 2 years	Agricul-ture	Land-holding peasant	Unmarried		
27	Inhabited for generations	Illiteracy	Agricul-ture	Land-holding peasant	Married		No religion
1	Inhabited for generations	Illiteracy	Agricul-ture	Land-holding peasant	Married		No religion
1	Inhabited for generations	Illiteracy	Agricul-ture	Land-holding peasant	Widowed		No religion
14	Inhabited for generations						
31	Inhabited for generations	Illiteracy					
8	Inhabited for generations	Illiteracy					
16	Inhabited for generations	Illiteracy	Agricul-ture	Land-holding peasant	Widowed		No religion
25	Inhabited for generations	Illiteracy	Agricul-ture	Land-holding peasant	Married		No religion
1	Inhabited for generations	Illiteracy	Agricul-ture	Land-holding peasant	Married		No religion
15	Inhabited for generations	Illiteracy	Agricul-ture	Land-holding peasant	Unmarried		No religion

(*cont.*)

	The no. of household	Jia	Household	Relationship with the household head	Surname	First name	Sex	The Chinese zodiac	Birthdate Year	Month
Ganquan Xiang 7th Bao				The second grandson	曹	安安	Male	Ox	1937	9
Ganquan Xiang 7th Bao				The second daughter	曹	洩哇	Female	Monkey	1920	6
Ganquan Xiang 7th Bao				Cohabitant	李	条儿	Female	Rooster	1908	8
Ganquan Xiang 7th Bao				Cohabitant	曹	登雪	Male	Rat	1898	2
Ganquan Xiang 7th Bao	820	7	7	The household head	曹	仲廷	Male	Snake	1892	4
Ganquan Xiang 7th Bao				The household head's wife	易	玉子	Female	Snake	1892	7
Ganquan Xiang 7th Bao				Son	曹	张姓	Male	Ox	1925	2
Ganquan Xiang 7th Bao				Son's wife	易	过哇	Female	Ox	1925	2
Ganquan Xiang 7th Bao	821	7	8	The household head	曹	金哇	Male	Dog	1921	4
Ganquan Xiang 7th Bao				The household head's wife	曹	灰哇	Female	Tiger	1926	3
Ganquan Xiang 7th Bao				Son	曹	休哇	Male	Rooster	1945	4
Ganquan Xiang 7th Bao				Daughter	曹	交哇	Female	Dog	1946	8
Ganquan Xiang 7th Bao	822	7	9	The household head	曹	等哇	Male	Goat	1932	1

Date	The beginning time of inhabitation in this county	Educational level	Profession sector	Occupation	Marital status	Physical disability	Religion
1	Inhabited for generations	Illiteracy					
17	Inhabited for generations	Illiteracy	Agriculture	Land-holding peasant	Divorced		
1	Inhabited for generations	Illiteracy	Agriculture	Land-holding peasant	Widowed		
9	Inhabited for generations	Illiteracy	Agriculture	Land-holding peasant	Divorced		
11	Inhabited for generations	Illiteracy	Agriculture	Land-holding peasant	Married		No religion
1	Inhabited for generations	Illiteracy	Agriculture	Land-holding peasant	Married		No religion
28	Inhabited for generations	Illiteracy	Agriculture	Land-holding peasant	Married		No religion
28	Inhabited for generations	Illiteracy	Agriculture	Land-holding peasant	Married		No religion
8	Inhabited for generations	Illiteracy	Agriculture	Land-holding peasant	Married		No religion
27	Inhabited for generations	Illiteracy	Agriculture	Land-holding peasant	Married		No religion
28	Inhabited for generations						
8	Inhabited for generations						
20	Inhabited for generations	Illiteracy	Agriculture	Land-holding peasant	Unmarried		No religion

(*cont.*)

	The no. of house-hold	Jia	House-hold	Relation-ship with the house-hold head	Surname	First name	Sex	The Chinese zodiac	Birthdate	
									Year	Month
Ganquan Xiang 7th Bao				Mother	马	氏	Female	Dragon	1891	8
Ganquan Xiang 7th Bao	823	7	10	The house-hold head	曹	仲库	Male	Tiger	1889	9
Ganquan Xiang 7th Bao				The house-hold head's wife	董	爱哇	Female	Monkey	1896	1
Ganquan Xiang 7th Bao				Son	曹	登明	Male	Horse	1918	11
Ganquan Xiang 7th Bao				The second son	曹	王代	Male	Goat	1931	7
Ganquan Xiang 7th Bao				The third son	曹	登祥	Male	Rooster	1933	12
Ganquan Xiang 7th Bao				The fourth son	曹	跟哇	Male	Rabbit	1939	9
Ganquan Xiang 7th Bao				Daughter	曹	香女	Female	Horse	1942	12
Ganquan Xiang 7th Bao	824	7	10	The house-hold head	曹	登林	Male	Tiger	1914	6
Ganquan Xiang 7th Bao				The house-hold head's wife	贾	瓜儿	Female	Pig	1923	7
Ganquan Xiang 7th Bao				Son	曹	未未	Male	Tiger	1938	7
Ganquan Xiang 7th Bao	825	7	11	The house-hold head	曹	仲义	Male	Dog	1885	10
Ganquan Xiang 7th Bao				The house-hold head's wife	冯	氏	Female	Dog	1886	1

Date	The beginning time of inhabitation in this county	Educational level	Profession sector	Occupation	Marital status	Physical disability	Religion
22	Inhabited for generations	Illiteracy	Agricul-ture	Land-holding peasant	Divorced		No religion
19	Inhabited for generations	Illiteracy	Agricul-ture	Land-holding peasant	Married		No religion
15	Inhabited for generations	Illiteracy	Agricul-ture	Land-holding peasant	Married		No religion
23	Inhabited for generations	Illiteracy	Agricul-ture	Land-holding peasant	Unmarried		No religion
20	Inhabited for generations	Illiteracy	Agricul-ture	Land-holding peasant	Unmarried		No religion
7	Inhabited for generations	Illiteracy	Agricul-ture	Land-holding peasant			
20	Inhabited for generations	Illiteracy					
10	Inhabited for generations						
4	Inhabited for generations	Illiteracy	Agricul-ture	Land-holding peasant	Married		No religion
15	Inhabited for generations	Illiteracy	Agricul-ture	Land-holding peasant	Married		No religion
25	Inhabited for generations	Illiteracy					
2	Inhabited for generations	Illiteracy	Agricul-ture	Land-holding peasant	Married		No religion
14	Inhabited for generations	Illiteracy	Agricul-ture	Land-holding peasant	Married		No religion

(*cont.*)

	The no. of household	Jia	Household	Relationship with the household head	Surname	First name	Sex	The Chinese zodiac	Birthdate Year	Mont
Ganquan Xiang 7th Bao				Son	曹	登平	Male	Rabbit	1915	4
Ganquan Xiang 7th Bao				Son's wife	冯	改哇	Female	Pig	1910	2
Ganquan Xiang 7th Bao				Grandson	冯	家	Male	Horse	1942	7
Ganquan Xiang 7th Bao				The second son	曹	登永	Male	Rooster	1921	6
Ganquan Xiang 7th Bao				The second son's wife	张	仓女	Female	Monkey	1932	12
Ganquan Xiang 7th Bao	826	7	12	The household head	曹	仲奎	Male	Goat	1919	5
Ganquan Xiang 7th Bao				The household head's wife	辛	见娃	Female	Rooster	1921	7
Ganquan Xiang 7th Bao				Mother	辛	氏	Female	Goat	1894	9
Ganquan Xiang 7th Bao				Cohabitant	蕁	顺	Male	Goat	1928	9
Ganquan Xiang 7th Bao	827	7	13	The household head	曹	仲全	Male	Pig	1898	10
Ganquan Xiang 7th Bao				The household head's wife	文	常娃	Female	Horse	1905	11
Ganquan Xiang 7th Bao				Son	曹	进财	Male	Horse	1930	5
Ganquan Xiang 7th Bao				Son's wife	文	连秀	Female	Rabbit	1927	8

Date	The beginning time of inhabitation in this county	Educational level	Profession sector	Occupation	Marital status	Physical disability	Religion
20	Inhabited for generations	Illiteracy	Agriculture	Land-holding peasant	Married		No religion
28	Inhabited for generations	Illiteracy	Agriculture	Land-holding peasant	Married		No religion
18	Inhabited for generations						
26	Inhabited for generations	Illiteracy					
12	Inhabited for generations	Illiteracy					
3	Inhabited for generations	Illiteracy	Agriculture	Land-holding peasant	Married		No religion
5	Inhabited for generations	Illiteracy	Agriculture	Land-holding peasant	Married		No religion
3	Inhabited for generations	Illiteracy	Agriculture	Land-holding peasant	Widowed		No religion
2	Inhabited for generations	Illiteracy	Agriculture	Land-holding peasant	Widowed		No religion
18	Inhabited for generations	Illiteracy	Agriculture	Land-holding peasant	Married		No religion
1	Inhabited for generations	Illiteracy	Agriculture	Land-holding peasant	Married		No religion
20	Inhabited for generations	Illiteracy	Agriculture	Land-holding peasant	Married		No religion
25	Inhabited for generations	Illiteracy	Agriculture	Land-holding peasant	Married		No religion

(*cont.*)

	The no. of household	Jia	Household	Relationship with the household head	Surname	First name	Sex	The Chinese zodiac	Birthdate	
									Year	Month
Ganquan Xiang 7th Bao				Grandson	曹	双喜	Male	Dog	1946	6
Ganquan Xiang 7th Bao				Granddaughter	曹	后女	Female	Pig	1947	5
Ganquan Xiang 7th Bao				Younger brother	曹	仲收	Male	Tiger	1914	2
Ganquan Xiang 7th Bao				Younger brother's wife	张	凤罔	Female	Horse	1918	4
Ganquan Xiang 7th Bao				Nephew	曹	等女	Female	Dog	1946	5
Ganquan Xiang 7th Bao	828	7	14	The household head	曹	船娃	Male	Ox	1913	8
Ganquan Xiang 7th Bao				The household head's wife	张	香香	Female	Dragon	1921	11
Ganquan Xiang 7th Bao				Son	曹	刘家	Male	Ox	1937	10
Ganquan Xiang 7th Bao				The second son	曹	丑娃	Male	Horse	1942	11
Ganquan Xiang 7th Bao	829	7	15	The household head	曹	款娃	Male	Rooster	1921	6
Ganquan Xiang 7th Bao				The household head's wife	张	义娃	Female	Ox	1925	12
Ganquan Xiang 7th Bao				Father	曹	如彦	Male	Dog	1897	6
Ganquan Xiang 7th Bao				Mother	张	连娃	Female	Dog	1898	10

Date	The beginning time of inhabitation in this county	Educational level	Profession sector	Occupation	Marital status	Physical disability	Religion
15	Inhabited for generations						
13	Inhabited for generations						
12	Inhabited for generations	Illiteracy	Agricul-ture	Land-holding peasant	Married		
21	Inhabited for generations	Illiteracy	Agricul-ture	Land-holding peasant	Married		
27	Inhabited for generations						
30	Inhabited for generations	Illiteracy	Agricul-ture	Land-holding peasant	Married		No religion
7	Inhabited for generations	Illiteracy	Agricul-ture	Land-holding peasant	Married		No religion
8	Inhabited for generations	Illiteracy					
13	Inhabited for generations						
5	Inhabited for generations	Illiteracy	Agricul-ture	Land-holding peasant	Married		No religion
1	Inhabited for generations	Illiteracy	Agricul-ture	Land-holding peasant	Married		No religion
21	Inhabited for generations	Illiteracy	Agricul-ture	Land-holding peasant	Married		No religion
13	Inhabited for generations	Illiteracy	Agricul-ture	Land-holding peasant	Married		No religion

(*cont.*)

	The no. of household	Jia	Household	Relationship with the household head	Surname	First name	Sex	The Chinese zodiac	Birthdate	
									Year	Month
Ganquan Xiang 7th Bao				Son	曹	柏哇	Male	Rat	1936	6
Ganquan Xiang 7th Bao				Daughter	曹	女儿	Female	Rooster	1933	4
Ganquan Xiang 7th Bao				The second daughter	曹	四女	Female	Snake	1942	12
Ganquan Xiang 7th Bao				The third daughter	曹	谢家	Female	Rooster	1945	12
Ganquan Xiang 7th Bao	830	8	1	The household head	杨	保春	Male	Horse	1918	11
Ganquan Xiang 7th Bao				The household head's wife	魏	响响	Female	Snake	1917	2
Ganquan Xiang 7th Bao				Daughter	杨	见女	Female	Snake	1941	2
Ganquan Xiang 7th Bao	831	8	2	The household head	杨	法春	Male	Snake	1917	3
Ganquan Xiang 7th Bao				The household head's wife	侯	能能	Female	Horse	1918	5
Ganquan Xiang 7th Bao				Father	杨	纪龙	Male	Snake	1892	6
Ganquan Xiang 7th Bao				Mother	龙	氏	Female	Ox	1888	3
Ganquan Xiang 7th Bao				Son	杨	王旺	Male	Snake	1941	3
Ganquan Xiang 7th Bao				The second son	杨	忠信	Male	Rabbit	1939	8

Date	The beginning time of inhabitation in this county	Educational level	Profession sector	Occupation	Marital status	Physical disability	Religion
8	Inhabited for generations	Illiteracy					
14	Inhabited for generations	Illiteracy					
20	Inhabited for generations						
7	Inhabited for generations						
14	Inhabited for generations	Illiteracy	Agriculture	Land-holding peasant	Married		No religion
15	Inhabited for generations	Illiteracy	Agriculture	Land-holding peasant	Married		No religion
11	Inhabited for generations	Illiteracy					
9	Inhabited for generations	Illiteracy	Agriculture	Land-holding peasant	Married		No religion
16	Inhabited for generations	Illiteracy	Agriculture	Land-holding peasant	Married		No religion
21	Inhabited for generations	Illiteracy	Agriculture	Land-holding peasant	Married		No religion
5	Inhabited for generations	Illiteracy	Agriculture	Land-holding peasant	Married		No religion
6	Inhabited for generations	Illiteracy					
21	Inhabited for generations	Illiteracy					

(*cont.*)

	The no. of household	Jia	Household	Relationship with the household head	Surname	First name	Sex	The Chinese zodiac	Birthdate	
									Year	Mont[h]
Ganquan Xiang 7th Bao				Daughter	杨	肖肖	Female	Dog	1946	12
Ganquan Xiang 7th Bao	832	8	3	The household head	杨	纪昌	Male	Monkey	1896	6
Ganquan Xiang 7th Bao				The household head's wife	雍	歪歪	Female	Dragon	1903	9
Ganquan Xiang 7th Bao				Son	杨	冯家	Male	Goat	1931	7
Ganquan Xiang 7th Bao				Daughter	杨	英英	Female	Pig	1935	5
Ganquan Xiang 7th Bao				The second daughter	杨	三女	Female	Snake	1941	8
Ganquan Xiang 7th Bao				The third daughter	杨	代女	Female	Ox	1938	2
Ganquan Xiang 7th Bao	833	8	4	The household head	陈	建邦	Male	Dragon	1916	3
Ganquan Xiang 7th Bao				The household head's wife	魏	二女	Female	Tiger	1921	11
Ganquan Xiang 7th Bao				Father	陈	明	Male	Ox	1898	5
Ganquan Xiang 7th Bao				Mother	魏	氏	Female	Horse	1893	8
Ganquan Xiang 7th Bao				Younger brother	陈	万顺	Male	Goat	1919	4
Ganquan Xiang 7th Bao				Younger brother's wife	尹	四顺	Female	Ox	1925	10

Date	The beginning time of inhabitation in this county	Educational level	Profession sector	Occupation	Marital status	Physical disability	Religion
1	Inhabited for generations						
3	Inhabited for generations	Illiteracy	Agriculture	Land-holding peasant	Married		No religion
23	Inhabited for generations	Illiteracy	Agriculture	Land-holding peasant	Married		No religion
1	Inhabited for generations	Illiteracy	Agriculture	Land-holding peasant			
1	Inhabited for generations	Illiteracy	Agriculture	Land-holding peasant			
9	Inhabited for generations	Illiteracy					
1	Inhabited for generations	Illiteracy					
11	Inhabited for generations	Illiteracy	Agriculture	Land-holding peasant	Married		No religion
17	Inhabited for generations	Illiteracy	Agriculture	Land-holding peasant	Married		No religion
13	Inhabited for generations	Illiteracy	Agriculture	Land-holding peasant	Married		Buddhism
17	Inhabited for generations	Illiteracy	Agriculture	Land-holding peasant	Married		Buddhism
12	Inhabited for generations	Illiteracy	Agriculture	Land-holding peasant	Married		No religion
24	Inhabited for generations	Illiteracy	Agriculture	Land-holding peasant	Married		No religion

(*cont.*)

	The no. of household	Jia	Household	Relationship with the household head	Surname	First name	Sex	The Chinese zodiac	Birthdate	
									Year	Month
Ganquan Xiang 7th Bao				Nephew	陈	有存	Male	Dog	1946	5
Ganquan Xiang 7th Bao				The third younger brother	陈	足平	Male	Pig	1923	9
Ganquan Xiang 7th Bao				The third Younger brother's wife	王	佛存	Female	Pig	1923	8
Ganquan Xiang 7th Bao				The fourth younger brother	陈	元邦	Male	Dragon	1928	5
Ganquan Xiang 7th Bao	834	8	5	The household head	陈	惠	Male	Dog	1891	12
Ganquan Xiang 7th Bao				The household head's wife	董	志珍	Female	Goat	1894	7
Ganquan Xiang 7th Bao				Younger brother	陈	杰	Male	Goat	1900	5
Ganquan Xiang 7th Bao				Younger brother's wife	尹	虽叩	Female	Rooster	1908	8
Ganquan Xiang 7th Bao				Nephew	陈	仲邦	Male	Pig	1935	11
Ganquan Xiang 7th Bao				The second nephew	陈	人子	Male	Monkey	1944	8
Ganquan Xiang 7th Bao				Niece	陈	郭郭	Female	Rooster	1933	8
Ganquan Xiang 7th Bao	835	8	6	The household head	陈	斌	Male	Goat	1894	5

Date	The beginning time of inhabitation in this county	Educational level	Profession sector	Occupation	Marital status	Physical disability	Religion
1	Inhabited for generations						
21	Inhabited for generations	Illiteracy	Agriculture	Land-holding peasant	Married		No religion
11	Inhabited for generations	Illiteracy	Agriculture	Land-holding peasant	Married		No religion
20	Inhabited for generations	Primary school 2 years	Agriculture	Land-holding peasant	Unmarried		No religion
18	Inhabited for generations	Illiteracy	Agriculture	Land-holding peasant	Married		No religion
9	Inhabited for generations	Illiteracy	Agriculture	Land-holding peasant	Married		No religion
1	Inhabited for generations	Illiteracy	Agriculture	Land-holding peasant	Married		No religion
1	Inhabited for generations	Illiteracy	Agriculture	Land-holding peasant	Married		No religion
11	Inhabited for generations	Primary school 2 years					
29	Inhabited for generations						
31	Inhabited for generations	Illiteracy					
4	Inhabited for generations	Sishu 3 years	Agriculture	Land-holding peasant	Widowed		No religion

(*cont.*)

	The no. of household	Jia	Household	Relationship with the household head	Surname	First name	Sex	The Chinese zodiac	Birthdate	
									Year	Month
Ganquan Xiang 7th Bao				Son	陈	荣华	Male	Monkey	1920	8
Ganquan Xiang 7th Bao				Son's wife	梁	董娃	Female	Rat	1924	9
Ganquan Xiang 7th Bao				Granddaughter	陈	桂英	Female	Monkey	1944	2
Ganquan Xiang 7th Bao				The second granddaughter	陈	秀英	Female	Dog	1947	2
Ganquan Xiang 7th Bao				Daughter	陈	香香	Female	Pig	1935	4
Ganquan Xiang 7th Bao				Older brother	陈	忠	Male	Horse	1893	8
Ganquan Xiang 7th Bao				Nephew's wife	王	氏	Female	Dog	1909	6
Ganquan Xiang 7th Bao				Niece	陈	青英	Female	Tiger	1938	6
Ganquan Xiang 7th Bao	836	8	7	The household head	石	彦昌	Male	Monkey	1921	1
Ganquan Xiang 7th Bao				The household head's wife	苟	品娃	Female	Goat	1919	6
Ganquan Xiang 7th Bao				Father	石	生珍	Male	Rat	1886	2
Ganquan Xiang 7th Bao				Mother	张	氏	Female	Rat	1886	11
Ganquan Xiang 7th Bao				Son	石	田福	Male	Rooster	1945	11

Date	The beginning time of inhabitation in this county	Educational level	Profession sector	Occupation	Marital status	Physical disability	Religion
21	Inhabited for generations	Illiteracy	Agriculture	Land-holding peasant	Married		No religion
8	Inhabited for generations	Illiteracy	Agriculture	Land-holding peasant	Married		No religion
9	Inhabited for generations						
8	Inhabited for generations						
23	Inhabited for generations	Illiteracy					
2	Inhabited for generations	Illiteracy	Agriculture	Land-holding peasant	Widowed		
9	Inhabited for generations	Illiteracy	Agriculture	Land-holding peasant	Widowed		
29	Inhabited for generations	Illiteracy					
18	Inhabited for generations	High school 2 years	Agriculture	Land-holding peasant	Married		No religion
3	Inhabited for generations	Illiteracy	Agriculture	Land-holding peasant	Married		No religion
10	Inhabited for generations	Illiteracy	Agriculture	Land-holding peasant	Married		No religion
9	Inhabited for generations	Illiteracy	Agriculture	Land-holding peasant	Married		No religion
9	Inhabited for generations						

(*cont.*)

	The no. of house-hold	Jia	House-hold	Relation-ship with the house-hold head	Surname	First name	Sex	The Chinese zodiac	Birthdate	
									Year	Mont
Ganquan Xiang 7th Bao				Younger brother	石	彦杰	Male	Pig	1922	12
Ganquan Xiang 7th Bao				Younger brother's wife	安	玄娃	Female	Pig	1922	12
Ganquan Xiang 7th Bao				Nephew	石	赠赠	Female	Snake	1941	4
Ganquan Xiang 7th Bao				The second younger brother	石	彦明	Male	Dragon	1928	10
Ganquan Xiang 7th Bao				Younger brother's wife	魏	三香	Female	Rooster	1921	3
Ganquan Xiang 7th Bao	837	8	8	The house-hold head	石	彦财	Male	Snake	1916	3
Ganquan Xiang 7th Bao				The house-hold head's wife	龙	桂香	Female	Rat	1927	6
Ganquan Xiang 7th Bao				Father	石	生珠	Male	Snake	1892	7
Ganquan Xiang 7th Bao				Mother	刘	氏	Female	Goat	1894	3
Ganquan Xiang 7th Bao				Son	石	双福	Male	Rooster	1945	4
Ganquan Xiang 7th Bao				Daughter	石	双女	Female	Pig	1947	3
Ganquan Xiang 7th Bao				Younger brother	石	彦文	Male	Rat	1925	1
Ganquan Xiang 7th Bao				Younger brother's wife	李	川女	Female	Horse	1929	6

Date	The beginning time of inhabitation in this county	Educational level	Profession sector	Occupation	Marital status	Physical disability	Religion
25	Inhabited for generations	Sishu 2 years	Agriculture	Land-holding peasant	Married		No religion
28	Inhabited for generations	Illiteracy	Agriculture	Land-holding peasant	Married		No religion
12	Inhabited for generations	Illiteracy					
8	Inhabited for generations	Primary school 2 years	Agriculture	Land-holding peasant	Married		No religion
26	Inhabited for generations	Illiteracy	Agriculture	Land-holding peasant	Married		No religion
1	Inhabited for generations	Sishu 2 years	Agriculture	Land-holding peasant	Married		No religion
5	Inhabited for generations	Illiteracy	Agriculture	Land-holding peasant	Married		No religion
18	Inhabited for generations	Illiteracy	Agriculture	Land-holding peasant	Married		No religion
14	Inhabited for generations	Illiteracy	Agriculture	Land-holding peasant	Married		No religion
28	Inhabited for generations						
1	Inhabited for generations						
23	Inhabited for generations	Primary school 1 year	Agriculture	Land-holding peasant	Married		
2	Inhabited for generations	Illiteracy	Agriculture	Land-holding peasant	Married		

(*cont.*)

	The no. of household	Jia	Household	Relationship with the household head	Surname	First name	Sex	The Chinese zodiac	Birthdate Year	Month
Ganquan Xiang 7th Bao	838	8	9	The household head	王	德忠	Male	Goat	1919	11
Ganquan Xiang 7th Bao				The household head's wife	侯	早娃	Female	Rabbit	1928	7
Ganquan Xiang 7th Bao				Mother	张	春娃	Female	Goat	1899	2
Ganquan Xiang 7th Bao				Younger brother	王	安定	Male	Dog	1922	12
Ganquan Xiang 7th Bao				The second younger brother	王	凤娃	Male	Tiger	1926	9
Ganquan Xiang 7th Bao				Younger brother's wife	魏	三女	Female	Dragon	1928	3
Ganquan Xiang 7th Bao				Younger sister	王	如意	Female	Rooster	1933	2
Ganquan Xiang 7th Bao	839	8	10	The household head	王	自平	Male	Tiger	1901	2
Ganquan Xiang 7th Bao				The household head's wife	苟	顺娃	Female	Tiger	1914	11
Ganquan Xiang 7th Bao				Son	王	仁拜	Male	Monkey	1932	6
Ganquan Xiang 7th Bao				Son's wife	田	强娃	Female	Snake	1929	8
Ganquan Xiang 7th Bao				The second son	王	有拜	Male	Snake	1941	3
Ganquan Xiang 7th Bao				Daughter	王	苏信	Female	Rat	1936	2

Date	The beginning time of inhabitation in this county	Educational level	Profession sector	Occupation	Marital status	Physical disability	Religion
16	Inhabited for generations	Illiteracy					
27	Inhabited for generations	Illiteracy					
11	Inhabited for generations	Illiteracy					
7	Inhabited for generations	Illiteracy					
29	Inhabited for generations	Illiteracy					
22	Inhabited for generations						
9	Inhabited for generations						
31	Inhabited for generations	Illiteracy	Agricul-ture	Land-holding peasant	Married		No religion
3	Inhabited for generations	Illiteracy	Agricul-ture	Land-holding peasant	Married		No religion
31	Inhabited for generations	Illiteracy	Agricul-ture	Land-holding peasant	Married		No religion
1	Inhabited for generations	Illiteracy	Agricul-ture	Land-holding peasant	Married		No religion
4	Inhabited for generations	Illiteracy					
7	Inhabited for generations						

(*cont.*)

	The no. of household	Jia	Household	Relationship with the household head	Surname	First name	Sex	The Chinese zodiac	Birthdate	
									Year	Month
Ganquan Xiang 7th Bao				The second daughter	王	迎信	Female	Tiger	1938	2
Ganquan Xiang 7th Bao	840	8	11	The household head	吴	子良	Male	Snake	1917	9
Ganquan Xiang 7th Bao				The household head's wife	张	竣竣	Female	Goat	1931	9
Ganquan Xiang 7th Bao				Older brother	吴	登弟	Male	Monkey	1907	9
Ganquan Xiang 7th Bao				Older brother's wife	王	乖儿	Female	Monkey	1907	8
Ganquan Xiang 7th Bao				Nephew	吴	军军	Male	Goat	1931	7
Ganquan Xiang 7th Bao				Nephew's wife	莹	二娃	Female	Ox	1931	10
Ganquan Xiang 7th Bao				The second nephew	吴	跑娃	Male	Horse	1942	11
Ganquan Xiang 7th Bao				The second older brother	吴	子明	Male	Ox	1913	8
Ganquan Xiang 7th Bao				Son's wife	董	来女	Female	Pig	1910	12
Ganquan Xiang 7th Bao				Niece	吴	产娃	Female	Rat	1936	6
Ganquan Xiang 7th Bao				The second niece	吴	韩娃	Female	Rooster	1945	6
Ganquan Xiang 7th Bao	841	8	12	The household head	王	满春	Male	Rabbit	1915	4

Date	The beginning time of inhabitation in this county	Educational level	Profession sector	Occupation	Marital status	Physical disability	Religion
22	Inhabited for generations						
5	Inhabited for generations	Sishu 2 years	Agriculture	Land-holding peasant	Married		No religion
29	Inhabited for generations	Illiteracy	Agriculture	Land-holding peasant	Married		No religion
1	Inhabited for generations	Illiteracy	Agriculture	Land-holding peasant	Married		No religion
1	Inhabited for generations	Illiteracy	Agriculture	Land-holding peasant	Married		No religion
26	Inhabited for generations	Illiteracy	Agriculture	Land-holding peasant	Married		No religion
1	Inhabited for generations	Illiteracy	Agriculture	Land-holding peasant	Married		No religion
23	Inhabited for generations						
30	Inhabited for generations	Illiteracy	Agriculture	Land-holding peasant	Married		No religion
5	Inhabited for generations	Illiteracy	Agriculture	Land-holding peasant	Married		No religion
1	Inhabited for generations	Illiteracy					
23	Inhabited for generations						
15	Inhabited for generations	Illiteracy	Agriculture	Land-holding peasant	Married		No religion

(*cont.*)

	The no. of household	Jia	Household	Relationship with the household head	Surname	First name	Sex	The Chinese zodiac	Birthdate Year	Month
Ganquan Xiang 7th Bao				The household head's wife	刘	栋娃	Female	Tiger	1914	5
Ganquan Xiang 7th Bao				Father	王	万章	Male	Horse	1893	4
Ganquan Xiang 7th Bao				Mother	赵	氏	Female	Tiger	1889	2
Ganquan Xiang 7th Bao				Son	王	赠仁	Male	Dog	1934	3
Ganquan Xiang 7th Bao				The second son	王	结义	Male	Horse	1942	2
Ganquan Xiang 7th Bao				Daughter	王	混混	Female	Tiger	1938	2
Ganquan Xiang 7th Bao				The second daughter	王	登明	Female	Monkey	1944	3
Ganquan Xiang 7th Bao				The third daughter	王	朵朵	Female	Rooster	1945	11
Ganquan Xiang 7th Bao				Younger brother	王	张遂	Male	Dragon	1928	2
Ganquan Xiang 7th Bao				Younger brother's wife	侯	家娃	Female	Ox	1925	12
Ganquan Xiang 7th Bao				Cohabitant	王	万宝	Male	Monkey	1899	2
Ganquan Xiang 7th Bao				Cohabitant	侯	兔娃	Female	Rabbit	1915	10
Ganquan Xiang 7th Bao				Cohabitant	张	女	Female	Dog	1934	5

Date	The beginning time of inhabitation in this county	Educational level	Profession sector	Occupation	Marital status	Physical disability	Religion
1	Inhabited for generations	Illiteracy	Agriculture	Land-holding peasant	Married		No religion
16	Inhabited for generations	Illiteracy	Agriculture	Land-holding peasant	Married		No religion
18	Inhabited for generations	Illiteracy	Agriculture	Land-holding peasant	Married		No religion
31	Inhabited for generations	Illiteracy					
22	Inhabited for generations						
20	Inhabited for generations	Illiteracy					
15	Inhabited for generations						
26	Inhabited for generations						
29	Inhabited for generations	Illiteracy	Agriculture	Land-holding peasant	Married		
28	Inhabited for generations	Illiteracy	Agriculture	Land-holding peasant	Married		No religion
11	Inhabited for generations	Illiteracy	Agriculture	Land-holding peasant	Married		No religion
1	Inhabited for generations	Illiteracy	Agriculture	Land-holding peasant	Married		No religion
16	Inhabited for generations	Illiteracy	Agriculture	Land-holding peasant			

(*cont.*)

	The no. of household	Jia	Household	Relationship with the household head	Surname	First name	Sex	The Chinese zodiac	Birthdate Year	Month
Ganquan Xiang 7th Bao				Cohabitant	王	草娃	Female	Goat	1943	8
Ganquan Xiang 7th Bao	842	8	12	The household head	王	得财	Male	Rabbit	1903	2
Ganquan Xiang 7th Bao				The household head's wife	冯	东连	Female	Pig	1910	10
Ganquan Xiang 7th Bao				Younger brother	王	得学	Male	Goat	1906	3
Ganquan Xiang 7th Bao				Younger brother's wife	童	補娃	Female	Tiger	1914	17
Ganquan Xiang 7th Bao				Niece	王	董董	Female	Tiger	1939	1
Ganquan Xiang 7th Bao				The second daughter	王	班班	Female	Dog	1946	5
Ganquan Xiang 7th Bao	843	9	1	The household head	王	万珍	Male	Monkey	1895	7
Ganquan Xiang 7th Bao				The household head's wife	王	根哇	Female	Monkey	1895	4
Ganquan Xiang 7th Bao				Son	王	团团	Male	Horse	1918	3
Ganquan Xiang 7th Bao				Son's wife	董	交哇	Female	Monkey	1920	4
Ganquan Xiang 7th Bao				Grandson	王	廷廷	Male	Goat	1943	7
Ganquan Xiang 7th Bao				The second son	王	团俊	Male	Dog	1934	7

Date	The beginning time of inhabitation in this county	Educational level	Profession sector	Occupation	Marital status	Physical disability	Religion
4	Inhabited for generations						
30	Inhabited for generations	Illiteracy	Agriculture	Land-holding peasant	Married		No religion
9	Inhabited for generations	Illiteracy	Agriculture	Land-holding peasant	Married		No religion
27	Inhabited for generations	Illiteracy	Agriculture	Land-holding peasant	Married		No religion
6	Inhabited for generations	Illiteracy	Agriculture	Land-holding peasant	Married		No religion
3	Inhabited for generations	Illiteracy					
19	Inhabited for generations						
3	Inhabited for generations	Illiteracy	Agriculture	Land-holding peasant	Married		No religion
8	Inhabited for generations	Illiteracy	Agriculture	Land-holding peasant	Married		No religion
10	Inhabited for generations	Illiteracy					
1	Inhabited for generations						
2	Inhabited for generations						
11	Inhabited for generations	Illiteracy					

(*cont.*)

	The no. of household	Jia	Household	Relationship with the household head	Surname	First name	Sex	The Chinese zodiac	Birthdate Year	Month
Ganquan Xiang 7th Bao	844	9	2	The household head	王	得义	Male	Goat	1906	6
Ganquan Xiang 7th Bao				The household head's wife	曹	胡家	Female	Goat	1906	10
Ganquan Xiang 7th Bao				Son	王	刘均	Male	Goat	1931	9
Ganquan Xiang 7th Bao				The second son	王	吴均	Male	Pig	1936	1
Ganquan Xiang 7th Bao				Daughter	王	坡哇	Female	Rabbit	1939	11
Ganquan Xiang 7th Bao				The second daughter	王	吉唤	Female	Dog	1946	7
Ganquan Xiang 7th Bao				Younger brother	王	羊娃	Male	Goat	1919	2
Ganquan Xiang 7th Bao				Younger brother's wife	吴	桃女	Female	Rooster	1921	4
Ganquan Xiang 7th Bao	845	9	3	The household head	王	在福	Male	Dragon	1879	1
Ganquan Xiang 7th Bao				Son	王	苟苟	Male	Pig	1932	3
Ganquan Xiang 7th Bao				The second son	王	治才	Male	Monkey	1932	3
Ganquan Xiang 7th Bao				Daughter	王	宋哇子	Female	Rabbit	1927	4
Ganquan Xiang 7th Bao				Older brother	王	成代儿	Male	Ox	1874	4

Date	The beginning time of inhabitation in this county	Educational level	Profession sector	Occupation	Marital status	Physical disability	Religion
21	Inhabited for generations	Illiteracy	Agriculture	Land-holding peasant	Married		No religion
31	Inhabited for generations	Illiteracy	Agriculture	Land-holding peasant	Married		No religion
26	Inhabited for generations	Illiteracy	Agriculture	Land-holding peasant			
23	Inhabited for generations	Illiteracy					
19	Inhabited for generations	Illiteracy					
29	Inhabited for generations						
6	Inhabited for generations	Illiteracy	Agriculture	Land-holding peasant	Married		
1	Inhabited for generations	Illiteracy	Agriculture	Land-holding peasant	Married		
11	Inhabited for generations	Illiteracy	Agriculture	Land-holding peasant	Widowed		No religion
19	Inhabited for generations	Illiteracy	Agriculture	Land-holding peasant	Unmarried		No religion
8	Inhabited for generations	Illiteracy	Agriculture	Land-holding peasant	Unmarried		No religion
1	Inhabited for generations	Illiteracy	Agriculture	Land-holding peasant	Unmarried		
10	Inhabited for generations	Illiteracy	Agriculture	Land-holding peasant	Widowed		

(*cont.*)

	The no. of household	Jia	Household	Relationship with the household head	Surname	First name	Sex	The Chinese zodiac	Birthdate	
									Year	Mont
Ganquan Xiang 7th Bao	846	9	4	The household head	王	万魁	Male	Dragon	1916	3
Ganquan Xiang 7th Bao				The household head's wife	曹	早娃	Female	Rat	1924	4
Ganquan Xiang 7th Bao				Mother	赵	氏	Female	Pig	1887	2
Ganquan Xiang 7th Bao				Daughter	王	董女	Female	Monkey	1944	1
Ganquan Xiang 7th Bao				The second daughter	王	柱哇	Female	Dog	1946	10
Ganquan Xiang 7th Bao				Older brother	王	万昌	Male	Rat	1912	11
Ganquan Xiang 7th Bao				Older brother's wife	陈	曹女	Female	Horse	1918	11
Ganquan Xiang 7th Bao				Niece	王	雪雪	Female	Pig	1935	11
Ganquan Xiang 7th Bao				The second niece	王	习气	Female	Goat	1943	5
Ganquan Xiang 7th Bao				The third niece	王	承承	Female	Rooster	1944	12
Ganquan Xiang 7th Bao				Cohabitant	汪	女子	Female	Dragon	1945	5
Ganquan Xiang 7th Bao	847	9	5	The household head	王	维学	Male	Pig	1898	3
Ganquan Xiang 7th Bao				The household head's wife	赵	氏	Female	Dragon	1891	9

Date	The beginning time of inhabitation in this county	Educational level	Profession sector	Occupation	Marital status	Physical disability	Religion
3	Inhabited for generations	Illiteracy	Agriculture	Land-holding peasant	Married		No religion
1	Inhabited for generations	Illiteracy	Agriculture	Land-holding peasant	Married		No religion
4	Inhabited for generations	Illiteracy	Agriculture	Land-holding peasant	Widowed		No religion
27	Inhabited for generations						
17	Inhabited for generations						
8	Inhabited for generations	Illiteracy	Agriculture	Land-holding peasant	Married		
6	Inhabited for generations	Illiteracy	Agriculture	Land-holding peasant	Married		
22	Inhabited for generations	Illiteracy					
31	Inhabited for generations						
6	Inhabited for generations						
2	Inhabited for generations	Illiteracy	Unemployed		Widowed		No religion
12	Inhabited for generations	Illiteracy	Agriculture	Land-holding peasant	Married		No religion
2	Inhabited for generations	Illiteracy	Agriculture	Land-holding peasant	Married		No religion

(*cont.*)

	The no. of household	Jia	Household	Relationship with the household head	Surname	First name	Sex	The Chinese zodiac	Birthdate	
									Year	Month
Ganquan Xiang 7th Bao				Son	王	万廷	Male	Pig	1910	3
Ganquan Xiang 7th Bao				Son's wife	苏	寸寸	Female	Horse	1918	9
Ganquan Xiang 7th Bao				Grandson	王	里时	Male	Horse	1942	7
Ganquan Xiang 7th Bao				Granddaughter	王	俊女	Female	Rooster	1945	11
Ganquan Xiang 7th Bao				Granddaughter	王	交交	Female	Dragon	1940	11
Ganquan Xiang 7th Bao				The second son	王	午申	Male	Monkey	1920	10
Ganquan Xiang 7th Bao				The second son's wife	张	团哇	Female	Rabbit	1927	3
Ganquan Xiang 7th Bao				The third son	王	万岐	Male	Monkey	1932	3
Ganquan Xiang 7th Bao				The fourth son	王	吴姓	Male	Rooster	1934	1
Ganquan Xiang 7th Bao	848	9	6	The household head	王	维义	Male	Tiger	1901	12
Ganquan Xiang 7th Bao				The household head's wife	曹	党女	Female	Ox	1909	8
Ganquan Xiang 7th Bao				Son	王	万鼎	Male	Ox	1937	6
Ganquan Xiang 7th Bao				The second son	王	仓儿	Male	Dog	1946	2

Date	The beginning time of inhabitation in this county	Educational level	Profession sector	Occupation	Marital status	Physical disability	Religion
20	Inhabited for generations	Illiteracy	Agriculture	Land-holding peasant	Married		No religion
2	Inhabited for generations	Illiteracy	Agriculture	Land-holding peasant	Married		No religion
22	Inhabited for generations						
3	Inhabited for generations						
2	Inhabited for generations						
22	Inhabited for generations	Illiteracy	Agriculture	Land-holding peasant	Married		
9	Inhabited for generations	Illiteracy	Agriculture	Land-holding peasant	Married		
26	Inhabited for generations	Primary school 2 years	Agriculture	Land-holding peasant	Unmarried		
10	Inhabited for generations	Illiteracy					
23	Inhabited for generations	Illiteracy	Agriculture	Land-holding peasant	Married		No religion
2	Inhabited for generations	Illiteracy	Agriculture	Land-holding peasant			No religion
1	Inhabited for generations	Illiteracy					
20	Inhabited for generations						

(*cont.*)

	The no. of household	Jia	Household	Relationship with the household head	Surname	First name	Sex	The Chinese zodiac	Birthdate	
									Year	Month
Ganquan Xiang 7th Bao				Daughter	王	义娃	Female	Monkey	1932	7
Ganquan Xiang 7th Bao				The second daughter	王	月月	Female	Goat	1943	4
Ganquan Xiang 7th Bao	849	9	7	The household head	陈	凯	Male	Snake	1892	3
Ganquan Xiang 7th Bao				Son	陈	昷邦	Male	Dragon	1916	7
Ganquan Xiang 7th Bao				Son's wife	辛	开娃	Female	Horse	1918	9
Ganquan Xiang 7th Bao				The second son	陈	昷昌	Male	Dog	1934	4
Ganquan Xiang 7th Bao				Daughter	陈	七娃	Female	Tiger	1938	8
Ganquan Xiang 7th Bao	850	9	8	The household head	辛	长	Male	Ox	1913	8
Ganquan Xiang 7th Bao				The household head's wife	梁	讲讲	Female	Tiger	1914	1
Ganquan Xiang 7th Bao				Mother	庞	氏	Female	Dog	1885	7
Ganquan Xiang 7th Bao				Son	辛	崴成	Male	Pig	1947	11
Ganquan Xiang 7th Bao				Daughter	辛	乖哇	Female	Snake	1941	9
Ganquan Xiang 7th Bao	851	9	9	The household head	杨	忠	Male	Ox	1900	12

Date	The beginning time of inhabitation in this county	Educational level	Profession sector	Occupation	Marital status	Physical disability	Religion
22	Inhabited for generations	Illiteracy	Agriculture	Land-holding peasant			
20	Inhabited for generations						
11	Inhabited for generations	Illiteracy	Agriculture	Land-holding peasant	Widowed		No religion
1	Inhabited for generations	Illiteracy	Agriculture	Land-holding peasant	Married		No religion
1	Inhabited for generations	Illiteracy	Agriculture	Land-holding peasant	Married		No religion
5	Inhabited for generations	Illiteracy					
1	Inhabited for generations	Illiteracy					
6	Inhabited for generations	Illiteracy	Agriculture	Land-holding peasant	Married		No religion
27	Inhabited for generations	Illiteracy	Agriculture	Land-holding peasant	Married		No religion
27	Inhabited for generations	Illiteracy	Agriculture	Land-holding peasant	Widowed		No religion
16	Inhabited for generations						
24	Inhabited for generations	Illiteracy					
1	Inhabited for generations	Illiteracy	Agriculture	Land-holding peasant	Married		No religion

(*cont.*)

	The no. of household	Jia	Household	Relationship with the household head	Surname	First name	Sex	The Chinese zodiac	Birthdate	
									Year	Month
Ganquan Xiang 7th Bao				The household head's wife	董	柱娃	Female	Rooster	1908	7
Ganquan Xiang 7th Bao				Mother	王	氏	Female	Horse	1889	8
Ganquan Xiang 7th Bao				Son	杨	世春	Male	Horse	1930	12
Ganquan Xiang 7th Bao				Daughter	杨	春女	Female	Rooster	1933	12
Ganquan Xiang 7th Bao				The second daughter	杨	贡哇	Female	Rat	1937	1
Ganquan Xiang 7th Bao	852	9	10	The household head	杨	恕	Male	Goat	1905	1
Ganquan Xiang 7th Bao				The household head's wife	张	苟哇	Female	Dragon	1917	1
Ganquan Xiang 7th Bao				Son	杨	葵卫	Male	Dragon	1940	11
Ganquan Xiang 7th Bao				Daughter	杨	慧女	Female	Horse	1942	1
Ganquan Xiang 7th Bao				The second daughter	杨	全女	Female	Dog	1947	1
Ganquan Xiang 7th Bao	853	9	11	The household head	辛	芳	Male	Rat	1912	8
Ganquan Xiang 7th Bao				The household head's wife	雍	撞娃	Female	Tiger	1926	8
Ganquan Xiang 7th Bao				Mother	张	氏	Female	Rat	1887	3

Date	The beginning time of inhabitation in this county	Educational level	Profession sector	Occupation	Marital status	Physical disability	Religion
30	Inhabited for generations	Illiteracy	Agriculture	Land-holding peasant	Married		No religion
16	Inhabited for generations	Illiteracy	Agriculture	Land-holding peasant	Widowed		No religion
4	Inhabited for generations	Illiteracy	Agriculture	Land-holding peasant			
29	Inhabited for generations	Illiteracy	Agriculture	Land-holding peasant			
3	Inhabited for generations	Illiteracy					
18	Inhabited for generations	Illiteracy	Agriculture	Land-holding peasant	Married		No religion
4	Inhabited for generations	Illiteracy	Agriculture	Land-holding peasant	Married		No religion
16	Inhabited for generations	Illiteracy					
28	Inhabited for generations						
1	Inhabited for generations						
23	Inhabited for generations	Illiteracy	Agriculture	Land-holding peasant	Married		No religion
23	Inhabited for generations	Illiteracy	Agriculture	Land-holding peasant	Married		No religion
1	Inhabited for generations	Illiteracy	Agriculture	Land-holding peasant	Widowed		No religion

(*cont.*)

	The no. of household	Jia	Household	Relationship with the household head	Surname	First name	Sex	The Chinese zodiac	Birthdate	
									Year	Month
Ganquan Xiang 7th Bao	854	9	12	The household head	辛	凯	Male	Rat	1899	12
Ganquan Xiang 7th Bao				The household head's wife	梁	焦娃	Female	Horse	1909	6
Ganquan Xiang 7th Bao				Son	辛	赵姓	Male	Monkey	1932	7
Ganquan Xiang 7th Bao				The second son	辛	王娃	Male	Tiger	1938	10
Ganquan Xiang 7th Bao				The third son	辛	石娃	Male	Horse	1942	9
Ganquan Xiang 7th Bao				The fourth son	辛	吴娃	Male	Dog	1946	2
Ganquan Xiang 7th Bao				Daughter	辛	秀娃	Female	Dog	1934	11
Ganquan Xiang 7th Bao	855	9	13	The household head	陈	瑞	Male	Horse	1894	1
Ganquan Xiang 7th Bao				The household head's wife	何	条於	Female	Goat	1906	10
Ganquan Xiang 7th Bao				Son	陈	根怀	Male	Pig	1923	6
Ganquan Xiang 7th Bao				Son's wife	陈	两娃	Female	Dog	1922	12
Ganquan Xiang 7th Bao				Grandson	陈	孔哇	Male	Monkey	1944	10
Ganquan Xiang 7th Bao				The second son	陈	怀哇	Male	Rooster	1933	5

Date	The beginning time of inhabitation in this county	Educational level	Profession sector	Occupation	Marital status	Physical disability	Religion
2	Inhabited for generations	Illiteracy	Agriculture	Land-holding peasant	Married		No religion
8	Inhabited for generations	Illiteracy	Agriculture	Land-holding peasant	Married		No religion
17	Inhabited for generations	Primary school 1 year	Agriculture	Land-holding peasant	Unmarried		No religion
27	Inhabited for generations	Primary school 2 years					
13	Inhabited for generations						
18	Inhabited for generations						
7	Inhabited for generations	Illiteracy	Agriculture	Land-holding peasant			No religion
5	Inhabited for generations	Illiteracy	Agriculture	Land-holding peasant	Married		No religion
24	Inhabited for generations	Illiteracy	Agriculture	Land-holding peasant	Married		No religion
9	Inhabited for generations	Illiteracy	Agriculture	Land-holding peasant	Married		No religion
3	Inhabited for generations	Illiteracy	Agriculture	Land-holding peasant	Married		No religion
1	Inhabited for generations						
4	Inhabited for generations	Illiteracy					

(*cont.*)

	The no. of household	Jia	Household	Relationship with the household head	Surname	First name	Sex	The Chinese zodiac	Birthdate	
									Year	Montl
Ganquan Xiang 7th Bao				The third son	陈	张姓	Male	Goat	1943	6
Ganquan Xiang 7th Bao				Cohabitant	文	忠	Male	Pig	1898	8
Ganquan Xiang 7th Bao				Cohabitant	根	早	Male	Horse	1930	10
Ganquan Xiang 7th Bao	856	9	14	The household head	陈	文地	Male	Horse	1905	2
Ganquan Xiang 7th Bao				The household head's wife	马	桃桃	Female	Horse	1908	6
Ganquan Xiang 7th Bao				Father	陈	富	Male	Horse	1881	1
Ganquan Xiang 7th Bao				Daughter	陈	春娃	Female	Pig	1935	5
Ganquan Xiang 7th Bao				The second daughter	陈	甘人	Female	Goat	1943	12
Ganquan Xiang 7th Bao				Uncle	陈	俊	Male	Rat	1900	1
Ganquan Xiang 7th Bao				Uncler's wife	尹	吓娃	Female	Dragon	1903	4
Ganquan Xiang 7th Bao				Uncle's son	陈	对娃	Male	Rabbit	1927	10
Ganquan Xiang 7th Bao				Uncle's second son	陈	跑哇	Male	Goat	1932	1
Ganquan Xiang 7th Bao				Uncle's third son	陈	张对	Male	Rabbit	1940	1

Date	The beginning time of inhabitation in this county	Educational level	Profession sector	Occupation	Marital status	Physical disability	Religion
30	Inhabited for generations						
10	Inhabited for generations	Illiteracy	Agriculture	Land-holding peasant	Widowed		
9	Inhabited for generations	Illiteracy	Agriculture	Land-holding peasant	Unmarried		
16	Inhabited for generations	Illiteracy	Agriculture	Land-holding peasant	Married		No religion
20	Inhabited for generations	Illiteracy	Agriculture	Land-holding peasant	Married		No religion
6	Inhabited for generations	Illiteracy	Agriculture	Land-holding peasant	Widowed		No religion
20	Inhabited for generations	Illiteracy					
12	Inhabited for generations						
2	Inhabited for generations	Illiteracy	Agriculture	Land-holding peasant	Married		
5	Inhabited for generations	Illiteracy	Agriculture	Land-holding peasant	Married		
18	Inhabited for generations	Illiteracy	Agriculture	Land-holding peasant	Unmarried		
29	Inhabited for generations	Illiteracy	Agriculture	Land-holding peasant			
10	Inhabited for generations	Illiteracy					

(*cont.*)

	The no. of household	Jia	Household	Relationship with the household head	Surname	First name	Sex	The Chinese zodiac	Birthdate Year	Month
Ganquan Xiang 7th Bao				Uncle's fourth son	陈	腰儿	Male	Goat	1943	8
Ganquan Xiang 7th Bao				Uncle's daughter	陈	腰女	Female	Pig	1947	3
Ganquan Xiang 7th Bao	857	9	15	The household head	陈	芳	Male	Goat	1906	7
Ganquan Xiang 7th Bao				Daughter	陈	长义	Female	Dragon	1940	5
Ganquan Xiang 7th Bao	858	10	1	The household head	辛	刚	Male	Pig	1898	8
Ganquan Xiang 7th Bao				The household head's wife	黄	双哇	Female	Horse	1905	8
Ganquan Xiang 7th Bao				Son	辛	永才	Male	Horse	1918	1
Ganquan Xiang 7th Bao				Son's wife	张	桃姐	Female	Pig	1910	8
Ganquan Xiang 7th Bao				The second son	辛	苟苟	Male	Horse	1942	5
Ganquan Xiang 7th Bao				Daughter	辛	虽女	Female	Rooster	1945	1
Ganquan Xiang 7th Bao	859	10	2	The household head	杨	赵拜	Male	Tiger	1927	2
Ganquan Xiang 7th Bao				The household head's wife	张	对娃	Female	Dragon	1928	11
Ganquan Xiang 7th Bao				Mother	张	那明	Female	Snake	1899	2

Date	The beginning time of inhabitation in this county	Educational level	Profession sector	Occupation	Marital status	Physical disability	Religion
4	Inhabited for generations						
1	Inhabited for generations						
28	Inhabited for generations	Illiteracy	Agriculture	Land-holding peasant	Married		No religion
28	Inhabited for generations	Illiteracy	Agriculture	Land-holding peasant	Married		No religion
9	Inhabited for generations	Illiteracy	Agriculture	Land-holding peasant	Married		No religion
29	Inhabited for generations	Illiteracy	Agriculture	Land-holding peasant	Married		No religion
10	Inhabited for generations	Illiteracy	Agriculture	Land-holding peasant	Married		No religion
6	Inhabited for generations	Illiteracy	Agriculture	Land-holding peasant	Married		No religion
29	Inhabited for generations						
2	Inhabited for generations						
6	Inhabited for generations	Illiteracy	Agriculture	Land-holding peasant	Married		No religion
22	Inhabited for generations	Illiteracy	Agriculture	Land-holding peasant	Married		No religion
8	Inhabited for generations	Illiteracy	Agriculture	Land-holding peasant	Widowed		No religion

(*cont.*)

	The no. of household	Jia	Household	Relationship with the household head	Surname	First name	Sex	The Chinese zodiac	Birthdate Year	Month
Ganquan Xiang 7th Bao				Younger brother	杨	官官	Male	Rooster	1933	6
Ganquan Xiang 7th Bao	860	10	3	The household head	马	生金	Male	Pig	1898	9
Ganquan Xiang 7th Bao				The household head's wife	田	召儿	Female	Tiger	1897	10
Ganquan Xiang 7th Bao				Mother	田	氏	Female	Ox	1946	1
Ganquan Xiang 7th Bao				Son	马	世祥	Male	Tiger	1938	6
Ganquan Xiang 7th Bao	861	10	4	The household head	陈	举	Male	Monkey	1920	11
Ganquan Xiang 7th Bao				The household head's wife	杨	林林	Female	Snake	1917	6
Ganquan Xiang 7th Bao				Cohabitant	竣	山	Male	Ox	1900	8
Ganquan Xiang 7th Bao	862	10	5	The household head	李	春耀	Male	Tiger	1914	5
Ganquan Xiang 7th Bao	863	10	6	The household head	魏	登全	Male	Rat	1924	2
Ganquan Xiang 7th Bao				The household head's wife	赵	菊香	Female	Goat	1919	8
Ganquan Xiang 7th Bao				Father	魏	怀道	Male	Pig	1886	10
Ganquan Xiang 7th Bao				Mother	杨	明镜	Female	Rabbit	1890	8

Date	The beginning time of inhabitation in this county	Educational level	Profession sector	Occupation	Marital status	Physical disability	Religion
5	Inhabited for generations	Illiteracy	Agriculture	Land-holding peasant	Unmarried		No religion
25	Inhabited for generations	Illiteracy	Agriculture	Land-holding peasant	Married		No religion
1	Inhabited for generations	Illiteracy	Agriculture	Land-holding peasant	Married		No religion
26	Inhabited for generations	Illiteracy	Agriculture	Land-holding peasant	Widowed		No religion
15	Inhabited for generations	Illiteracy					
19	Inhabited for generations	Illiteracy	Agriculture	Land-holding peasant	Married		No religion
11	Inhabited for generations	Illiteracy	Agriculture	Land-holding peasant	Married		No religion
13	Inhabited for generations	Illiteracy	Agriculture	Land-holding peasant	Widowed		No religion
9	Inhabited for generations	Primary school graduated			Married		No religion
14	Inhabited for generations	Illiteracy	Agriculture	Land-holding peasant	Married		No religion
1	Inhabited for generations	Illiteracy	Agriculture	Land-holding peasant	Married		No religion
19	Inhabited for generations	Illiteracy	Agriculture	Land-holding peasant	Married		No religion
2	Inhabited for generations	Illiteracy	Agriculture	Land-holding peasant	Married		No religion

(*cont.*)

	The no. of household	Jia	Household	Relationship with the household head	Surname	First name	Sex	The Chinese zodiac	Birthdate	
									Year	Month
Ganquan Xiang 7th Bao				Daughter	魏	柏林	Female	Dog	1946	10
Ganquan Xiang 7th Bao	864	10	7	The household head	杨	纪文	Male	Ox	1900	2
Ganquan Xiang 7th Bao				The household head's wife	柏	金姐	Female	Rabbit	1902	7
Ganquan Xiang 7th Bao				Father	杨	生海	Male	Ox	1877	7
Ganquan Xiang 7th Bao				Son	杨	贵春	Male	Dragon	1940	7
Ganquan Xiang 7th Bao				Daughter	杨	全秀	Female	Rat	1936	3
Ganquan Xiang 7th Bao				The second daughter	杨	佐哇	Female	Rabbit	1939	11
Ganquan Xiang 7th Bao	865	10	8	The household head	赵	元思	Male	Goat	1906	5
Ganquan Xiang 7th Bao				The household head's wife	文	魏魏	Female	Rooster	1908	3
Ganquan Xiang 7th Bao				Mother	苟	氏	Female	Rooster	1907	1
Ganquan Xiang 7th Bao				Son	赵	羊家	Male	Dragon	1928	5
Ganquan Xiang 7th Bao				The second son	赵	二顺	Male	Pig	1935	5
Ganquan Xiang 7th Bao				The third son	赵	三顺	Male	Dragon	1940	10

Date	The beginning time of inhabitation in this county	Educational level	Profession sector	Occupation	Marital status	Physical disability	Religion
14	Inhabited for generations						
25	Inhabited for generations	Illiteracy	Agricul-ture	Land-holding peasant	Married		No religion
30	Inhabited for generations	Illiteracy	Agricul-ture	Land-holding peasant	Married		No religion
1	Inhabited for generations	Illiteracy	Agricul-ture	Land-holding peasant	Widowed		No religion
12	Inhabited for generations	Illiteracy					
8	Inhabited for generations	Illiteracy					
26	Inhabited for generations	Illiteracy					
7	Inhabited for generations	Illiteracy	Agricul-ture	Land-holding peasant	Married		No religion
25	Inhabited for generations	Illiteracy	Agricul-ture	Land-holding peasant	Married		No religion
5	Inhabited for generations	Illiteracy	Agricul-ture	Land-holding peasant	Widowed		No religion
1	Inhabited for generations	Illiteracy	Agricul-ture	Land-holding peasant	Unmarried		No religion
23	Inhabited for generations	Illiteracy	Agricul-ture	Land-holding peasant			
7	Inhabited for generations	Illiteracy					

(*cont.*)

	The no. of household	Jia	Household	Relationship with the household head	Surname	First name	Sex	The Chinese zodiac	Birthdate	
									Year	Mont
Ganquan Xiang 7th Bao				The fourth son	赵	平儿	Male	Rooster	1915	9
Ganquan Xiang 7th Bao				Daughter	赵	香女	Female	Snake	1941	8
Ganquan Xiang 7th Bao				Older brother	元	凯	Male	Rooster	1896	8
Ganquan Xiang 7th Bao	866	10	9	The household head	魏	登像	Male	Ox	1879	2
Ganquan Xiang 7th Bao				The household head's wife	王	氏	Female	Monkey	1884	1
Ganquan Xiang 7th Bao	867	10	10	The household head	胡	生义	Male	Dog	1909	11
Ganquan Xiang 7th Bao				The household head's wife	胡	鼠哇	Female	Snake	1912	8
Ganquan Xiang 7th Bao				Mother	吴	氏	Female	Pig	1870	9
Ganquan Xiang 7th Bao				Son	胡	羊信	Male	Tiger	1938	12
Ganquan Xiang 7th Bao				The second son	胡	蒲信	Male	Monkey	1944	6
Ganquan Xiang 7th Bao				Daughter	胡	凤英	Female	Dog	1946	4
Ganquan Xiang 7th Bao	868	10	11	The household head	魏	怀转	Male	Pig	1899	1
Ganquan Xiang 7th Bao				The household head's wife	王	劝信儿	Female	Rabbit	1907	10

Date	The beginning time of inhabitation in this county	Educational level	Profession sector	Occupation	Marital status	Physical disability	Religion
17	Inhabited for generations						
23	Inhabited for generations	Illiteracy					
2	Inhabited for generations	Illiteracy	Agriculture	Land-holding peasant	Widowed		No religion
5	Inhabited for generations	Illiteracy	Agriculture	Land-holding peasant	Married		No religion
17	Inhabited for generations	Illiteracy	Agriculture	Land-holding peasant	Married		No religion
7	Inhabited for generations	Illiteracy	Agriculture	Land-holding peasant	Married		No religion
1	Inhabited for generations	Illiteracy	Agriculture	Land-holding peasant	Married		No religion
10	Inhabited for generations	Illiteracy	Agriculture	Land-holding peasant	Widowed		No religion
3	Inhabited for generations	Illiteracy					
23	Inhabited for generations						
4	Inhabited for generations						
6	Inhabited for generations	Illiteracy	Agriculture	Land-holding peasant	Married		No religion
1	Inhabited for generations	Illiteracy	Agriculture	Land-holding peasant	Married		No religion

(*cont.*)

	The no. of household	Jia	Household	Relationship with the household head	Surname	First name	Sex	The Chinese zodiac	Birthdate	
									Year	Month
Ganquan Xiang 7th Bao				Mother	雍	氏	Female	Ox	1874	7
Ganquan Xiang 7th Bao				Son	魏	满财	Male	Dog	1946	11
Ganquan Xiang 7th Bao				Daughter	魏	马女	Female	Tiger	1938	3
Ganquan Xiang 7th Bao				The second daughter	魏	崴女	Female	Horse	1942	4
Ganquan Xiang 7th Bao	869	10	12	The household head	赵	元海	Male	Rat	1899	3
Ganquan Xiang 7th Bao				Son	赵	均娃	Male	Dragon	1929	9
Ganquan Xiang 7th Bao				The second son	赵	二均	Male	Goat	1923	9
Ganquan Xiang 7th Bao				The third son	赵	三义	Male	Dog	1934	4
Ganquan Xiang 7th Bao	870	10	13	The household head	赵	元仓	Male	Dog	1910	1
Ganquan Xiang 7th Bao				The household head's wife	陈	春林	Female	Snake	1905	3
Ganquan Xiang 7th Bao				Cohabitant	李	氏	Female	Horse	1881	7
Ganquan Xiang 7th Bao	871	10	14	The household head	张	文选	Male	Dragon	1916	6
Ganquan Xiang 7th Bao				The household head's wife	赵	曾女	Female	Rabbit	1915	8

Date	The beginning time of inhabitation in this county	Educational level	Profession sector	Occupation	Marital status	Physical disability	Religion
1	Inhabited for generations	Illiteracy	Agriculture	Land-holding peasant	Widowed		No religion
14	Inhabited for generations						
19	Inhabited for generations	Illiteracy					
29	Inhabited for generations						
4	Inhabited for generations	Illiteracy	Agriculture		Married		No religion
1	Inhabited for generations	Illiteracy	Agriculture	Land-holding peasant	Married		No religion
2	Inhabited for generations	Illiteracy	Agriculture	Land-holding peasant	Unmarried		No religion
17	Inhabited for generations	Illiteracy	Agriculture	Land-holding peasant			
3	Inhabited for generations	Illiteracy	Agriculture	Land-holding peasant	Married		No religion
12	Inhabited for generations	Illiteracy	Agriculture	Land-holding peasant	Married		No religion
8	Inhabited for generations	Illiteracy	Agriculture	Land-holding peasant	Widowed		No religion
1	Inhabited for generations	Illiteracy	Agriculture	Land-holding peasant	Married		No religion
22	Inhabited for generations	Illiteracy	Agriculture	Land-holding peasant	Married		No religion

(*cont.*)

	The no. of household	Jia	Household	Relationship with the household head	Surname	First name	Sex	The Chinese zodiac	Birthdate	
									Year	Month
Ganquan Xiang 7th Bao				Son	张	郭才	Male	Horse	1942	5
Ganquan Xiang 7th Bao				Daughter	张	兵团	Female	Tiger	1938	6
Ganquan Xiang 7th Bao				The second daughter	张	丙娃	Female	Dog	1946	3
Ganquan Xiang 7th Bao	872	10	15	The household head	赵	士明	Male	Rat	1935	10
Ganquan Xiang 7th Bao				Son	赵	志福	Male	Ox	1925	10
Ganquan Xiang 7th Bao				The second son	赵	志禄	Male	Rabbit	1927	2
Ganquan Xiang 7th Bao	873	11	1	The household head	赵	士连	Male	Rooster	1908	4
Ganquan Xiang 7th Bao				The household head's wife	赵	平儿	Female	Goat	1906	5
Ganquan Xiang 7th Bao				Father	赵	元森	Male	Goat	1882	1
Ganquan Xiang 7th Bao				Son	赵	芝芳	Male	Monkey	1932	7
Ganquan Xiang 7th Bao				Daughter	赵	永水	Female	Ox	1937	5
Ganquan Xiang 7th Bao				The second daughter	赵	抢抢	Female	Goat	1943	4
Ganquan Xiang 7th Bao				The third daughter	赵	赵	Female	Pig	1947	5

Date	The beginning time of inhabitation in this county	Educational level	Profession sector	Occupation	Marital status	Physical disability	Religion
7	Inhabited for generations						
1	Inhabited for generations	Illiteracy					
6							
17	Inhabited for generations	Illiteracy	Agricul-ture	Land-holding peasant	Widowed		No religion
29	Inhabited for generations	Illiteracy	Agricul-ture	Land-holding peasant	Unmarried		No religion
11	Inhabited for generations	Illiteracy	Agricul-ture	Land-holding peasant	Unmarried		No religion
16	Inhabited for generations	Illiteracy	Agricul-ture	Land-holding peasant	Married		No religion
22	Inhabited for generations	Illiteracy	Agricul-ture	Land-holding peasant	Married		No religion
10	Inhabited for generations	Sishu 3 years	Agricul-ture	Land-holding peasant	Widowed		No religion
17	Inhabited for generations	Primary school 5 years	Agricul-ture	Land-holding peasant	Unmarried		No religion
26	Inhabited for generations	Illiteracy	Agricul-ture				
17	Inhabited for generations						
20	Inhabited for generations						

(*cont.*)

	The no. of household	Jia	Household	Relationship with the household head	Surname	First name	Sex	The Chinese zodiac	Birthdate	
									Year	Month
Ganquan Xiang 7th Bao				Nephew	赵	韩韩	Male	Snake	1929	5
Ganquan Xiang 7th Bao				Nephew's wife	冯	香娃	Female	Dragon	1928	8
Ganquan Xiang 7th Bao				Cohabitant	平	平	Male	Snake	1929	2
Ganquan Xiang 7th Bao				Cohabitant	赵	孟欠儿	Female	Dragon	1928	5
Ganquan Xiang 7th Bao	874	11	2	The household head	赵	元堂	Male	Dog	1909	8
Ganquan Xiang 7th Bao				The household head's wife	王	银娃	Female	Goat	1907	1
Ganquan Xiang 7th Bao				Son	赵	步日	Male	Rat	1936	12
Ganquan Xiang 7th Bao				The second son	赵	文财	Male	Dragon	1940	10
Ganquan Xiang 7th Bao				Cohabitant	学	道	Male	Ox	1876	7
Ganquan Xiang 7th Bao				Cohabitant	元	满	Male	Monkey	1895	1
Ganquan Xiang 7th Bao	875	11	3	The household head	杨	正春	Male	Dragon	1916	12
Ganquan Xiang 7th Bao				The household head's wife	王	娘女	Female	Dragon	1916	2
Ganquan Xiang 7th Bao				Mother	张	氏	Female	Monkey	1883	6

Date	The beginning time of inhabitation in this county	Educational level	Profession sector	Occupation	Marital status	Physical disability	Religion
21	Inhabited for generations	Illiteracy	Agriculture	Land-holding peasant	Married		
1	Inhabited for generations	Illiteracy	Agriculture	Land-holding peasant	Married		
15	Inhabited for generations	Illiteracy	Agriculture	Land-holding peasant	Married		
1	Inhabited for generations	Illiteracy	Agriculture	Land-holding peasant	Married		No religion
25	Inhabited for generations	Illiteracy	Agriculture	Land-holding peasant	Married		No religion
11	Inhabited for generations	Illiteracy	Agriculture	Land-holding peasant	Married		No religion
23	Inhabited for generations	Illiteracy					
11	Inhabited for generations	Illiteracy					
2	Inhabited for generations	Illiteracy	Agriculture	Land-holding peasant	Widowed		
22	Inhabited for generations	Illiteracy	Agriculture	Land-holding peasant	Widowed		
9	Inhabited for generations	Sishu 2 years	Agriculture	Land-holding peasant	Married		No religion
29	Inhabited for generations	Illiteracy	Agriculture	Land-holding peasant	Married		No religion
5	Inhabited for generations	Illiteracy	Agriculture	Land-holding peasant	Widowed		No religion

(*cont.*)

	The no. of house-hold	Jia	House-hold	Relation-ship with the house-hold head	Surname	First name	Sex	The Chinese zodiac	Birthdate	
									Year	Month
Ganquan Xiang 7th Bao				Uncle's wife	田	氏	Female	Rooster	1885	1
Ganquan Xiang 7th Bao				Son	杨	三斤	Male	Rabbit	1939	2
Ganquan Xiang 7th Bao				The second son	杨	吴姓	Male	Dog	1946	1
Ganquan Xiang 7th Bao				Daughter	杨	笔墨	Female	Horse	1942	7
Ganquan Xiang 7th Bao				Younger brother	杨	全春	Male	Rat	1924	4
Ganquan Xiang 7th Bao				Younger brother's wife	魏	明儿	Female	Monkey	1920	5
Ganquan Xiang 7th Bao	876	11	4	The house-hold head	赵	元山	Male	Pig	1910	10
Ganquan Xiang 7th Bao				The house-hold head's wife	赵	三明	Female	Goat	1919	4
Ganquan Xiang 7th Bao				Mother	贾	氏	Female	Rooster	1884	7
Ganquan Xiang 7th Bao				Son	赵	郭财	Male	Snake	1942	4
Ganquan Xiang 7th Bao				Daughter	赵	周娃	Female	Ox	1937	5
Ganquan Xiang 7th Bao				The second daughter	赵	信娃	Female	Horse	1932	12
Ganquan Xiang 7th Bao				The second daughter	赵	崴信	Female	Rooster	1945	12

Date	The beginning time of inhabitation in this county	Educational level	Profession sector	Occupation	Marital status	Physical disability	Religion
9	Inhabited for generations	Illiteracy	Agriculture	Land-holding peasant	Widowed		No religion
29	Inhabited for generations	Primary school 2 years					
3	Inhabited for generations						
20	Inhabited for generations						
6	Inhabited for generations	Illiteracy	Agriculture	Land-holding peasant	Married		
7	Inhabited for generations	Illiteracy	Agriculture	Land-holding peasant	Married		
2	Inhabited for generations	Illiteracy	Agriculture	Land-holding peasant	Married		No religion
22	Inhabited for generations	Illiteracy	Agriculture	Land-holding peasant	Married		No religion
6	Inhabited for generations	Illiteracy	Agriculture	Land-holding peasant	Widowed		No religion
21	Inhabited for generations						
21	Inhabited for generations	Illiteracy					
19	Inhabited for generations						
11	Inhabited for generations						

(*cont.*)

	The no. of household	Jia	Household	Relationship with the household head	Surname	First name	Sex	The Chinese zodiac	Birthdate	
									Year	Month
Ganquan Xiang 7th Bao	877	11	5	The household head	赵	学林	Male	Pig	1898	9
Ganquan Xiang 7th Bao				The household head's wife	陈	爱娃	Female	Dog	1897	10
Ganquan Xiang 7th Bao				Mother	苏	氏	Female	Monkey	1871	3
Ganquan Xiang 7th Bao				Son	赵	赠哇	Male	Dog	1922	11
Ganquan Xiang 7th Bao				The second son	赵	俊哇	Male	Goat	1931	12
Ganquan Xiang 7th Bao				Daughter	赵	义香	Female	Snake	1941	4
Ganquan Xiang 7th Bao	878	11	6	The household head	魏	有泰	Male	Snake	1917	9
Ganquan Xiang 7th Bao				The household head's wife	张	挑女	Female	Monkey	1920	5
Ganquan Xiang 7th Bao				Father	魏	崇彦	Male	Dragon	1903	11
Ganquan Xiang 7th Bao				Mother	侯	氏	Female	Dog	1897	9
Ganquan Xiang 7th Bao				Daughter	魏	罗润	Female	Monkey	1942	9
Ganquan Xiang 7th Bao				Younger brother	魏	有山	Male	Dog	1922	8
Ganquan Xiang 7th Bao				Younger brother's wife	束	香	Female	Rooster	1921	11

Date	The beginning time of inhabitation in this county	Educational level	Profession sector	Occupation	Marital status	Physical disability	Religion
16	Inhabited for generations	Illiteracy	Agriculture	Land-holding peasant	Married		No religion
19	Inhabited for generations	Illiteracy	Agriculture	Land-holding peasant	Married		No religion
2	Inhabited for generations	Illiteracy	Agriculture	Land-holding peasant	Widowed		No religion
29	Inhabited for generations	Illiteracy	Agriculture	Land-holding peasant	Unmarried		No religion
23	Inhabited for generations	Illiteracy	Agriculture	Land-holding peasant	Unmarried		No religion
11	Inhabited for generations	Illiteracy					
18	Inhabited for generations	Sishu 5 years	Agriculture	Land-holding peasant	Married		No religion
26	Inhabited for generations	Illiteracy	Agriculture	Land-holding peasant	Married		No religion
4	Inhabited for generations	Sishu 2 years	Agriculture	Land-holding peasant	Married		No religion
22	Inhabited for generations	Illiteracy	Agriculture	Land-holding peasant	Married		No religion
19	Inhabited for generations						
24	Inhabited for generations	Sishu 2 years	Agriculture	Land-holding peasant	Married		
9	Inhabited for generations	Illiteracy	Agriculture	Land-holding peasant	Married		

(*cont.*)

	The no. of house-hold	Jia	House-hold	Relation-ship with the house-hold head	Surname	First name	Sex	The Chinese zodiac	Birthdate	
									Year	Mont
Ganquan Xiang 7th Bao				Niece	魏	照娃	Female	Monkey	1944	6
Ganquan Xiang 7th Bao				The second daughter	魏	照润	Female	Pig	1947	2
Ganquan Xiang 7th Bao				The second younger brother	魏	有安	Male	Tiger	1926	11
Ganquan Xiang 7th Bao				Younger brother's wife	高	婴儿	Female	Dragon	1928	10
Ganquan Xiang 7th Bao				Younger sister	魏	桃义	Female	Goat	1931	11
Ganquan Xiang 7th Bao				Cohabitant	老	李	Male	Pig	1886	2
Ganquan Xiang 7th Bao	879	11	7	The house-hold head	魏	登海	Male	Dog	1897	4
Ganquan Xiang 7th Bao				The house-hold head's wife	张	满娃	Female	Pig	1910	7
Ganquan Xiang 7th Bao				Son	魏	寅十	Male	Rooster	1934	2
Ganquan Xiang 7th Bao				Son's wife	刘	安明	Female	Rat	1924	7
Ganquan Xiang 7th Bao	880	11	8	The house-hold head	魏	荣章	Male	Rooster	1896	10
Ganquan Xiang 7th Bao				The house-hold head's wife	童	金瑕儿	Female	Rooster	1896	9

Date	The beginning time of inhabitation in this county	Educational level	Profession sector	Occupation	Marital status	Physical disability	Religion
5	Inhabited for generations						
1	Inhabited for generations						
22	Inhabited for generations	Junior middle school 1 year	Agriculture	Land-holding peasant	Married		
27	Inhabited for generations	Illiteracy	Agriculture	Land-holding peasant	Married		No religion
7	Inhabited for generations	Illiteracy	Agriculture	Land-holding peasant	Unmarried		No religion
30	Inhabited for generations	Illiteracy	Agriculture	Land-holding peasant	Widowed		No religion
25	Inhabited for generations	Illiteracy	Agriculture	Land-holding peasant	Married		No religion
7	Inhabited for generations	Illiteracy	Agriculture	Land-holding peasant	Married		No religion
12	Inhabited for generations	Illiteracy	Agriculture	Land-holding peasant	Married		No religion
1	Inhabited for generations	Illiteracy	Agriculture	Land-holding peasant	Married		No religion
25	Inhabited for generations	Illiteracy	Agriculture	Land-holding peasant	Married		No religion
8	Inhabited for generations	Illiteracy	Agriculture	Land-holding peasant	Married		No religion

(*cont.*)

	The no. of household	Jia	Household	Relationship with the household head	Surname	First name	Sex	The Chinese zodiac	Birthdate Year	Month
Ganquan Xiang 7th Bao				Son	魏	船船	Male	Snake	1929	12
Ganquan Xiang 7th Bao				The second son	魏	团团	Male	Monkey	1932	12
Ganquan Xiang 7th Bao				Daughter	魏	尖女	Female	Rabbit	1939	10
Ganquan Xiang 7th Bao	881	11	9	The household head	魏	崇厚	Male	Snake	1918	1
Ganquan Xiang 7th Bao				The household head's wife	张	兵兵	Female	Dragon	1916	9
Ganquan Xiang 7th Bao				Son	魏	有林	Male	Dog	1934	11
Ganquan Xiang 7th Bao				Son's wife	董	爱英	Female	Dragon	1928	9
Ganquan Xiang 7th Bao				Older brother	魏	拜拜	Male	Rabbit	1916	1
Ganquan Xiang 7th Bao				Older brother's wife	曹	香儿	Female	Ox	1925	9
Ganquan Xiang 7th Bao				Niece	魏	郝女	Female	Dragon	1940	3
Ganquan Xiang 7th Bao	882	11	10	The household head	魏	登满	Male	Horse	1893	12
Ganquan Xiang 7th Bao				The household head's wife	张	玄玄	Female	Tiger	1901	11
Ganquan Xiang 7th Bao				Son	魏	崇堂	Male	Rooster	1934	1

Date	The beginning time of inhabitation in this county	Educational level	Profession sector	Occupation	Marital status	Physical disability	Religion
22	Inhabited for generations	Illiteracy	Agriculture	Land-holding peasant	Unmarried		No religion
9	Inhabited for generations	Illiteracy	Agriculture	Land-holding peasant	Unmarried		No religion
4	Inhabited for generations	Illiteracy					
29	Inhabited for generations	Illiteracy	Agriculture	Land-holding peasant	Married		No religion
1	Inhabited for generations	Illiteracy	Agriculture	Land-holding peasant	Married		No religion
29	Inhabited for generations	Illiteracy	Agriculture	Land-holding peasant	Married		No religion
1	Inhabited for generations	Illiteracy	Agriculture	Land-holding peasant	Married		No religion
29	Inhabited for generations	Illiteracy	Agriculture	Land-holding peasant	Married		No religion
1	Inhabited for generations	Illiteracy	Agriculture	Land-holding peasant	Married		No religion
16	Inhabited for generations	Illiteracy					
8	Inhabited for generations	Illiteracy	Agriculture	Land-holding peasant	Married		No religion
2	Inhabited for generations	Illiteracy	Agriculture	Land-holding peasant	Married		No religion
7	Inhabited for generations	Primary school 2 years	Agriculture	Land-holding peasant	Unmarried		No religion

(*cont.*)

	The no. of house-hold	Jia	House-hold	Relation-ship with the house-hold head	Surname	First name	Sex	The Chinese zodiac	Birthdate	
									Year	Month
Ganquan Xiang 7th Bao				Daughter	魏	中女	Female	Tiger	1938	3
Ganquan Xiang 7th Bao				The second daughter	魏	见见	Female	Tiger	1940	9
Ganquan Xiang 7th Bao	883	11	11	The house-hold head	司	彦文	Male	Dragon	1916	7
Ganquan Xiang 7th Bao				The house-hold head's wife	魏	丙娃	Female	Tiger	1914	3
Ganquan Xiang 7th Bao				Son	司	韩成	Male	Monkey	1944	2
Ganquan Xiang 7th Bao				The second son	司	二成	Male	Pig	1945	4
Ganquan Xiang 7th Bao				Daughter	司	朵娃	Female	Pig	1913	12
Ganquan Xiang 7th Bao				The second daughter	司	杨娃	Female	Snake	1941	9
Ganquan Xiang 7th Bao	884	11	12	The house-hold head	魏	登千	Male	Rat	1887	4
Ganquan Xiang 7th Bao				The house-hold head's wife	荀	氏	Female	Pig	1910	3
Ganquan Xiang 7th Bao				Son	魏	崇礼	Male	Monkey	1907	12
Ganquan Xiang 7th Bao				Son's wife	王	条明	Female	Horse	1905	10
Ganquan Xiang 7th Bao				Grandson	魏	邦成	Male	Rooster	1945	3

Date	The beginning time of inhabitation in this county	Educational level	Profession sector	Occupation	Marital status	Physical disability	Religion
9	Inhabited for generations	Illiteracy					
10	Inhabited for generations	Illiteracy					
2	Inhabited for generations	Illiteracy	Agriculture	Land-holding peasant	Married		No religion
25	Inhabited for generations	Illiteracy	Agriculture	Land-holding peasant	Married		No religion
20	Inhabited for generations						
1	Inhabited for generations						
5	Inhabited for generations	Illiteracy					
13	Inhabited for generations	Illiteracy					
20	Inhabited for generations	Illiteracy	Agriculture	Land-holding peasant	Married		No religion
2	Inhabited for generations	Illiteracy	Agriculture	Land-holding peasant	Married		No religion
9	Inhabited for generations	Illiteracy	Agriculture	Land-holding peasant	Married		No religion
5	Inhabited for generations	Illiteracy	Agriculture	Land-holding peasant	Married		No religion
16	Inhabited for generations						

(*cont.*)

	The no. of household	Jia	Household	Relationship with the household head	Surname	First name	Sex	The Chinese zodiac	Birthdate	
									Year	Mont[h]
Ganquan Xiang 7th Bao				Granddaughter	魏	召娃	Female	Rooster	1933	11
Ganquan Xiang 7th Bao				The second daughter	魏	崴女	Female	Rabbit	1939	8
Ganquan Xiang 7th Bao	885	11	13	The household head	魏	崇荣	Male	Dog	1910	1
Ganquan Xiang 7th Bao				The household head's wife	张	雪梅	Female	Horse	1905	11
Ganquan Xiang 7th Bao				Son	魏	郭成	Male	Ox	1937	5
Ganquan Xiang 7th Bao				Daughter	魏	丑娃	Female	Dog	1934	3
Ganquan Xiang 7th Bao				The second daughter	魏	崴丑	Female	Snake	1941	3
Ganquan Xiang 7th Bao	886	11	14	The household head	魏	崇富	Male	Horse	1918	8
Ganquan Xiang 7th Bao				The household head's wife	杨	郝娃	Female	Rooster	1921	5
Ganquan Xiang 7th Bao				Mother	刘	翠儿	Female	Goat	1894	5
Ganquan Xiang 7th Bao				Daughter	魏	桂英	Female	Goat	1943	4
Ganquan Xiang 7th Bao	887	11	15	The household head	魏	普哇	Male	Dragon	1928	2
Ganquan Xiang 7th Bao				The household head's wife	侯	绑娃	Female	Ox	1925	6

Date	The beginning time of inhabitation in this county	Educational level	Profession sector	Occupation	Marital status	Physical disability	Religion
31	Inhabited for generations	Illiteracy	Agriculture	Land-holding peasant			
12	Inhabited for generations	Illiteracy					
1	Inhabited for generations	Illiteracy	Agriculture	Land-holding peasant	Married		No religion
15	Inhabited for generations	Illiteracy	Agriculture	Land-holding peasant	Married		No religion
6	Inhabited for generations	Illiteracy					
31	Inhabited for generations	Illiteracy					
30	Inhabited for generations	Illiteracy					
29	Inhabited for generations	Illiteracy	Agriculture	Land-holding peasant	Married		No religion
14	Inhabited for generations	Illiteracy	Agriculture	Land-holding peasant	Married		No religion
1	Inhabited for generations	Illiteracy	Agriculture	Land-holding peasant	Widowed		No religion
30	Inhabited for generations						
23	Inhabited for generations	Illiteracy	Agriculture	Land-holding peasant	Married		No religion
30	Inhabited for generations	Illiteracy	Agriculture	Land-holding peasant	Married		No religion

(*cont.*)

	The no. of household	Jia	Household	Relationship with the household head	Surname	First name	Sex	The Chinese zodiac	Birthdate	
									Year	Month
Ganquan Xiang 7th Bao				Mother	刘	拜娃	Female	Dog	1897	5
Ganquan Xiang 7th Bao				Grandmother	王	氏	Female	Pig	1886	3
Ganquan Xiang 7th Bao				Older brother	魏	崇秀	Male	Rat	1924	7
Ganquan Xiang 7th Bao				Older brother's wife	杨	赵娃	Female	Rat	1924	3
Ganquan Xiang 7th Bao				Nephew	魏	牛福	Male	Rooster	1945	11
Ganquan Xiang 7th Bao				Uncle	魏	吴儿	Male	Ox	1900	4
Ganquan Xiang 7th Bao	888	12	1	The household head	魏	仲章	Male	Monkey	1907	3
Ganquan Xiang 7th Bao				The household head's wife	张	闹娃	Female	Rat	1912	4
Ganquan Xiang 7th Bao				Father	魏	老管	Male	Horse	1943	2
Ganquan Xiang 7th Bao				Son	魏	招财	Male	Dog	1923	9
Ganquan Xiang 7th Bao				Son's wife	王	交哇	Female	Dragon	1928	2
Ganquan Xiang 7th Bao				The second son	魏	八斤	Male	Ox	1937	10
Ganquan Xiang 7th Bao				Daughter	魏	肖肖	Female	Ox	1937	8

Date	The beginning time of inhabitation in this county	Educational level	Profession sector	Occupation	Marital status	Physical disability	Religion
1	Inhabited for generations	Illiteracy	Agriculture	Land-holding peasant	Widowed		No religion
1	Inhabited for generations	Illiteracy	Agriculture	Land-holding peasant	Widowed		No religion
7	Inhabited for generations	Illiteracy	Agriculture	Land-holding peasant	Married		No religion
1	Inhabited for generations	Illiteracy	Agriculture	Land-holding peasant	Married		No religion
10	Inhabited for generations						
1	Inhabited for generations	Illiteracy	Agriculture	Land-holding peasant	Widowed		
3	Inhabited for generations	Illiteracy	Agriculture	Land-holding peasant	Married		No religion
10	Inhabited for generations	Illiteracy	Agriculture	Land-holding peasant	Married		No religion
25	Inhabited for generations	Illiteracy	Agriculture	Land-holding peasant	Widowed		No religion
10	Inhabited for generations	Illiteracy	Agriculture	Land-holding peasant	Married		No religion
1	Inhabited for generations	Illiteracy	Agriculture	Land-holding peasant	Married		No religion
21	Inhabited for generations	Illiteracy					
1	Inhabited for generations	Illiteracy					

(*cont.*)

	The no. of household	Jia	Household	Relationship with the household head	Surname	First name	Sex	The Chinese zodiac	Birthdate	
									Year	Month
Ganquan Xiang 7th Bao				The second daughter	魏	赦赦	Female	Monkey	1945	2
Ganquan Xiang 7th Bao	889	12	2	The household head	魏	有泰	Male	Rooster	1908	2
Ganquan Xiang 7th Bao				The household head's wife	刘	香儿	Female	Ox	1913	6
Ganquan Xiang 7th Bao				Son	魏	皆用	Male	Dog	1934	9
Ganquan Xiang 7th Bao				The second son	魏	长王儿	Male	Ox	1943	4
Ganquan Xiang 7th Bao				Daughter	魏	如意	Female	Dragon	1940	6
Ganquan Xiang 7th Bao				The second daughter	魏	足足	Female	Horse	1942	10
Ganquan Xiang 7th Bao				The third daughter	魏	玉梅	Female	Dog	1947	1
Ganquan Xiang 7th Bao	890	12	3	The household head	魏	良良	Male	Dog	1922	4
Ganquan Xiang 7th Bao				The household head's wife	冯	桃敏哇	Female	Rat	1924	8
Ganquan Xiang 7th Bao				Mother	王	氏	Female	Goat	1882	1
Ganquan Xiang 7th Bao	891	12	4	The household head	魏	石哇	Male	Rat	1912	1
Ganquan Xiang 7th Bao				The household head's wife	魏	绑哇	Female	Tiger	1926	9

Date	The beginning time of inhabitation in this county	Educational level	Profession sector	Occupation	Marital status	Physical disability	Religion
7	Inhabited for generations						
25	Inhabited for generations	Illiteracy	Agriculture	Land-holding peasant	Married		No religion
10	Inhabited for generations	Illiteracy	Agriculture	Land-holding peasant	Married		No religion
17	Inhabited for generations	Illiteracy	Agriculture	Land-holding peasant			No religion
10	Inhabited for generations						
17	Inhabited for generations	Illiteracy					
13	Inhabited for generations						
1	Inhabited for generations						
1	Inhabited for generations	Illiteracy	Agriculture	Land-holding peasant	Married		No religion
7	Inhabited for generations	Illiteracy	Agriculture	Land-holding peasant	Married		No religion
17	Inhabited for generations						
3	Inhabited for generations	Illiteracy	Agriculture	Land-holding peasant	Married		No religion
18	Inhabited for generations	Illiteracy	Agriculture	Land-holding peasant	Married		No religion

(*cont.*)

	The no. of household	Jia	Household	Relationship with the household head	Surname	First name	Sex	The Chinese zodiac	Birthdate	
									Year	Month
Ganquan Xiang 7th Bao				Father	魏	登顺	Male	Rooster	1884	7
Ganquan Xiang 7th Bao				Mother	张	像哇	Female	Dragon	1903	3
Ganquan Xiang 7th Bao				Younger sister	魏	代成	Female	Rooster	1933	9
Ganquan Xiang 7th Bao				The second younger sister	魏	谢成	Female	Rat	1946	10
Ganquan Xiang 7th Bao	892	12	5	The household head	何	丙哇	Female	Dragon	1903	8
Ganquan Xiang 7th Bao				The household head's husband	魏	登泰	Male	Rooster	1908	5
Ganquan Xiang 7th Bao				Son	魏	掌世	Male	Dog	1946	5
Ganquan Xiang 7th Bao				Daughter	魏	创哇	Female	Rooster	1933	5
Ganquan Xiang 7th Bao				The second daughter	魏	茂味	Female	Snake	1941	11
Ganquan Xiang 7th Bao	893	12	6	The household head	魏	崇万	Male	Snake	1904	10
Ganquan Xiang 7th Bao				The household head's wife	荀	条姐	Female	Rooster	1908	1
Ganquan Xiang 7th Bao				Mother	党	氏	Female	Dragon	1879	6
Ganquan Xiang 7th Bao				Son	魏	有荣	Male			

Date	The beginning time of inhabitation in this county	Educational level	Profession sector	Occupation	Marital status	Physical disability	Religion
24	Inhabited for generations	Illiteracy	Agriculture	Land-holding peasant	Married		No religion
24	Inhabited for generations	Illiteracy	Agriculture	Land-holding peasant	Married		No religion
7	Inhabited for generations	Illiteracy	Agriculture	Land-holding peasant			No religion
9	Inhabited for generations						
7	Inhabited for generations	Illiteracy	Agriculture	Land-holding peasant	Married		No religion
11	Inhabited for generations	Illiteracy	Agriculture	Land-holding peasant	Married		No religion
24	Inhabited for generations						
30	Inhabited for generations	Illiteracy	Agriculture	Land-holding peasant			No religion
2	Inhabited for generations						
2							
29							
12							

(*cont.*)

	The no. of household	Jia	Household	Relationship with the household head	Surname	First name	Sex	The Chinese zodiac	Birthdate	
									Year	Month
Ganquan Xiang 7th Bao				Son's wife	童	乖哇	Female			
Ganquan Xiang 7th Bao				The second son	魏	有聿	Male			
Ganquan Xiang 7th Bao				The third son	魏	安成	Male			
Ganquan Xiang 7th Bao	894	12	7	The household head	魏	登雪	Male	Horse	1894	1
Ganquan Xiang 7th Bao				The household head's wife	王	腰连	Female	Snake	1892	3
Ganquan Xiang 7th Bao				Son	魏	崇孝	Male	Snake	1918	1
Ganquan Xiang 7th Bao				Son's wife	杨	称娃	Female	Pig	1910	6
Ganquan Xiang 7th Bao				Grandson	魏	崴家	Male	Horse	1942	5
Ganquan Xiang 7th Bao				The second grandson	魏	成俊	Male	Dog	1946	6
Ganquan Xiang 7th Bao	895	12	8	The household head	魏	永常	Male	Pig	1909	2
Ganquan Xiang 7th Bao				The household head's wife	赵	当当	Female	Snake	1917	9
Ganquan Xiang 7th Bao				Daughter	魏	桂女	Female	Pig	1935	3
Ganquan Xiang 7th Bao				Uncle	魏	玉林	Male	Goat	1906	7

Date	The beginning time of inhabitation in this county	Educational level	Profession sector	Occupation	Marital status	Physical disability	Religion
17	Inhabited for generations	Illiteracy	Agriculture	Land-holding peasant	Married		No religion
14	Inhabited for generations	Illiteracy	Agriculture	Land-holding peasant	Married		No religion
16	Inhabited for generations	Illiteracy	Agriculture	Land-holding peasant	Married		No religion
28	Inhabited for generations	Illiteracy	Agriculture	Land-holding peasant	Married		No religion
25	Inhabited for generations						
4	Inhabited for generations						
24	Inhabited for generations	Illiteracy	Agriculture	Land-holding peasant	Married		No religion
30	Inhabited for generations	Illiteracy	Agriculture	Land-holding peasant	Married		No religion
9							
4	Inhabited for generations	Illiteracy	Agriculture	Land-holding peasant	Widowed		No religion

(*cont.*)

	The no. of household	Jia	Household	Relationship with the household head	Surname	First name	Sex	The Chinese zodiac	Birthdate Year	Month
Ganquan Xiang 7th Bao	896	12	9	The household head	魏	崇贵	Male	Monkey	1907	5
Ganquan Xiang 7th Bao				The household head's wife	苟	礼娃	Female	Ox	1913	9
Ganquan Xiang 7th Bao				Father	魏	登玉	Male	Snake	1892	3
Ganquan Xiang 7th Bao				Son	魏	升升	Male	Goat	1931	9
Ganquan Xiang 7th Bao				The second son	魏	丑丑	Male	Rooster	1933	10
Ganquan Xiang 7th Bao				The third son	魏	永永	Male	Pig	1935	10
Ganquan Xiang 7th Bao				The fourth son	魏	天运	Male	Dog	1945	1
Ganquan Xiang 7th Bao				Daughter	魏	凤英	Female	Ox	1938	1
Ganquan Xiang 7th Bao	897	12	10	The household head	魏	登明	Male	Ox	1888	8
Ganquan Xiang 7th Bao				The household head's wife	王	氏	Female	Rat	1887	3
Ganquan Xiang 7th Bao				Son	魏	崇平	Male	Goat	1919	11
Ganquan Xiang 7th Bao				Son's wife	杨	桃女	Female	Ox	1925	8
Ganquan Xiang 7th Bao				Grandson	魏	條成	Male	Dog	1946	10

Date	The beginning time of inhabitation in this county	Educational level	Profession sector	Occupation	Marital status	Physical disability	Religion
1	Inhabited for generations	Illiteracy	Agriculture	Land-holding peasant	Married		No religion
1	Inhabited for generations	Illiteracy	Agriculture	Land-holding peasant	Married		No religion
22	Inhabited for generations	Sishu 2 years	Agriculture	Land-holding peasant	Widowed		No religion
15	Inhabited for generations	Illiteracy	Agriculture	Land-holding peasant	Unmarried		No religion
23	Inhabited for generations	Primary school 2 years	Agriculture	Land-holding peasant			
28	Inhabited for generations	Illiteracy					
4	Inhabited for generations						
14	Inhabited for generations	Illiteracy					
1	Inhabited for generations	Illiteracy	Agriculture	Land-holding peasant	Married		No religion
14	Inhabited for generations	Illiteracy	Agriculture	Land-holding peasant	Married		No religion
27	Inhabited for generations	Illiteracy	Agriculture	Land-holding peasant	Married		No religion
25	Inhabited for generations	Illiteracy	Agriculture	Land-holding peasant	Married		No religion
30	Inhabited for generations						

(*cont.*)

	The no. of household	Jia	Household	Relationship with the household head	Surname	First name	Sex	The Chinese zodiac	Birthdate Year	Mont
Ganquan Xiang 7th Bao	898	12	11	The household head	魏	登虎	Male	Tiger	1926	5
Ganquan Xiang 7th Bao				The household head's wife	杨	能能	Female	Horse	1930	3
Ganquan Xiang 7th Bao				Father	魏	怀芳	Male	Dragon	1879	2
Ganquan Xiang 7th Bao				Mother	成	氏	Female	Monkey	1883	9
Ganquan Xiang 7th Bao	899	12	12	The household head	魏	崇昌	Male	Tiger	1914	8
Ganquan Xiang 7th Bao				The household head's wife	庞	旦娃	Female	Ox	1914	1
Ganquan Xiang 7th Bao				Mother	庞	氏	Female	Rooster	1882	12
Ganquan Xiang 7th Bao				Son	魏	郭时	Male	Rooster	1945	4
Ganquan Xiang 7th Bao				Daughter	魏	家女	Female	Dragon	1940	3
Ganquan Xiang 7th Bao				The second daughter	魏	马女	Female	Snake	1941	4
Ganquan Xiang 7th Bao	900	12	13	The household head	魏	有德	Male	Dog	1910	1
Ganquan Xiang 7th Bao				The household head's wife	张	秦娃	Female	Ox	1913	5
Ganquan Xiang 7th Bao				Son	魏	宽用	Male	Horse	1942	11

Date	The beginning time of inhabitation in this county	Educational level	Profession sector	Occupation	Marital status	Physical disability	Religion
12	Inhabited for generations	Illiteracy	Agriculture	Land-holding peasant	Married		No religion
3	Inhabited for generations	Illiteracy	Agriculture	Land-holding peasant	Married		No religion
5	Inhabited for generations	Illiteracy	Agriculture	Land-holding peasant	Married		No religion
22	Inhabited for generations	Illiteracy	Agriculture	Land-holding peasant	Married		No religion
25	Inhabited for generations	Illiteracy	Agriculture	Land-holding peasant	Married		No religion
3	Inhabited for generations	Illiteracy	Agriculture	Land-holding peasant	Married		No religion
18	Inhabited for generations	Illiteracy	Agriculture	Land-holding peasant	Widowed		No religion
14	Inhabited for generations						
19	Inhabited for generations	Illiteracy					
18	Inhabited for generations	Illiteracy					
23	Inhabited for generations	Illiteracy	Agriculture	Land-holding peasant	Married		No religion
4	Inhabited for generations	Illiteracy	Agriculture	Land-holding peasant	Married		No religion
16	Inhabited for generations						

(*cont.*)

	The no. of household	Jia	Household	Relationship with the household head	Surname	First name	Sex	The Chinese zodiac	Birthdate	
									Year	Month
Ganquan Xiang 7th Bao				Daughter	魏	利娃	Female	Goat	1933	9
Ganquan Xiang 7th Bao				The second daughter	魏	斌团	Female	Pig	1935	7
Ganquan Xiang 7th Bao				The third daughter	魏	三团	Female	Tiger	1941	6
Ganquan Xiang 7th Bao	901	13	1	The household head	魏	有全	Male	Goat	1919	2
Ganquan Xiang 7th Bao				The household head's wife	王	成女	Female	Tiger	1914	3
Ganquan Xiang 7th Bao				Daughter	魏	福连	Female	Ox	1937	9
Ganquan Xiang 7th Bao				The second daughter	魏	进用	Female	Rooster	1945	10
Ganquan Xiang 7th Bao				Younger brother	魏	才才	Male	Tiger	1938	4
Ganquan Xiang 7th Bao				Younger brother's wife	董	对娃	Female	Rabbit	1927	3
Ganquan Xiang 7th Bao	902	13	2	The household head	魏	崇芳	Male	Ox	1901	2
Ganquan Xiang 7th Bao				The household head's wife	苏	落	Female	Monkey	1907	5
Ganquan Xiang 7th Bao				Son	魏	谢應	Male	Rabbit	1927	8
Ganquan Xiang 7th Bao				Son's wife	刘	款娃	Female	Ox	1925	2

Date	The beginning time of inhabitation in this county	Educational level	Profession sector	Occupation	Marital status	Physical disability	Religion
8	Inhabited for generations	Illiteracy					
4	Inhabited for generations	Illiteracy					
2	Inhabited for generations	Illiteracy					
6	Inhabited for generations	Illiteracy	Agricul-ture	Land-holding peasant	Married		No religion
31	Inhabited for generations	Illiteracy	Agricul-ture	Land-holding peasant	Married		No religion
26	Inhabited for generations	Illiteracy					
13	Inhabited for generations						
3	Inhabited for generations	Illiteracy			Married		No religion
6	Inhabited for generations	Illiteracy	Agricul-ture	Land-holding peasant	Married		No religion
1	Inhabited for generations	Illiteracy	Agricul-ture	Land-holding peasant	Married		No religion
31	Inhabited for generations	Illiteracy	Agricul-ture	Land-holding peasant	Married		No religion
29	Inhabited for generations	Illiteracy	Agricul-ture	Land-holding peasant	Married		No religion
28	Inhabited for generations	Illiteracy	Agricul-ture	Land-holding peasant	Married		No religion

(*cont.*)

	The no. of household	Jia	Household	Relationship with the household head	Surname	First name	Sex	The Chinese zodiac	Birthdate Year	Month
Ganquan Xiang 7th Bao				The second son	魏	有才	Male	Goat	1931	9
Ganquan Xiang 7th Bao	903	13	3	The household head	魏	有义	Male	Tiger	1926	11
Ganquan Xiang 7th Bao				The household head's wife	王	？娃	Female	Tiger	1926	9
Ganquan Xiang 7th Bao				Father	魏	国正	Male	Horse	1893	4
Ganquan Xiang 7th Bao				Mother	赵	根儿	Female	Snake	1904	6
Ganquan Xiang 7th Bao				Son	魏	德用	Male	Rooster	1945	3
Ganquan Xiang 7th Bao				Younger sister	魏	顺女	Female	Dog	1934	3
Ganquan Xiang 7th Bao				The second younger sister	魏	全女	Female	Dog	1946	7
Ganquan Xiang 7th Bao	904	13	4	The household head	魏	转哇	Male	Dog	1909	3
Ganquan Xiang 7th Bao				The household head's wife	张	响娃	Female	Dragon	1903	6
Ganquan Xiang 7th Bao				Daughter	魏	邦娃子	Female	Rabbit	1939	4
Ganquan Xiang 7th Bao				Cohabitant	反	魏	Male	Rabbit	1915	3
Ganquan Xiang 7th Bao	905	13	5	The household head	魏	崇义	Male	Pig	1910	4

Date	The beginning time of inhabitation in this county	Educational level	Profession sector	Occupation	Marital status	Physical disability	Religion
24	Inhabited for generations	Illiteracy	Agriculture	Land-holding peasant	Unmarried		No religion
11	Inhabited for generations	Illiteracy	Agriculture	Land-holding peasant	Married		No religion
8	Inhabited for generations	Illiteracy	Agriculture	Land-holding peasant	Married		No religion
1	Inhabited for generations	Illiteracy	Agriculture	Land-holding peasant	Married		No religion
4	Inhabited for generations	Illiteracy	Agriculture	Land-holding peasant	Married		No religion
14	Inhabited for generations						
20	Inhabited for generations	Illiteracy	Agriculture	Land-holding peasant	Unmarried		No religion
3	Inhabited for generations						
15	Inhabited for generations	Illiteracy	Agriculture	Land-holding peasant	Married		No religion
22	Inhabited for generations	Illiteracy	Agriculture	Land-holding peasant	Married		No religion
25	Inhabited for generations	Illiteracy					
22	Inhabited for generations	Illiteracy	Agriculture	Land-holding peasant	Widowed		No religion
9	Inhabited for generations	Illiteracy	Agriculture	Land-holding peasant	Married		No religion

(*cont.*)

	The no. of household	Jia	Household	Relationship with the household head	Surname	First name	Sex	The Chinese zodiac	Birthdate Year	Month
Ganquan Xiang 7th Bao				The household head's wife	尹	柱娃	Female	Pig	1910	4
Ganquan Xiang 7th Bao				Son	魏	良良	Male	Pig	1935	7
Ganquan Xiang 7th Bao				The second son	魏	二顺	Male	Rabbit	1939	11
Ganquan Xiang 7th Bao				The third son	魏	三顺	Male	Snake	1941	3
Ganquan Xiang 7th Bao				The fourth son	魏	董才	Male	Dog	1946	1
Ganquan Xiang 7th Bao				Daughter	魏	佐成	Female	Rooster	1945	7
Ganquan Xiang 7th Bao				The second daughter	魏	世世	Female	Pig	1935	4
Ganquan Xiang 7th Bao	906	13	6	The household head	魏	登辉	Male	Horse	1893	10
Ganquan Xiang 7th Bao				The household head's wife	柏	歪儿	Female	Rat	1909	1
Ganquan Xiang 7th Bao				Son	魏	张才	Male	Rat	1936	11
Ganquan Xiang 7th Bao				The second son	魏	跟财	Male	Pig	1947	2
Ganquan Xiang 7th Bao				Daughter	魏	款女	Female	Monkey	1932	4
Ganquan Xiang 7th Bao				The second daughter	魏	味味子	Female	Rooster	1945	10

Date	The beginning time of inhabitation in this county	Educational level	Profession sector	Occupation	Marital status	Physical disability	Religion
8	Inhabited for generations	Illiteracy	Agriculture	Land-holding peasant	Married		No religion
18	Inhabited for generations	Illiteracy	Agriculture	Land-holding peasant	Unmarried		No religion
25	Inhabited for generations	Primary school 2 years					
1	Inhabited for generations						
15	Inhabited for generations						
12	Inhabited for generations						
26	Inhabited for generations	Illiteracy					
31	Inhabited for generations	Illiteracy	Agriculture	Land-holding peasant	Married		No religion
15	Inhabited for generations	Illiteracy	Agriculture	Land-holding peasant	Married		No religion
7	Inhabited for generations	Illiteracy	Agriculture	Land-holding peasant			No religion
13	Inhabited for generations						
15	Inhabited for generations	Illiteracy	Agriculture	Land-holding peasant			No religion
20	Inhabited for generations						

(*cont.*)

	The no. of household	Jia	Household	Relationship with the household head	Surname	First name	Sex	The Chinese zodiac	Birthdate	
									Year	Month
Ganquan Xiang 7th Bao				Cohabitant	魏	登隆	Male	Pig	1933	2
Ganquan Xiang 7th Bao	907	13	7	The household head	田	生荣	Male	Pig	1910	4
Ganquan Xiang 7th Bao				The household head's wife	刘	俊娃	Female	Rabbit	1915	4
Ganquan Xiang 7th Bao				Daughter	田	银银	Female	Snake	1941	9
Ganquan Xiang 7th Bao				The second daughter	田	音音	Female	Rooster	1945	4
Ganquan Xiang 7th Bao	908	13	8	The household head	刘	生	Male	Rooster	1914	2
Ganquan Xiang 7th Bao				The household head's wife	张	雪儿	Female	Snake	1904	11
Ganquan Xiang 7th Bao				Son	刘	顺娃	Male	Rooster	1933	9
Ganquan Xiang 7th Bao				The second son	刘	绑绑	Male	Dog	1946	8
Ganquan Xiang 7th Bao				Concubine	郭	赠娃	Female	Pig	1910	1
Ganquan Xiang 7th Bao	909	13	9	The household head	刘	千	Male	Rabbit	1903	2
Ganquan Xiang 7th Bao				The household head's wife	耕	清	Female	Rat	1913	1
Ganquan Xiang 7th Bao				Son	刘	长张	Male	Rat	1936	9

Date	The beginning time of inhabitation in this county	Educational level	Profession sector	Occupation	Marital status	Physical disability	Religion
8	Inhabited for generations	Illiteracy	Agriculture	Land-holding peasant			No religion
2	Inhabited for generations	Illiteracy	Agriculture	Land-holding peasant	Married		No religion
1	Inhabited for generations	Illiteracy	Agriculture	Land-holding peasant	Married		No religion
4	Inhabited for generations						
7	Inhabited for generations						
5	Inhabited for generations	Illiteracy	Agriculture	Land-holding peasant	Married		No religion
29	Inhabited for generations	Illiteracy	Agriculture	Land-holding peasant	Married		No religion
23	Inhabited for generations	Illiteracy	Agriculture	Land-holding peasant			
31	Inhabited for generations						
26	Inhabited for generations	Illiteracy	Agriculture	Land-holding peasant			
2	Inhabited for generations	Illiteracy	Agriculture	Land-holding peasant	Married		No religion
11	Inhabited for generations	Illiteracy	Agriculture	Land-holding peasant	Married		No religion
22	Inhabited for generations	Primary school 2 years					

(*cont.*)

	The no. of household	Jia	Household	Relationship with the household head	Surname	First name	Sex	The Chinese zodiac	Birthdate Year	Month
Ganquan Xiang 7th Bao	910	13	10	The household head	刘	华	Male	Tiger	1926	2
Ganquan Xiang 7th Bao				Mother	李	氏	Female	Goat	1882	7
Ganquan Xiang 7th Bao				Older brother	刘	生全	Male	Pig	1910	6
Ganquan Xiang 7th Bao				Older brother's wife	苟	冬冬	Female	Goat	1906	3
Ganquan Xiang 7th Bao				Nephew	刘	均均	Male	Snake	1942	1
Ganquan Xiang 7th Bao				The second son	刘	二均娃	Male	Monkey	1944	1
Ganquan Xiang 7th Bao				Niece	刘	检娃	Female	Goat	1943	4
Ganquan Xiang 7th Bao	911	13	11	The household head	刘	應	Male	Dragon	1903	7
Ganquan Xiang 7th Bao				The household head's wife	苟	能能	Female	Dragon	1903	11
Ganquan Xiang 7th Bao				Son	刘	焦哇	Male	Dog	1934	10
Ganquan Xiang 7th Bao				Son's wife	张	战娃	Female	Horse	1928	3
Ganquan Xiang 7th Bao				The second son	刘	冯长	Male	Horse	1942	3
Ganquan Xiang 7th Bao	912	13	12	The household head	刘	思选	Male	Horse	1905	9

Date	The beginning time of inhabitation in this county	Educational level	Profession sector	Occupation	Marital status	Physical disability	Religion
26	Inhabited for generations	Illiteracy	Agriculture	Land-holding peasant	Married		No religion
12	Inhabited for generations	Illiteracy	Agriculture	Land-holding peasant	Widowed		No religion
29	Inhabited for generations	Illiteracy	Agriculture	Land-holding peasant	Married		No religion
30	Inhabited for generations	Illiteracy	Agriculture	Land-holding peasant	Married		No religion
25	Inhabited for generations						
25	Inhabited for generations						
7	Inhabited for generations						
5	Inhabited for generations	Illiteracy	Agriculture	Land-holding peasant	Married		No religion
2	Inhabited for generations	Illiteracy	Agriculture	Land-holding peasant	Married		No religion
9	Inhabited for generations	Illiteracy	Agriculture	Land-holding peasant	Married		No religion
23	Inhabited for generations	Illiteracy	Agriculture	Land-holding peasant	Married		No religion
21	Inhabited for generations						
3	Inhabited for generations	Illiteracy	Agriculture	Land-holding peasant	Married		No religion

(*cont.*)

	The no. of household	Jia	Household	Relationship with the household head	Surname	First name	Sex	The Chinese zodiac	Birthdate	
									Year	Month
Ganquan Xiang 7th Bao				The household head's wife	李	令娃	Female	Goat	1906	3
Ganquan Xiang 7th Bao				Son	刘	郭生儿	Male	Dragon	1928	2
Ganquan Xiang 7th Bao				The second son	刘	羊羊	Male	Horse	1930	7
Ganquan Xiang 7th Bao				Son's wife	刘	居娃	Female	Tiger	1926	5
Ganquan Xiang 7th Bao				Granddaughter	刘	交用	Female	Rooster	1945	8
Ganquan Xiang 7th Bao				The third son	刘	羊生娃	Male	Rooster	1933	5
Ganquan Xiang 7th Bao				Daughter	刘	棋哇	Female	Snake	1941	11
Ganquan Xiang 7th Bao	913	13	13	The household head	刘	万义	Male	Monkey	1920	8
Ganquan Xiang 7th Bao				The household head's wife	曹	均均	Female	Snake	1929	4
Ganquan Xiang 7th Bao				Older brother	刘	顺哇	Male	Horse	1918	3
Ganquan Xiang 7th Bao				Son's wife	王	登娃	Female	Rooster	1921	3
Ganquan Xiang 7th Bao				Younger sister	刘	三长	Female	Rooster	1933	2
Ganquan Xiang 7th Bao	914	14	1	The household head	魏	崇满	Male	Horse	1893	5

Date	The beginning time of inhabitation in this county	Educational level	Profession sector	Occupation	Marital status	Physical disability	Religion
15	Inhabited for generations	Illiteracy	Agriculture	Land-holding peasant	Married		No religion
25	Inhabited for generations	Illiteracy	Agriculture	Land-holding peasant			No religion
7	Inhabited for generations	Illiteracy	Agriculture	Land-holding peasant	Married		
25	Inhabited for generations	Illiteracy	Agriculture	Land-holding peasant	Married		
17	Inhabited for generations						
12	Inhabited for generations	Illiteracy	Agriculture	Land-holding peasant			
20	Inhabited for generations						
21	Inhabited for generations	Illiteracy	Agriculture	Land-holding peasant	Married		No religion
5	Inhabited for generations	Illiteracy	Agriculture	Land-holding peasant	Married		No religion
17	Inhabited for generations	Illiteracy	Agriculture	Land-holding peasant	Married		No religion
13	Inhabited for generations	Illiteracy	Agriculture	Land-holding peasant	Married		No religion
30	Inhabited for generations	Illiteracy	Agriculture	Land-holding peasant			No religion
2	Inhabited for generations	Illiteracy	Agriculture	Land-holding peasant	Married	No	No religion

(*cont.*)

	The no. of household	Jia	Household	Relationship with the household head	Surname	First name	Sex	The Chinese zodiac	Birthdate	
									Year	Month
Ganquan Xiang 7th Bao				The household head's wife	赵	里儿	Female	Snake	1894	1
Ganquan Xiang 7th Bao				Son	魏	大招娃	Male	Rabbit	1915	12
Ganquan Xiang 7th Bao				Son's wife	刘	洛娃	Female	Goat	1919	11
Ganquan Xiang 7th Bao				The second son	魏	崴招娃	Male	Pig	1923	12
Ganquan Xiang 7th Bao				The second son's wife	张	开娃	Female	Dog	1921	3
Ganquan Xiang 7th Bao				Grandson	魏	中信	Male	Goat	1943	6
Ganquan Xiang 7th Bao				The third son	魏	三招娃	Male	Goat	1931	6
Ganquan Xiang 7th Bao				Daughter	魏	招姐娃	Female	Rat	1936	3
Ganquan Xiang 7th Bao	915	14	2	The household head	魏	顺娃	Male	Dog	1922	4
Ganquan Xiang 7th Bao				Father	魏	怌清	Male	Dog	1885	3
Ganquan Xiang 7th Bao				Mother	张	氏	Female	Monkey	1883	8
Ganquan Xiang 7th Bao	916	14	3	The household head	赵	董义	Male	Pig	1924	1
Ganquan Xiang 7th Bao				The household head's wife	董	跑娃	Female	Pig	1923	6

Date	The beginning time of inhabitation in this county	Educational level	Profession sector	Occupation	Marital status	Physical disability	Religion
17	Inhabited for generations	Illiteracy	Agriculture	Land-holding peasant	Married	No	No religion
16	Inhabited for generations	Illiteracy	Agriculture	Land-holding peasant	Married	No	No religion
1	Inhabited for generations	Illiteracy	Agriculture	Land-holding peasant	Married	No	No religion
30	Inhabited for generations	Illiteracy	Agriculture	Land-holding peasant	Married	No	No religion
29	Inhabited for generations	Illiteracy	Agriculture	Land-holding peasant	Married	No	No religion
17	Inhabited for generations						
28	Inhabited for generations	Illiteracy	Agriculture	Land-holding peasant	Married	No	No religion
30	Inhabited for generations	Illiteracy	Agriculture	Land-holding peasant	Unmarried	No	No religion
28	Inhabited for generations	Illiteracy	Agriculture	Land-holding peasant	Unmarried		No religion
13	Inhabited for generations	Illiteracy	Agriculture	Land-holding peasant	Married		No religion
11	Inhabited for generations	Illiteracy	Agriculture	Land-holding peasant	Married		No religion
7	Inhabited for generations	Illiteracy	Agriculture	Land-holding peasant	Married		No religion
30	Inhabited for generations	Illiteracy	Agriculture	Land-holding peasant	Married		No religion

(*cont.*)

	The no. of house-hold	Jia	House-hold	Relation-ship with the house-hold head	Surname	First name	Sex	The Chinese zodiac	Birthdate	
									Year	Month
Ganquan Xiang 7th Bao				Father	赵	学会	Male	Tiger	1889	12
Ganquan Xiang 7th Bao				Mother	王	条儿	Female	Rooster	1896	2
Ganquan Xiang 7th Bao				Older brother	赵	任义	Male	Goat	1919	12
Ganquan Xiang 7th Bao				Older brother's wife	杨	党党	Female	Horse	1918	4
Ganquan Xiang 7th Bao				Younger brother	赵	洛义	Male	Monkey	1932	9
Ganquan Xiang 7th Bao				Younger sister	赵	女	Female	Horse	1930	3
Ganquan Xiang 7th Bao	917	14	4	The house-hold head	魏	登礼	Male	Tiger	1914	5
Ganquan Xiang 7th Bao				The house-hold head's wife	董	祥姐子	Female	Pig	1912	1
Ganquan Xiang 7th Bao				Son	魏	得用	Male	Monkey	1944	3
Ganquan Xiang 7th Bao				The second son	魏	丢信	Male	Dog	1946	8
Ganquan Xiang 7th Bao				Uncle	魏	来禾	Male	Snake	1870	5
Ganquan Xiang 7th Bao				Daughter	魏	节用	Female	Snake	1941	8
Ganquan Xiang 7th Bao	918	14	5	The house-hold head	魏	崇美	Male	Horse	1905	10

Date	The beginning time of inhabitation in this county	Educational level	Profession sector	Occupation	Marital status	Physical disability	Religion
23	Inhabited for generations	Illiteracy	Agriculture	Land-holding peasant	Married		No religion
6	Inhabited for generations	Illiteracy	Agriculture	Land-holding peasant	Married		No religion
1	Inhabited for generations	Illiteracy	Agriculture	Land-holding peasant	Married		No religion
1	Inhabited for generations	Illiteracy	Agriculture	Land-holding peasant	Married		No religion
2	Inhabited for generations	Illiteracy	Agriculture	Land-holding peasant	Unmarried		No religion
25	Inhabited for generations	Illiteracy	Agriculture	Land-holding peasant	Unmarried		No religion
8	Inhabited for generations	Illiteracy	Agriculture	Land-holding peasant	Married		No religion
30	Inhabited for generations	Illiteracy	Agriculture	Land-holding peasant	Married		No religion
1	Inhabited for generations	Illiteracy	Agriculture	Land-holding peasant			
11	Inhabited for generations	Illiteracy	Agriculture	Land-holding peasant			
6	Inhabited for generations	Illiteracy	Agriculture				
24	Inhabited for generations	Illiteracy	Agriculture	Land-holding peasant	Widowed	No	No religion
5	Inhabited for generations	Illiteracy	Agriculture	Land-holding peasant	Married	No	No religion

(*cont.*)

	The no. of household	Jia	Household	Relationship with the household head	Surname	First name	Sex	The Chinese zodiac	Birthdate	
									Year	Mont[h]
Ganquan Xiang 7th Bao				The household head's wife	王	清哇	Female	Pig	1910	3
Ganquan Xiang 7th Bao				Mother	辛	氏	Female	Rabbit	1876	6
Ganquan Xiang 7th Bao				Son	魏	林姓	Male	Horse	1930	5
Ganquan Xiang 7th Bao				Daughter	魏	撒女	Female	Rooster	1933	2
Ganquan Xiang 7th Bao				The second daughter	魏	耕撒	Female	Rabbit	1939	6
Ganquan Xiang 7th Bao				The third daughter	魏	三女	Female	Goat		
Ganquan Xiang 7th Bao				The fourth daughter	魏	谢女	Female	Dog	1946	6
Ganquan Xiang 7th Bao				Younger brother	魏	款哇	Male	Rooster	1921	10
Ganquan Xiang 7th Bao	919	14	6	The household head	魏	进才	Male	Horse	1918	3
Ganquan Xiang 7th Bao				The household head's wife	魏	长哇	Female	Horse	1918	9
Ganquan Xiang 7th Bao				Mother	李	苏儿	Female	Snake	1892	3
Ganquan Xiang 7th Bao	920	14	7	The household head	安	全福	Male	Rabbit	1915	3
Ganquan Xiang 7th Bao				The household head's wife	田	哇	Female	Goat	1919	7

Date	The beginning time of inhabitation in this county	Educational level	Profession sector	Occupation	Marital status	Physical disability	Religion
4	Inhabited for generations	Illiteracy	Agriculture	Land-holding peasant	Married	No	No religion
1	Inhabited for generations	Illiteracy	Agriculture	Land-holding peasant	Widowed	No	No religion
1	Inhabited for generations	Illiteracy	Agriculture	Land-holding peasant	Unmarried	No	No religion
6	Inhabited for generations	Illiteracy	Agriculture	Land-holding peasant		No	No religion
18	Inhabited for generations	Illiteracy				No	No religion
	Inhabited for generations	Illiteracy				No	No religion
4	Inhabited for generations					No	No religion
31	Inhabited for generations					No	No religion
14	Inhabited for generations	Illiteracy	Agriculture	Land-holding peasant	Married	No	No religion
6	Inhabited for generations	Illiteracy	Agriculture	Land-holding peasant	Married	No	No religion
9		Illiteracy	Agriculture	Land-holding peasant	Married	No	No religion
17	Inhabited for generations	Illiteracy	Agriculture	Land-holding peasant	Married	No	No religion
2	Inhabited for generations	Illiteracy	Agriculture	Land-holding peasant	Married	No	No religion

(*cont.*)

	The no. of household	Jia	Household	Relationship with the household head	Surname	First name	Sex	The Chinese zodiac	Birthdate	
									Year	Month
Ganquan Xiang 7th Bao				Father	安	崔明	Male	Monkey	1882	4
Ganquan Xiang 7th Bao				Mother	安	氏	Female	Tiger	1908	
Ganquan Xiang 7th Bao				Son	安	拜拜	Male	Horse	1942	9
Ganquan Xiang 7th Bao				The second son	安	柏儿	Male	Monkey	1944	10
Ganquan Xiang 7th Bao				Daughter	安	菜哇	Female	Rat	1936	7
Ganquan Xiang 7th Bao				The second daughter	安	肖女	Female	Tiger	1938	7
Ganquan Xiang 7th Bao				The third daughter	安	三女	Female	Goat	1943	8
Ganquan Xiang 7th Bao				The fourth daughter	安	柏女	Female	Dog	1946	6
Ganquan Xiang 7th Bao	921	14	8	The household head	王	润润子	Male	Horse	1905	3
Ganquan Xiang 7th Bao				The household head's wife	瞿	双足儿	Female	Horse	1905	9
Ganquan Xiang 7th Bao				Son	王	崴团	Male	Ox	1937	7
Ganquan Xiang 7th Bao	922	14	9	The household head	韩	建成	Male	Horse	1919	1
Ganquan Xiang 7th Bao				The household head's wife	杨	必祥	Female	Rat	1912	3

Date	The beginning time of inhabitation in this county	Educational level	Profession sector	Occupation	Marital status	Physical disability	Religion
2	Inhabited for generations	Illiteracy	Agriculture	Land-holding peasant	Married	No	No religion
	Inhabited for generations	Illiteracy	Agriculture	Land-holding peasant	Married	No	No religion
15	Inhabited for generations	Illiteracy	Agriculture			No	No religion
30	Inhabited for generations	Illiteracy	Agriculture			No	No religion
9	Inhabited for generations	Illiteracy	Agriculture			No	No religion
25	Inhabited for generations	Illiteracy	Agriculture			No	No religion
2	Inhabited for generations	Illiteracy	Agriculture			No	No religion
23	Inhabited for generations	Illiteracy	Agriculture			No	No religion
2	Inhabited for generations	Illiteracy	Agriculture	Land-holding peasant	Married	No	No religion
22	Inhabited for generations	Illiteracy	Agriculture	Land-holding peasant	Married	No	No religion
23	Inhabited for generations	Illiteracy				No	
8	Inhabited for generations	Illiteracy	Agriculture	Land-holding peasant	Married		No religion
3	Inhabited for generations	Illiteracy	Agriculture	Land-holding peasant	Married		No religion

(*cont.*)

	The no. of household	Jia	Household	Relationship with the household head	Surname	First name	Sex	The Chinese zodiac	Birthdate	
									Year	Month
Ganquan Xiang 7th Bao				Son	韩	长哇	Male	Ox	1937	9
Ganquan Xiang 7th Bao				The second son	韩	魏哇	Male	Horse	1943	2
Ganquan Xiang 7th Bao				Daughter	韩	崴女	Female	Dog	1946	9
Ganquan Xiang 7th Bao	923	14	10	The household head	杨	举	Male		1894	10
Ganquan Xiang 7th Bao				The household head's wife	李	考儿	Female	Tiger	1901	4
Ganquan Xiang 7th Bao				Mother	马	氏	Female	Ox		
Ganquan Xiang 7th Bao				Son	杨	明明	Male	Dog	1909	4
Ganquan Xiang 7th Bao				Son's wife	杨	尹娃	Female	Rat	1924	6
Ganquan Xiang 7th Bao				The second son	杨	秃女子	Male	Dragon	1941	1
Ganquan Xiang 7th Bao	924	14	11	The household head	杨	均	Male	Horse	1905	6
Ganquan Xiang 7th Bao				The household head's wife	柏	田田子	Female	Rabbit	1915	
Ganquan Xiang 7th Bao				Son	杨	丑哇	Male	Rat	1936	2
Ganquan Xiang 7th Bao				The second son	杨	无进	Male	Horse	1942	10

Date	The beginning time of inhabitation in this county	Educational level	Profession sector	Occupation	Marital status	Physical disability	Religion
27	Inhabited for generations	Illiteracy					
4	Inhabited for generations						
10	Inhabited for generations						
26	Inhabited for generations	Illiteracy	Agriculture	Land-holding peasant	Married	No	No religion
20	Inhabited for generations	Illiteracy	Agriculture	Land-holding peasant	Married	No	No religion
	Inhabited for generations	Illiteracy	Agriculture	Land-holding peasant	Married	No	No religion
1	Inhabited for generations	Illiteracy	Agriculture	Land-holding peasant	Married	No	No religion
12	Inhabited for generations	Illiteracy	Agriculture	Land-holding peasant	Married	No	No religion
21	Inhabited for generations						
26	Inhabited for generations	Illiteracy	Agriculture	Land-holding peasant	Married	No	No religion
	Inhabited for generations	Illiteracy	Agriculture	Land-holding peasant	Married	No	
25	Inhabited for generations	Illiteracy	Agriculture			No	
14	Inhabited for generations					No	

(*cont.*)

	The no. of house-hold	Jia	House-hold	Relation-ship with the house-hold head	Surname	First name	Sex	The Chinese zodiac	Birthdate Year	Mont[h]
Ganquan Xiang 7th Bao	925	14	12	The house-hold head	杨	魁	Male	Ox	1900	5
Ganquan Xiang 7th Bao				The house-hold head's wife	崔	娘连儿	Female	Rooster	1908	4
Ganquan Xiang 7th Bao				Son	杨	吉娃	Male	Goat	1931	2
Ganquan Xiang 7th Bao				Son's wife	杨	进女	Female	Monkey	1932	5
Ganquan Xiang 7th Bao				Younger brother	杨	雨儿	Male	Rooster	1908	8
Ganquan Xiang 7th Bao				Younger brother's wife	崔	冻娃儿	Female	Tiger	1914	11
Ganquan Xiang 7th Bao				Nephew	杨	庞子	Male	Horse	1942	11
Ganquan Xiang 7th Bao				Niece	杨	冬庆	Female	Ox	1937	12
Ganquan Xiang 7th Bao				The second niece	杨	苏女子	Female	Monkey	1944	6
Liulin Xiang 1st Bao	926	1	2	The house-hold head	潘	希寿	Male	Rat	1912	8
Liulin Xiang 1st Bao				The house-hold head's wife	李	贵香	Female	Snake	1917	3
Liulin Xiang 1st Bao				Son	潘	禄儿	Male	Dragon	1940	7
Liulin Xiang 1st Bao				Daughter	潘	春燕	Female	Tiger	1938	3

Date	The beginning time of inhabitation in this county	Educational level	Profession sector	Occupation	Marital status	Physical disability	Religion
1	Inhabited for generations	Illiteracy	Agriculture	Land-holding peasant	Married	No	
16	Inhabited for generations	Illiteracy	Agriculture	Land-holding peasant	Married	No	
8	Inhabited for generations	Illiteracy	Agriculture	Land-holding peasant	Married	No	
11	Inhabited for generations	Illiteracy	Agriculture	Land-holding peasant	Married	No	
27	Inhabited for generations	Illiteracy	Agriculture	Land-holding peasant	Married		No religion
24	Inhabited for generations	Illiteracy	Agriculture	Land-holding peasant	Married		No religion
9	Inhabited for generations					No	
7	Inhabited for generations	Illiteracy				No	
2	Inhabited for generations					No	
25	Inhabited for generations	Lower primary school 4 years	Agriculture	Land-holding peasant	Married	No	No religion
27	Inhabited for generations	Illiteracy	Agriculture	Land-holding peasant	Married		
31	Inhabited for generations	Illiteracy					
10	Inhabited for generations	Illiteracy					

Continued in Volume 2, DOI:10.1163/9789004727335_002

www.ingramcontent.com/pod-product-compliance
Lightning Source LLC
LaVergne TN
LVHW010636300926
849161LV00005B/960